Proceedings of the First

Construction History Society

Conference

Queens' College, University of Cambridge, 11-12th April 2014

Edited by

James W.P. Campbell

Wendy Andrews

Nicholas Bill

Karey Draper

Patrick Fleming

Yiting Pan

Formatting and layout by Yiting Pan
First printed by Short Run Press, Bittern Road, Exeter, EX2 7LW

Proceedings of the First Construction History Society Conference
edited by James W.P. Campbell,
Wendy Andrews, Nicholas Bill, Karey Draper, Patrick Fleming, and Yiting Pan

The First Construction History Society Conference

Organised by:

The Construction History Society

Hosted by:

Queens' College, University of Cambridge
&
The Department of Architecture, University of Cambridge

Acknowledgements

The preparation of any conference takes an enormous amount of time and effort on behalf of a large number of people. I am hugely grateful to all those who have assisted in the planning of this, the First Construction History Society Conference. In particular I would like to thank Yiting Pan who has worked so hard to put these proceedings together, a task that has taken her away from her PhD. David Yeomans as chair of the Scientific Committee took on the unenviable task of collecting all the abstracts together and distributing them to the various members of the scientific committee for review. I am as always grateful for his sage advice and I would like to thank him and all the other members of that committee for their efforts in reviewing all these proposals. Once the papers were in the editorial committee took over and went painstakingly through every paper. Any errors remaining are my fault not theirs. I would like to thank all those members of the committee of the Construction History Society who give their time so generously to the society without any form of remuneration. They are Antonio Becchi, Michael Driver, Michael Heaton, Niall Bird, Nick Beech, Alan Palmer, Michael Tutton, Nina Baker, Andrew Jackson, Treve Rosoman, and the journal editors Hermann Schlimme and Bill Addis. Without their support the Society would not function. Lastly I would like to thank all those staff of the Architecture Department and Queens' College, in the University of Cambridge for hosting this conference, without whose aid none of this would be possible.

James W P Campbell
2014

Contents

Preliminaries

Session papers

Contents

Preliminaries

James W.P. Campbell

Introduction

The editing of Conference Proceedings is obviously not like the writing a book. When you call for abstracts you never know what you are going to receive and how much interest the conference is going to elicit. This is particularly true of a new conference. It is not unreasonable at this point to ask why this is the first such conference that the society has held.

The Construction History Society is the oldest society in the world devoted to the study of the history of construction. It was set up in the 1980s with the aim of providing a common focus for individuals from various academic disciplines and backgrounds. They were united by an interest in how the technical, social and economic factors in the building trade had shaped the world around us. Among its first members were archaeologists, bringing their observations on building techniques of the past from first hand recording. There were architectural historians who were exploring building accounts and physical fabric and seeking to understand the world of the building craftsmen. There were economic historians looking at the wage and material price trends in building records who were also keen understand their context. There were social historians who were fascinated by working conditions and the history of the struggle for workers' rights. There were engineers seeking to understand the history of their own profession, which was (and still is) rarely taught in engineering schools. And then there were members of the building industry, often retired, who were keen to share their experiences and explore the subject further.

Construction History was — and still is — deliberately loosely defined. The journal provided a valuable space for the publication of articles on subjects that did not fit easily elsewhere. Initially it was just a national society, organising trips to see buildings and regular meetings and providing a journal and a newsletter. The Construction History Society now boasts a truly international membership and it is no longer the only society. The Spanish society (*Sociedad Española de Historia de la Construcción*) was founded after the first international congress in Madrid in 2003 and has been highly successful. Following on from that success the Construction History Society hosted the Second International Congress in Queens' College in 2006 and this lead to an affiliated society in America (the Construction History Society of America), which it helped to found in 2008. More recently new independent societies have been founded in France and Germany.

Since 2003 the Spanish society under the leadership of Professor Santiago Huerta has held regular national conferences. These started from comparatively modest beginnings and have grown into major events, held very two years, with hundreds of papers. It is in response to this success that we decided three years ago to start our own series of regular conferences. Queens' College, Cambridge was an obvious venue. The college had successfully hosted the International Congress in 2006 and, as I continue to enjoy a Fellowship here, the college allows us to host our events for free. It is a pleasant venue and because it has buildings from all periods and a spectacular wooden bridge it is a rather appropriate one. It is the intention to maintain this relationship for the foreseeable future, or at least until the members decide they would rather go elsewhere. Our hope is that this can become a regular event. Attempts in recent years to run visits and trips had been increasingly disappointing. The membership has changed and fewer are able to take the time off work to attend visits or simply live too far away. It is thus felt that a single annual event providing overnight accommodation might work better and that if it is successful it could be combined with visits and other elements that might be even more attractive to members.

This set of proceedings also boasts another first for the Construction History Society: our own International Standard Book Number (ISBN). The Construction History Society has of course had an International Standard Serial Number (ISSN) for the journal for decades. The issuance of a series of ISBNs now allows us to publish books and we would welcome suggestions from potential authors for volumes that we might want to produce. Our focus would be on titles that are of use as sources of reference for our members: those working in or on construction history (bibliographies, glossaries, directories etc.).

There are many ways to arrange papers in a set of proceedings. Large conferences lend themselves to thematic groupings. In a small conference this becomes more difficult as there are often relatively few papers on a subject: subject groupings with just one paper are not really groupings at all. For this reason it seemed simplest to stick to an alphabetical listing. The lack of an index will no doubt disappoint some readers, but the pressure to produce these proceedings to be available at the conference sadly meant that there was no time for all the work that would have involved. An electronic index will be produced and placed on the website for download at a future date with the option to print on demand.

The 47 papers presented here show the breadth of what we mean by "construction history." They cover a very board range of issues and periods. Some readers will notice two areas are missing. The first is structural calculations on existing buildings. When construction history first emerged as subject there were many papers showing computer modelling of existing structures demonstrating that they stood up or that they should not stand up, although they clearly did. This kind of paper has thankfully largely disappeared. The methods used by designers in the past to calculate whether their buildings would be safe are of course of interest. New methods for calculating the safety of existing structures do have their place, but they belong in structural engineering journals rather than in construction history. The other papers that have been excluded are those providing accounts of new methods of repairing old buildings or of conservation projects. This is simply building conservation, for which there are already plenty of other outlets. Of course conservation projects can — and often do — provide new information on historic building techniques. The presentation of these in their historical context is of great interest, but building conservation *per se* is not.

The number of new societies and increasing number of conferences taking place across the world suggest that construction history is a subject that is growing in popularity. For someone who is excited about the subject this is a positive development. All over world growing populations and expanding cities put increasing pressures on historical preservation. Much evidence disappears daily. Moreover architecture has a habit of "reinventing the wheel". Our ancestors often lived in a low-energy, low-impact manner. As these papers demonstrate, mankind is endlessly creative, and there is much we can learn from all periods and all countries, whether it be archaeological digs or oil rigs.

James W.P. Campbell

Chairman

Construction History Society

April 2014

The contribution made by the Journal Construction History towards establishing the history of construction as an academic discipline

Bill Addis
Co-editor of Construction History

Summary

This short paper takes a look back at papers published in Construction History, the peer-reviewed journal of the Construction History Society, presenting the range of subjects covered under headings such as main theme, period of history, location of subject, type of building or other structure and aspect of the building design and construction process. In an important sense, the 179 papers published in the journal have greatly helped to define the scope of the subject during the 29 years since its first publication. Finally, this paper comments on the state of development of the history of construction in Britain, and in the wider context of the EU, towards being a fully-fledged academic discipline in the way that history of science now is.

Introduction

Construction History is the international, peer-reviewed journal of the Construction History Society. It was first published in 1985, two years after the formation of the Construction History Group which later changed its name to the Construction History Society. Mark Swenarton was the first editor assisted by Simon Pepper as Abstracts Editor and Robert Thorne as Reviews editor. Mark asked Chris Powell to join him as co-editor for issues three to five (1987-89) before he stepped down to concentrate on Architecture Today. Robert Thorne and Chris Powell took over as editor, with Simon Pepper on Abstracts and Robert and Chris remained editors until their retirement in 2012. Since that time, Nick Bullock (who stepped down in 2013), Bill Addis and, most recently, Hermann Schlimme have been co-editors.

Papers published in Construction History

The papers published in Construction History during the last 28 years give an indication of what subjects have been aired under the overarching scope of the emerging-discipline. During this period there has been no attempt by either the Committee of the Construction History Society or the editors of Construction History to define careful what subjects might fall within the scope of "Construction History", and what not.

The journal was initially published annually, 27 issues appearing between 1985 and 2012. From volume 28, published in 2013 the number of issues per years has increased to two a year with additional special issues, reflecting the growing interest in the field. In these 28 Volumes (31 issues) 179 papers have been published.

The contribution made by the Journal Construction History towards establishing the history of construction as an academic discipline

Thematic analysis of papers published in Construction History

The tables in the following sections show the results of an analysis of the titles of the papers according to 7 categories:

- Main theme of papers
- Period of history considered
- Location of subject
- Type of building or other structure
- Stage within the design / construction process
- Building element or design issue
- Construction material

Methodology

The analysis was carried out by the author, based on the titles of the papers published and pre-existing knowledge of the content of many of these papers built up over 25 years, as well as pre-existing knowledge of many of the authors and their interests in construction history. It was not based on a recent, detailed reading of each paper.

Main theme of construction history

Table 1 shows the numbers of papers written on various main themes of construction history. The most popular subject has been, not surprisingly perhaps, the design and construction of buildings and engineering structures. It is interesting that all the remaining subject areas have attracted a very similar degree of interest.

Tab. 1. Main theme of the papers

	No.	%
Design and construction	74	37
Historiography	15	8
Sources	7	4
Books, bibliography	12	6
Science, theory	6	3
Innovation, technology transfer	17	9
History of construction firms	9	5
History of other organisations	5	3
Biography	18	9
Policy	5	3
Economics, commercial issues	14	7
Labour, trades, professions	16	8
Training, education	2	1
Total	200	

Period of construction history

Table 2 shows that the dominant interest of authors has been the nineteenth and twentieth centuries. This does reflect the construction history of Britain. Surveys of a similar Italian publication would probably show greater interest in ancient and Renaissance themes and less in the 19th century. In France the interest would probably be more focused on the Enlightenment and nineteenth centuries; in Germany the focus of attention would probably be the nineteenth and twentieth centuries, while in Spain the mediaeval / Renaissance would probably dominate.

Tab. 2. Period of history considered in papers

	No.	%
Ancient	2	1
Mediaeval / renaissance	11	6
1600 – 1800	31	16
1800 – 1850	47	25
1850 – 1914	39	21
1914 – 1945	34	18
1945 –	26	14
Total	190	

Location of subjects and structures considered in papers

Although the *Construction History* is now firmly established as an international journal, its origins lie in the UK. It this therefore no surprise that the geographical locations of the subjects considered in papers are mostly in the UK. Nevertheless, a good number of the EU subjects were written about in the first 20 years of publication. With English as the language of publication, the majority of other papers have dealt with themes in English-speaking countries, mainly the USA and Australasia.

Tab. 3. Location of subjects and structures considered in papers

	No.	%
UK	64	42
EU	43	28
USA	25	17
Russia	4	3
Latin America	5	4
China	3	2
Other	7	4
Total	151	

Type of construction considered in papers

Nearly half of the published papers have dealt with the design and construction of particular types of buildings. The distribution is shown in Table 4. Of these papers, about a third were devoted to named individual buildings while the remainder dealt more generally with a class of structure such as housing or railways.

Tab. 4. Type of construction considered in papers

	No.	%
Churches	14	16
Housing	22	25
Education	5	5
High rise	9	11
Bridges	12	15
Docks, dams, industrial	8	10
Urban development	5	7
Military	5	7
Railways	4	5
Other	11	15
Total	74	
A named structure	22	36

Stage or aspect of the construction process considered in papers

The vast majority of papers dealing with the process of delivering buildings have concentrated on their design and construction. Table 5 shows that a significant number have also been concerned with off-site manufacturing processes associated with the design and construction of structures. Management and contractual issues have attracted comparatively less interest. For instance, the work of the quantity survey (cost consultant) has yet to be addressed by an author.

Tab. 5. Stage or aspect of the construction process considered in papers

	No.	%
Design & construction	69	69
Industrialisation, prefabrication	6	6
Manufacturing processes	16	16
Construction management	8	8
Contracts	1	1
Total	100	

Stage or aspect of the construction process considered in papers

The majorly of those papers that dealt with specific elements of buildings were devoted to the structure of buildings and, particularly, their roof structures. Building services have attracted only a few papers and building façades, little more. Other aspects of building engineering, including acoustics, fire resistance and provision of a healthy internal environment have attracted very little interest at all.

Tab. 5. Building element or aspect of design/construction considered in papers

	No.	%
Structure	21	39
Roof	16	30
Building services	6	11
Façade, envelope	9	17
Acoustics	0	0
Fire resistance	1	2
Health	1	2
Total	54	

Construction materials considered in papers

Building is, fundamentally, about materials and how they have been used in construction. About a third of all papers address various construction materials, directly or indirectly, and these were quite evenly spread across the main materials that have been used.

Table 6. The construction material considered in papers

	No.	%
Masonry / stone	13	17
Brick, fired clay	11	14
Timber	16	21
Iron, steel	16	21
Cement, concrete	12	16
Aluminium	3	4
Other	6	8
Total	77	

Is the history of construction now an established discipline?

Two aspects of this question need to be considered: what is the nature of an academic discipline, and whether a supposed discipline is demonstrably separate from other, related disciplines.

The characteristics of an academic discipline

Before considering how or whether the papers in *Construction History* have contributed to establishing the history of construction as a discipline, it is appropriate to summarise what might be needed to confirm whether a certain subject is a "discipline" or not.

For the purposes of this paper, it is assumed that a discipline is an area of study and research at a university which has a name that gives it a discrete identity.

This would suggest that there should be at least on department or academic unit that includes the name of the discipline, and that there should be at least one member of staff at professorial level (in the UK, or a chair in most other countries). There should be some research, funded by a funding agency, being undertaken within the field of activity, and evidence of related publications such as books, conference papers and peer-reviewed papers. There should also be research students working towards doctorates (PhD) and some students at postgraduate level and/or undergraduate level who study publicly-advertised courses or course modules in the subject area. A discipline would also be expected to support formal, national and international congresses, and at least one peer-review periodical.

Table 7 shows a crude assessment of how the history of construction meets these criteria, in the UK and in the wider context of the European Union.

Tab. 7. Characteristics of a discipline and suggested "scores" in UK and EU

	UK	EU
1 or more University-based Research unit / Department	✗	✓
1 or more Professorial chair	✗	✓
Private research students	✓	✓
Funded research students	✗	✓
Books	✓	✓
Occasional meetings / workshops	✓	✓
Conferences with published proceedings	✓	✓
1 or more Peer-review Journal	✓	✗
Occasional lectures to students	✓	✓
Undergraduate taught courses / modules	✗	✓
Post-graduate taught courses / modules	✗	✓

There is likely to be much discussion about this simple assessment since there are many ways of understanding what is meant by "a course or module in construction history", especially when comparisons are made across national borders and different education systems. Likewise when it comes to defining "funded research", both at post-graduate and post-doctoral levels. Some research undoubtedly addresses construction history, but most often under the name of "architectural history" or architectural conservation. The assessment has been made mainly on the basis of whether the name "construction history" is used to describe courses or research programmes. A much more detailed study would expose these differences and nuances.

Differentiation of Construction History as a discipline

Of equal significance to the nature of a discipline is whether or not a supposed field of interest be considered as a discipline separate from its near neighbours.

An analysis of neighbouring subjects enables an assessment to be made as to how distinct the subject of construction history is from related fields of interest. One can imagine a Venn diagram showing areas for two disciplines, and an area of overlap in which a certain paper might be judged to fall, for example, in either the history of construction or the history of architecture.

It should also be noted that possible overlap between disciplines is, of course, highly subjective. A paper discussing the design and construction of a Victorian textile mill, for example, might be considered by different people as construction history, architectural history, engineering history, industrial archaeology, engineering heritage or architectural heritage. Indeed, papers on construction history have been, and can still be found in journals devoted to these other disciplines.

The following table shows a rough assessment of the overlap between pairs of subject areas, based on the titles of papers published in Construction History.

The data were determined by asking the following question:

> If *Construction History* did not exist, in which journal in a related discipline, would it be most likely to be published?

To a large extent, the answer to this question is what would have been the case 30 years ago, before *Construction History* had been founded.

The data illustrated in Figure 8 suggest the following points:

- There is already a good separation between the history of construction and the various strands of heritage studies – conservation, restoration, etc.
- It would be worthwhile marketing *Construction History* to authors who currently publish papers in journals devoted to the following fields of interest:
    - architectural history (e.g. Journal of Architectural Historians),
    - history of technology (proceeding of the Newcomen Society, History of Technology, Technology and Culture),
    - history of building and civil engineering (e.g. Engineering History and Heritage from the ICE), and
    - economic and business history.

Tab. 8. Crossover with other disciplines

	No.	%
History of architecture	71	26
History of urbanism	16	6
History of technology	62	22
History of building engineering	43	15
History of civil engineering	24	9
Industrial archaeology	1	0
Architectural heritage	3	1
Civil engineering heritage	0	0
Economic / business history	50	18
Military history	8	3
Total	278	

Access to papers published in Construction History

A full list of the papers published in *Construction History* can be downloaded from the CHS Website.

The contribution made by the Journal Construction History towards establishing the history of construction as an academic discipline

Since the rise of the internet the Construction History Society has actively sought to to link the Journal to the modern world of access via the Internet. Since 2011 the full text of the journal has been available on Proquest and since 2013 it has also been available on JSTOR. In both cases the last two years of journals are embargoed. The Journal is thus available through not one but two separate digital publishing outlets, which together are the most commonly used science and humanities resources.

The subject of impact factors, has only begun to be addressed recently by the Society. There use in the arts subjects is still hotly disputed, although they are widely accepted in the sciences. Increasingly however Universities are looking to these indices when considering academics for promotion. The Journal *Construction History* is being submitted to Thomson Reuter's Web of Science. This is now almost a requisite of an academic journal, since only those papers published in Journals registered with Reuters are accepted by many Universities as counting towards the various forms of assessing the research quality of University departments and their staff.

Conclusion

The very existence of *Construction History* as a peer-review Journal has done much to help establish the subject as a discipline.

The wide range of subjects addressed by authors has done much to establish what the discipline of construction is, and what it is not.

The availability of the Journal via both Proquest and JSTOR will improve access to its full content, and the inclusion of the Journal in the Web of Science will attract more potential authors.

Nevertheless, the existence of the journal, by itself, does not mean that the history of construction is now established as an academic discipline. This will only be achieved when various different countries have established professorial chairs and research units dedicated to the history of construction (by name). Alongside this step forward is the beginning and continued growth in funded research in the field. The UK is beginning to make progress in this respect but it lags behind Germany, Italy, France and Spain.

It should probably be concluded from these various observations that construction history has not yet fully achieved the state of academic discipline, either in the UK or in the EU as a whole. It has not yet reached the status of, for example, History of Science which is now widely established and to be found in many Universities. It is of some consolation that it took well over half a century for History of Science to become established; the history of construction is scarcely 30 years old.

Session Papers

The early German production and use of modern hydraulic binders: between English influences and the search for a scientific approach, 1817-1839

Salvatore Aprea
Theory and History of Architecture Laboratory, Swiss Federal Institute of Technology in Lausanne (EPFL), Switzerland

Introduction

A general need to improve building standards strongly characterizes the German "art of building" in the first decades of the nineteenth century. As a consequence, the interest towards technical aspects of buildings increases and the publication of several technical treaties and periodicals is an evidence of that. The preface to the first issue of the periodical *Sammlung nützlicher Aufsätze und Nachrichten, die Baukunst betreffend* published by the Preußisches Ober-Bau Departement from 1797 to 1806 can be considered as a manifesto of the technical approach to architecture. Art and science should merge together in architecture otherwise "the artist risks weirdness", asserts one of the focal concepts of this preface.[1]

It is in such a context that the interest in hydraulic binders increases and it was driven by the need for more efficient and durable mortars in hydraulic engineering as well as for plasters and artificial stones. The ability to produce and apply efficient hydraulic binders in Germany seems to have lagged behind France and England.[2] Several French and British chemists conducted significant research on hydraulic lime and mortar from the eighteenth century, while master builders, masons, brick makers, entrepreneurs or just inventors managed to improve methods for producing and applying them. One of those inventors is the British clergyman James Parker who patents a new calcareous hydraulic binder that is known as Parker's or Roman cement. Unlike lime, which is normally used close to the site of its production, Roman cement can be packaged and transported far away, as any another manufacturing product. It is an efficient hydraulic binder as well as a great manufacturing opportunity, arousing therefore the interest of master builders and of all those who were involved in the process of industrialisation in Germany.

Investigations on Roman cement

From 1806 to 1814, the embargo known as "Continental Blockade" prevented relations and exchanges between the United Kingdom and European countries. After 1814, British trade resumed and, in 1817, a Roman cement dealer named Benjamin Dodson sets up in Hamburg[3]. Dodson worked on behalf of a London manufacturer and beside cement he also supplies "the tools that allow the use of cement" - probably moulds.[4] Not only trade resumed after 1814 but also the so-called "Technologische Reise". The expression "Technologische Reise" describes journeys undertaken by German technicians, businessmen or civil servants to investigate new industrial products and manufacturing processes abroad. England was the main goal of these journeys. Most travellers were interested in mechanical things but some attention was also paid to new building techniques and materials.

It must have been quite impossible for German travellers not to notice the Roman cement plasters and decorations that spread all over in London after the lapse of the Parker's patent in 1810. It is probably thanks to such a traveller that prince Leopold of Bavaria received a report about Roman cement by the London department of building. The prince handed the report over to Christoph Arzberger (1772-1822), a mathematician and civil servant of high rank in Coburg. Arzberger analysed the report and wrote a short article that was published in the *Kunst- und Gewerbe- Blatt des polytechnischen Vereins für das Königreich Bayern* in 1819[5]. Arzberger describes Roman cement as a fine milled powder made of clayish limestone that has been burnt in brick kilns and is suitable for composing mortar to use in hydraulic engineering as well as for plaster and for moulded decorations. Arzberger also presumes that Roman cement could be reproducible in Germany by burning stones composed of lime and clay, the so-called calcareous marls.

Another significant investigation on Roman cement is that of [Johann Heinrich] Krahmer, master builder at the Prussian Oberbaudeputation.[6] Krahmer went on a technological journey to England in 1821, during which he visits the house of Charles Francis, owner of the greatest Roman cement factory of that time, together with his associated John Bazley White. Upon his return to Berlin, Krahmer reported to the Oberbaudeputation about what he has seen and learnt in England. He synthetically describes the composition and the properties of Roman cement, and lists all the applications that he has observed in Francis's garden: several plaster works on walls, staircases, pavements, basins of fountains and many moulded decorations (Figs 1-2). Krahmer also reports to have seen some gothic style moulded decorations in the newly restored New College Chapel in Oxford. While on one hand Krahmer appears to be impressed by the quality of the Roman cement mortar and plaster he has observed, on the other he complains about the impossibility to get detailed information about the raw materials and the manufacturing process of cement. As Arzberger before, Krahmer also believed that it could be possible to produce Roman cement in Germany after having investigated its composition through chemical analysis. Consequently, he brought from England to Berlin two casks of Roman cement and handed them over to Johann Albert Eytelwein (1764-1849), head of the Oberbaudeputation, and to Peter Christian Wilhelm Beuth (1781-1853), head of the Technische Deputation and of the Verein zur Beförderung des Gewerbfleißes in Preußen (Gewerbverein).

The two institutions led by Beuth were committed to developing industry in Prussia. In 1822 the Gewerbverein promotes a competition divided into several sections one of which concerns the invention of a mortar similar to cement and suitable as artificial stone to produce moulded objects and decorations.[7] This competition is probably inspired by Krahmer's experience in London but, because of the lack of literature and any other knowledge or experience about Roman cement, participants are invited to read German and French literature about hydraulic lime, and to refer to a previous attempt at producing artificial stone that was carried out in Berlin about ten years before by the master builder Paul Ludwig Simon (1771-1815).[8]

While the competition does not yield the expected results, it nevertheless contributed to raising the interest in hydraulic binders and to spreading the notoriety of Roman cement in other German states. In 1823, the king of Württemberg, Wilhelm I, asks his industrial spy in England, Friedrich Schmidt, to search for information about Roman cement focusing on its quality, its price and its suitability to plaster buildings in Stuttgart.[9] In June 1823, Schmidt sends the king two reports in which he asserts that Roman cement is an appropriate binder for plaster, and describes briefly how to produce it. Just like Arzberger and Krahmer before, even Schmidt believes that Roman cement could be produced in Germany and furthermore he thinks that cement manufacturing could become an important field of the rising industry in Wurttemberg. He therefore suggested searching for clayish limestone along rivers but, unfortunately, the search, that actually takes place around 1823, does not give the expected results. The available amount of clayish limestone was judged too small and no cement manufacturing was undertaken.

Despite attempts and efforts, around 1825 no hydraulic binder similar to Roman cement had been produced in Germany. Roman cement was, however, imported in small amounts, and some masters of building begin to employ it to plaster walls or just to seal masonry joints that are exposed to water or damp. Around 1826, Krahmer himself experimented with the use of Roman cement in a suburban villa nearby Berlin.[10] He plastered some exterior walls and staircases, some pavements in the house basement, and finally he seals the joints of a dock along the Spree, the Berlin River. Roman cement was also used to seal the joints of the Friedrich-Wilhelm Canal sluices and in 1827 to seal the joints of the new lighthouse of Kap Arkona on the Rügen Island. The definitive plans of the lighthouse are elaborated by August Adolph Günther (1779-1842) probably in collaboration with Karl Friedrich Schinkel (1781-1841) around 1825.[11]

In 1826, Schinkel went on a journey to France and England together with his friend Beuth. In London they visit the building site of the first Thames tunnel conceived by the engineer Marc Isambard Brunel (1769-1849) who personally led them around. In his travel diary, Schinkel drafts two sections of the tunnel structure and notices that the main building materials are bricks and Roman cement mortar.[12] One year later, in the magazine of the Gewerbverein, Beuth publishes an article about the construction of the Thames tunnel in which he writes on the huge quantity of Roman cement that Brunel has been using (Fig. 3).[13]

Schinkel's journey to England actually happens during the construction of the Kap Arkona lighthouse but it would be difficult to prove that Schinkel himself suggested the use of Roman cement to finish the lighthouse joints. On the contrary it is an evidence that, after the journey of Schinkel and Beuth, the Gewerbverein begins to plan a new competition, which should be clearly focused on the reproduction of Roman cement. To make the task easier, the Interior Minister Friedrich von Schuckmann bought two casks of Roman cement from the factory of Francis and White and gives them to the Gewerbverein in order to conduct experiments and better understand the properties of Roman cement.[14] The experiments take place around 1828 and aimed to verify if Roman cement mortar set under water, if it binds bricks and if it works efficiently as a plaster for walls and water containers (Fig. 4). The results are not extraordinary as the plaster of water containers cracks.[15] Furthermore the Gewerbverein technicians judge Roman cement too expensive for hydraulic engineering while they consider normal lime efficient enough to plaster buildings. Roman cement therefore appears suitable just to seal founding walls against rising damp, to build masonry vaults that are exposed to rain and to plaster wet walls. The following year, the Gewerbverein set aside the idea of a competition focused on the reproduction of Roman cement and asked for the development of a hydraulic mortar that should have all properties identified as necessary during the tests on Roman cement mortar, i.e. the mortar should set under water as well as at air without cracks, it should be suitable for hydraulic engineering as well as for ordinary masonry and it should cost no more than six time the price of ordinary lime mortar.[16]

Like the previous one, even this competition had no winner. The efforts undertaken to discover a binder with similar properties to those of Roman cement appeared fruitless, partly because the exact knowledge of the manufacturing process remained unknown and partly because the suitable raw materials were still to be found.[17]

The cement production in the northwest regions

Since the late 1820s some calcareous cements similar to Roman cement have been produced in the northwest regions, from the Frisian coast to the city of Kassel. That resulted from the initiatives of private entrepreneurs who could easily learn about British manufacturing processes, owing to the intense relationships that these regions maintained with England. Initially, even raw materials were sometimes imported from England, as it happens for the cement production that Claus Shipmann undertook in Hamburg in 1829.[18] Two years later Reinhard Woltmann (1757-1837), a hydraulic engineer in Hamburg, wrote a report about the Shipmann cement, asserting that this cement is suitable to plaster buildings and can be used for hydraulic engineering or the construction of concrete walls.[19] Neither concrete walls nor concrete mixed with cement were common in Germany in the 1830s when concrete was still composed with lime and *Traß* or brick powder, and was used in few cases just to build foundations for hydraulic engineering.

In the first half of the 1830s, after Shipmann's cement mill, several others are founded in the above-mentioned area.[20] Two are located in the Frisian villages of Carolinsiel and Emden, one in Buxtehude near Hamburg, one in Hameln in the south of Hannover, one in Barbis near the Harz Mountains and another one in Kassel.

The cement mills in Hameln and the one in Kassel had a particularly good reputation. The engineer Georg Theodor Wendelstadt (1790-1860) and the banker Adolph Meyer (1807-1866) found the first one in 1833.[21] The factory is located along the Weser River, next to the future building site of the suspended bridge of Hameln. There is a connection between the foundation of the cement mill and the project for the suspended bridge in Hameln.[22] Wendelstadt himself was charged with the project for the bridge, which provided him the opportunity to learn more about Roman cement thanks to studies of previous works and personal contacts with other specialists. An important reference for Wendelstadt's project was the Menai Suspension Bridge built by Thomas Telford (1757-1834) between 1819 and 1826. The pillars of which are made of stones and mortar partly composed with hydraulic lime, while the joints are sealed with Roman cement mortar.[23] Furthermore, around 1833-34, Wendelstadt orders some strength tests on metal chains that take place in Kassel, in the foundry of Carl Anton Henschel (1780-1861) who had just been to England to study railways and did meet Brunel, the engineer of the Thames tunnel. At that time, Brunel was testing the force of adhesion between cement mortar and bricks through the experimental construction of an impressive half arch (Fig. 5).[24] Henschel learnt about the potential of combining cement mortar and bricks - as the construction of his new foundry some years later demonstrates - and he could have talked to Wendelstadt about Roman cement. The project for the bridge in Hameln was approved in 1836 and the bridge was built between 1836 and 1839, using huge quantities of cement produced by the factory of Wendelstadt and Meyer for the masonry of pillars and for the anchorages of the metal braces into the earth (Fig. 6).

Around the time the Hameln cement mill was founded, the entrepreneur Ernst Martin Koch (1786-1860) began producing cement in Kassel. The fame of Koch's cement was primarily due to two events. The first one concerns the publication in 1833 of an instruction booklet that is probably the first German manual entirely dedicated to cement. The second event concerns the building of the dome for the above-mentioned new foundry of Henschel in 1837, for which Koch delivers the cement.[25] Henschel himself designed the dome that is made of hollow circular frustum-shaped bricks bonded with cement mortar. Henschel did not use any centring to build the dome, he just exploits the property of cement mortar to set in few minutes and strongly adhere to bricks, as he had probably learnt from Brunel in London.

The scientific approach to the production of hydraulic lime in Bavaria

While in the northwest and central regions some entrepreneurs launch the production of cements similar to the Roman cement, in Bavaria some master builders begin to produce and use hydraulic lime. British influences are mostly absent in this case as the use of the expression "hydraulic lime" instead of the term "cement" demonstrates. Moreover, the manufacturing process of hydraulic lime in Bavaria derives from the implementation of a scientific theory rather than from the imitation of a foreign manufacturing process and the role of the Bavarian State in supporting scientific research as well as the production and use of hydraulic lime is much more important than the role of private entrepreneurs.

Towards to end of the 1820s, Johann Nepomuk von Fuchs (1774-1856), a Bavarian chemist, was charged by the director of the ministerial building department Von Purgel to study lime and mortar.[26] Based on precedent researches carried out mostly by French and German chemists since the end of the eighteenth century, Fuchs defines a theory to explain the chemical processes that concern clay and lime during lime burning and during mortar hardening. The practical implications of Fuchs's theory are no less important than the chemical aspects of it. Fuchs encourages master builders to produce hydraulic lime using the Bavarian marls and asserts that hydraulic lime is suitable both for hydraulic engineering and for conventional buildings. In 1829, Fuchs read a dissertation about the results of his research at a sitting of the Polytechnischer Verein für das Königreich Bayern arising partly interest and partly the opposition of those who believe that the best hydraulic lime ever is the Roman cement and that Bavarians should acquire in England the principles of cement manufacturing rather that investing in researches.[27] Some others even believe that the production of hydraulic lime in Bavaria would be impossible because of the lack of coke that is used to burn Roman cement in England.[28]

The opposition to Fuchs's theory soon became outdated thanks to the first practical experiences with Bavarian hydraulic lime. The main promoter of Fuchs's theory implementation is Friedrich Panzer (1794-1854), an engineer at the Bavarian public administration. Panzer himself experimented with the production of hydraulic lime from marls, undertook strength tests and uses it to compose mortar and plaster for a barn that he built in Würzburg in 1830, and for a bridge over the River Main that he built in the village of Schonungen in 1831 (Figs 7-8). Several Bavarian master builders and masons follow the example of Panzer and begin to burn and use hydraulic lime to compose mortar and plaster.[29] Beyond the building activities, Panzer also wrote essays. In 1831, he published the first that is entirely focused on hydraulic lime and mortar.[30] In the preface Panzer affirms that the existence of a scientific theory is the inescapable condition to spread knowledge and practices about hydraulic lime. He wrote: "For long time hydraulic lime was known in Italy and Holland, later on in England and France, but the lack of scientific justification did not allow the further spread of it".[31] Panzer's efforts deeply change the approach to the question of hydraulic binder in Bavaria and even in near Württemberg. This is clearly expressed in an article published on the *Polytechnisches Journal* of Stuttgart in 1831, which states the following: "until now hydraulic lime has been considered a rare natural product and it has been thought that the best hydraulic lime ever was available only in England. Lime of the same quality is conversely available in abundance in all Bavaria. Until now, it was neither known nor used because of the lack of a foundational theory".[32]

In 1832, Fuchs's theoretical research and Panzer's practical activities were highly recognised by the *Hollandsche Maatschappij van Wetenschappen* (Dutch Society of Sciences) as well as by the *Polytechnisches Verein für das Königreich Bayern*. The first institution awarded Fuchs the first prize for a competition aimed to discover the composition and properties of hydraulic cements, and to define the chemical process that takes place during the solidification of cements. The second institution awarded Panzer the first prize for a competition aimed to reward "that engineer, master builder or mason who, following the instructions given by Dr. Fuchs ..., discovers a domestic hydraulic lime" suitable to prepare a mortar able to resist to water and damp.[33] After that, the *Polytechnisches Verein* also decided to award Fuchs a medal in recognition of his contributions to the scientific theory of mortars and to the improvement of the art of building.

The full recognition of Fuchs and Panzer's works is also an acknowledgement of the Bavarian scientific approach to the question of hydraulic lime as well as a hard critic of those who had claimed the necessity to search abroad for the knowledge and practice to produce hydraulic lime. This point is clearly expressed in an article published in the *Kunst- und Gewerbe- Blatt des polytechnischen Vereins für das Königreich Bayern* in 1832. The author finds it regrettable that after the publication of Fuchs's theory "someone was sent at the expense of the State to acquire the method for producing hydraulic lime in England. A place, where there had been no theoretical developments on the subject. Where hydraulic lime was produced from only one type of raw material and by following a quite empirical method. Where nobody was able to provide any explanation based on general principles".[34]

Conclusions

The cement that was produced in the north and the hydraulic lime that is produced in the south of Germany are two very similar hydraulic binders, even identical in some cases. Cement of that time can be considered a good hydraulic lime. "Cement properties are those of a highest quality hydraulic lime" states in 1835 B. W. Rodewick, a cement producer in Emden.[35] Three years later, Carl Wilhelm Dempp, who taught at the Königliche Baugewerbeschule in Münich, confirmed this point by writing: "the art of building distinguishes ... aerial lime from the hydraulic one that is also named cement".[36]

The approach to the manufacturing of these binders was very different in the North and in the South, resulting from different cultural backgrounds. The approach followed in Bavaria was representative of the efforts carried out in some regions of Germany to find an autonomous character for the German *Baukunst* that should be based on the exaltation of scientific and technical aspects. The search for establishing theories and general principles also characterises the future development of German hydraulic binders and enables Germany to become, some years later, the first country to develop a set of norms for the production and the delivery of cement.

References

[1] Mitglieder des Königl. Preuss. Ober - Bau - Departments, "Vorrede", *Sammlung nützlicher Aufsätze und Nachrichten, die Baukunst betreffend*, vol. 1, 1797, pp. IV.

[2] Regardless of the political form of Great Britain, German authors of the first half of the eighteenth century mainly refer to England in technical literature about cement. Actually England was the main aim of their interest.

[3] *Hamburgisches Address – Buch für das Jahr 1817*. Hamburg: Hermann am Fischmarkt, 1817, p. 70; *Hamburgisches Address – Buch für das Jahr 1818*. Hamburg: Hermann am Fischmarkt, 1818, p. 74; *Hamburgisches Address – Buch für das Jahr 1819*. Hamburg: Hermann am Fischmarkt, 1817, p. 76.

[4] Ch. Arzberger, "Römisches Cement", *Kunst- und Gewerbe- Blatt des polytechnischen Vereins für das Königreich Bayern*, vol. 5, no. 43, 1819, col. 600

[5] Arzberger, (Note 3), coll. 598-600.

[6] T. Brunsch, "Die historische Verwendung zementgebundener Kunststeine im Außenraum im 19. und frühen 20. Jahrhundert unter besonderer Berücksichtigung Berlins und Brandenburgs" (Ph.D thesis, Technisches Universität of Berlin, 2006), pp. 56-8. [J. H.] Krahmer, "Bemerkungen über die Dauer und Festigkeit des in England erfundenen Roman-Cement. Nebst Nachrichten von einigen andern Cementen", *Journal für die Baukunst*, vol. 3, 1830, pp. 476-83.

[7] "Preisaufgaben des Vereins: Vierte Preisaufgabe", *Verhandlungen des Vereins zur Beförderung des Gewerbfleißes in Preußen*, vol. 1, 1822, p. 22.

[8] Geheimes Staatsarchiv Preußischer Kulturbesitz, I.HA Rep. 93B, Ministerium für öffentliche Arbeiten, Nr. 1394, nn.

[9] H. Albrecht, "Vom Caementum zu Zement" in H. Albrecht, (ed.) *Kalk und Zement in Württemberg : Industriegeschichte am Südrand der Schwabischen Alb*, Ubstadt-Weiher: Verlag Regionalkultur, 1991, pp. 117-230.

[10] Krahmer, (Note 6) pp. 476-478.

[11] R. Strecke, "Schinkel und der Leuchtturm auf Kap Arkona", *Jahrbuch Preußischer Kulturbesitz*, vol. 32, Berlin: Gebr. Mann Verlag, 1996, pp. 297-319.

[12] R. Wegner, M. Kühn, (eds.), *Karl Friedrich Schinkel. Die Reise nach Frankreich und England : Das Tagebuch*, München: Deutscher Kunstverlag, 1990, pp. 46, 125.

[13] P. Ch. W. Beuth, "Der Tunnel unter der Themse", *Verhandlungen des Vereins zur Beförderung des Gewerbfleißes in Preußen*, vol. 6, 1827, pp. 286-94.

[14] F. von Schuckmann, "Letter to the Gewerbverein, 6th April 1828" in Anon. "Ueber die mit Roman Cement, aus der Fabrik von Francis, White und Francis, in London, angestellten Versuche", *Verhandlungen des Vereins zur Beförderung des Gewerbfleißes in Preußen*, vol. 8, 1829, p. 123.

[15] Ibid. pp. 123-126.

[16] "Preisaufgaben für die Jahre 1829 und 1830: Dritte Preisaufgabe", *Verhandlungen des Vereins zur Beförderung des Gewerbfleißes in Preußen*, vol. 8, 1829, p. 31.

[17] Von Schuckmann, (Note 15). More than ten years later, in the Gewerbverein Protocol, it is stated that "the theory for producing cement is well known and many cements had already been produced in Prussia" but what still misses is "a place where great quantity of suitable raw materials cold be found", Anon., "Angelegenheit des Vereins: Protokoll April 1841", *Verhandlungen des Vereins zur Beförderung des Gewerbfleißes in Preußen*, vol. 20, 1841, p. 62.

[18] J. Cohn, *Ueber die Wichtigkeit der Cemente*. Breslau: Druck der Stadt- und Universitäts-Buchdruckerei von Grass, Barth & Comp., 1855, p. 10.

[19] O. Ch. F. Reinhold, "Sammlung practischer Erfahrungen und Vorschriften, Cemente, Mörtel und Bétons betreffend", *Journal für die Baukunst*, vol. 15, 1841, pp. 94-96.

[20] Ibid., p. 89.

[21] Niedersächsisches Landesarchiv - Hauptstaatsarchiv Hannover, Hann. 80, Nr. 15903, "Privilegium Exclusivum" released by Wilhelm IV March 27, 1833 Fr. von Reden, "Bericht über die von dem Gewerbe-Vereine für das Königreich Hannover in den Monaten Mai und Juni 1835 veranstaltete erste Ausstellung inländischer gewerblicher Erzeugnisse", *Mittheilungen des Gewerbe-Vereins für das Königreich Hannover*, 7. Lieferung, 1835, coll. 528-529.

[22] E. M. Hermann, "'Leicht wie ein Gedanke'. Erste Große deutsche Kettenbrücke: 1836/39 bei Hameln erbaut", *Die Weser*, vol. 44, no. 1, 1970, pp. 48-49. M. Mende, "'Not only a Matter of Taste but ... of the Laws of Mechanics': The Adoption of British Models in Nineteenth-Century Continental Suspension Bridge Design", *Journal of Design History*, vol. 6, 1993, pp.86-88. H. Knobe, *Hamelner Wasserbauwerke an der Weser*, Bielefeld: Verlag für Regionalgeschichte, 2003, pp. 114-121.

[23] T. Telford, *An historical and descriptive account of the suspension bridge constructed over the Menai Strait in North Wales: with a brief notice of the Conway bridge*. London: Ibotson and Palmer, Strand, 1828, p. 24.

[24] Ch. Rohault de Fleury, "Nachricht von einem Versuch des Herrn Brunel in London, Brücken ohne Gerüst zu wölben", *Journal für die Baukunst*, vol. 10, 1836, pp. 170-173 (originally published as "Voûtes en ciment hydraulique", *Annales des ponts et chausses*, 1835, pp. 77-80).

[25] The building is still in existence and it is used as conference hall of the University of Kassel.

[26] J. N. Fuchs, "Ueber Kalk und Mörtel", *Beilage zum Kunst- und Gewerbe-Blatt des polytechnischen Vereins für das Königreich Bayern*, vol. 15, 1829, coll. 1-56.

[27] "Auszug aus den Protokollen der Sitzungen in den Monaten April, Mai und Juni 1829", *Kunst- und Gewerbe-Blatt des polytechnischen Vereins für das Königreich Bayern*, vol. 15, no. 29, 1829, col. 392.

[28] Anon., "Würdigung der Arbeit des Professors und Akademikers Fuchs über Kalk und Mörtel", *Kunst- und Gewerbe-Blatt des polytechnischen Vereins für das Königreich Bayern*, vol. 18, 1832, coll. 758-759.

[29] F. Panzer, "An die Redaktion des Kunst- und Gewerbe-Blattes", *Kunst- und Gewerbe-Blatt des polytechnischen Vereins für das Königreich Bayern*, vol. 10, 1832, coll. 111-144.

[30] F. Panzer, *Anleitung über die Bereitung des Mörtels aus hydraulischem Kalke in Beziehung auf die Auffindung des vorzüglich hierzu sich eignenden Mergels, so wie über das Verfahren, welches bei der Anwendung des hydraulischen Mörtels, sowohl bei Landgebäuden, als auch bei Bauten im Wasser zu beobachten ist*, München: Joseph Lindauer'sche Buchhandlung, 1832 (1st edition published in Würzburg, 1831).

[31] Panzer, *Anleitung*, (Note 32), p. 3.

[32] Anon., "Ueber Aufbewahrung des Getreides in Silo's und die Benuzung des hydraulischen Mörtels zur Erbauung derselben", *Polytechnisches Journal*, vol. 39, 1831, p. 428.

[33] "Preisaufgaben des polytech. Vereins in Bayern", *Polytechnische Journal*, vol. 40, 1831, p. 388.

[34] Anon., (Note 30), col. 759.

[35] Reinhold, (Note 21), p. 90.

[36] C. W. Dempp, "Ueber die Benutzung des Mergelkalkes zu hydraulischem Kalk im Allgemeinen, und dessen Anwendung in Bayern insbesondere", *Allgemeine Bauzeitung*, vol. 3, 1838, p. 442.

Form v function: the structural development of early fifteenth-century hammer-beam roofs

Robert Beech
Department of History of Art, University of Birmingham, UK

The earliest hammer-beam roofs were carpentered in the south of England at the turn of the thirteenth century. Whether to provide clear circulation in a busy kitchen, (Bishop's Palace, Chichester), or to provide an open area in an aisled hall (Pilgrims' Hall, Winchester), the intent behind the new construction was to create a space uncluttered by vertical posts. [1] Yet despite its ingenuity and utility, patrons seemed little interested in the new construction, and only five hammer-beam roofs are known to predate the 1390s. By contrast, the fifteenth century saw an explosion in hammer-beam construction. Carpenters constructed 161 such roofs in Norfolk and Suffolk alone. This paper examines what propelled this sudden upsurge in construction, and considers subsequent developments in hammer-beam carpentry.

The Influence of Westminster Hall

Hugh Herland (d. *c.*1411) completed the hammer-beam roof of Westminster Hall for Richard II probably by 1398. The first major event to be enacted in this arena for kingly display was a revolution: Richard's own deposition in 1399. The nobles and prelates of the realm were required to attend, and the proceedings confirming the deposition were the best attended of any medieval assembly. [2] Revolutionary also was Herland's new roof, innovative and unique in both construction and ornament, the first to feature hammer beams carved as angels. Before the great and good of the land, therefore, and after a century of indifference, the hammer-beam roof's formal and ornamental potential was unveiled for the first time.

It was in East Anglia where Herland's formal and ornamental seeds found fertile soil. Fuelled by the wealth of the cloth trade, the fifteenth century saw an intensive campaign of church-building and enlargement in Norfolk and Suffolk. Indeed, the towns of Norwich and Ipswich must at times have resembled enormous building sites. [3] To gauge the extent in the region of the influence of Westminster Hall immediately after its construction, the author has isolated fourteen examples of roofs with hammer beams carpentered in the first third of the fifteenth century. Few East Anglian hammer-beams have undergone dendrochronology. The dates presented in Table 1 have been determined from the study of primary and secondary sources, and a first hand examination of the construction of the roofs and details in the associated masonry.

BUILDING	CONSTRUCTION	SPAN	PITCH	DATE	CERTAINTY	EVIDENCE
DEBENHAM, Suff., St Mary Magdalene	TB/HB no collar	22ft 6ins / 6.86m	shallow	1403 ± 6	***	tree-ring
NORWICH, Norf. St Giles	HB/AB no collar	19ft / 5.79m	shallow	early C15	**	Structure Doc 1 / Doc 2
KINGS LYNN, Norf., St Nicholas	TB/HB	32 ft / 9.75m	shallow	*c*. 1400-19	***	Doc 1 / Doc 2
GREAT SHELFORD, Cambs, St Mary	TB/AB/HB	23ft 8in / 7.21m	V. shallow	1400-11	**	Structure Doc 2
WESTERFIELD, Suff., **St Mary Magdelene**	HB/AB collar / KP	19ft / 5.79m	steep	early C15	*	Structure Doc 2
BEESTON-next-MILEHAM, Norf. St Mary	HB/HB no collar	22ft 11in / 7m	shallow	*c*. 1410	**	Doc 1
WINGFIELD, Suff., St Andrew	CR/HB collar, soulaces	19ft / 5.8m	shallow	*c*. 1415	**	Doc 1 / Doc 2 structure
WIGGENHALL, Norf., St Mary Magdelen	TB/HB no collar	21ft 6in / 6.55m	shallow	1419 ± 16	***	tree-ring
BILDESTON, Suff., St Mary	TB / HB no collar	21ft 7in / 6.57m	steep	*c*. 1420	**	Structure Doc 2
BARDWELL, Suff., St Peter & St Paul	HB/HB no collar	26ft 6in / 8.08m	steep	*c*. 1421	**	Doc 1
ELY CATHEDRAL N. & S. transepts	HB/HB no collar	*c*. 33ft / 10.06m	shallow	1426 – 27	***	tree-ring
MILDENHALL, Suff., St Mary	TB/HB	22ft 6in / 6.86m	V. shallow	1420-30	**	Doc 2
GREAT CRESSINGHAM, Norf., St Michael	HB no collar	17ft 4in / 5.28 m	shallow	*c*. 1430.	**	Structure Doc 1 / Doc 2
BURY ST EDMUNDS, Suff., St Mary	HB/AB collar	26ft 9in / 8.15m	shallow	*c*. 1430	**	Doc 1 / Doc 2

Table 1.

Key:
CONSTRUCTION: principal transverse frames given first. Abbreviations: TB: tie-beam; HB: hammer-beam; AB: arch brace; KP: king post; CR: single-framed common rafter roof
PITCH: steep: above 45°; shallow: below 45°; v. shallow: c. 24° or below.
*CERTAINTY: of date. *: possible; **: probable; ***: certain.*
EVIDENCE: Evidence for date. 'Structure': construction connotes early date. 'Doc 1': Primary documents; 'Doc 2': secondary documents.

Of note is that seven of the fourteen examples are not structural hammer-beam roofs. Five of these seven, for example King's Lynn, are tie-beam roofs with hammer beams inserted in the intermediate frames. Wingfield is an exception, where the nave contains a common rafter roof with collars and soulaces (Fig. 1). It is a type of construction belonging to a carpentry tradition then centuries old, functioning happily for generations without the carpenter finding it necessary to include hammer beams. And yet here they are.

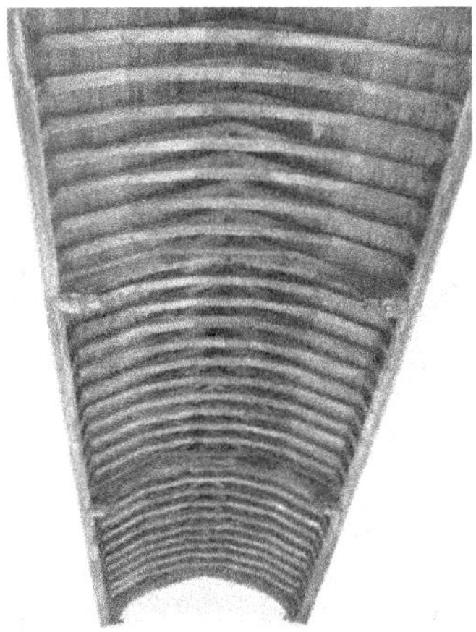

Fig.1. Wingfield Church, nave roof, looking east. (All photographs / illustrations by the author)

The owner of Wingfield Church in the early fifteenth century was the Second Earl of Suffolk, Michael de la Pole (d.1415). De la Pole was present at Westminster Hall during the momentous events of the autumn of 1399 [4], and here, in his family church in the early years of the fifteenth century, we find, arch-braced to common rafters, angel-hammer beams inserted for visual effect. The evidence from the seven non-structural hammer-beams is compelling: in the two decades following the completion of Westminster Hall, patrons were eager to specify angel-hammer beams irrespective of structural need.

Turning to the early roofs in which the hammer-beam framing is structural, a distinct typology emerges. None of the Norfolk carpenters chose to frame in the normally desirable collar, while their Suffolk counterparts chose to include collars in two of their three roofs. The sample in the table is small but indicative, and when the chronology is scaled-up to include Norfolk and Suffolk roofs throughout the later medieval period, a remarkable picture emerges. Of fifty-two medieval single hammer-beams in Norfolk, forty-one have *no* collars. In Suffolk, of sixty-six single hammer-beam roofs, collars are present in all but two. The two collarless examples are just few miles south of the county border, at Bardwell and Whickam Skeith. What could account for this remarkable structural dichotomy? [5]

It is likely that post-Westminster Hall hammer-beam construction in East Anglia developed from two distinct structural scions located in major towns. Norfolk carpenters chose to follow the example of St Giles, Norwich, and perhaps later the transepts of Ely Cathedral. In these roofs collars are absent; the principals are arch-braced to a wall post, and rise to a king pendant (Fig. 2).

Form v function: the structural development of early fifteenth-century hammer-beam roofs

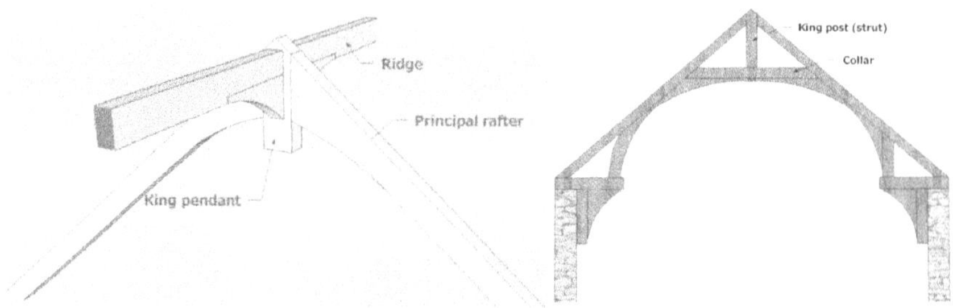

Fig. 2. (left)Generic king-pendant construction. Fig. 3. (right) Generic Suffolk hammer-beam frame. Hammer post and arch brace are sometimes merged, as on the left.

Suffolk carpenters chose to follow the example of St Mary's, Bury St Edmunds, and its more traditional arch-braced collar construction. [6] Indeed, the arch-braced collar, usually with a king strut, became somewhat unimaginative Suffolk construction in virtually all of the county's subsequent hammer-beams (Fig. 3). Though differing structurally from each other and from Westminster, Herland's influence is still present in these three influential early roofs: all possess ornamental recumbent angel-hammer beams.

In omitting collars, early fifteenth-century Norfolk carpenters were more structurally daring than their Suffolk counterparts and, significantly, proved themselves to be no slavish copiers of Herland, who included horizontal transverse timbers in his masterpiece. Indeed, it is remarkable how few of the Norfolk roofs do not suffer from rafter sag. [7] Is Norfolk practice a pursuit of form heedless to potential structural catastrophe? The authors of these roofs certainly were interested in form, in creating a clear longitudinal vista uncluttered by collars leading the eye towards the most sacred part of the church. It was not, however, the chancel, where the mysteries of the Mass were enacted, that was the principal object of this vision-channelling device. Above the chancel arch would be the emblem of redemption, the holy rood, and behind, the unambiguous portrayal of glorious salvation or monstrous damnation: the Doom. A rhythm of receding, uninterrupted gothic arches would serve perfectly to frame these devotional cynosures, and how much better if this vista was further channelled by a twin rank of observing hammer-beam angels. Here one senses the influence of Westminster Hall. The deific Richard II must have been delighted at the way Herland's receding trefoils created a tabernacled, niche-like space at the high end of his new hall, drawing the eyes of his subjects, and framing the monarch in angelic glory. [8]

Nonetheless, structure was not neglected by the form-fixated Norfolk carpenters. Indeed, the majority of these collarless roofs have survived the centuries remarkably intact, testament that the craftsman adapted his carpentry successfully to withstand the demands of form. We can understand how he achieved stability by looking at the splendid roofs over the transepts of Ely Cathedral. These are prime examples of collarless construction and, exacerbating structural challenges, are of fairly low pitch and exceptionally wide span. The massive masonry of the cathedral walls must prevent spread, but what prevents the rafters from sagging? All is conventional in the principal's lower framing: an arch-braced hammer-post provides the stiffening here. In the upper framing the carpenter solved the problem of how to strengthen the rafter by utilising a king pendant (see Fig. 2). Crucially, the king pendant provides a timber to which the principal can now be arch-braced, thus effectively increasing its section, stiffening it, and allowing the collar to be omitted. Embryonic king pendants are framed into the very early fifteenth-century hammer-beam at St Giles, Norwich. Remarkably, in some cases, as at Beeston-next-Mileham and Great Cressingham, rather than arch-brace the principal, the carpenter flared out the upper section of the rafter, so that 'arch brace' and principal are one timber.

The king pendant has a further, less structural, function in providing a point to which a ridge may be framed. Unlike modern roofs, the ridge in medieval roofs was primarily ornamental, indeed, the ridge at Beeston is located simply by a short, un-pegged stub tenon. [9] Usually moulded, the ridge piece served to mask the unsightly terminations of the common rafters, and frequently it was arch-braced to the pendant. Framed thus, it introduced a longitudinal arcaded element at the apex which often alludes to similar arcading produced by structurally superfluous wall post to wall plate arch-bracing.

The Waning of Westminster Influence: Pendant Hammer Post Construction

Collarless construction and the ornamental ridge piece in the upper framing show that fifteenth-century Norfolk carpenters were abandoning an exclusively structural approach to roof carpentry as a means purely to keep out the rain. Across the county border, Suffolk carpenters were turning their attention to the lower hammer-beam framing in a quest, not for form, but for more ornament. In the trilateral area formed by connecting the East Anglian towns of Ipswich, Bury St Edmunds and Gestingthorpe, there seems to have been a minor industry in the production of pendant-post hammer-beam roofs. To explain this type of configuration: in conventional hammer-beam construction the hammer post is tenoned vertically into the hammer beam. In pendant construction the hammer beam is tenoned horizontally into the hammer post (Fig. 4).

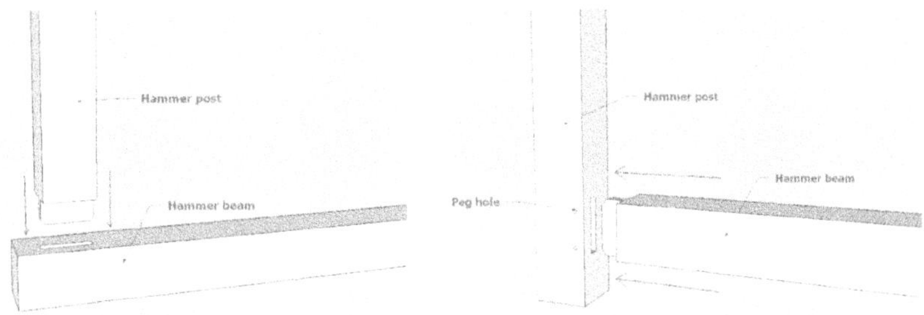

Fig. 4. Left: conventional framing, right: pendant post construction.

It is a construction of negative structural implications. Pendant configuration means that on square timbers of a section of, for example, eight inches (210 mm), any shearing force the beam is intended to carry rests on a 1 ½ inch (40 mm) tenon rather than its full 8-inch width in conventional construction. Further, under certain lateral wind-loading conditions the mortice and tenon joint may now be in tension (conventionally it sat happily in compression), and any forces this triangulation (with the principal rafter) is meant to control depends merely on the one-inch pegs locating the joint and the negligible resistance to shear of the short-grain tenon. Nonetheless, carpenters persevered with this structurally suspect construction, the pendant post favoured even in the august royal setting of Eltham Palace (*c.*1475-79). Why?

Form v function: the structural development of early fifteenth-century hammer-beam roofs

The pursuit of ornament was the driver. By now we are in the mid fifteenth century, and the influence of Westminster Hall has waned. The notion of the hammer-beam roof as a vehicle solely for angel-hammer beams has been abandoned and other visual motifs gain favour. Indeed, pendant-post construction, unless the roof is a double hammer-beam, precludes the recumbent angels of Westminster Hall. One of these motifs was the ornamental wooden 'boss.' [10] During the fifteenth-century, ornate bosses, often foliate in nature, became a common feature of East Anglian roofs. These were frequently mounted at the intersection of the purlin with principle and intermediate rafters, and on the underside of king pendants. Wymondham Abbey in Norfolk (mid-fifteenth century) contains some particularly extravagant examples. Pendant-post framing allowed the carpenter to incorporate similar ornament at a lower, more visually dominant level below the hammer beam.

If the suggestion of a descended boss as a spur to pendant-post construction seems facile, then consider another ornamental catalyst. The date of the hammer-beam roof in Tostock Church, Suffolk, probably corresponds with the remodelling of its nave in the mid fifteenth century. If so, then Tostock is the earliest pendant-post hammer-beam roof in England. [11] Tostock is also significant because it displays a further ornamental advantage of the pendant post. In its lower section, the carpenter created a niche, and in that canopied space he carved a saint (all decapitated during iconoclastic purges). Such figures are common denizens of the wall posts of East Anglian roofs, but the pendant post supplied added locations for these devotional figures.

A further twist in the framing at Tostock demonstrates a final stage in an astonishing turnaround in carpenters' priorities during the later Middle Ages: from the rational and prosaic, to the ascendance of form and ornament. In contrast to the conventional pedant-post framing, at Tostock the carpenter rotated the hammer post by 45° and framed it arris-first into the hammer beam (Fig. 5). [12]

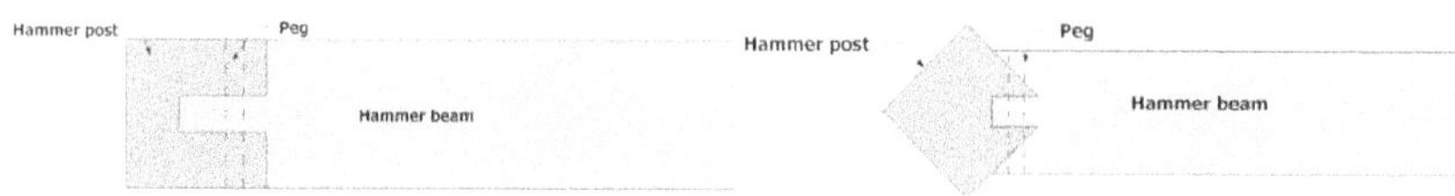

Fig. 5. Left: conventional pendant post arrangement (in plan); right: rotated, arris-framed pendant (presence of tenon in right is assumed)

Such construction meant that the layout and scribing procedures at the framing yard were rendered considerably more problematic. The carpenter had to cut a vee at the end of the hammer beam, scribed to the form of the rotated hammer post, and still create a tenon within that vee. A mortice then had to be chopped into the arris of a hammer post presumably secured by chocks. Furthermore, such construction also compromises the strength of the joint. To engage any substantial section of timber in the hammer post, the peg locating the joint can only be positioned near the end of the hammer beam tenon and its oblique shoulders, and thus the risk of the end-grain shearing under any tensile force is greatly increased. Nonetheless, the Tostock carpenter was willing to compromise structure and increase framing difficulties in order to accommodate a sculptural programme. An arris-framed hammer post allows the carver greater depth of timber in which to create his niche and to carve a more roundly modelled, three-dimensional occupant. At Tostock, if the conventional, non-rotated, configuration had been maintained, the hammer-beam tenon would have protruded somewhere in the vicinity of the saint's nose.

With constructions such as Tostock, hammer-beam framing was losing much of its structural function. One is left wondering if the hammer post is assisting the principal, or whether the hammer-beam framing is now suspended from the rafter.

Concluding Remarks

The phenomenal expansion in hammer-beam construction following Westminster Hall was no less than a revolution in roof carpentry. Westminster Hall, however, remained unique in structure, a masterpiece never replicated by medieval carpenters. Its influence lay in its unprecedented form and ornament which went on to inspire future patrons as they embarked on their church-building campaigns. They saw that the motifs of this secular hall, the receding arcuated forms, the angel-hammer beams, could be interpreted and applied with greater potency to an ecclesiastical space. Thus, angel-hammer beams suddenly appeared in church roofs regardless of structural need. The structural hammer-beam became a theme upon which innovative craftsmen improvised. Norfolk carpenters adapted structure primarily in the pursuit of form; Suffolk carpenters in the pursuit of ornament. As the century unfolded, craftsmen, displaying themselves no mere banausic workmen, developed a discriminating aesthetic acumen. Structure, though seldom neglected, became secondary, and although developing divergent structural traditions, East Anglian carpenters maintained a common principal objective: visual delight in the service of ritual and salvation.

References

[1] As argued by J. T. Smith, "Medieval Roofs: A Classification", *Archaeological Journal*, Vol. 115 (1958), p. 123.

[2] C. Given-Wilson, *Chronicles of the Revolution, 1397-1400*. Manchester: Manchester University Press, 1993, pp.124-125, 165, 171-172; C Given-Wilson, *The Parliament Rolls of Medieval England 1275-1504*, Vol. 8, Boydell Press / National Archives, 2005, pp.3, 9; J. S. Roskell, *The House of Commons 1386-1421*, Vol. 1, Alan Sutton, 1992, p.33.

[3] Nicolas R. Armor, *Late Medieval Ipswich: Trade and Industry*. Woodbridge: Boydell Press, 2011, p.206; L. J. Redstone, *Ipswich Through the Ages*. Ipswich: 1969, pp.73-74.

[4] Given-Wilson, *Rolls* (note 2), pp.3, 9; Roskell, *Commons* (note 2), p.33.

[5] H. Munro Cautley noted the lack of collars in Norfolk hammer-beams, *Suffolk Churches*. Woodbridge: Boydell Press, 5th ed., 1982, p.89.

[6] Bury lacks an orthodox king strut; the carpenter elected to frame twin struts in this position.

[7] For the function of the collar in medieval roofs: David Yeomans, *How Structures Work*. Chichester: Wiley-Blackwell, 2009, pp.184-186.

[8] Richard's heightened perception of his divinity: Nigel Saul, *Richard II*. Yale University Press, 1997, pp.449-450, 453; 'The Kingship of Richard II', in Goodman & Gillespie (eds.), *Richard II The Art of Kingship*. Oxford: OUP, 1999, pp.40, 46, 49. Also, the panel painting 'The Wilton Diptych', *c*.1395, London, National Gallery.

[9] Personal communication with architect Ruth Blackman of Birdsall, Swash & Blackman Ltd, Beeston.

[10] These ornamental timber 'bosses' have no structural relationship to the masonry boss.

[11] Date: English Heritage List for England, see website <http://list.english-heritage.org.uk/resultsingle.aspx?uid=1352395> [accessed on 14 November 2013]

[12] The only other example of this technique identified by the author is at Cotton Church, Suffolk.

Some examples of structural innovation in the construction of English churches in the long nineteenth century

Paul W.R. Bell
Jampel Davison & Bell, UK

Summary

The nineteenth century in England was a time of reaction in architecture, particularly church architecture. The Gothic Revival was a reversion to traditional patterns in liturgy, in decoration and in construction. A large part of the work of the architectural profession was the design of churches, while non-architects produced secular buildings, employing all the technical innovations that had become available during the Industrial Revolution. However there was cross-over between these two worlds, often when maverick builders with their own area of expertise were able to influence or interfere with the design of a particular church. There were also Architects, who were prepared to be more innovative than the mainstream. Cost played a part, when a cheaper method of achieving the desired architectural effect was acceptable. So advances in technology were able to leak into even this most conservative area of construction.

The examples of structural innovation are categorised and examined by the material used, dealing with laminated timber, with iron and with concrete. Where possible, connections are made between the builders and engineers working elsewhere in these materials and their use in the church buildings illustrated.

Introduction

The long nineteenth century (1789-1914) was a time of enormous change in England. Innovation, a concept promoted now, often by those who don't have to do the innovating themselves, was then spontaneous and rampant. It is believed by some [1] that the rate of innovation peaked in the year 1873 and is now declining. Certainly it is difficult to imagine more profound innovations than the change from the courier on horseback to the telegram/telephone and the introduction of the steam and internal combustion engines.

Innovation involves doing something new and original. It is the application of an invention or process. It means doing something in a different way with an implication that it is an improvement on the existing way of doing things. However innovation does not always end happily. Since it is new it carries risk. It can fail and some of the examples in this paper are examples of such failure or at least of technological cul-de-sacs.

The survival rate of church buildings can differ to some extent from that of buildings in general. They are subject to the usual forces and risks such as the weather, fire and warfare. Arson is sometimes a hazard due to their accessibility and, at least in the past, to sectarianism. They also become redundant due to population movements and declining congregations. However they are less susceptible than other buildings to commercial pressure and the system of ecclesiastical and planning jurisdiction in modern times places constraints on alteration.

Some examples of structural innovation in the construction of English churches in the long nineteenth century

The Romantic Movement in England in literature, art and music was a reaction to the Industrial Revolution. The Gothic Revival in church architecture, springing from the Romantic Movement, looked back to the medieval as the model. In its origin with Pugin it was an attempt to use the past in the hope of shaping the moral state of the present. Churches were expected to conform to the pre-Reformation paragon in their liturgy, in their planning and in their construction. The word church encompasses the various degrees of Anglican, Catholic, Presbyterian, Methodist and other denominations, all subject to the social changes of the time and with their different approaches, which were, however, tending to move in a similar direction architecturally.

One could assume it would be unlikely that structural innovation would take place in the construction of English churches in the nineteenth century. However this was not the case. There were churches built by men of industry who wanted to have their own way in using new materials and methods. There were churches built by rogue architects pursuing their individualistic goals. Later there were architects prepared to be innovative within the mainstream. And there was the general problem of the impossibility of returning to medieval methods five hundred years later. Churches were therefore at the point of collision between the forces of technological progress and architectural reaction.

Engineers have a tradition of categorising by structural material. This tradition has been followed in the organisation of this paper.

Laminated Timber

Laminated timber is a method of building up substantial timber members from smaller sections in order to overcome the limitations of size and span for single pieces of timber. In beams it is most useful to use horizontal laminations. The method was developed early in the nineteenth century in Bavaria, France and England especially for arch bridges [2 and 3]. The laminations, being relatively thin, could be bent to achieve the form of the arch, and then connected together. It was also used for roofs, as in the original roof of Kings Cross station (1851), later replaced by wrought iron, and in the nearby surviving German Gym (1864).

The problem for nineteenth century engineers was that they did not have modern glues to transfer the large "horizontal" shear forces between the laminations, which must be resisted for the separate pieces to act together as a single member. It is unclear to what extent in the early part of the century they understood the magnitude of these forces. The standard work of the time, Tredgold's "Elementary Principles of Carpentry", first published in 1820, describes the technique but in relation to thick laminations on a large scale and without a rigorous analysis of the forces. Brunel was certainly aware of the importance of dealing with these forces and conducted experiments in 1841 with different types of mechanical connector [4].

The examples found in the roofs of churches are on a small scale compared to the laminated timber railway bridges being built at the same time. Animal glues, pegs, trenails, screws and bolts were used to connect the laminations but none of these are very convincing to modern eyes.

Laminated timber in the nineteenth century could be considered as an innovation that became a cul-de-sac. It was superseded by iron by 1865 and later by steel both for bridges and roofs. In the mid to late twentieth century the machinery and glues became available that allowed it to function successfully, although usually for appearance rather than for structural/economic reasons.

Holy Trinity, Cambo, Northumberland (1842) J. & B. Green

This church is a classic example of men of industry using their expertise in a small church building. The father and son team of John and Benjamin Green were successful architect/engineer/contractors building substantial laminated timber railway viaducts in the north-east of England in the 1830s. Their Ouseburn and Willington viaducts were major structures. They also built station roofs of laminated timber. Unsurprisingly the bridges succumbed to decay within 30 years as gaps between laminations let water into the structures. The station roofs eventually went the same way. The small church at Cambo survives as does another at Horsley (1844). Both are in remote parts of Northumberland. The roofs are barrel vaults (Fig.1) with laminated timber arches. At Cambo there is no access to the space between the later ceiling and the pitched roof above.

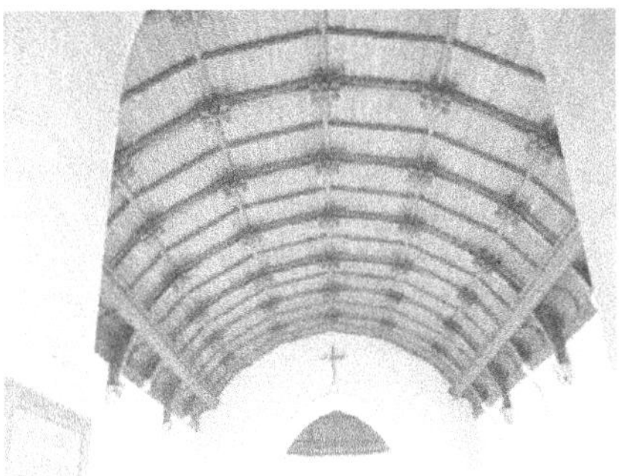

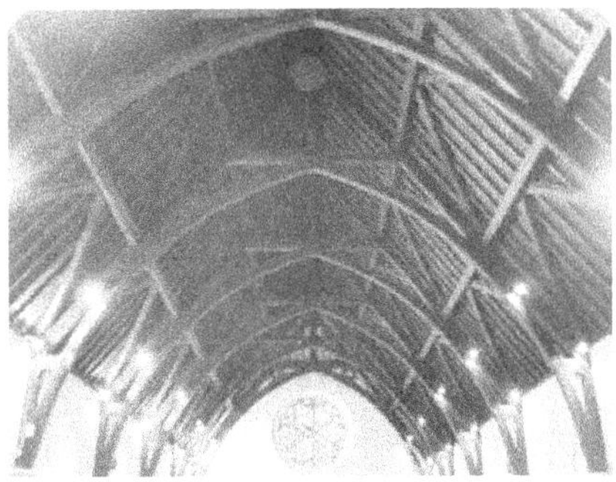

Fig. 1. (left) Holy Trinity, Cambo. Fig. 2. (right) St. Luke, Formby.

St. Luke, Formby, Lancashire (1852) W. Culshaw

There is a cluster of churches with laminated timber roofs in the Liverpool area built in the 1850s. The use of laminated timber made possible a vaulted roof with an appearance that might please contemporary ecclesiological taste at a reasonable price. It is difficult to know if the technique was chosen by the architect or by the builder. Clearly there was a reservoir of technical knowledge in this area of the country that the local architects were able to draw on. Lancashire at the time of the expansion of the coal and cotton industries had the apparent need for churches that could be built quickly and cheaply. St. Luke has a splendid nave roof (Fig. 2), unappreciated in Pevsner, who comments "Somewhat dreary interior". The main arches have bolts through the laminations, admittedly at wide centres. The curved braces also appear to be laminated.

Holy Trinity, St. Helens, Lancashire (1857) W. & J. Hay

Holy Trinity, St. Helens, is another quite similar example (Fig. 3). It has the additional challenge of a crossing, where the nave and transepts meet. The laminated arches can be seen close to in the enclosed west gallery. The walls of this building are built of crazy-paved copper slag, an additional innovation.

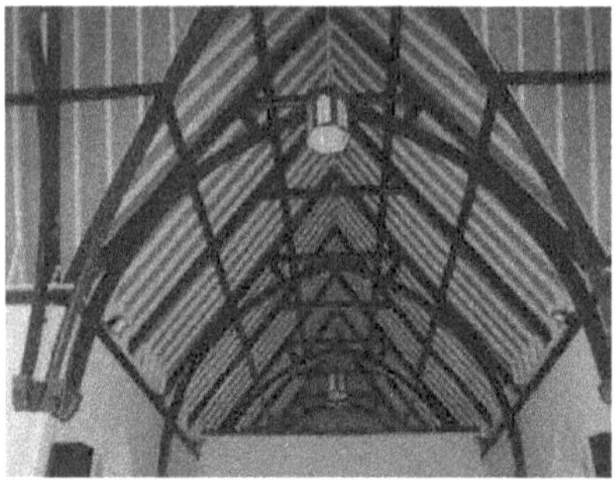
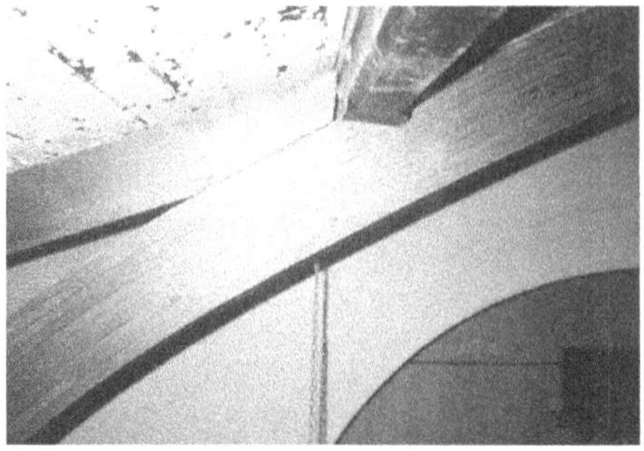

Fig. 3. (left) Holy Trinity, St Helens. Fig. 4. (right) St. Paul's, Deformed Arch.

St. Paul's Presbyterian Church, Isle of Dogs, London (1859) T.E. Knightley

The paradox of the Victorian age is summed up in this one small building - a Presbyterian church with an Italian Romanesque façade and a laminated timber roof. The foundation stone was laid by John Scott Russell, builder of the Great Eastern, launched in 1858, in the nearby shipyard. The fabricators of the most advanced technological artefact of the time worshipped in a building whose façade was a pastiche of Pisa Cathedral. There is the possibility of connecting the innovative roof structure with the technological giant Brunel, with his knowledge of laminated timber, who would have been working close by.

The roof is supported by three laminated timber arches. The timber beams are made up of 11 laminations connected by pairs of screws. The arches are deformed (Fig. 4) and have delaminated in places. The arches alone have to support the roof, without the benefit of in-plane forces. The point loads of the clerestory and purlins have induced bending effects not present if the arch was more evenly loaded. The church has survived despite its use for some years as a warehouse. It was rescued and restored in the 1990s after which it became "The Space" - a performing arts centre. The repairs included putting in steel dowels to carry the horizontal shear.

What conclusions can one draw from these laminated roofs? A sceptic might say that these arches are not laminated timber but just a series of bent planks fixed together in a dubious way without consideration of the shear forces. It may be significant that all the examples are of arches, where laminating is a cheap way of producing the curved members. It would be helpful to know how the laminations are connected, which is difficult to discover by visual observation, especially at a distance. The laminated arches may not be highly stressed since the boarded roofs will have a capacity to act as inclined deep beams. Load can be shed by the arches if they deflect, due perhaps to an inability to transfer shear. The buttresses at the arch positions provide horizontal restraint to either the arch or the whole roof.

Certainly there is a cluster of examples in Lancashire; there is at least one more, St Mary's, Grassendale, making a total of three. They are much more substantial structures than the Northumberland and London examples. The Cambo and Horsley churches have a clear connection to engineers with considerable knowledge of the technique. The London church has a possible connection. Engineers such as Vignoles and Locke were building laminated timber bridges in the north-west at the same time the Lancashire churches were being built and future research might find a link.

Other laminated timber roofs in churches may be overlooked since the laminations are not obvious if painted. A non-engineer would not necessarily appreciate that they are of technological significance.

Iron

The development of construction in iron in its various forms was a fundamental aspect of the Industrial Revolution - first cast iron, then wrought iron and finally steel.

The early cast iron churches in Liverpool were truly innovative. St James, Toxeth, has a very early use of cast iron columns (1774), which is contemporary with the cast iron columns in the early mills. The churches built in Liverpool by John Cragg in 1814-16 are a clear example of a man of industry imposing his ideas on the architect (Rickman) and the construction of a church. John Cragg was the proprietor of the Mersey Iron Foundry and a keen promoter of cast iron. His three churches are well known – St. George, Everton, St. Michael-in-the-Hamlet and St. Philip, Hardman Street. They are well documented elsewhere [5] so are not described in detail in this paper.

Cragg developed cast iron into a complete form of system-building for these churches. Only the outer walls of the churches are masonry, the rest is a kit of cast iron parts. Why was the system not adopted elsewhere? The conventional answer is that cast iron was unacceptable to the ecclesiastical authorities and their architects but these churches are so early that they pre-date the ecclesiological and architectural debates to come.

The central figure in those debates, in the struggle between iron construction and "architecture" in the mid-nineteenth century, was John Ruskin [6]. Ruskin's attitude towards the use of iron in buildings hardened over time. The key building is the Oxford Museum where he was so influential. During the 1850s he could accept that iron had a place but by 1877 he was saying "... it was all over with the building (Oxford) as soon as Woodward allowed iron". Ruskin and his contemporaries distinguished between wrought and cast iron, rather in the way that modern architects have distinguished between in-situ and precast concrete. Wrought iron, produced by craftsmen and hammered into shape was acceptable. Cast iron, poured into a mould and therefore capable of mass production was unacceptable.

Although they could not by then be described as truly structurally innovative the achievements of the mid-century architects using cast iron columns in their churches are remarkable. The innovative element was in the height and detailing of the columns. These are the rogue architects, who took the flamboyant aspects of Gothic to the extreme to the general disapproval of contemporary and later critics. They were unaffected by Ruskin's pangs of conscience and used cast iron to great effect. Two particularly free spirits were Bassett Keeling and Teulon [7].

St. George, Campden Hill (1864) E. Bassett Keeling

Few of Bassett Keeling's buildings have survived and those that have, like St George, have often been radically altered. Not all of its iron columns remain. The modern engineer would be concerned for their stability, particularly during construction. Bassett Keeling had a reputation for being able to complete buildings on time and within budget and cast iron could have been useful in this respect. The Italian Gothic favoured by the rogues implied the use of small columns and cast iron would have been attractive for that reason.

Some examples of structural innovation in the construction of English churches in the long nineteenth century

St. Mary, Ealing (1866) S.S.Teulon

Teulon's cast iron columns are slender, with beautiful detailing at the gallery supports and the capitals (Fig. 5). The use of cast iron allowed new spatial possibilities beyond what could be achieved with masonry.

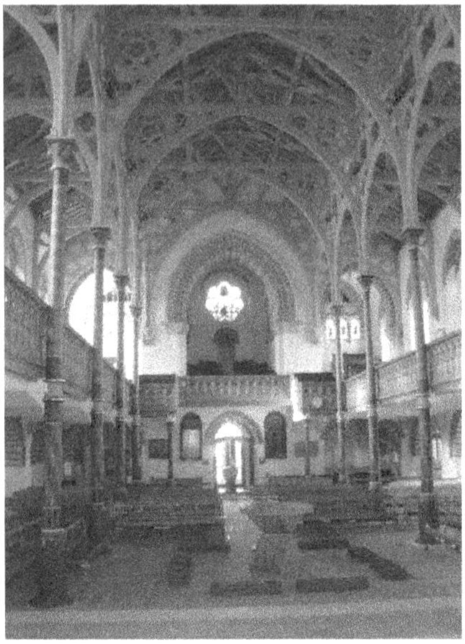

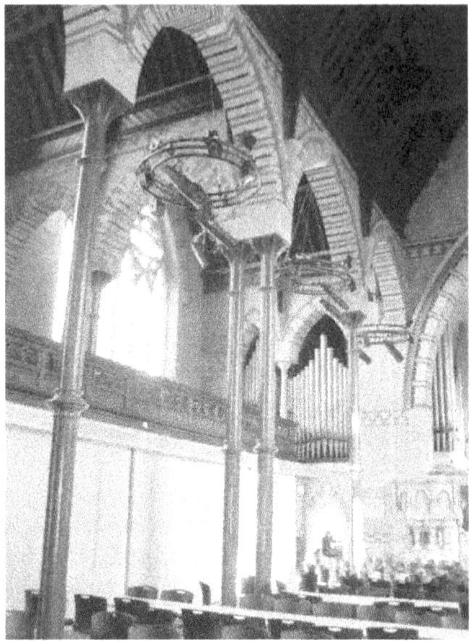

Fig. 5. (left) St Mary, Ealing. Fig. 6. (right) St Jude, Kensington.

St. Jude, Kensington (1867) G. & H. Godwin

The detailing of the cast iron in this church is superb. The columns are unrestrained (Fig. 6). Again the modern engineer wonders at their stability, especially during construction, and at their ability to cope with the lateral thrust from the masonry arches they support. The bold capitals look as if they have been designed by an engineer.

It would be interesting to know to what extent the work of Viollet-Le-Duc had an influence on these rogue Gothic architects. He put forward a convincing case for the use of cast iron in Gothic buildings and the use by the rogues of cast iron columns has echoes of his ideas. Their churches bear comparison with churches such as St Augustin in Paris, based directly on his work.

Another area to be explored is the use of steel in churches in the late nineteenth century. For example, Holy Redeemer, Clerkenwell (1887) by J.D. Sedding is said to have a steel frame [8]. If so, this would be an early use of the material.

Concrete

To find examples of concrete churches it is necessary to stretch the definition of the long nineteenth century up to the First World War. The great era of church building in England was over by then. There had been some odd earlier concrete examples such as St. Mary, Westley (1836), built in blockwork.

Lethaby used mass concrete for the vaulted nave of All Saints, Brockhampton (1901) under the thatched roof. But, unreinforced, concrete has no significant advantage over masonry. It is only when it began to be reinforced that new forms became possible.

St. Sidwell, Exeter (1901), built by Cottancin using his patent system, is a true example of innovation. It is well documented [9] so is not described in detail here. Cottancin's system was fragile and lacked a coherent theoretical rationale. There were questions over the stability of his design for St. Jean-de-Montmartre and Cottancin was bankrupted when acting as the contractor for St. Sidwell. This is another example of an innovation that failed.

Clare College Mission, Rotherhithe (1911) Simpson & Ayrton

The enigma concerning this concrete church (Fig. 7) is its possible connection with Sir Owen Williams, the outstanding English architect/engineer exponent of reinforced concrete [10]. The architects of the church were to work with him later on the British Empire Exhibition.

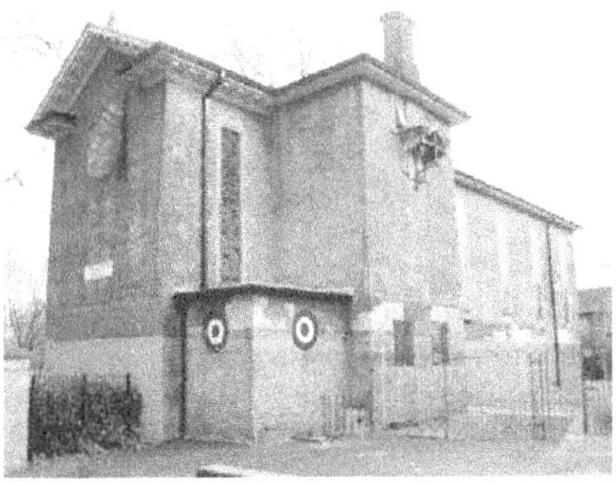

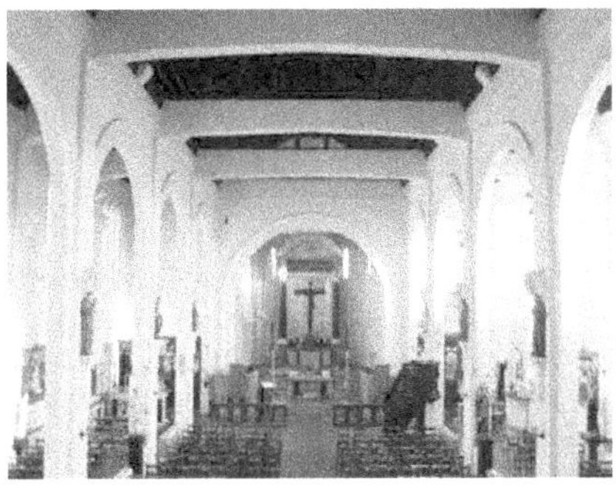

Fig. 7. (left) Clare College Mission. Fig. 8. (right) St. John & St. Mary, Goldthorpe.

St John & St Mary Magdalene, Goldthorpe (1914) Alfred Nutt

This church, although not a structural tour-de-force, begins to show what might be achieved in reinforced concrete (Fig. 8).

The real innovations in reinforced concrete had to wait for the work of August Perret in France after the First World War, culminating in Notre Dame du Raincy [11].

Conclusion

It is not easy to draw general conclusions from such different examples of innovation. A first conclusion is that technological innovation has a momentum of its own. It overcomes whatever barriers might appear to be in its way.

A second conclusion is that innovation can fail if the technology is not up to the demands made upon it at the time. Laminated timber falls into this category. Its large scale use for bridges was not successful, it survives in small churches where it was not highly stressed and was protected from decay. In addition to the technological deficit there was a theory deficit in the understanding of horizontal shear stress.

The early use of reinforced concrete can also be considered an unsuccessful innovation in some respects. Systems such as Cottancin's were failures. There was also a misplaced faith in the durability of concrete with inadequate cover and deleterious ingredients leading to subsequent problems.

One definite conclusion is that more fieldwork is needed. It could produce interesting results, particularly in the area of laminated timber. There are certainly a number of possible connections between builders, engineers and architects, currently rather speculative, which need to be confirmed. Engineers with an interest in construction history have tended to concentrate on engineering structures and secular buildings and neglected churches. It might have been assumed that churches have less to offer but this paper shows that not to be the case.

References

[1] J. Huebner, "A Possible Declining Trend for Worldwide Innovation", *Technological Forecasting and Social Change,* vol. 72, Issue 8, October 2005, pp.980-986.

[2] L.G. Booth, "The Development of Laminated Timber Arch Structures in Bavaria, France and England in the early nineteenth century", *Journal of the Institute of Wood Science*, 1971, 5, pp. 3-16.

[3] L.G. Booth, "Laminated Timber Arch Railway Bridges in England and Scotland", *Transactions of the Newcomen Society,* vol.44, 1971-2, pp.1-21.

[4] L.G. Booth, "Timber Works", in Sir A. Pugsley, (ed.) *The Works of Isambard Kingdom Brunel.* London: ICE, 1976, pp.107-135.

[5] Mario Rinke, Joseph Schwartz (ed.), *Before Steel.* Zurich: Verlag Niggli, 2010.

[6] Michael W. Brooks, *John Ruskin and Victorian Architecture*. London, Thames and Hudson, 1989.

[7] James S. Curl, *Victorian Architecture - Diversity & Invention*. Reading: Spire, 2007.

[8] Alastair A. Jackson, "The Development of Steel Framed Buildings in Britain 1880-1905", *Construction History,* vol.14, 1998, pp. 21-40.

[9] G.J. Edgell, The Remarkable Structures of Paul Cottancin, *The Structural Engineer,* vol. 63A, 7, 1985, pp. 201-207.

[10] David Cottam, *Sir Owen Williams*. London, The Architectural Association, 1986.

[11] Peter Collins, *Concrete: The Vision of a New Architecture. A Study of Auguste Perret and his Precursors*. London: Faber and Faber, 1959.

'Fabric wardens' and the organisation of parish church construction in the late Middle Ages

Gabriel Byng
Clare College, University of Cambridge, UK

Introduction

Between the artistic skills of medieval architects and the glamorous munificence of their patrons, the managers of late medieval and early modern church construction are a quotidian and mysterious group. Although the recent researches of Beat Kumin, Eamon Duffy, Katherine French and numerous others have thoroughly surveyed the administration of parishes in this period, they have only rarely looked beyond the financing of church construction.[1] Historians have largely focussed their attention on the work of churchwardens, who only occasionally ran building work, and the manor court. Clive Burgess stands as an important exception to this rule, and has suggested the importance of the 'masters' of the parish in running construction, a point which will be further explored in this paper.[2] The limitations of churchwardens' accounts and court records points to an important lacuna in our understanding of the place of parish church building in contemporary society and the 'politics of the parish'.[3] At its heart is the question of the identity of the individuals or groups who conducted major communal projects, including construction, but this also throws up numerous other lines of enquiry: what was their relationship to the mason, the parish, and other local authorities, especially the parish masters? How much discretion did they have, how were they chosen, and who was eligible for membership?

The paucity of documentation makes providing incontrovertible answers to these questions all but impossible, especially in the context of this brief paper. The first part is intended to identify examples where the conduct of parish church construction was given not to the churchwardens but to 'fabric wardens', dedicated temporary officials appointed by the parish assembly or masters. This term is chosen for convenience and, it must be emphasised, was not universally employed in the sources being examined. Evidence regarding their identity, status and appointment is suggested by a broad range of sources, across churchwardens' accounts, municipal records, wills, and inscriptions, although often only in brief, previously overlooked, references. The second part of this paper will be concerned with two exceptional manuscripts that provide important clues to, and, indeed, might be the only surviving indications of, the managerial approaches used by the bodies in charge of church building. They are the neat annual accounts made for audit by fabric wardens that survive for two important building sites. Although both have been in publication for well over a century, one, for work at Bodmin, has been used almost exclusively to provide insights into the fabric of the building and its funding, not its management, while the other, at Hedon, Yorkshire, has been altogether disregarded.

This paper will affirm that the 'fabric wardens' of parochial building projects were senior members of the parish, often former churchwardens, who brought years of experience in guilds, parish office, or local government to their roles, as well as considerable personal wealth, often used to provide credit for the work. They were closely involved in the daily execution of projects and, although appointed by senior authorities and kept in check using audits and the threat of replacement, they probably acted with significant independence. The use of fabric wardens had the advantage of providing dedicated managers, with proven experience and ability, as well as a clear organisational structure, accountable both for the project's execution and the compilation of accounts for audit. It freed other bodies from having to take on onerous duties and assured the parish the work would be undertaken thoroughly. Occasionally, of course, the churchwardens were sufficiently able, trusted or had large enough incomes that they were the appropriate choice for a construction project. However, projects run by the churchwardens, of which many well-known examples survive, or by individuals are beyond the remit of this paper, as are questions of finance and contracting.[4]

Evidence for 'fabric wardens'

Numerous sources, not limited to churchwardens' accounts, suggest the appointment of fabric wardens to run building work. Testators sometimes named individuals who may have acted as fabric wardens: Richard Turnour fulfilled a grant he had made for a new steeple to 'Mr Parke of London, merchant, and Mr Lamb of Leeds, and others of the parish' of Sutton Valence, Kent, in 1528; as Parke and Lamb were not masons it seems likely that they were fabric wardens.[5] More common was the delegation of such work to executors, but as these are usually the gifts of individuals rather than of the parish they are beyond the remit of this paper.[6] Inscriptions also suggest the use of dedicated overseers, as at Long Melford where they note donors 'of whose goods, John Clopton, Master; Robert Coteler and Thomas Elys did this arch make and glaze, and the roof over the porch' and 'of whose goods the said Kateryn, John Clopton, Master Willem Qwaytis and John Smyth, did these VI arches new repair'. It is likely that Clopton, Coteler, Elys, Qwaytis and Smyth acted, perhaps at different times, as fabric wardens (as well as patrons). If correct, Katherine, wife of Roger Moryell, would be the only female fabric warden I have identified.

Little information regarding the appointment of fabric wardens survives, but churchwardens' accounts and municipal records do provide some insight. In villages and many urban parishes fabric wardens might, like churchwardens, be appointed by the parish assembly or the parish masters, but for projects run and financed by municipal authorities they would fall under the ambit of local government. Some details of the appointment of fabric wardens for the building of Totnes tower in Devon survive in the Mayor's Court rolls (Fig. 1).[7] In January 1449, the court appointed six wardens ('*procuratores*', often translated as 'factors'), a 'receiver of the pence' to run the Sunday collection for the work (an office discharged, in later years, by one of the wardens), and a master mason, Roger Growden. In October 1449, two men were appointed as 'overseers of the workmen of the belfry and of quarrymen', apparently separate from the wardens and presumably as permanent employees of the works. The wardens may have been divided (as is suggested in January 1451) into two supervisors and four superintendents; as no further information survives regarding accounting procedure, it is reasonable to surmise that the receiver acted as accountant for the work, acting independently from the superintendents and submitting his records to the mayor's court for audit.

The appointment process for the Totnes tower wardens is not recorded, although officers seem to have been appointed biannually at the January and October meetings of the court (and in the April session in 1449). Almost half served only one term, but in 1450 the number of wardens was reduced to four who were then reappointed throughout 1451 with two additional members, forming a stable core that had been absent in 1449. Only one served throughout the period: William Rowe, Burhed's neighbour, who would become a long-serving mayor from 1452. Others fulfilled more junior positions as receivers, boundary wardens, tax assessors, chapel wardens, and fisheries wardens, while two were fined for rejecting the role of provost; some do not appear to have held office. This was comparable to other municipal building works, as at Beverley Bar, where twelve men were appointed '*custodes*' for the work.[8] However, the work could be given to existing officers: in Norwich an assembly appointed a temporary fabric committee of 24 men in 1407 to recruit builders and organise fundraising for the new guildhall, which placed the two chamberlains in charge (1408-15).[9] Another committee ran a local levy in 1408.

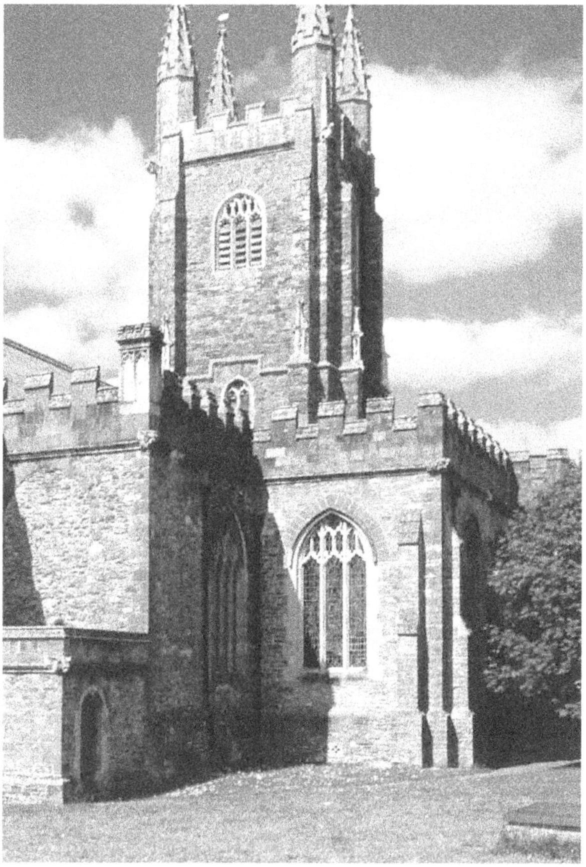

Fig. 1. The tower of Totnes, Devon (Photograph: Manfred Heyde).

An important contrast can be made with the appointment of fabric wardens by the parish, rather than by local government, to run building work and keep accounts at St Mary-at-Hill, London. An appendix to the churchwardens' accounts in 1499-1500, includes money given to a former warden and auditor, Thomas Colyns, to pay him back for his expenditure on the new tower and aisles, which he notes was 'paid in my reckoning... as it appears more plainly in my account'.[10] In 1501-02 Colyns was paid for his travel to Maidstone to 'bind Mawnde the mason in to perform his covenants' and for the cost of making the contract for work on the windows. Even more explicitly, in 1512-13, the churchwardens noted 'that John Allthorpe and Stephen Sanderson have promised to take charge and keep reckoning to pay all such workmen as shall make the battlements of our church... and Mr Parson is to assist them with his good diligence and wisdom... and Thomas Monders is chosen by the said parish to wait upon the said Stephen and Allthorpe in their absence and at their commandment for the furtherance of the same work'.[11] The men had proved themselves in several offices, including that of churchwarden, as well as being elder members of the community and personally wealthy. Allthorpe was an auditor in

1508-09, when he adds a memorandum himself, in 1509-10, and with Sanderson in 1513-14; Allthorpe was churchwarden in 1501-03, and Sanderson in 1505-07 and 1516-17. Sanderson paid 13s 4d for burial in the chapel in 1523-24. Although Colyns and Allthorpe and Sanderson kept accounts, they have not survived.

More opaque are the churchwardens' accounts at Saffron Walden, where, in 1486, the wardens noted that the collectors for the rood light 'delivered [their money] to William Middleton and John Nicholls to Thomas Spurgeon, receivers for the new work of the south aisle'.[12] It is likely that 'receiver' in this case was one office in a committee of fabric wardens, probably in charge of fundraising, as at Totnes. The men were all senior parishioners, members of large local families, appearing regularly in property deeds, all former churchwardens (in 1484, 1485 and 1485 respectively) and had already worked as receivers or ran the Easter collection.[13] They were, respectively, a yeoman, a draper and a dyer. Nevertheless, it was the churchwardens who travelled to Cambridge to contract for the work, although no details of building work made it into the accounts.

Surviving fabric wardens' accounts

These examples have augmented our understanding of the appointment and qualifications of fabric wardens. More significant, however, are the survival of two sets of accounts kept by fabric wardens, for the rebuilding of Bodmin, Cornwall (in a decaying book, covering 1469-72; Fig. 2),[14] and for the tower at St Augustine, Hedon, Yorkshire (in five damaged rolls, covering 1428-38, although building had begun earlier; Fig. 3).[15] It is likely that other examples survive, unknown to this author, but there may have been far less cause to preserve them than the accounts of the churchwardens: ordinarily such documents were no doubt discarded once they had served their purpose and the building was finished. Burgess has suggested that churchwardens' accounts survive as commemorative records of their tenure, as well as useful legal records, a purpose that may explain the survival of the Hedon or Bodmin accounts too.[16]

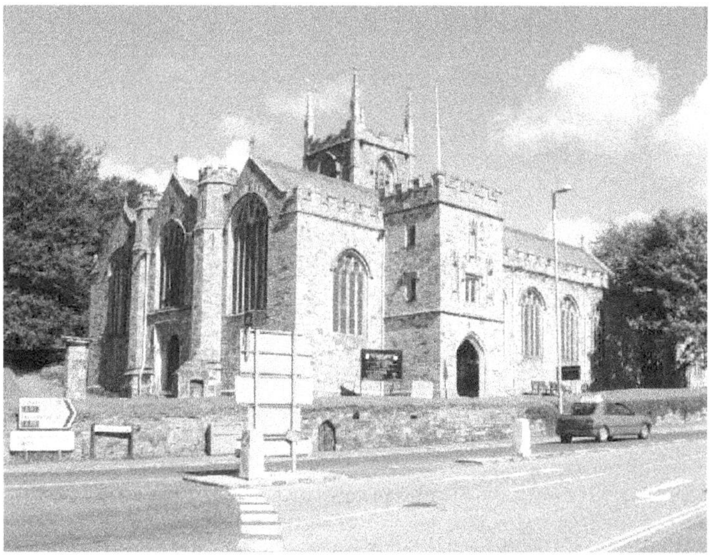

Fig. 2. Bodmin church, Cornwall (Photograph: Janine Forbes).

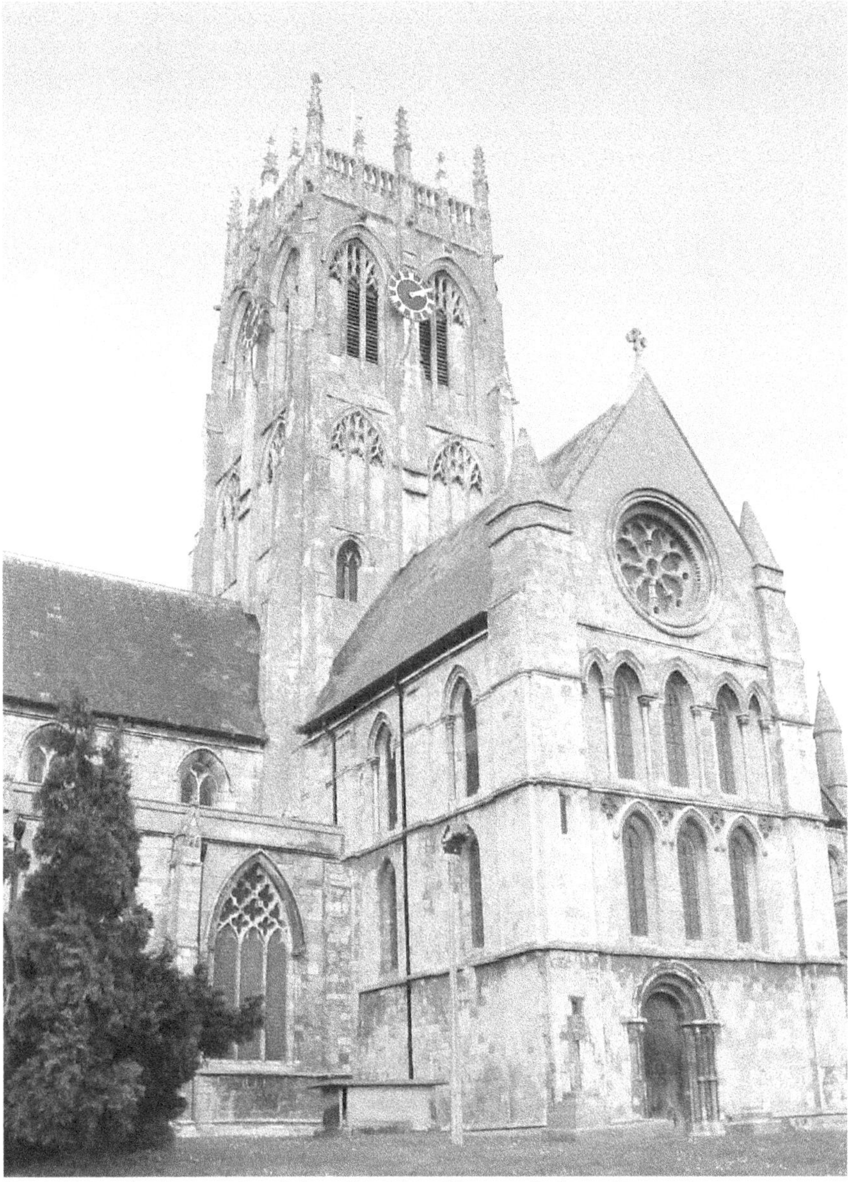

Fig. 3. Hedon church tower, East Riding of Yorkshire (Photograph: James Hopgrove).

The Bodmin and Hedon accounts, despite their fragmentary survival, provide useful insights into the work of parochial fabric wardens. Both were kept by important figures from the urban hierarchy. At Bodmin their title is never given (except, once, as 'receiver of the billing'), it is only noted that the accounts were kept 'for the church fabric'.[17] Nevertheless, their keepers' activities were those of fabric wardens. There were two over the course of construction, each working alone: Thomas Jerman

(October 1469-October 1471), a member of the Guild of St Eloy, a smiths' guild,[18] and Thomas Lucomb (October 1471-June 1472), who served as mayor of the town on four occasions (three times before the building work, and one afterwards), acted as a witness and was collector of the King's taxes.[19] Their positions within the civic hierarchy appear to have recommended them to take on the management of the building work on behalf of the town. They also contributed significant sums personally (Jerman almost £8; Lucomb almost £7), indicating both wealth and generosity. Lucomb lent almost £7, subsidised the glazing of the gable window as well as having his arms carved in the ceiling.[20] At Hedon, the wardens worked in pairs and were explicitly named as 'wardens of the fabric of the new belltower'.[21] The names of five survive: John Thorkleby and William West (Michaelmas 1428-30); followed by Robert Baty and John Ellerton (1434-35); and John Ellerton and William Chapman (1437-38). Baty probably gave up the position to become a bailiff.[22] The wardens had graduated through local government and the churchwardenship. Robert Baty and John Ellerton had served together as churchwardens in 1428-30, like Colyns, Allthorpe and Sanderson at St Mary-at-Hill. Baty and Ellerton served as bailiffs in 1422-23 and 1444-45,[23] and Chapman and Ellerton as chamberlains.[24] Baty and Ellerton were involved in land sales and acted as witness.[25] The town ran collections by street (as did Bodmin) and Chapman had worked as a collector in 1428-29.

The selection of the wardens at Bodmin is entirely opaque, and it is unknown whether the procedure is closer to that of St Mary-at-Hill or Totnes. The position was probably chosen annually in October but reappointment was possible (Jerman served at least two years) as was replacement (in Jerman's case by Lucomb). The extra set of accounts prepared by Jerman, summarising his income over the course of his entire tenure, and his replacement by a more senior peer suggest that his management may have been found wanting by the parish masters, assembly, or municipal authorities, and he was deposed. Lucomb, already mayor of the town on three occasions, may have been selected to take the work in hand as a safe pair of hands. If the account book was to be commemorative then Lucomb's formal Latin style may also have been preferred to Jerman's English. At Hedon, the wardens were appointed annually and changed regularly, so that no individual served for the entire period of construction, but the accounts are too fragmentary to determine their typical term of office: Thorkleby and West for at least two consecutive years; Ellerton possibly for four years; and Baty and Chapman perhaps just for one each. If others had served as wardens in 1430-34 and 1435-37, as is likely, then the ranks of collectors, bailiffs, churchwardens and chamberlains in the town would have provided a pool of experienced volunteers. William Rihill, for example, was collector (1428-30), procurator of a chantry (1429-30) and bailiff (1431).[26] Continuity in oversight may have rested with the town's government or a subcommittee as wardens moved between different local offices and the fabric wardenship, rarely occupying the latter for more than a couple of years.

The accounting procedure

The Bodmin and Hedon accounts provide surprisingly sharp contrasts, given that similar organisational approaches were adopted in both cases. At Hedon, the wardens followed similar practice to the churchwardens: their accounts begin with an introduction naming the wardens and the date; followed by an account of arrears (none in the extant examples); receipts from collections, gifts, and wardens; and expenses on stone, 'necessary expenses' and wages. Over the course of construction the format and wording of the accounts changes little; indeed the accounting procedure is so well established that the same subheadings are used even when there is nothing to record. It is of no surprise to discover that some of those running the work were former churchwardens and it seems likely the system was deliberately borrowed, having proved itself capable. These are clearly the neat annual editions of the accounts prepared for audit, quite possibly conducted by the same group and at the same time as that of the churchwardens' accounts, on Michaelmas, from when the accounts are dated and a festival often used by churchwardens as the start of the accounting year. Like churchwardens, the fabric wardens kept paper accounts (*parcellae*) during construction, which were checked against the annual accounts at audit.[27] The collectors also kept 'indentures', possibly agreements between them and the wardens or them and the donors as to how each would donate.

Although the Bodmin accounts were evidently also copied out and rationalised for audit or commemoration they do not follow the format of churchwardens' accounts and are often highly confusing. There are many missing sums in the accounts (e.g. one subtotal is for £70 but less than £5 of expenditure is recorded), probably abbreviated for the audit, and one section appears to be a revised summary of the whole accounts.[28] Unlike Hedon, they show the personal imprint of individual wardens, suggesting they were able to act with greater discretion. Jerman's accounts (written in the first person singular in English with occasional Latin), from 1469 to Michaelmas 1471, are divided between income and expenditure, grouped under relevant subsections, so that guild donations, sales, payments for wax and oblations are grouped separately; while major pieces of task work are divided from repeat payments for transport or materials. Jerman tends to use different forms for different types of income: 'item of' for donations and '*de*' for guilds. His accounts are probably a neat copy of two separate annual accounts, which were possibly in the form of the Hedon accounts: references are made to sums received 'this year'; a memorandum notes gifts given after Michaelmas 1470, probably after the start of his second term of office; while a subtitle in the expenses is for Easter 1471 (half way through his second year of office). Subsections always finish with their own total, but one is followed by what is probably an annual total ('*tot. fol.*') of £57. Gifts from the same guilds appear twice in the receipts, suggesting they were annual, and the names of churchwardens change (William Mason early in the accounts; David Witfen and Martin Hogge later). If these accounts were copied from two sets of annual accounts (perhaps of the churchwarden/Hedon form) the reason may have been commemorative, as is suggested by the use of formal Latin headings (e.g. '*Hic liber dico…*'). An alternative may be suggested, however: if Jerman left under a cloud he may have been required to compile a summary of his tenure to prove that his sums added up and to hand over to Lucomb.

Lucomb's accounts (in Latin and without pronouns), for 4 October 1471 to June 1472, headed 'Last Year', are shorter and more formal, with two lists of income and expenditure, and no subheadings.[29] They are followed, under the heading 'Next Year', by an undated 'memorandum of the accounts' (in Latin and without pronouns) by Lucomb and Jerman, although presumably written by the former.[30] These also consist of two short lists of income and expenditure, largely on timber, transport and payment to a carpenter. Most unusual of all is the next set of accounts, by Jerman, summarising various families of income for 1469-Michaelmas 1471, from sales, wax, gifts, guilds, the 'five stewards', collections and levies (in English and in the first person singular). This document has the appearance of a summary of the monies taken over Jerman's time in office, prepared from the full accounts perhaps to commemorate the contributions of the guilds or in response to concerns over his management: it ends with the total received, the total spent and the reckoning of the sum clear (about £2). There follows a list of guild debts and the current ownership of various donations or moneys owed (in Latin).

Responsibility and delegation

The decision to use individual fabric wardens at Bodmin seems curious: a major building project was demanding in time and expertise, and a second warden would help reduce the likelihood of mistake or corruption. Indeed, as mentioned above, there is an undated 'memorandum of the accounts' headed by both Jerman and Lucomb, although probably written by Lucomb. There is some evidence of delegation: the churchwardens raise and donate money; senior guild officers occasionally sell materials or donate sums; and others carry out organisational tasks. During Jerman's tenure as fabric warden, Lucomb received money from the churchwarden for the work; while Thomas Bere, who worked with the churchwardens, housed the carpenters and made a journey to fetch lime. Richard Trote was paid for riding to find a carver, 'to labour to have the timber home' and a relative, Bartholemew, who was mayor, worked with Jerman in St Eloy's Guild, and gave half a window. John Bare took a letter to carpenters in Devon; while the Hancocks, particularly the two Johns (one a mason 'and his fellows', the other a smith), were members of St Eloy's Guild, donated sums, handled part-payments and regularly worked on site.

The warden relieved himself of several tasks by contracting out several units of various sizes: under Jerman, Richard Richowe 'and his fellows' were paid £28 for the pillars, over £17 for the north and south walls, and almost £14 for the porch and pillars; Samuel Carpenter received almost £10 for task work and £7 for work on the roof 'as it appears by [an] account between the said Sam and me'; and William Carpenter received a part-payment of £3 for task work. Samuel Carpenter received £14 for task work under Lucomb, while Jerman and Lucomb's joint account includes £15 to William Carpenter in part-payment. In all some two-fifths of total recorded expenditure went on contractual work. Six windows were also glazed by parishioners, apparently paying for the work directly. However, much work was clearly run directly by Jerman or Lucomb: both their accounts, but particularly Jerman's, record detailed expenditure on many small tasks, involving transport, materials and labour (individual sums could be as low as 1d, as, for example, for the purchase of a single nail). He constructed a wooden lodge for seasoning the timber, and paid for drink, gloves and paper work. Although the accounts are clearly being prepared for audit, with many small payments noted as one (e.g. Witford was paid for 42 days work, probably seven weeks), the warden paid others for small individual tasks, from a day's work, to 'help' on a project or payment for drink. Richowe, the main contractor, was also paid on thirteen other occasions (with 'his fellows') for small tasks.

The fabric wardens at Hedon managed the minutiae of building work, buying stone, lime, sand, tiles, iron and timber, and paying wages. The purchases are not itemised, presumably as the accounts were being prepared for audit, but no contract is mentioned and the wardens seem to have been directly involved in making purchases. The fabric wardens delegated substantially to the churchwardens, who paid several masons' wages and bought stone. The churchwardens' contributions were worth large sums of money, in one year accounting for almost four-fifths of the fabric wardens' total income. The fabric wardens also seem to have been able to call on a large band of volunteers from local government, most notably for the collections that were run on a street by street basis by pairs of collectors, often serving for several years. The mayor, John Burhed, took on constructing a new quay for unloading barges.

Conclusion

Late medieval church building was not invariably run by the churchwardens, as is often claimed or implied by parish historians. Indeed, some seven examples have been given in this paper of parochial projects probably carried out by groups of senior parishioners appointed by the parish or town authorities. These bodies were highly variable in their operation, varying from individual wardens (Bodmin) to large committees where individuals were given specialised tasks to execute (Totnes). The seniority of the members, probably higher than that of the churchwardens, suggests a considerable degree of independence, but the authorities that appointed the wardens could be relatively interventionist, as at Totnes. The use of audits, contracts, meetings, and, ultimately, dismissal, ensured oversight and decision making remained with the parish's leading authorities. The examples in this paper are largely taken from towns but there is little reason to suppose that such arrangements were unknown in rural areas: certainly villages had well-structured social hierarchies and sophisticated managerial systems by the fourteenth century. Similarly, although the examples in this paper are from the fifteenth and early sixteenth century it is plausible that fabric wardens were used earlier still. They provided dedicated, experienced and trustworthy managers to oversee a taxing but important job, leaving existing authorities to continue with their ordinary duties. Nevertheless, considerable delegation of responsibilities was often necessary. Fabric wardens kept accounts for audit, often probably modelled on those of the churchwardens, and, remarkably, two examples survive, at Bodmin and Hedon. These show the highly detailed level of daily control required of fabric wardens: purchasing and recording small units of labour and materials, directing work and monitoring progress. This kind of attentive and thorough management was and is critical for the successful completion of building projects, as much as the contributions of funders and contractors. These responsibilities were, nevertheless, the preserve of a minority of parishioners, whose identity must be integrated into our understanding of how the relationship between patron, community, and craftsmen reflected parish society and politics.

References

[1] Beat A Kumin, *The Shaping of a Community: The Rise and Reformation of the English Parish, C. 1400-1560*. Aldershot: Scolar Press, 1996; Katherine L. French and Gary G. Gibbs, *The Parish in English Life, 1400-1600*, Manchester: Manchester University Press, 1997; Katherine L French, *The People of the Parish: Community Life in a Late Medieval English Diocese*. Philadelphia: University of Pennsylvania Press, 2001; Eamon Duffy, *The Voices of Morebath: Reformation and Rebellion in an English Village*. New Haven: Yale University Press, 2001.

[2] Clive Burgess, "Pre-Reformation Churchwardens' Accounts and Parish Government: Lessons from London and Bristol", *The English Historical Review* 117, no. 471, 1 April 2002, pp.306 –332; Beat Kumin, "Late Medieval Churchwardens' Accounts and Parish Government: Looking beyond London and Bristol", *The English Historical Review* 119, no. 480, 1 February 2004, pp. 87 –99; Clive Burgess, "The Broader Church? A Rejoinder to 'Looking Beyond'", *The English Historical Review* 119, no. 480, 2 January 2004, pp.100–116.

[3] Keith Wrightson, "The Politics of the Parish in Early Modern England", in P. Griffiths and S. Hindle (eds.), *The Experience of Authority in Early Modern England*. London: Palgrave Macmillan, 1996.

[4] For the contracting, finance and churchwarden-led projects see my PhD thesis "Planning and Paying for Parish Church Construction in the Later Middle Ages", at Cambridge University; for the only surviving private accounts of a gentry run project see: Gabriel Byng, "The Construction of the Tower at Bolney Church", *Sussex Archaeological Collections* 151, 2013.

[5] A. Hussey, (ed.) *Testamenta Cantiana (East Kent)*. London: Mitchell, Hughes and Clarke, 1906, p.333.

[6] Most exceptionally, William Wykeham charged his with finding lime, sand and quarries, and paying the contractors, for the nave of Winchester Cathedral: J. Nichols, (ed.) *A Collection of the Wills Now Known to Be Extant of the Kings and Queens of England*. London: Printed by J. Nichols, printer to the Society of Antiquaries, 1780, pp. 339, 766.

[7] Devon History Centre 1579A-0/9/19-21; C. F. Rea, "Building of Totnes Parish Church", *Transactions of the Devon Association* 57, 1925, p.282; Hugh Robert Watkin, *The History of Totnes Priory & Medieval Town, Devonshire*. Torquay: the author, 1914, pp.395–425; Percy Russell, *The Good Town of Totnes*. Exeter: Devonshire Association for the Advancement of Science, Literature and Art, 1964, pp.34–42.

[8] George Poulson, *Beverlac*, vol. 1, London: Printed for G. Scaum, Beverley, and sold by Longman, Rees, Orme, Brown and Green, 1829, pp.118–72.

[9] Richard Howlett, "A Fabric Roll of Norwich Guildhall", *Norfolk Archaeology* XV, no. 2, 1903, pp.174–89.

[10] London Metropolitan Archives P69/MRY/B/005/MS01239/001/001, f. 181; Henry Littlehales, (ed.) *The Medieval Records of a London City Church (St. Mary at Hill) A.D. 1420-1559*. London:

Published for the Early English Text Society by K. Paul, Trench, Trübner & Co., 1904, p.239; this quote is used by Burgess in "Lessons" (Note 2), p.326.

[11] London Metropolitan Archives P69/MRY/B/005/MS01239/001/002, ff. 328-40; Littlehales, *St. Mary at Hill*, (Note 10) p. 284.

[12] Essex Record Office D/DBy Q18, f. 140.

[13] Essex Record Office D/ACR1 f. 38; TNA:PRO PROB 11/18/272; PROB 11/37/451; PROB 11/42B/699.

[14] Cornwall Record Office BBOD/244; J. J. Wilkinson, "Receipts and Expenses in the Building of Bodmin Church, A.D. 1469 to 1472", *Camd. N. S.* 14, 1875, pp.1–41.

[15] J. R. Boyle, *The Early History of the Town and Port of Hedon*. Hull: A. Brown & sons, Simpkin, Marshall, Hamilton, Kent, 1895, Note x.

[16] Burgess, "Lessons" (Note 2), pp. 314–5.

[17] '*pro fabrica Ecclesie*'.

[18] Cornwall Record Office ART/2/5/7; Wilkinson, "Bodmin", (Note 14) p. 38.

[19] Cornwall Record Office BTRU/18; AR/1/812; J. Wallis, *The Bodmin Register*. Bodmin: Liddell and son, 1827, p.277.

[20] Wilkinson, "Bodmin", (Note 14) pp. 35, 24, 31.

[21] '*custodem fabrice nove campanilis*'.

[22] East Riding of Yorkshire Archives DDCC/45/29.

[23] East Riding of Yorkshire Archives DDHE/19/1, pp. 192, 199.

[24] East Riding of Yorkshire Archives DDHE/19/1, pp. 250, 245; DDCC/45/20 and 29; West Yorkshire Archive Service SpSt/4/11/112/2.

[25] East Riding of Yorkshire Archives DDCC/45/23, 25-6; DDHE/19/1, f.26; DDCC/81/13.

[26] East Riding of Yorkshire Archives DDCC/45/25.

[27] See two entries in 1429-30: '[...]*tempus compoti prout patet per parcellas huius compoti super hunc compotum*[...]'; '*prout patet per parcellas super hujus compotum ostensas et examinatas*'.

[28] This point is also made in: Norman John Greville Pounds, *A History of the English Parish: The Culture of Religion from Augustine to Victoria*. Cambridge: Cambridge University Press, 2000, p.401.

[29] '*Ultimo anno*'.

[30] '*Proximo Anno*'; '*memorandum de compoto*'.

Strategies for the accord of Gothic and Classical stone construction systems in 16th- and 17th-century buildings in Northern Spain

Alberto Calderón-González
Universidad Politécnica de Madrid, Spain

Introduction

It is an increasingly attested fact that Gothic architecture, rather than fading away into irrelevance, kept developing through the whole sixteenth century and remained in existence much longer. Thus, the introduction of Classical aesthetics – together with a new constructive logic based on surface definition – far from causing the disappearance of the line-defined Gothic system, meant that both would reach their peak simultaneously, coexisting well into the seventeenth century.

Although purely stylistic overlapping and incidental juxtaposition abound, deliberate conjunctions of both systems are rarer and have generally been overlooked. This paper will focus on these, showing how formal concerns become inextricable from construction-related issues, particularly in the need to address the accord along contact surfaces of two stereotomy systems with oft-conflicting logics, work strategies and singular points.

A case study will be undertaken of buildings in Northern Spain showcasing these features. These demonstrate a range of solutions, with specific technical and aesthetic implications which will be discussed. Strategies of varying complexity have been recognized, ranging from simple instances of vault juxtaposition to the remarkable case of the Villasidro parish church, in which the search for formal continuity through the use of shared vault elements triggered a learning process, with multiple solutions being tried and perfected over time.

The coexistence of Gothic and Classicism

Traditional historiography viewed stylistic evolution as a linear succession of events, with newer styles replacing older, already 'decadent' ones. In this light, late Gothic was described as contrived and out of fashion, and it was considered natural that it should be replaced, swiftly and uniformly, by Renaissance in the early sixteenth century.

Development of the concept of 'Gothic survival' [1] started to put this into question, acknowledging that Gothic had somehow continued to exist well beyond its conventional time frame, even into the eighteenth century, although as an outdated form. Later research [2] has elaborated on this in two directions: establishing that Gothic construction continued to advance and develop all over Europe well into the sixteenth century, yielding some of its greatest achievements; and also attesting that all throughout the seventeenth and, in some cases, the eighteenth centuries it often remained a relevant option.

In Spain, Javier Gómez Martínez [3] was the first to shed light on the sophistication reached by Gothic architecture in the sixteenth century and its continuation well afterwards. Enrique Rabasa [4] and José

Carlos Palacios [5], among others, have followed in this direction, researching into the construction techniques associated with it.

Nonetheless, it is also a fact that Spain was, together with France, at the forefront of the Classical stone construction system, being home to one of the first stereotomy treatises [6] on record and paramount examples of Renaissance stonecutting. This implies a simultaneous currency of Gothic and Renaissance at both the formal and technical levels; and therefore a high degree of mutual influence, with Renaissance design being influenced by Gothic construction techniques and vice versa [7], and simultaneous use of Gothic and Classical strategies in many buildings.

These usually employ Gothic vaulting for the larger part of the building, limiting Classical stonemasonry to 'prestige' elements (apses, chapels or domed lantern towers); this can be justified by a greater constructive expense [8] that often prevented it from being used pervasively outside specific regions or high-profile buildings.

Different approaches to stone construction

The strategies of Gothic and Classical stone construction are fundamentally different. [9] Form definition in Gothic vault construction is line-based: the vault is defined by a set of lines physically expressed as intersecting ribs, disregarding the surface shape which comes only as a result and is concealed by the prominence of these. Rib intersections become singular points, with the tas-de-charge and bosses requiring specific definition and concentrating design effort. The former are the common springing of ribs, made of solid horizontal courses defined through their beds by the use of templates, with the last of these being tapered for a better support of the ribs; the latter are rib intersections, commonly conceived from a vertical axis and thus usually carved from a horizontal extrados [10], cutting a branch at a different angle for each concurring rib.

Design of Classical vaulting, instead, generally focuses first on defining the geometric shape of the vault surface, (often of an ideal nature, or else an easier to define substitute), and then trying to solve the design of its elements. In this way, all parts of the surface have similar prominence, and emphasis on singular points and lines is greatly reduced, even if these may still be used as tools for geometric control. [11] The importance given to surfaces and their geometric development implies that voussoirs are usually carved from the intrados, and bed geometries derive from it.

A conjunction of Gothic and Renaissance: multiple spatial possibilities

The substantial difference between Gothic and Classical stone construction made physical continuity between both systems difficult to put into practice. This had to be taken into account when designing both into one building; therefore, while elaborate examples of purposeful combination of the Gothic and Classical construction systems -La Santa Espina, Valladolid (Fig. 1) or La Vid, Burgos- are relatively common, they tend to arrange different types of vaulting at different levels to avoid contact between them. Besides requiring proficiency of the masons in both systems (which was fairly common, judging by the contents of stonecutting treatises or the work history of many masters), construction of such buildings did not pose any novel issue.

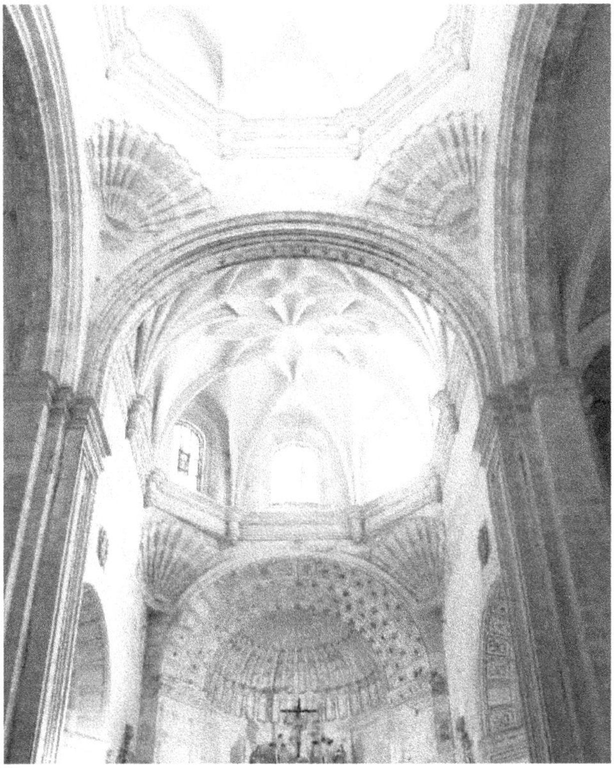

Fig. 1. La Santa Espina, Valladolid. Chancel vaulting (Author: Paula Guillot).

In some cases, however, physical continuity was not avoided, and Gothic and Classical vaulting were designed into a formal continuum through common springing heights and similar arch radii. This was less frequent, for it meant that the issue of system accord had to be addressed. Nevertheless, a significant number of cases have been discovered. A relatively large, geographically close sample of these is located in Burgos province in Northern Spain [12], on which this work will therefore focus (examples from other, specified, regions will be included when necessary); these cases are actually quite varied in their spatial configuration.

The most common of these employed Classical vaulting for the church apse, which could be a half-dome or a semi-octagonal vault, supported by either squinches over a square or octagon, pendentives over a square, or simply a semi-circular wall. Examples include Espinosa de los Monteros (half-octagon on squinches), Ajo, Cantabria, and Olmillos de Sasamón (half-dome over a semi-circular wall), or Las Bernardas, Salamanca (half-dome on a combination of squinches and pendentives).

In some cases, a more elaborate chancel was designed comprising a central Gothic vault with a Classical apse and two barrel vaults forming a small transept. This was common for enlarging a previous, smaller building, as in Ledesma, Salamanca (the apse a pointed half-dome on a semi-circular wall) and Moradillo de Sedano (the apse a half-dome on octagonal pendentives).

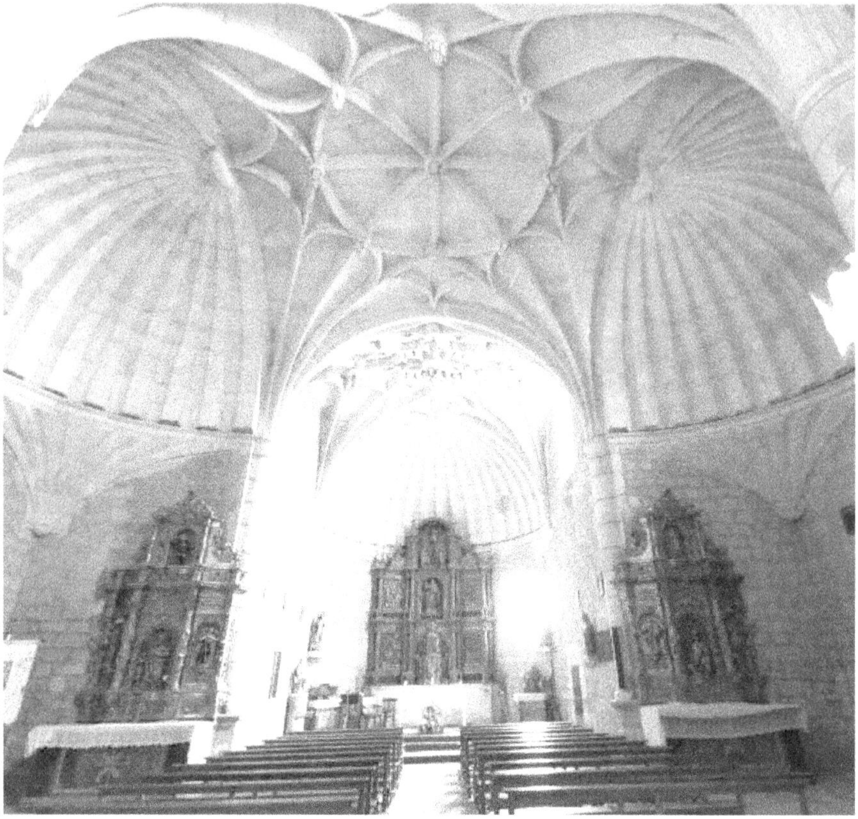

Fig. 2. Villasidro, Burgos. General vault layout.

A third option was Classical vaulting of lateral spaces acting as either a transept or side chapels. This happens in Sedano with half-domes on irregular pendentives; in Rubena, two juxtaposed transepts (half-domes on squinches) were added in this fashion. The church in Villasidro is a combination of this case and the first one: a three-bay Gothic nave with Classical apse (half-dome on octagonal pendentives) and transept featuring half-domes on square pendentives (Fig. 2). Though a rather small building, it is a particularly noteworthy example of formal interaction of both systems creating a more fluid space.

Exceptionally, in the Villaveta church Classical vaulting was not applied to any apse or chapel, but to the connection between the nave and the much smaller choir through what amounts to a massive, truncated squinch.

A range of solutions

Master masons approached the issue of designing the elements of contact between both vaults through various methods. Overall, solutions can be grouped into three general strategies: juxtaposition of both vaults, concentric superposition and, in some rare and very specific instances, actual use of a common arch (Fig. 3). [13]

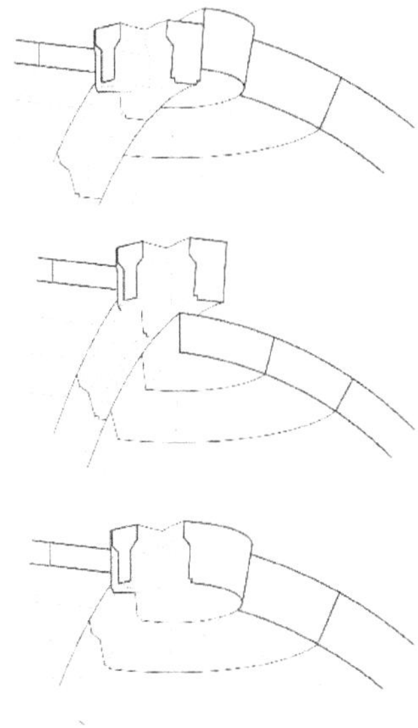

Fig. 3. Strategies for vault accord: juxtaposition, superposition, shared arch.

Vault juxtaposition

Juxtaposition of both vaults appears to have been the most frequent and widespread of these strategies. [14] Its main advantage is that both vaults become totally independent not only construction-wise, but also geometrically. Neither stereotomy, nor arch radii, nor even springing heights need coincide. Such liberties do not preclude a careful design, but they allow for a wide range in the quality of solutions, as demonstrated by the cases of Rubena and Moradillo de Sedano.

In spite of their spatial originality, the double-transept half-domes in the church in Rubena -built in two stages, the former after 1580 and the latter well within the seventeenth century [15] - exhibit a highly uneven construction: vaults spring at inconsistent heights unrelated to the wall rib, half-domes are not semi-circular in either plan or elevation and even arch cross-sections are awkwardly fluctuating. Wall rib intrados radii are smaller than those of the half-domes, which helps disguise execution errors and design imprecisions. [16]

In contrast, the Moradillo de Sedano church, while using the same basic strategy, shows a much more careful design. Its 1616 [17] chancel, although a combination of constructive systems, manages to form a coherent whole (Fig. 4). Attention to details, such as pendentives continuing the ashlar courses of the wall and ribs interlocking with pillar mouldings in the tas-de-charge (which requires displacing the rib springing points on the ground plan), emphasize the precision of the design. In this vein, when juxtaposing the apse and crossing vaults, half-dome and rib radii appear to be the same, but the rib shows moulding on both sides, thus marking a subtle separation from the apse through a line of shadow and highlighting the well-executed joint instead of concealing it.

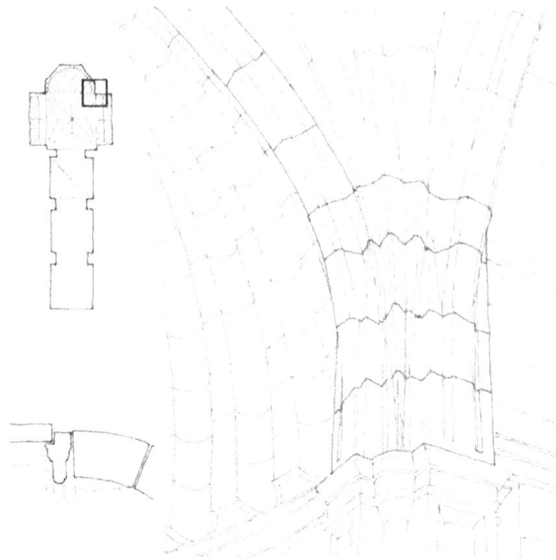

Fig. 4. Moradillo de Sedano. Juxtaposed springing and detail of arch moulding.

Vault superposition

An alternative strategy to vault juxtaposition was concentric superposition of both vaults, resting the Gothic transverse rib on the extrados of the Classical vault. In this way, vault stereotomies remain independent. However, arch radii do not; or at least, disagreements between them become much more visible. Furthermore, arch concentricity requires a horizontal space reserve for the springing of the Classical vault within the rib intrados, translating to a wall setback equivalent to the section of the former. Although this might appear obvious when looking at the finished building, it had to be considered from the very moment of plan design, where it would have been much less evident. Failure to do so in the cases studied led to design complications that otherwise could have been easily avoided; this was the case in Villaveta and Espinosa de los Monteros.

The large squinch in Villaveta (possibly earlier than other examples [18]) was clearly meant to be concentric with the transverse rib, as their upper two-thirds fit perfectly. However, the two oblique walls supporting the squinch are directly adjacent to the pillars supporting the Gothic vaulting, so that no space reserve was left for the springing of its arch face. In order to solve this inconvenience, its lower courses were tapered into a wedge, so that the outer arch of the squinch departs visibly from the general circular shape at a given point (Fig. 5).

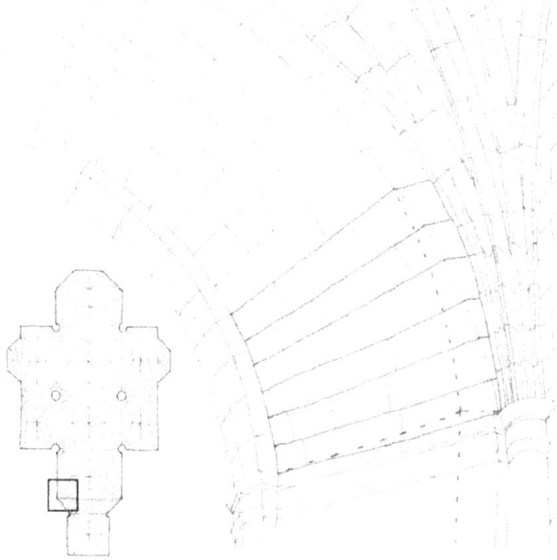

Fig. 5. Villaveta. Connection squinch, showing theoretical and actual arch shapes.

In the semi-octagonal apse vault at Espinosa de los Monteros, apparently finished by 1572 [19], the issues would have been similar, but the solution was quite the opposite; instead of tapering the inner vault, a cut was made into the Gothic tas-de-charge to make room for it. Furthermore, both arches do not fit properly, even in the upper ranges where nothing would prevent it, and the inner vault shows evidence of deformation. This might indicate construction of the apse at an earlier stage and subsequent structural movements before work began in the nave vaults [20], which would also explain why the Gothic tas-de-charge had to adapt to the half-octagon shell, rather than otherwise. [21]

In the already mid seventeenth century [22] Sedano church, contact between superposed arches was avoided altogether: the arch radii of transept vaults are significantly smaller than those of the crossing vault, leaving a stretch of wall between both (Fig. 6). This inelegant solution is, however, probably not due to an intention of avoiding construction problems, but rather to the long and haphazard history of the building, to which the crossing vault was added well after the apse, choir and transept were finished. The aforementioned wall stretches continue beyond the vaulting and support the nave roof; this would have been useful for closing the church before construction of the crossing vault began.

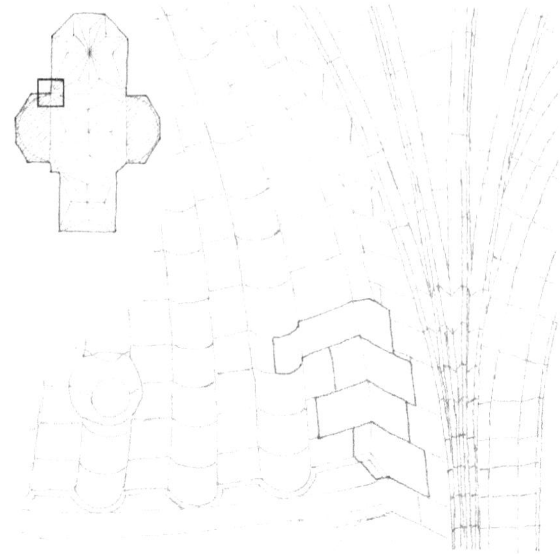

Fig. 6. Sedano. Springing, showing a stretch of wall separating both vaults.

Villasidro: a unique case of conjunction through shared elements

The Villasidro parish church, apparently finished in 1603 [23], departs markedly from the previous strategies. Instead of avoiding the difficulties of conjunction, they are faced directly. No separation is left between vaults: Gothic infilling rests directly on the apse and transept half-domes, so that their edges function as wall ribs for the lierne vaults, creating a spatial continuity that stands out from the previous examples.

Such a situation leaves some of the constructive elements in a contradictory situation. The springing of the half-dome, according to Classical designs, would typically consist of conical, sloped courses; however, the courses of the Gothic tas-de-charge would have to be horizontal. Less significantly, the half-dome keystone with springing liernes would also have been approached from two opposite strategies.

A close look at the vaults in this building reveals not only that these issues did surface, but also that they were addressed in more than one way. In the apse vault, the pieces of the tas-de-charge continue into the half-dome, but their courses do not coincide with those of the latter, which are evidently conical as can be seen through the fluting on its surface; they interlock in an irregular way (Fig. 7). Thus, it appears that the tas-de-charge was prolonged into the apse surface and then adapted ad-hoc.

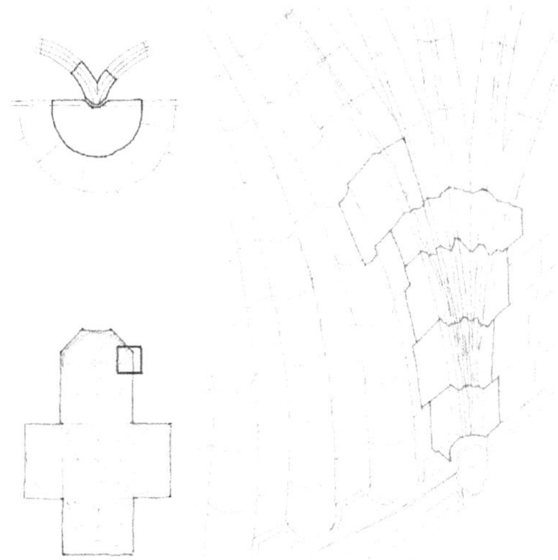

Fig. 7. Villasidro. Shared arch: apse solution.

In the transept vaulting, the fluting instead shows that the first two courses are horizontal; their levels coincide with those of the tas-de-charge (Fig. 8). This makes it possible to build the Classical dome and Gothic ribs as continuous courses without awkward interlocking. Although inconsistent arch radii (which coincide in the upper two thirds but diverge oddly in the springing due to a slightly stilted half-dome, possibly meant to adjust arch heights) somewhat deteriorate the final result, this solution is still a clear improvement over the former. It employs an already existing design strategy of making the first courses of a dome horizontal to reduce centring expenditure (a technique with examples as notable as the Escorial domes [24]), and adapts it to fit new needs.

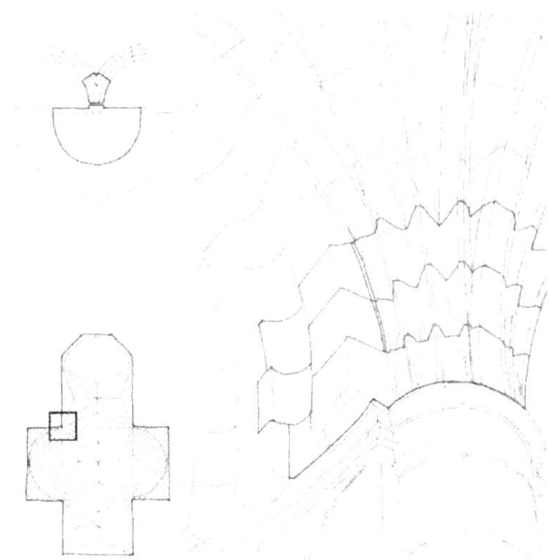

Fig. 8. Villasidro. Shared arch: transept solution.

Keystone designs are also different. The very large apse dome keystone does not incorporate lierne springing; it leaves a socket, concealed by a decorative volute, into which they fit. The transept vaults, instead, reduce the dome keystone, and add to it a short, easy-to-carve section of straight rib, so that lierne bifurcation still happens in a separate piece, but continuity between the liernes and the half-dome is emphasized.

The Villasidro vaulting has a close parallel in the apse at Olmillos de Sasamón church. This nearby, stylistically similar building (although larger and less uniform, showing evidence of works between 1540 and 1610 at least [25]) also includes an apse half-dome continuous with the Gothic nave, in a design that suggests a likely common authorship.

In this case, the vault was painted over and stone courses and joints are therefore much harder to distinguish. Close examination of the springing seems to show discontinuity between the half-dome and the tas-de-charge; if this were so, the shared arch would be part of the half-dome only, and would have to be cut somewhat abruptly to fit with the tas-de-charge.

This, although visually acceptable, seems an unsatisfactory solution, requiring an awkward tas-de-charge design. Therefore, and considering a possibly slightly earlier date for the Olmillos vaulting [26], it is plausible to think of it as the earlier attempt at integration, which would later lead to the more refined solutions in Villasidro. Among these, the apse vault would be earlier, achieving a single but imperfectly integrated springing; the final solution, realizing the accord of vault courses, would be the one given in the transept vaulting.

Conclusions

The cases studied here bear witness to the resourcefulness of sixteenth and seventeenth-century master masons. They come to show that even minor masters working in relatively low-profile projects were willing to experiment and face new problems, and could draw solutions -although not always wholly successful- from their own ingenuity, perfecting them over time, instead of falling back on established procedures. The line of research carried out in Olmillos and Villasidro is a clear example of a modest task undertaken with a high degree of formal and technical ambition, and of the different strategies tested in order to address difficulties arising as a result.

With the odd exception, construction of all the assessed buildings was carried out over the last decades of the sixteenth century and the first half of the seventeenth century. Therefore, they do not speak of a transitional situation between Gothic and Classicism, but rather of a conjunction taking place when both construction systems had already been well established and incorporated into the common knowledge of master masons.

Acknowledgements

This work would not have been possible without the support of several people. Particular credit should be given to Miguel Sobrino González for suggesting some of the cases of study, (and especially for his discovery of the Villasidro church, which prompted research leading to the publication of this paper [27]); and to Enrique Rabasa Díaz and Carmen Pérez de los Ríos for their invaluable help in reviewing the paper and general support and advice.

References

[1] J. Corblet, "L' Architecture du Moyen Âge jugée par les écrivains des deux derniers siècles", *Revue de l'Art Chrétien*, 1860, p. 60; J.H. Parker: "On the Late, or Debased, Gothic Buildings of Oxford", in *ABC of Gothic Architecture*, Oxford: J. Parker, 1881, pp. 219-256; P. Frankl, "The survival of the Gothic style", in *Gothic architecture*, New Haven: Yale University Press, [1962] 2000.

[2] H. Rousteau-Chambon, *Le gothique des temps modernes*. Paris: Éditions A. et J. Picard, 2003; N. Nussbaum, *German Gothic Church Architecture*. New Haven: Yale University Press, 2000.

[3] J. Gómez Martínez, *El gótico español de la Edad Moderna*. Valladolid: Universidad de Valladolid, 1998.

[4] E. Rabasa Díaz, "Principios y construcción de las bóvedas de crucería", *Loggia*, vol. 20, 2007, pp. 86-97.

[5] J.C. Palacios Gonzalo, *La cantería medieval: la construcción de la bóveda gótica española*. Madrid: Editorial Munilla-Lería, 2009.

[6] MS 12.719, Biblioteca Nacional de España and MS R.10, Biblioteca de la E.T.S. de Arquitectura de Madrid, A. de Vandelvira, 'Libro de traças de cortes de piedras', 1575-1580 (facs. in G. Barbé-Coquelin de Lisle, *Tratado de Arquitectura de Alonso de Vandelvira*, Albacete: Caja de Ahorros, 1977).

[7] See notes 3, 4 and 5

[8] Acknowledged since as far back as the 18th century: A.-F. Frézier, *La théorie et la pratique de la coupe de pierres* (3 vols.), Strasbourg/Paris: Charles-Antoine Jombert, 1737-39 (facs. in Nogent-le-Roy: L.A.M.E., 1980), vol. II, p. 102.

[9] E. Rabasa Díaz, *Forma y construcción en piedra. De la cantería medieval a la estereotomía del siglo XIX*, Madrid: Ediciones Akal, 2000.

[10] R. Willis, "On the construction of the vaults of the Middle Ages", *Transactions of the Institute of British Architects*, vol.1, [1842] 1910, pp. 22-24.

[11] Sometimes more significantly than might be assumed (A. López Mozo, E. Rabasa Díaz, M. Sobrino González, "La línea en el control material de la forma", pp. 743-754 in S. Huerta et al. (eds.), *Actas del Séptimo Congreso Nacional de Historia de la Construcción*. Madrid: Instituto Juan de Herrera, 2011).

[12] This should not be construed as an exceptional concentration of actual occurrences, as no systematic listing of cases has yet been compiled.

[13] Grid-crossing vaults, which approximate the desired surface through parallel and meridian ribs, were sometimes used similarly; however, as their constructive logic is rather different, they will not be discussed here.

[14] Outside Burgos, Ajo (Cantabria), Ledesma and Las Bernardas Convent (Salamanca) and Logrosán (Badajoz), among others, appear to have followed this strategy.

[15] A.C. Ibáñez Pérez, "Arquitectura burgalesa del siglo XVI", *Historia 16*, vol. 91, 1993, p. VII.

[16] This technique was also used in other, better executed buildings (Ajo, Las Bernardas).

[17] An addition to a 12th-century nave, built in 1616 by a mason called Cotero (F.J. Gómez Oña, *Las mil y una iglesias de la diócesis de Burgos*, Burgos: Editorial Monte Carmelo, 2011, pp.296-297).

[18] S. Andrés Ordax, *La provincia de Burgos*, León: Ediciones Lancia, 1991, p. 122. It was begun in 1526 and underwent significant work in 1529-38; still, the tower was not completed until 1606, which could betray a more coherent late-16th-century chronology for the choir and squinch.

[19] R. Gutiérrez-Solana and H. de Diego, "El proceso constructivo de la iglesia parroquial de Santa Cecilia en Espinosa de los Monteros (Burgos)", pp. 485-491 in A. Graciani et al. (eds.), *Actas del Tercer Congreso Nacional de Historia de la Construcción*. Madrid: Instituto Juan de Herrera, 2000, p. 488.

[20] Construction was riddled with structural problems, and suffered frequent interruptions (R. Gutiérrez-Solana and H. de Diego, "El proceso constructivo de la iglesia parroquial de Santa Cecilia en Espinosa de los Monteros (Burgos)", in A. Graciani et al. (eds.), *Actas*, pp. 486-489).

[21] Vault deformation and the awkward springing design might cast some doubt on whether both arches were originally supposed to be concentric. However, presence of carved decoration in the half-octagon arch face indicates that it was, at the very least, intended to be visible.

[22] J. L. Gutiérrez Hurtado, *Sedano, Villa y Honor*. Burgos: Caja de Burgos, 1997, pp. 142-147. Transept vaults were finished in 1649, and the crossing was not begun until 1659. Construction of this relatively modest building, already "in a reasonably advanced state" in 1566, thus spanned over a century.

[23] This date, carved over its entrance, remains the main source so far for dating the building.

[24] A. López Mozo, "La cúpula de El Escorial: geometría, estereotomía y estabilidad", pp. 763-776 in S. Huerta et al. (eds.), *Actas del Sexto Congreso Nacional de Historia de la Construcción*, Madrid: Instituto Juan de Herrera, 2009, pp. 769-771.

[25] I. Rilova Pérez, *Olmillos de Sasamón. Villa, Iglesia y Fortaleza*, Burgos: Adeco Camino, 1997, p. 155.

[26] The Olmillos chancel had been vaulted by García de Arce in 1589. (Rilova, *Olmillos* (Note 20), p. 159). If the Villasidro church was indeed completed in 1603, vaulting would have been the last stage of construction, and thus slightly later than, or at least simultaneous to, the Olmillos apse, though precise dating remains difficult.

[27] M. Sobrino González, "Piedra labrada en la arquitectura burgalesa del Renacimiento", in E. J. Rodríguez Pajares (ed.), *El arte del Renacimiento en el territorio burgalés*. Burgos: Universidad Popular, 2008, pp. 55-56.

Pier Luigi Nervi vs Fazlur Khan: the developing of the outrigger system for skyscrapers

Luciano Cardellicchio
School of Architecture, University of Kent, UK

Introduction

The structural development of tall buildings has been a continuously evolving process. There is a distinct structural history of tall buildings similar to the history of their architectural styles.[1] In this structural history it might be interesting to investigate how one of the most popular system used nowadays to build super tall skyscrapers, the outrigger structure, was designed for the first time. The paper focuses on how this particular structural system was first developed, trying to link the experience of two outstanding engineers who worked after the Second World War, the Italian Pierluigi Nervi (1891–1979) and the Bangledeshi-American Fazlur Rahman Khan (1929–1982).

In the outrigger structures, a lateral load-resisting system is extended from the conventional core to the building perimeter columns through the outriggers that connect them. This basic configuration often requires perimeter super columns or other structural components at the outrigger levels, and these elements of this system are sometimes integrated into the aesthetic of the building.

A very early example of outrigger structure is the Place Victoria Office Tower of 1965 in Montreal designed by Pier Luigi Nervi and Luigi Moretti (Fig.1). It was also used by Fazlur Khan in BHP House in Melbourne, completed in the 1972 and in the First Wisconsin Center of 1973 in Milwaukee. However, major applications of this structural system can be seen in the contemporary skyscrapers such as the Jin Mao Building in Shanghai and the Taipei 101 Tower in Taipei.

Skyscraper is a type of building where the structure has to efficiently match all the severe economical requirements needed. Indeed, a good structural scheme for a high-rise building has to employ the minimum amount of material in order to maximize the marketable space at each level and to contain the construction costs. The ability to form and shape a high-rise building is strongly influenced by the structural system. This influence becomes progressively more significant as the height of the building increases. Buildings up to 20 storeys can be shaped without undue influence of the structure. Between 20 and 40 storeys, one must begin to identify a specific structural system, its composition and its efficiency, and the flexibilities for shaping offered by the system. Structures that are 40-60 storeys tall will have more specific restrictions regarding asymmetry of profile and plan. Towers beyond 60 storeys are more decidedly affected by the structure. As every common structure, skyscrapers are subject to two main load categories: the gravitational loads (dead loads such as self weight of the structure, walls and cladding and live load such as people, furniture, vehicles) and lateral loads (wind, earthquakes). How these two main sets of loads affect the structural behavior of a skyscraper is extremely different than a low-rise structure. Indeed, the increase in structural material required for a tall building is largely due to the additional strengthening and stiffening necessary to resist lateral forces. In other words, the primary structural concern for a high-rise building is to develop sufficient lateral stiffness.

Pier Luigi Nervi vs Fazlur Khan: the developing of the outrigger system for skyscrapers

In the early 1960s, the Bangle-American engineer, Fazlur R. Khan, father of the modern skyscrapers structure, developed a method for assessing the differential between the material quantity associated with support of gravity loads only and the one required by actual load condition (gravity plus lateral loads). He named this differential with the expression "premium for height". Taller buildings could not compete economically with shorter buildings if engineers did not find a way to control, or most of all to reduce, this structural premium without increasing the size of structural components"[2]. Thus, in a skyscraper structure the main difficulty is to support the vertical load with a minimum increase of the structural element necessary to support the gravitational load. The research made by Khan in the early 1960s was oriented towards the shaping of tall buildings in order to make them act as pure cantilevers against lateral load. In the process he developed several new structural systems, most of them distinguishable by a full integration between structure and façade.

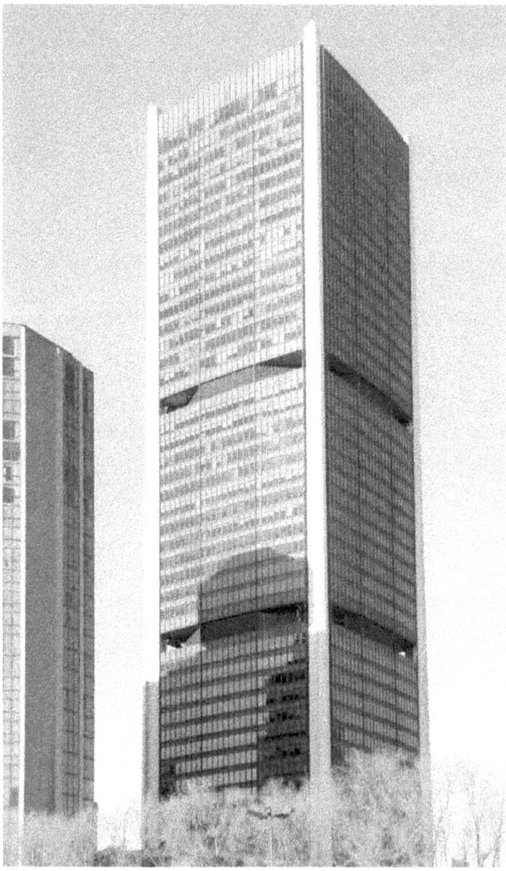

Fig. 1. Luigi Moretti, Pier Luigi Nervi, Stock Exchange Tower, Montreal, 1965. (Photo by the author)

Before that, the most predominant steel system of the 1950s and early 1960s was the plane frame system, which consisted of plane frames (rigidly connected columns and beams all in essentially one plane) arranged in two directions constituting a bay arrangement. The frame resists lateral loads primarily by the bending of member depths. Khan understood this frame behavior thoroughly, especially its inefficiency and limitations. As the bay dimensions increased or the structure grew taller, disproportionately heavier members were required, which reduced the potential offered by this system for such applications. Khan considered this system to be the weakest and placed it at the beginning of a range of possible systems. At the other end of the range, he put a pure cantilever, which utilized the entire three-dimensional form, as the strongest.

In 1969, Fazlur Khan classified structural systems for tall buildings relating to their heights with considerations for efficiency in the form of "Heights for Structural Systems" diagrams [3]. This marked the beginning of a new era of skyscraper revolution in terms of multiple structural systems. Later, he upgraded these diagrams by way of modifications[4]. He developed these schemes for both steel and concrete. He introduced this chart in order to define a sequence of systems, each of them with particular limitations. This set the stage for the "system approach" in the design of tall buildings, whereby the selection of an appropriate system was a significant design event. He argued that the process of optimization of the structure rested with the appropriate system selection; member optimization was secondary. New systems were placed into the chart as they were developed. During that period, the structural emergence in high-rise architectural expression was primarily aiming at integrating façade and structure, almost into a single entity.

Today, the substantial passing of pure stereometry in contemporary skyscrapers and the development of new and complex figurative solutions in terms of design, usually demand a hybrid approach combining different structural solutions, which can be adapted case by case to match better the requirements of the architectural form. Therefore, today the classification of structural solutions in relation to height presented by Khan, offering a clear diversification between different types of structure, cannot be integrated and updated. A quick and rough analysis shows that the recent high-rise structures have embodied more structural solutions previously associated to separate classifications: for example, the Spiral Tower in Nagoya, Japan (170 meters) has a structure consisting of an inner trusses tube with a triangular grid on the outside; the Swiss Re Tower of London (180 meters) by Sir Norman Foster is supported by a Rigid Frame steel inside and a triangular grid on the outside; the Al Sharq Tower in Dubai (367 meters) of Skidmore Owings & Merrill should have provided a structure composed of pre-stressed pipe system arranged in a Bundled Tube, braced by cables on the outside and concrete shear walls on the inside; the Shanghai World Financial Center (492 meters) has a Embedded Core, stabilized by four mega-columns and mega-diagonals on the outside, connected to the core with outrigger elements; the London Shard has at least three inner systems (Vierendeel Trusses, an "hat" truss and several outrigger elements) to increase the stiffness against lateral loads while producing the sharp outline.

Among all these hybrid systems in use, the outrigger structure has become one of the most popular solutions for super-tall buildings of more than one hundred storeys. Outrigger systems have been historically used by sailing ships to help resist the wind forces in their sails, making the tall and slender masts stable and strong. The core in a tall building is analogous to the mast of the ship, with outriggers acting as the spreaders and the exterior columns like the stays. As for the sailing ships, outriggers reduce the overturning moment in the core that would otherwise act as pure cantilever, and transfer the reduced moment to the outer columns through the outriggers connecting the core to the columns. These elements are configured as cantilevered structures that protrude from the core, made up of shear walls, towards the external structure. Outrigger components are generally allocated inside service floors among groups of regular floors in order to reduce the commercial loss of viable space. This system is able to stabilize the core in the horizontal direction, reducing greatly the bending moment at the bottom of the vertical shear walls, transferring a part to the elements in the facade that, depending on the motion of bending will be compressed or placed in tension. The structural principle can be detailed in several sub-systems, such as the Outrigger trusses, as in the Shanghai World Financial Center where four super columns at perimeter shape the profile of the building, or such as Outrigger shear walls in reinforced concrete as in the Burj

Khalifa in Dubai (828 meters), the currently world's tallest building (2013). Although loss of marketable surface occurs in the levels where the outrigger elements are inserted, the structure involves substantial benefits in the floor plan layouts and in the architectural vocabulary. Indeed, the use of this system releases the façade cladding from load bearing components (except the ones that have to be connected with the outriggers), so that the overlapping parts can have bulges, convexity, concavity, depression, and tapers. The visionary and expressive power of this structural scheme was brought to life for the first time by Pier Luigi Nervi in the Stock Exchange Tower (190 meters) in Montreal designed by Luigi Moretti.

When completed, on May 1st, 1965, the Montreal Stock Exchange Tower with its 190 meters was the tallest reinforced concrete building in the world. But it is not only the type of structure which clearly distinguishes the work of Moretti and Nervi from the original model of the American steel frame skyscraper: there are far more significant differences. It is the architect himself who highlights the main differences in one of the reports compiled during the long designing process. Indeed, he does not even hesitate to indicate as sources of inspiration of the project a few flaws of the Seagram Building, universally regarded as the archetype of skyscraper perfection. For Moretti, however, the "shape of the prism without taper" and "uncatchable image of the structure and his behavior" are deficiencies which cannot be countered by the "glass walls beautifully sized"[5]. Indeed, in the main symbols of modernity for a high-rise building, which are the undifferentiated and indefinite stereometry and the texture of the curtain wall, Moretti sees the mechanical translation of the logic of industrial production. For him, that is a radical depersonalization of what is incompatible with the conception of architecture as an artistic creation of the individual: architecture must be unique and unrepeatable. It is not simply the preference for the reinforced concrete structure, then, that we are interested in investigating in this paper, but most of all how the particular structural conformation, tailored in the design process to strengthen the identity and originality of one building, has been developed from a customized engineering approach to a standard and popular structural scheme for high-rise building. The tower in Montreal is supported by a complex and innovative structural skeleton, the efficiency of which is entrusted to a static system called by Pier Luigi Nervi the "main-resistant"[6]. This system is three-dimensional, and allocated inside and outside of the squared floor plans of about forty meters per each side. In fact, a central core of two baffles of reinforced concrete crossing diagonally is connected by trusses (only in correspondence of three units) to the four pilasters arranged at the edges of the volume of the building. The structural intuition relies on the role of crutches on the corner posts, in order to stabilize the core cross against horizontal forces. As we said the main internal stress that the structure of a tower needs to reduce, in order to increase the efficiency of the structure without increasing the number of structural components, is the bending moment at the core base. Here the stiffness of the core is guaranteed by connecting the main core of the building with the four mega-columns at the building edges through four deep diagonal trusses (which Nervi had built in reinforced concrete to ensure a monolithic structural behavior). In detail, the angular pylons, almost free from bending stiffness due to their slenderness, behave like pendulums (either compressed or tensed), depending on the tower horizontal motion under vertical load, increasing effectively the bending inertia of the core and consequently reducing the moment at the base of the tower. Regarding the gravity loads they are carried out to the foundation through an ordinary concrete frame system which is made up by the four angular columns completed with two intermediate pillars on each side of the squared floor plan. Therefore it is possible to assume that the conformation and the static behavior of the entire structure are quite standard when supporting just the crossed ribs floor slabs. Although the structural solution was tailored on the tower in order to accomplish the will of the designer, not all the parts of the super structure collaborate with the image of the building. Indeed, the elevation of the tower shows only some components of the mega bracing system. This aspect is quite important in understanding the relationship between structure and architecture in this building. The balance between the structural elements exposed and the bracing components which are hidden and also the glazed cladding of the facades is quite sophisticated. The four corner pilasters, tapered and conspicuously detached from the glazing skin with a deep shadow gap, are crucial to the building appearance. More precisely, in order to emphasize the image of the pillars, the other eight intermediate pillars are concealed behind the curtain wall. Consequently, to avoid separate deformation, the external portion of the structure exposed to the significant daily temperature changes of Montreal, are protected by a shell made by white concrete panels[7]. The outer shape of angular pylons visible is independently detached from the inner structural section of pillars. Significantly bigger in terms of dimensions, the external cladding for the pillars provides a cavity for the passage of conditioning pipes and increases the impression of the static effort as the will of the architect. According to Moretti "the centralization of the loads in just few elements of the structure" to give a "high sense of stillness" to the building[8]. Essentially with a similar figurative strategy, the three floors, where the rigid connections between the nucleus and angular pillars are allocated, must be perceived in negative, as net horizontal cuts in the volume of the tower, interrupting the continuity of external surface. Therefore in order to release these fractures from all the structural components designated in them, the large trusses are not visible and the perimeter anti-torsion device, initially resolved by exposed structural wall-beams, is eventually reduced to two small ring beams, protruding only slightly more than a meter from the slab soffit, carefully hidden by the curtain wall. In addition, to better understand the delicate balance between architectural image and structural inventions, the three "cuts " do not all have the same height: the amplitude is reduced from bottom to top in order to create a perspective effect. This device is strikingly in contrast with the effects caused by lateral load, which are higher on the beams at the highest level, as confirmed in the calculation report[9]. Therefore, although structural expression seems to be a focal point of the building, it is important to highlight that the structure is exposed not to show off its efficiency but to contribute to the image of the building in an unpredictable way. The great achievement of the system is not only the powerful effectiveness of the four giant columns at the corners but also the way it releases the glass façade from any bearing load components. Thus, meanwhile Fazlur Khan was researching the way to achieve the "pure cantilevered behavior" designing a system where a dense grid of structural components concretely embody (and almost blended) the façade, Nervi and Moretti were working in the opposite direction, designing few linear super-structure parts (exposed and recognizable) and a non-standard glass cladding free by the main structural constraints. The priority of the overall figurative perception of the architecture compared to the functional and structural requirements is confirmed in a series of tiny and invisible deformations impressed on the architectural components as optical corrections. Indeed, the most original optical correction for the skyscraper is certainly the one about the taper and the convexity of the prism. To achieve that, Moretti, not happy with the compromise achieved initially with the project management, which agreed to have the sets back of the facades in just three shots concentrated at the technical floors, finally obtained, in a crucial meeting in Montreal May 19, 1962[10], the gradual reduction of the floor slab from plan to plan. In fact, the cantilever portion of the floor (about 15 cm thick), which has a polygonal profile that roughly outlines the slight curvature of the facades, is set back to about half an inch on each floor of the building. As mentioned, these transfigurations are managed by Moretti and Nervi not to show the structure but to emphasize the completeness, uniqueness and the figurative nature of the object. This approach, in contrast to the iterative solutions of American skyscrapers, leads the design strategy being named as "the unique form"[11], which has its archetype in the Pirelli skyscraper in Milan designed a few years earlier by Gio Ponti, and not surprisingly, with Pier Luigi Nervi as structural engineer in charge. However, although this Italian interpretation of the high-rise building based on a substantial customization of the structure led to the Stock Exchange tower, the intuition of Nervi had an impact well beyond the success of that building. Indeed the structural scheme has been developed over the years and nowadays what we call "outrigger" system represents one of the most popular solutions for super-tall skyscrapers. However, before being used in super-tall buildings the first steps in the development during the late 1960s and early 1970s involved the standardization of the system in order to make it replicable and adjustable for buildings of about 40 storeys. Ironically, the depersonalization of the architectural product so disliked by

Moretti has turned the original structural scheme created for the Stock Exchange Tower into a paradigmatic solution that could be replicated for skyscrapers with identical architectural features.

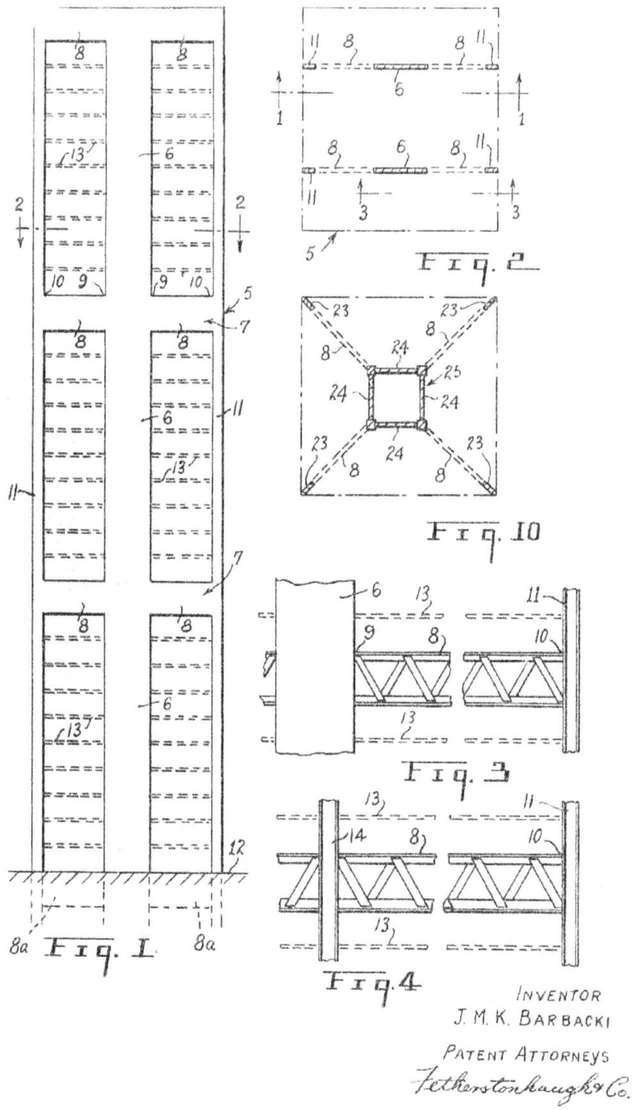

Fig. 2. Jacek M. K. Barbacki, Wind and Frame Earthquacke T-Frame system for building", drawing 1, patent n. 659142, CIPO. (Photo available on: cipo.gc.ca)

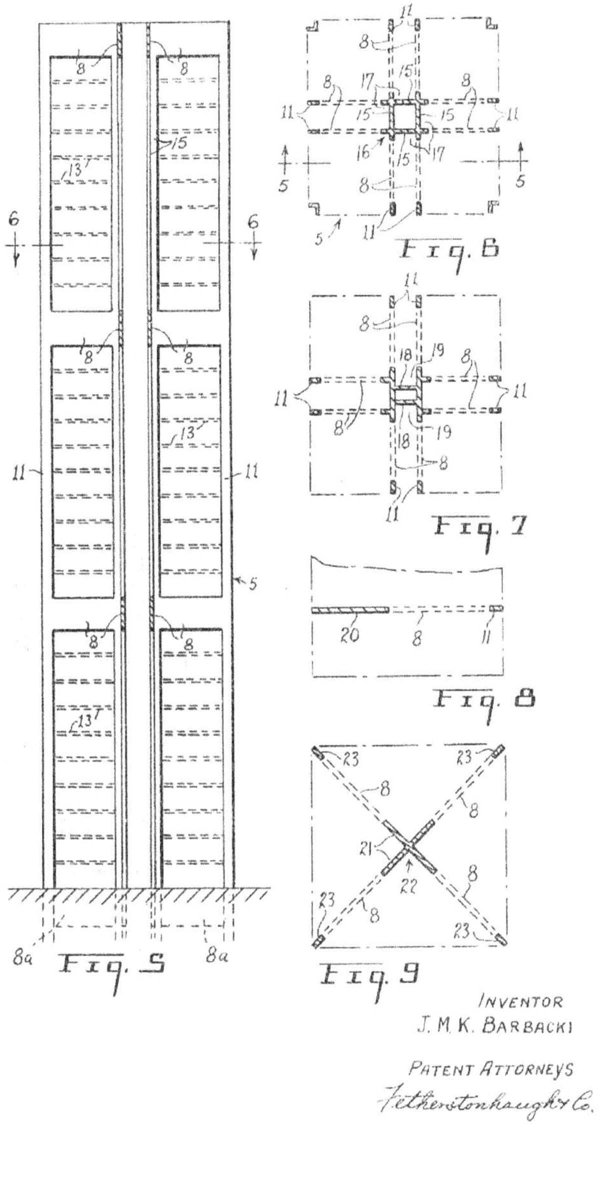

Fig. 3. Jacek M. K. Barbacki, Wind and Frame Earthquacke T-Frame system for building", drawing 2, patent n. 659142, CIPO. (Photo available on: cipo.gc.ca)

Pier Luigi Nervi vs Fazlur Khan: the developing of the outrigger system for skyscrapers

Pier Luigi Nervi developed the scheme for the Stock Exchange Tower collaborating with two Canadian engineering firm, Greenspoon, Freedlander & Dunne and D'Allemagne & Barbacki. The structural system was patented in Canada by Jacek M.K. Barbacki[12] in 1963 under the title "Wind and Frame Earthquake T-Frame system for buildings". In order to understand how the system was modified and consequently applied to other buildings over the following years it is necessary to analyze carefully the Canadian patent (Fig.2-3). The documents submitted for the patent showed a structural system made by components extrapolated from the Stock Exchange Tower, combined with different and more flexible and adjustable solutions. The patent proposal consists of a description where the main advantages of the system are highlighted and two pages of drawings showing how the same structural components can be assembled for different functional solutions. According to the description, the fundamental parts of the scheme are the vertical shear walls, the outer columns and the "horizontal frame" (the outrigger elements). [13] In addition, going beyond the Montreal experience, the report stressed that the T-Frame system "may be of any well known type of construction such as structural steel, reinforced concrete or other structural material or any combination of these types. For instance, the shear walls may be of reinforced concrete construction and the horizontal frames may be of steel, either solid beam or open truss construction."[14]. The drawings are two sections, six floor plans and two detailed enlarged vertical sections showing the connection between inner and outer structures. The floor plans describe six different ways of arranging the vertical shear walls, according to different architectural core layouts. In fact, "it is to be understood that the number of shear walls to be incorporated within the building will depend upon the size and shape of the building. For instance, in a building that is square in plan, only one shear wall may normally be required, whereas, in a building with a long face in plan, two or more shear walls may be installed" as it is shown in the figure number 2 of the drawings (Fig.2). In detail, the first two floor plans display shear walls parallel to the edges of the building. The figure number 8 (Fig.3) describes how to modify the system allocating the vertical walls at one side of the building with the horizontal frame spanning the floor to a column at the opposite exterior wall. The figure number 9 (Fig. 3) displays the original Stock Exchange Tower solution with the shear walls in form of a cross and with the horizontal components disposed diagonally towards columns at the corners of the building. Among the others, the most interesting shown layout is the one with the outrigger elements parallel to the side of the building. Indeed, there is no indication about how this system could be integrated in the building elevation especially because the outer main structure should be allocated in the middle portion of the façades. It is a significant change from the Nervi and Moretti approach who had put the outer super structure at the corners in order to combine structural improvement and architectural expression. Indeed, it should be noted how the system conceived in order to increase the formal independency of the façade from bearing load parts in Montreal, has been developed in the patent with options alluding to differing architectural outcomes. However, it is clear that any other external column allocation except the one with the columns at the corner would be weaker in terms of bending and torsional inertia[15]. From this assumption we can understand the development that Khan brought to this system, starting from the layout with the parallel outrigger ribs. Indeed, although the structural system was patented in Canada in the 1963, apparently Khan did not mention it in the chart of 1968 with the name of "outrigger" system. It is clear that Nervi and Barbacki came up with that system to improve the lateral stiffness of a framing "regular" building structure, while Khan's approach has always been oriented towards the "pure cantilever" behavior working on the stiffness of the building envelope. In fact, according to his research path, he developed his own version of outrigger system for two buildings around 44 storeys each, integrating the efficiency provided by linking inner shear walls with external structural components with the experienced tube structure. That system became the belt truss structure, employed for the BHP House in Melbourne (Fig.4), completed in the 1972, and for the First Wisconsin Center in Milwaukee (Fig.5), completed in 1974. The belt-truss is a structural component invented by Khan to reinforce the external wall structure at the specific levels where the outrigger elements are provided. The main aim was to improve the way in which the wind load was transferred to the foundations of the building.

Acting as a rigid belt around the building the orientation in plan of the outriggers was no longer an issue. The loss of bending and torsional inertia occurs from pointing the horizontal ribs to a central field of the facades and has been balanced with the belt-truss which is able to absorb the vertical load all along the exterior walls of the building and transfer it to the core at the outrigger levels. In more detail, "Khan suggested that at least two levels of horizontal outrigger trusses were needed to engage the perimeter wall: one around mid-height and one at the building top. Equally important, he suspected, was the stiffening of the exterior frame at these same levels to increase their participation in the interaction system. Without the stiffening, only a few columns of the frames would be directly engaged by the trusses, leaving the potential restraining influence of the exterior frames largely untapped: while the few engaged columns would work to anchor the core, the large majority of columns, further removed from outrigger connection points, would not"[16]. His idea, therefore, was to force the full width of each frame to participate in resisting the bending of the core structure by introducing storey-high "belt" trusses in the perimeter at each outrigger truss level. The two buildings structurally designed by Khan with the belt-truss system have development path extremely different.[17] However, when comparing the architectural expression of both it is easy to understand how the result led to a similar identical appearance.

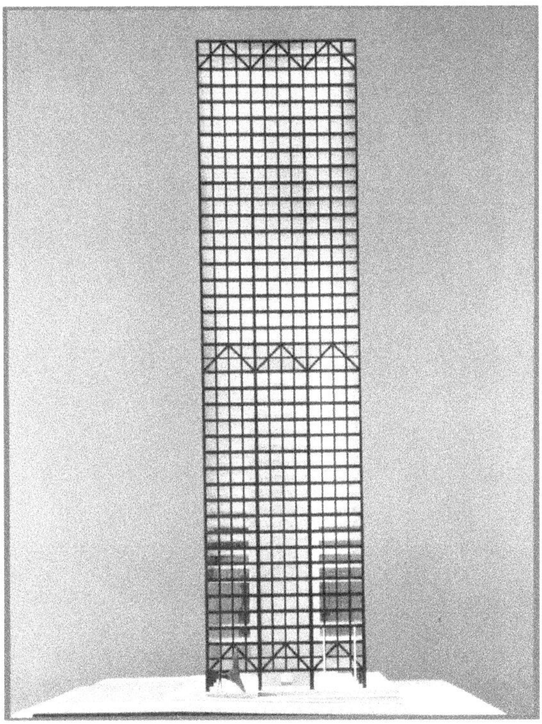

Fig. 4. Yuncken Freeman Architects with Fazlur Khan and Hal Iyengar (Engineers, S.O.M.), 140 William Street (formerly BHP House), Melbourne, 1972, model of the tower. (Picture by Wolfgang Sievers, available on:<http://nla.gov.au/nla.pic-vn4902728>)

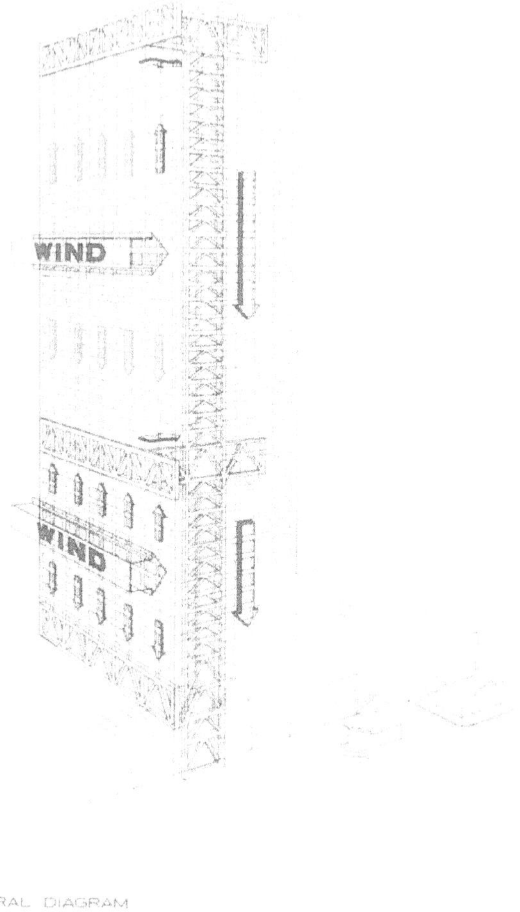

Fig. 5. Bruce Graham, James DeStefano (architects, S.O.M.), Fazlur Khan (Structural engineer, S.O.M.), First Wisconsin Center, Milwaukee, 1973, structural diagram. (Courtesy: James DeStefano)

Conclusion

Starting from the first experimentation, the system modified and developed by Khan is more replicable and adjustable for buildings with a boxy silhouette. According to Hal Iyengar, one of the engineers associated with Fazlur Khan on many projects, "the systems must be applicable over a broad cross section of structures and not just to particular solutions to particular buildings. The systems must be architecturally relevant, providing a basis and often a vocabulary for visual expression". [18] In these words is possible to recognize the outline of the American approach: it is the structural system that has to provide the architectural language of the building rather than being shaped as a non-standard support for the figurative artistic expression, as it was in the Montreal.

References

[1] M.M. Ali, P.J. Armstrong (eds.), *Architecture of Tall Buildings*. New York: McGraw-Hill, 1995, p 115.

[2] Y.S. Khan, *Engineering Architecture, the vision of Fazlur Khan*. New York: Norton, 2004, p. 69.

[3] F.R. Khan, "Recent structural systems in steel for high-rise buildings", *Proceedings of the British Constructional Steelwork Association Conference on Steel in Architecture*, London: British Constructional Steelwork Association, 1969.

[4] F.R. Khan, "Evolution of structural systems for high-rise buildings in steel and concrete", in J. Kozak (ed.), *Tall Buildings in the Middle and East Europe: Proceedings of the 10th Regional Conference on Tall Buildings-Planning, Design and Construction*. Bratislava: Czechoslovak Scientific and Technical Association, 1973.

[5] S. Poretti, G. Capurso, "Trasfigurazione di Strutture", in B. Reichlin, L. Tedeschi (eds.) *Luigi Moretti. Razionalismo e trasgressività tra barocco e informale*. Milano: Electa Mondadori, 2010, p.375.

[6] Ibid., p.375.

[7] J. Folch-Ribas, "Place Victoria, Précisions sur la Tour de la Bourse", *Royal Architectural Institut of Canada Journal*, vol. 42, October 1965, n.10, pp. 70-73.

[8] ACSRo (Archivio Centrale di Stato, Roma), Fondo SGIS, "Superfici di Solaio costruibili, ipotesi planimetriche, ipotesi strutturali ecc.", unsigned typewritten report.

[9] ACSRo (Archivio Centrale di Stato, Roma), Fondo SGIS, P.L. Nervi "Place Victoria St. Jacques Co. Relazione strutturale", typewritten calculation report.

[10] Poretti, Capurso, *Moretti*, (Note 5), p.378.

[11] G. Ponti, "Espressione dell'Edificio Pirelli in costruzione a Milano", *Domus*, 1956, n.316, pp.1-16.

[12] The patent was submitted the 21st June 1961 and approved on the 12th March 1963 with the number 659142 by the CIPO (Canadian Intellectual Property Office).

[13] J. Barbacki, "Wind and Frame Earthquake T-Frame system for buildings", CIPO Patent number 659142, report description, pp.1-9, available on:
<http://brevets-patents.ic.gc.ca/opic-cipo/cpd/eng/patent/659142/summary.html>

[14] Ibid., p.6.

[15] T. Trombetti, "Un nuovo equilibrio: Nervi e Moretti a Montreal", in *D'A. D'Architettura*, vol.34, 2007, pp.60-63.

[16] Khan, *Engineering Architecture*, (Note 2), p.176.

[17] Ibid., p.173.

[18] H. Iyengar, "Structural and Steel System", in D.P. Billington, M. Goldsmith (eds.), *Technique and Aesthetics in the desing of tall buildings, proceedings of the Fazlur R. Khan Session On Structural Expression in Buildings*, Institute for the Study of High-Rise Habitat, 1986, pp 57-69.

A digital edition of the didactic knowledge of construction: a critical edition of the courses of Antoine Desgodets, professor at the Académie Royale d'Architecture (Paris, 1719-1728)*

Robert Carvais
CNRS, Centre de théorie et analyse juridique (UMR 7074), Université Paris Ouest Nanterre La Défense, France

We all agree that the essence of architectural knowledge initially consists of a minimum of mathematical knowledge (geometry, proportion and arithmetic) to which must be added the art of representation (drawing, perspective, ornaments, orders, etc.). We simply tend to forget that three other disciplines were necessary to define and complete the status and practices of modern architects, implemented early in the 18th century: law, economy and the art of building. Modern architects had to know how to construct a building while also envisaging its conception. The two criteria of the Vitruvian architect reached their full potential: that of the building site (*fabrica*) and that of the creation (*ratiocinatio*). Renaissance works on architecture participated in liberating architecture from its original theoretical constraints and making it take root in the inescapable practice of construction. During this period of flourishing construction, linguistic and representational hurdles were overcome: the commentaries of Fra Giocondo, Daniele Barbaro, Guillaume Philandrier and Claude Perrault contributed significantly to this development. Progress in graphic representation through perspective and exact measurement thanks to the treatises of Alberti, Jean Pelerin, Serlio, Vignola and Hans Vredeman de Vries also favoured this revolutionary approach to architecture. During the 16th century, technical approaches to the art of building were developed with the writings of Phillibert de L'Orme and of Girard Desargues on stone cutting, of Salomon de Caus and Giovanni Poleni on hydraulics, and of Philippe de La Hire on mechanics, as well as innumerable treatises on fortifications by military engineers which inevitably touched on techniques of building. But few studies have investigated the legal and economic questions of construction, which are crucial to a complete understanding of the modern architect. These are fields which we will address in this paper.

To understand how the status of architects developed, the evolution of their knowledge needs to be examined. In France, the *Académie Royale,* established in 1671, taught architecture and trained the king's architects. The historiography of this institution is somewhat dated (with the exception of Baudez)[1]. The contents of the training are hardly known except for the few courses which were published. However, for a decade now there has been a renewal of works on its teachers: François Blondel[2], Philippe de la Hire[3], Jacques-François Blondel[4] Julien-David Leroy[5]; and on the essentially scientific disciplines taught at the *Académie*[6]. But how the precise and pioneering knowledge taught and delivered by the tenured teachers emerged remains to be discovered.

For a long time, the history of architecture was based on analyzing published treatises, rarely were there hand-written or unpublished courses. These, however, represent the first and fundamental part of knowledge passed on to future architects. Our contribution aims to present the digital edition of the courses taught at the *Académie Royale d' Architecture* by Antoine Desgodets from 1719 to 1728. We directed this research project financed by the *Agence Nationale de la Recherche* [National Agency for Research] from the end of 2007 to June 2013. It was the work of a multidisciplinary team comprised of twenty researchers from different disciplines: art historians, of course, but also philosophers, philologists, historians of science and techniques as well as law historians.

Preparing the publication of the courses took place in three successive stages. The first consisted of identifying the issues of the publication project, and in setting out its reasons, major themes and research hypotheses. The second stage consisted of exposing the procedures followed, the series of data collection, the constitution of the corpus and its treatment, in brief the material gathered and the methods used. The third stage dealt with comparing the initial results obtained with the initial hypotheses and with the state of the art in the different fields of knowledge transmitted by the teacher. What we will present here is only what pertained to the history of construction.

The stakes and issues of the project

The goal of our project was to measure the knowledge of architects at the end of the 17th century and at the beginning of the 18th century, based on the hand-written tradition of architectural writings at the very heart of the discipline's theoretical emergence dating back to the Vitruvian tradition of the 15th century. We know that architectural writings belong to the field of technical literature. They require more than the simple activity of passive, intellectual reading—assuming that mental mechanics do not influence action—but constitute veritable work instruments allowing architects to conceive and realize a building. This is even more true for the teaching material.

Framework of the project

Although the theoretical literature of architects, developed in the form of treatises and periodicals, has caught the attention of researchers in recent years[7], and a certain number of reflections on the practical knowledge of architects have been initiated[8], there is still a discursive part which represents the teaching delivered to the students has mostly been ignored. This is why it seemed important to dwell on the courses of Antoine Desgodets: these unpublished courses have continued to intrigue researchers in respect to their exact contents, their reception and their impact on the building of knowledge.

Three reasons impelled us to choose the oeuvre of Desgodets: First, Desgodets became a professor at the *Académie d'Architecture* in 1719, during the Regency, at a time when reforms to the dogma were being sought. Secondly, by choosing new themes to teach future architects, his teaching played a part in the reduction into art of the academic discipline, and this by means of one of the first standardisations of architectural knowledge. Lastly, the oeuvre of Desgodets came at a key moment for the growing professional legitimacy of architects; concurrently, it took into consideration the figure of the expert, an intermediary between the architect and the amateur who deals with the complementary professions of *toiseurs*, *vérificateurs*, etc.

For all these reasons, Antoine Desgodets's courses deserved to be reconstituted and delivered to the scientific community as texts that established the “status” of the modern architect. As opposed to Vitruvius' work, there is no original master text (no *pinceps* edition) of the scholastic oeuvre of Desgodets. A multidisciplinary team worked for five years to expose the significance of his teachings in those directions judged to be complementary: architectural theory, the execution of projects, their economic analysis and their legal aspects.

Creating institutions often reveals ruptures in the history of men and their knowledge. Coming from the building trades, architects were formed on work sites and in books. We know little about their academic training and why it was established. Although access to the courses given by the professors was free to 'any man of any age and condition whatsoever, who has a taste for Architecture,'[9] no provisions were made to preserve the content of the courses. Only posters announcing their times, four hours per week over two days, attested to their existence. Note taking of the professor's dictations or copies of the manuscript provides some insight into architectural knowledge at the beginning of the 18th century, or more precisely into what the *Académie Royale* and its professors judged as constituting knowledge for the architect. He, who would make the art of building his profession, would need to be versed in mathematics, drawing, and also understand the practical trades of building. The *Académie* envisaged exactly half of the courses to be given for 'practical geometry,' and half for 'lessons of different notions, teachings, rules and practice of architecture'[10] which must represent the principles of this art or the most necessary knowledge. The expression is sufficiently enigmatic and programmatic to make one wish to discover in concrete terms what it implied and the reasons for such a formulation.

Research themes

Three directions were considered for the project:

First, to reconstitute what at the beginning of the 18th century made up, in the eyes the Académie d'Architecture, the different kinds of knowledge architects were taught. It was a matter of understanding the disciplines that together defined this profession as a new science. Desgodets's courses are of various natures and involve rather disconnected fields, traditional for some, novel for others: the orders, *commodité*,[11] good taste, harmony, the *toisé* [art of measuring][12] and easements. Certain courses were directly connected to established disciplines such as *toisé* to economy and easements to law. Our initial method was to establish a master text by digitally comparing manuscripts. This minute process, framed by philological and encoding precautions,[13] would enable us to recover the hand-written manuscript, if one exists, or to reconstitute the closest approximation, whether it was notes taken by students attending the courses or copies used by practitioners. As a tool, the computer was indispensable for comparing the variations of different versions of notes for the same course. Equally, the regrets, hesitations, annotations and comments would bring us clarifications on the origin of the manuscripts. This first stage would help us measure the influence of the Vitruvian tradition of architectural science and its transformation. How did this period of institutionalization of architecture since the last quarter of the 17th century break with the Vitruvian past, in particular with the conceit of architecture as a building site practice and no longer as merely the theory of design?

Secondly, concerning the contents of architectural knowledge, our work was meant to facilitate the research on the filiations (or sources) of the different texts, before and after publication, given that the manuscripts were either preceded by a publication (*Les Edifices antiques de Rome*, 1682) or commented on in posthumous publications (*Loix, 1748 and Toisé, 1761)*. Moreover, this perspective would enable us to resolve certain imprecise points left by those who continued the work of Desgodets. Whom are we to believe, Quatremère de Quincy or Morisot, the first of whom claimed that the *Traité du toisé* by Desgodets was published by Ginet, or the second who claimed that it was published by Delespée? What do the preparatory course texts, which we found, teach us about the *Loix*? In itself the circulation of Desgodets's manuscripts is of invaluable help to deduce their uses. The same work of comparison would be carried out with the plates of drawings or isolated illustrations to substantiate their constancy or their modification, their origin or their novelty. The multiplicity of architectural practice (creative, constructive, learned or expert, etc.) emerged during this transitional period of developing knowledge. Revealing this diversity increased the understanding of the types of architectural writings, of their origins and filiations.

Finally, gathering this information, both texts and sources, would allow us to reconsider the status of architectural science at the turn of the 18th century as much in its traditional theoretical foundations as in the practices of the building site. From this point of view, the *Toisé* and the *Loix* should be analyzed both as construction techniques during the building period and as the final techniques in the service of expertise and post-construction practice. Such was our hypothesis. It was certainly a determining progress, which needed to be confirmed during our research: that science addresses the future of a building and not only its creation. Celebrated as the chief representative of classical architecture, was Desgodets more the innovator whom we did not expect in the field of technical disciplines? We could, for instance, examine the place of his work in the history of mathematical knowledge serving architecture, in connection with stone cutting, descriptive geometry and perspective, or even assess his influence on legal science. Thus, putting into perspective each of the themes taught with the state of knowledge in the concerned disciplines would allow us to examine and specify what should be understood by the birth of the disciplinary branches (law of construction, economy of the project, etc.).

Our research project would result in establishing the matrix of Antoine Desgodets's courses and the digital and critical edition of those courses. Putting them on line with open access should initiate thoughts in new directions, in particular on the constitution of different sorts of knowledge and their impact on the birth of the disciplines. But in the first place, what were our sources and how did we work?

Collected material and methods used

The work was carried out in seven stages sometimes overlapping for efficiency's sake. Each stage was the subject of debates and decisions taken during meetings of the project's scientific committee.

Defining the corpus

Based on an inventory after Desgodets's death,[14] the printer-publisher Jean Mariette listed the portfolios of prints and handwritten architectural treatises, which he then had valued and appraised. Unearthing these archived documents allowed us to take stock of the hand-written courses the professor owned at the time of his death. The value assigned to each of them gives an idea of what his work meant to the informed mind of the prominent and knowledgeable collector of all types of art, who was asked to assess the value of Desgodet's library.

The prints' estimated value was 505 *livres*

The four themes of the courses were estimated as follows:

- for the orders, 225 *livres*;
- for the *commodité,* 35 *livres*;
- for the *toisé*, 30 *livres*;
- for the easements, 40 *livres*.

Some of our initial hypotheses could not be confirmed. Among the elements mentioned in the project, we identified a different type of architecture course archived at the *Bibliothèque de l'Institut* and attributed to Desgodets, which upon analysis turned out to be the work of Courtonne. Moreover, according to Hélène Rousteau Chambon, Desgodets would have taught, as an assistant to Philippe de La Hire, stone cutting, drawing and perspective representation well before he was tenured, but no record of that has endured. The *toisé,* which was allegedly published by Charles François Delépée, his colleague at the *Académie*, was never recovered. Finally, after consultation, a manuscript discovered at the *Ecole Nationale des Ponts et Chaussées* turned out not to be by Desgodets.

On the other hand, we discovered a group of four courses by Desgodets imported to Sweden and conserved in a private library (Carl Johan Cronstedt, Fullerö, Stockholm). It was very useful to gauge the dissemination and impact of his courses throughout Europe. The *toisé* kept at the Bibliothèque Sainte Cornille in Compiègne was finally recovered after a long period when it was generally thought to have been destroyed by the bombings of WWI. It is also more than likely that we will later discover other manuscripts, such as the manuscript for the *'Traité des ordres d'architecture et de toutes leurs parties, expliqué en l'Académie roiale d'architecture par Mr Desgodets architecte des batimens du Roi et professeur de ladite Académie'* [Treatise on the orders and all their parts, explained at the Royal Academy of Architecture by Mr Desgodets architect of the King's buildings and professor at the said Academy] by J. Mansart de Jouy (*in-fol.*). This once belonged to Giuseppe Bossi and was sold at a February 1818 auction to the bookseller J. P. Giegler who later sold it to private buyers. Recently, the *Conservatoire National des Arts et Métiers* library acquired a new *toisé* course which has two particularities. It is a course dictated in 1751 by Louis-Adam Loriot, professor at the *Académie* from 1748 to 1762, thereby conclusively proving the uninterrupted use of Desgodet's course by the following generations. It also includes additions which in all likelihood were updates.

Overall, our corpus includes twenty-six manuscripts and seven printed books, for a total of thirty-three documents. Five manuscripts were critically edited versions to which we added other handwritten evidence for the more significant variations. A diplomatic transcription was carried out for these other manuscripts.

Dating the manuscripts

Some manuscripts were clearly dated, with either front page or colophon bearing a precise date. Others do not allow us to differentiate between the supposed date when the course was given from the date when the said course was copied. The discrepancy can be major. Others were approximately dated thanks to their watermarks.

Digitizing the manuscripts

Once the corpus was defined, a campaign was implemented to consult and reproduce all the manuscripts. We obtained the various authorizations required without too much trouble. Given the large number of manuscripts we chose to reproduce most documents ourselves. Some libraries did demand that their documents be digitized in-house, sometimes even by private companies to which they were linked, which generated some predictable and acceptable expenses. In some cases, we bought copies of a microfilm and had it digitized. The quality of these copies was generally rather poor but did not preclude transcription. For the library of the *Institut*, the municipal library of Nantes, and the Avery Library (Columbia), authorizations were harder to obtain and some of our acquaintances had to intervene. Only the Avery, given the fragility of their documents, especially the bindings, asked us to contribute to their restoration.

Transcribing the texts

In 2009, the question of transcribing and immediately modernizing the texts was posed. But most of the scientific committee supported the approach that rigorously respected the philological tradition allowing for the texts to be understood each in relation to the others.

We decided, however, to carry out both types of transcriptions. The first, which we will call 'classic', is a transcription true to the manuscript, according to the principles of the *Ecole Nationale des Chartes*. The second, which we will call 'modernized', is an adaptation of the first, which in addition emends spelling, accents and punctuation. Modernization would have allowed us to use the *Medite*[15] comparative software, which prevents interference when bringing versions together, but this was ultimately useless.

Comparing versions

Firstly, owing to the scientific committee's observations, we decided to edit scientifically just one course per subject-while offering clear justification for our choices. Secondly, and when necessary, this operation called for a comparison of the other existing manuscripts of the course thus revealing their variations (additions, suppressions, moving parts, etc.). To do that the initially chosen *Medite* software turned out to be too tricky to use. Instead, we ended up using the *Juxta* software, which is reliable and quite straightforward.[16] We compared illustrations manually. Upon completion of this work, we were able to establish a chart of the different versions of the courses. Each manuscript was the subject of a complete encoded description in XML-TEI format (*Extensible Markup Language – Text Encoding Initiative*).

Editing the chosen version

A team of editors was formed for each course. The scientific committee chose the XML-TEI standard to edit the courses in order to create records of the phenomena affecting the texts and with an eye to future publication on a web application. Besides affording durability and interoperability this international standard provides more versatility of expression than a simple transcription using word processing software (pre-encoding, encoding).

Obtaining broadcasting rights by implementing partnerships

Why would a public institution have the right to prevent us from broadcasting illustrations drawn from manuscripts, which, although they do belong to them, have entered into the public domain and are of use to the scientific community? Why should they be allowed to demand a financial contribution in return for their authorisation? We would like to offer these establishments a partnership based on reciprocity.

The first results measured in terms of our initial expectations and the state of the art

How is a specific knowledge formed and disseminated, and how in the course of time does it develop? We sought out and established the sources for the materials composing these courses, as well as their outcome: how did their publication play out, how did students receive them, how did subsequent teachers take over? This paper will introduce only a few results pertaining to the two 'constructive' courses. Of the two other courses, both more 'architectural', the one on the orders represents a real departure from tradition and the course on *commodité* is particularly innovative. Let us, however, examine the constructive courses on *toisé*, especially enduring, and on easements, which became a reference.

The treatise on toisé

Given from 1724 and 1728, the course on *toisé* by Antoine Desgodets is the first written trace of the passing on of this concrete knowledge within the framework of teaching. Its organisation bears witness to the transfer from spoken to written transmission, and towards theorisation. The course's layout no longer reflects the actual order of *toisé*, 'an order opposite to construction,' (as it still did in *l'Architecture Pratique* by Bullet in 1691) but rather an order thought out beforehand, in a pedagogic approach striving for comprehensiveness.

The links between architecture and science are here posited and debated. Initially conceived in a mostly mathematical approach, the methods of *toisé* first emerged in technical, military or surveying treatises. The first written record of *toisé* in the architectural sense only occurred in the early 17th century, in l'*Architecture Françoise* by Louis Savot. The latter, an amateur and not a technician, provides insight only useful to sponsors or clients but less so to builders. Architectural *toisé* was first truly formalised in the late 17th century by Pierre Bullet when in 1691 he published his l'*Architecture Pratique*, strictly used as a 'treatise on the *toisé* of buildings'.

Setting down the *toisé* on paper coalesced many issues, its methods varying widely from one region to another and especially between Paris and the rest of the country, or from one practitioner to the other. In the early 19th century, in his *Tableau détaillé de tous les ouvrages de bâtiment [Detailed chart of all the types of buildings]* Joseph-Marie Morisot reaffirmed the triple object of the 'science of *toisé*'. In the first instance allowing clients and contractors to foresee costs; second, estimating building costs for a building designed by an architect and the ensuing adjustments on the project; thirdly, estimating in retrospect the costs of building for payment. The field was at once related to design (and to creativity), law and economy. The course's layout, closely resembling a treatise, would suggest that Desgodets envisaged a wider circulation, for instance, among owners and contractors. For that reason the *toisé* was given a theoretical framework and streamlined; only its principles were exposed methodically and according to the major divisions of construction (masonry, framework, roofing, lightweight work, etc.).

The course inspired subsequent ones such as l'*Architecture Moderne* (1728 and 1764), the reissues of l'*Architecture Pratique* by Bullet, *Le Toisé* by Nicolas Ginet who, drawing from the Desgodets, also corrupted it. But the course remained unrivalled for a long time. Desgodets taught the *toisé* as it was used, but also criticised its incoherencies, uncertainties and difficulties for non-Parisians when dealing with the Custom of Paris. Although he expressed the need to reform these rules, he did not take on the task as he considered it the royal authority's prerogative. The chief value of the course was his endeavour to reform the hands-on method used. He never failed to point out misuses or problems and the means to mitigate or even solve them. Although the reform to which he aspired was carried out in the 19th century through the works of Morisot or Rondelet, in 1866 the *vérificateur* [professional quantity surveyor] Descouges wrote: 'The name of Desgodets has remained, and it is only fair. Today still his book is a precious and plentiful mine of useful information on all things pertaining to construction. It was consulted, fruitfully, by most of those who since then have written on urban planning and estimating construction.'[17]

The treatise on easements

Desgodets's course on easements consisted chiefly in commenting upon the sections of the Paris Custom on easements and the reports of experts, but he also dealt with other questions related to construction. The course was given to the students at the *Académie* between 1723 and 1725; it was meant for future architects to acquire sufficient legal knowledge to practise as experts just like contractors, who until May 1690, edict held the exclusive power of expertise.

We found only one copy of the course in its original version, not autographed, consisting of two complementary manuscripts held today at the *Bibliothèque du Sénat*. The first manuscript (S1) is the 'preliminary' and unabridged transcription by Desgodets of all the sections of the customs of France that dealt with easements and expertise, for each and every region and town. Then came the sections solving disputes over specific rights related to construction by drawing from the relevant customs (main manors or capon thefts, marks of justice and traces of gallows, dovecots, *franc alleu,*[18] *droit de champart,*[19] *retrait lignager,*[20] the difference between movable and immovable property, repairs by renters, usufructuaries, dowagers, or *bénéficières,*[21] or fortuitous cases of damage to constructions). The volume ends with a compilation of judgments by various courts of justice (parliaments, first instance courts, etc.) concerning architecture and/or construction and garnered from judgment commentators dealing with subjects various and sundry. In the second volume (S2), Desgodets explains the sections of the Paris Custom only, which by then already represents French law. This volume is divided into two parts: one on easements and expert reports and another on the sections scattered throughout the custom, which Desgodets deems integral to the new French law such as repairs, movable/immovable property, prescription, *retrait lignager.*

We chose to edit these two volumes and then to circulate online the text-stages subsequently published by Desgodets. These provide evidence of the legal knowledge of architects trained by Desgodets, as shown by their successive comments. These include those comments by architect-expert-*bourgeois* Martin Goupy in 1748 and 1768, the famous lawyers of the Parliament of Dijon in 1752 and 1799, lawyer Hugues Destrem in 1845[22], and by an anonymous author of Grenoble in 1855. Later we expect to publish the versions edited by the lawyer P. Lepage in 1808 and 1848 as well as by Ramelet, an architect from Besançon, in 1822. Both authors provided their texts with a systematic structure that differed completely from the initial text by Desgodets, thereby producing a veritable *mise en abime* of texts.

Regarding law, we were able to confirm both of our hypotheses. Desgodets widely drew from the existing legal literature to write his course including doctrine, jurisprudence, and customs. Likewise, the many successive editions of his course during the century when the *Code Civil* was implemented allow the general evolutions of law to be tracked. In its own way, the *Loix des bâtiments* would constitute a founding text of law just as, relatively speaking, the *De re aedificatoria* Alberti was for architecture.

The course became the reference point, reflecting legal trends during the Revolution as well as its adaptations after the *Code Civil*:[23]

> - Customs *versus* Roman law: as early as 1752 lawyers from Bourgogne used Roman law to comment on some editions of the *Loix*;
>
> - Paris law *versus* French law:[24] after 1804 the editions of *Loix* by lawyer P. Lepage[25] acknowledged Desgodets's far-sightedness in having focused his work on the Paris Custom. This was even more pronounced with Hugues Destrem who updated the Paris Custom a century after the *Loix* was published, even though only the *Code Civil* would have been enforced at the time.
>
> - The law of the provinces *versus* the uniformity of the *Code Civil*: in the 19th century lasting local uses gave Desgodets's course a second life through new local editions. For instance, in Besançon with the architect Ramelet, or in Grenoble where the Loix are published in a *Annuaire statistique de la cour*

impériale de Grenoble et du département de l'Isère [Statistical directory of the imperial court in Grenoble and of the Department of Isère] in 1855.

- French law *versus* foreign law[26]: many mentions of Desgodets can be spotted abroad *via* Lepage in Italy,[27] but also in Britain[28]. In fact, Desgodets initiated a movement in Europe in favour of writing a construction law. This was observed in Italy through the unfulfilled goals of architect Bernardo Antonio Vitonne (1704-1770) from the Academia di San Luca in Rome[29] as well as those which Francesco Milizia (1715-1798) did achieve in *Principi di architettura civile*[30].

Desgodets opened the debate on creating legal disciplines through new branches of law:

- The birth of these new branches does not seem to be limited to a particular public. This course describes how law was initially shaped by use, first by masons, then by architects, before reaching the world of lawyers. Could the new paradigm, which we have exposed, be applied to other fields of knowledge? In theory, would law be a field not exclusively practised by lawyers?

This would imply that:

- The need to create stems from a social need which, in this instance, would be to produce neighbourhood rules through the question of ownership, hence the importance of easements and joint ownership in constituting this law. More generally, this constant necessity could take shape in finding a common good.

- A discipline's autonomy often grows through its being taught, or given an academic treatment, a compilation or a code. These different means should be multiplied to be productive. Ultimately, the political authority will always set off the mechanism and have the upper hand on the final and effective creation.

* *

*

Any completed project deserves a future life, first because a project is never really finished and second because it encourages new endeavours. Our project opens unlimited perspectives for research. This work should allow the scientific community interested in the history of architecture, coming from all fields and backgrounds, to revisit its knowledge on the relationship between theory and practice. This is true not only for purely theoretical subjects such as the orders or the *commodité*, which could lend themselves to new approaches in practical simulation, but also for subjects usually considered as concrete, such as *toisé* and easements. Nonetheless, Desgodets, who called for reform of both these economic and legal fields, endowed them with theoretical aspects.

Our editorial project is nearly complete. We still need to publish all the versions of the treatises thus allowing researchers to access them freely. Moreover, new manuscripts probably remain to be found (cf. the *toisé* recently acquired by the Cnam, the rediscovered *Edifices antiques* annotated by Legrand and Molinos, etc.) and many archives still deserve to be sorted through. We plan on organising an international conference on Desgodets at the end of 2014. A paper publication of the critical edition should follow fairly soon after.

Above all, this project should encourage researchers to take on our approach to study similar teaching corpora in France, such as the manuscripts of Philippe de La Hire, or those of architecture and engineering professors, through courses from other regional or foreign schools. Courses are volatile manifestations of knowledge, which fortunately are rescued from oblivion by a multitude of intermediaries (students, professionals, colleagues, etc.). They need to be rediscovered and to that end the Digital Humanities are now a vital science.

References

* Antoine Desgodets, *Les cours d'Antoine Desgodets, une édition numérique*, R. Carvais (ed.), *2013*. < http://www.desgodets.net/>

[1] B. Baudez, *Architecture & tradition académique au temps des Lumières*. Rennes: Presses universitaires de Rennes, 2012.

[2] A. Gerbino, *François Blondel: Architecture, Erudition, and the Scientific Revolution*. London and New York: Routledge, 2010.

[3] A. Becchi, H. Rousteau-Chambon, et J. Sakarovitch (eds.), *Philippe de La Hire (1640-1718). Entre architecture et sciences*, Paris: Picard, 2013.

[4] A. Davrius, *Jacques-François Blondel (1708/9-1774): architecte des Lumières*. PhD. Paris: EPHE, 2011.

[5] Ch. D. Armstrong, *Julien-David Leroy and the Making of Architectural History*, London and New York: Routledge, 2012.

[6] B. Baudez, "Les mathématiques à l'Académie royale d'architecture dans la seconde moitié du XVIIIe siècle", *Revue de l'art*, 2011, pp.171-171, pp.9-15; H. Rousteau-Chambon, "L'enseignement des mathématiques dans la première moitié du XVIIIe siècle au sein de l'Académie royale d'architecture", *Studies on Voltaire and the Eighteenth century*, 06, 2008, pp.57-68; H. Rousteau-Chambon, *L'enseignement à l'Académie royale d'architecture. De Philippe de La Hire à Louis-Adam Loriot [1687-1762]*. HDR, Paris Est (to be published).

[7] Fr. Lemerle-Pauwels, Y. Pauwels, *Architectures de papier. La France et l'Europe, XVIe-XVIIe siècles*, Anvers, Brepols, 2013.

[8] J. –Ph. Garric, V. Nègre, and A. Thomine-Berrada (eds.), *La Construction savante. Les avatars de la littérature technique*. Paris: Picard, 2008.

[9] ". . . tout homme de quelque âge et condition qu'il soit, qui aura goût pour l'Architecture." From the letters patent creating the statutes of the *Académie d'architecture*, of February 1717, art. XXXVI.

[10] "géométrie-pratique," . . . "leçons de différentes notions, enseignements, règles et pratiques d'architecture." *Ibid.*, art. XXXV.

[11] The layout and proportions of churches and other buildings (*". . . la distribution et les proportions des églises et autres édifices."*) Antoine Desgodets, *Les cours d'Antoine Desgodets, une édition numérique, Traité de la commodité,* Robert Carvais, Hélène Rousteau-Chambon et Emmanuel Château (eds.), 2013. < http://www.desgodets.net/toise>

[12] It means, in mathematics, the science or art of measuring surfaces and solids, and of expressing their area or volume in parts of certain appropriate units: for example, in toise [six feet], if it is a surface; cubes, if it is a volume (*Dictionnaire de l'Académie Française*, 6th Edition, 1832-1835).

[13] Analysing watermarks, introductory notes, bindings, handwritings, etc.

[14] Archives nationales MC LIII, 245, inventory after Antoine Desgodets's death, June 17, 1728.

[15] This software, developed by researchers and engineers from the *Institut des Textes et Manuscrits Modernes* [Insitute of Modern Texts and Manuscripts] allows one to compare different versions of the same text, from the draft to the printed text. Ganascia, J.-G., Fénoglio I. and Lebrave, J.-L., "Manuscrits, genèse et documents numérisés. EDITE : une étude informatisée du travail de l'écrivain," in *Document numérique*, vol. 8, n°4/2004, pp. 91-111.

[16] <www.juxtasoftware.org> [accessed 30 May 2013]

[17] *"Le nom de Desgodets est resté, et ce n'est que justice. Son livre est encore aujourd'hui une mine précieuse et féconde en renseignements utiles sur toutes les choses relatives à la construction. Il a été consulté, avec fruit, par la plupart de ceux qui, depuis, ont écrit sur la voirie et l'estimation des travaux."*

[18] Land free from any feudal hierarchy.

[19] Rent in kind; initially, taken directly from the field before the tenant would harvest his crop.

[20] In case of alienation of a property from an estate by an outsider, the right for a kin of the lineage to ward off the former to reinstate the property into the family by refunding it to him.

[21] Owner/holder of a benefit: land concessions granted by the king to individuals holding a military office, or income from certain estates linked to an ecclesiastical function.

[22] H. Destrem, *Les Lois des bâtiments suivant la coutume de Paris par Desgodets, avec les notes de Goupy, nouvelle édition mise en rapport avec les lois et la jurisprudence modernes relatives aux bâtiments et construction*, Paris, 1845.

[23] Reference or model as in architect Ramelet's warning in his *lois relatives aux bâtiments* (1822*):"... j'ai imité Desgodets, comme on peut s'en convaincre à la lecture de sa préface et de son livre. Je ne pouvais suivre, en tout, un meilleur modèle, bien que je sois loin de l'avoir égalé,"* "... I imitated Desgodets, as one can tell when reading his preface and his book. I could not have followed, in all things, a better model, although I'm very far from ranking with him." R. Carvais, Comment traduire les *Loix des Bâtiments* après l'avènement du Code civil ? *in* R. Carvais and V. Nègre et al. (eds.) *Traduire l'architecture*. Paris. (to be published).

[24] On the uniformity of construction law for all the French people, read Madin, *Extrait de l'Essai sur les Loix des bâtiments*, Paris, An V, 1797.

[25] We know very little about this author. P. Lepage is cited on the title page as a "former lawyer", that is to say he served under the *Ancien Régime* [ie. before the Revolution]. He would have been specialised in the editing of reference books of *Ancien Régime* law.

[26] R. Carvais, "For a Comparative Study of Construction Laws", In R. Carvais, A. Guillerme, V. Nègre, and J. Sakarovitch (eds.), *Nuts & Bolts of Construction History. Culture, Technology and Society*, vol.3, t.1. Paris: Picard, 2012, pp. 475-487.

[27] The works of lawyer P. Lepage were translated into Italian, since the *Code Civil* ruled in Italy during the Napoleonic era. We found several editions of *Loix des bâtimens*, one edited in Rome in 1845, one edited in Naples, then another edited in Fuligno in 1854.

[28] J. Elmes, *A Practical Treatise on architectural Jurisprudence*, London, 1827; J. Kent, *Commentaries in American Law*, vol. 3. New York, 1928, pp.344-362.

[29] (B. A. Vitonne, *Istruzioni diverse concernenti l'officio dell'architetto civile*, Lugano, 1766, p. XIII.

[30] F. Milizia, *Principi di architettura civile*, t. III, chap. VIII, Bassano, 1785, pp. 298-315.

Building material migration: imported brick and localization in Taiwan in the seventeenth century

Yu-Sen Chou and Dimitris Theodossopoulos
ESALA, University of Edinburgh, UK

Summary

In certain circumstances, the evolution of architecture is highly connected with construction technique development, which in turn may be generated by external cultures. This paper gives insight to such developments by discussing how brick was established as a new imported building material into Taiwan in the 17th century under Dutch influence.

Taiwan's architectural system was essentially started by the Dutch from 1624, who built the first major permanent structures, three fortresses with associated burghs. Research into archives shows that the Dutch East India Company (VOC) brought for this purpose a new building material, brick, first by importing finished products from Southern China, then gradually moving to local production in Taiwan, which spread to several places within the island, even exporting to their strongholds in Japan. However, after the VOC left in 1662, the brick making industry supported by the VOC seems to have ended in Taiwan, as did the brick architecture.

Studies in this field have concentrated on the historic aspects of the VOC presence rather than their architecture. This study attempts for the first time a comprehensive dating of the key stages based on archival research. In addition, analysis of the masonry techniques of the scant remains serves to frame the technology transfer process within the context of Dutch construction of the time.

Introduction: a material of force: bricks imported by the Dutch East India Company (1624-1662)

For Taiwan the Seventeenth century was a crucial turning point in its history, from politics to architectural culture. Before that, Taiwan was an island with dozens of aboriginal tribes. Suddenly, Dutch, Spanish, Japanese and Chinese came to the island for their own interests. Among them, Dutch traders took the lead and became the greatest foreign influence on the island from 1624 to 1662. They also brought the new building material, brick, and changed the architectural landscape of Taiwan forever.

Up to that point, the building materials used in Taiwanese aborigines' housing were natural and local, such as timber, bamboo, grass and stone plates. The brick which migrated to Taiwan is a result of a series of historic events, highly related to the trading wars of the period.

The Dutch East India Company (VOC) showed a strong interest in trading in the Far East in the beginning of the 17th Century, following the Portuguese and Spanish. In 1622, to open business with China and Japan, twelve ships launched from Batavia city (now Jakarta), the VOC headquarters in Asia, to seek a trading base near China. After several battles , their target trading base switched from Macau to Pescadores (now Penghu Islands), an archipelago between Mainland China and Taiwan, and then finally to their third choice, Tayouan (the Anping area in Tainan City now) in Taiwan. In the following years, the VOC established its power on this island. As an extension of its power network, the Dutch introduced brick into Taiwan, initially importing materials for local production, then spreading to different places in Taiwan. Through this process, brick became a critical building material for the VOC to establish reliable structures, in particular the three main forts (Fort Zeelandia, Fort Provintia, Fort San Domingo) that safeguarded their trade presence (Fig. 1).

With this background, Taiwan in the 17th Century provides an excellent case study through which to discuss the issue of foreign building materials technical transfer in East Asia in the Age of Exploration, driven by a Western superpower at that time (the VOC). Discussion of this topic is still quite limited in current research, thus, this paper intends to explore the migration of bricks, an advanced building material in Taiwanat that time, from importation to localization under the VOC's influence, to understand the role of this particular building material in its historical context and how it was utilised by political powers to shape the local architectural landscape. This study attempts for the first time a comprehensive dating of the key stages based on archival research. In addition, analysis of the masonry techniques of the scant remains, combined with archivalresearch and fieldwork, frames the technology transfer process within the context of Dutch construction of the time.

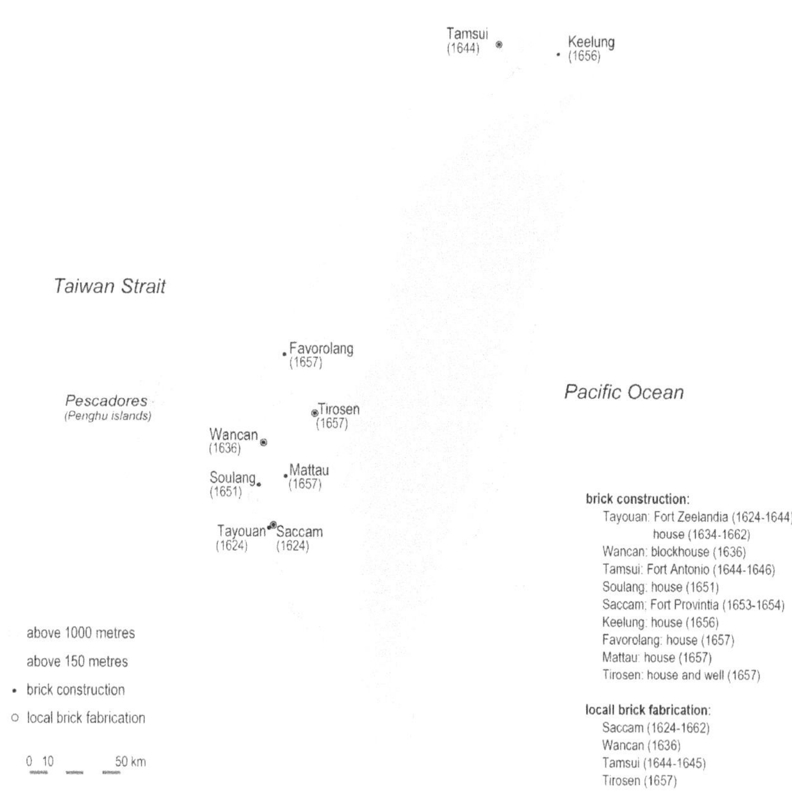

Fig. 1. Brick construction and production in Taiwan during Dutch rule (1624 - 1662) (author's drawing)

More than ballast - brick through the VOC network in Taiwan

The VOC created its empire by building its own network for trades, and brick played a significant part. Brick was commonly used as ballast for shipping in the 17th and 18th Centuries [1]; however, for the VOC in Taiwan in the 17th Century, it was also a major building material for which the company had a great demand, not only for construction, but from a certain perspective, also as a kind of defence weapon of their trading system. Brick, the material itself and its fabrication process, was imported as the VOC's power was expanding inTaiwan and spread to strategic points of its network to strengthen the presence of the VOC.

In order to understand the network of the VOC in Taiwan, two collections of VOC archives are crucial: "*Letters of the VOC-governors in Taiwan sent to Governor-General of Batavia*", Volume I (1622-1626) and II (1626-1629), which includes 29 letters written by the VOC-governors in Taiwan to give a general report of Taiwan affairs to its headquarter in Batavia, numbering three to four letters in one year; and "*Journal of Fort Zeelandia, Taiwan 1629-1662*", which is the daily record by the VOC secretary in Taiwan, providing detailed information about the company's activities. Through these two archive collections, an outline of the use of brick by the VOC in Taiwan can be defined.

Brick became an indispensable building material for the VOC in establishing permanent forts and company houses in Taiwan, so gathering supplies of bricks was one of the major works at the beginning of its business. On the one hand, the company imported bricks from its trading target China; on the other hand, the VOC introduced brick making techniques in order to produce them locally and to depend less on China. Both sources of brick existed during the 38 years of Dutch rule.

Start of imported and local production (1624-1635)

For defence requirements, Fort Zeelandia in Tayouan was the first major project for the VOC in Taiwan (Fig. 1). The building started in 1624 and the inner fort was finished in 1633. To obtain the reliable building material, the Dutch tried either importing from China or local production.

Importing bricks from China never stopped and it was a crucial operation in the first five years (1624-1629) of their rule of Taiwan. Table 1 records the flow of bricks fabricated, imported or stored and will be used to monitor the use of the material in critical periods. Less than half a year after the VOC moved to Tayouan, 14,000 bricks were imported from China in December 1624 for Fort Zeelandia. [2] Although local brick production appeared in the following year, imported bricks were still a crucial commodity for the company. In the archives, the Taiwan VOC-governor mentioned that bricks would continue being imported from Southeast China in 1625, 1626, and 1629 to reinforce the wall of Fort Zeelandia and build the warehouses and company houses in Tayouan [3]. After that, imported bricks were less often noted in the records. It could be considered that the imported bricks from China had lost importance for the VOC officers because the local production was growing, as discussed below.

Ensuring that brick production was local and sustainable was the one of the first goals of the VOC in Taiwan. [4] At the same time as importing the first batches of bricks, the VOC tried to import brick-making craftsmen from China in December 1624 and quickly acquired ten workers.. However, the performance of the first crew did not satisfy the acting Taiwan VOC-governor, Martinus Sonck, because of its low production speed. He may have replaced them with Dutch sailors who learned brick-making more quickly. [5] The demand for bricks for the VOC local authority was large and urgent, so they gathered bricks imported from China and produced them locally at the same time. As a result, the number of stored bricks increased quickly, from 14,000 in December 1624 to 111,000 in October 1625

(Table 1). [6] In the following years, Siaccam area (which changed its name to *Provintia* in the later period), a new town near Tayouan (Fig. 1), became crucial for local brick production. The local industry may have become strong enough to meet the demands of the Taiwan VOC in the 1630s. In 1634, the VOC introduced a tithe – a ten percent tax – on locally made bricks in Tayouan and Siaccam, including fired and unfired ones, to fund a brick company house; and forced all residents in Tayouan to rebuild their houses with bricks instead of the common thatched cottage, to prevent rebellions. [7] It also meant the local brick industry developed quickly. Notes about brick from China are absent in the archives after 1630, which suggests the Taiwan VOC officers were no longer worrying about the brick supply problem as they had in 1624 and 1625. Thisgives an indication that the brick making industry had taken root in Taiwan.

Tab. 1. Brick flow records in Taiwan (compiled by the author)

Date	Quantity	Origin	Destination	Project
05 Nov 1624	uncertain	China	Tayouan, Taiwan	Fort Zeelandia
12 Dec 1624	14,000	China	Tayouan, Taiwan	Fort Zeelandia
29 Oct 1625	110,000 (Partly imported)	China	Tayouan, Taiwan	Fort Zeelandia
14 Aug 1629	uncertain	China	Tayouan, Taiwan	Fort Zeelandia
18 May 1636 – 11 Jul 2016	uncertain (13 recordings)	Tayouan, Taiwan	Wancan, Taiwan	Blockhouse
15 Mar 1637	50,000	Tayouan, Taiwan	Firando, Japan	Office
01 Nov 1643	5,000	China	Tayouan, Taiwan	(no record)
10 Apr 1644	uncertain	Tayouan, Taiwan	Tamsui, Taiwan	Fort Anthonio
01 Jul 1644	10,000	China	Tayouan, Taiwan	(no record)
10 Jun 1645 – 16 Oct 1645	74,700 (total in 13 recordings)	China	Tayouan, Taiwan	(no record)
19 Mar 1646 – 16 Oct 1646	27,000 (total in 20 recordings)	China	Tayouan, Taiwan	(no record)
19 Mar 1646 – 16 Oct 1646	28,000 (total in 12 recordings)	China	Tayouan, Taiwan	(no record)

Transfer: from defence construction to houses (1636-1662)

After the brick making industry became established in Tayouan, the VOC created a network of building materials support systems. From 1636, apart from importing brick to the Tayouan area, the VOC also exported bricks to their expanding control points in Taiwan. From the two sets of archives, we learn that the VOC transported bricks to seven other areas in Taiwan in 38 years, from the material itself to fabrication techniques (Figure 1). These works often included defence concerns and strong military structures for forts or blockhouses, such as Wancan in 1636 and Tamsui in 1644. Those two cases demonstrate how the VOC in Taiwan kept moving bricks across its network and with practical materials first, then techniques to produce and construct, used bricks to strength that network.

Wancan, the major production centre for lime ash from 1624 [8], was also the first place to accept the transfer of local brick-making techniques from Tayouan. In 1636, the VOC decided to build a brick blockhouse there. To that purpose, the Taiwan VOC-governor Hans Putmans went to Wancan in March 1636 to investigate whether local production of brick was feasible. Once enough raw material was found, the project was planned to start on 14 April 1636. In the following two months, bricks were shipped from Tayouan to Wancan, as is shown in thirteen records, and the last shipment of brick to Wancan was on 11 July 1636. [9] There were brick construction activities in Wancan later, but there is no other record of shipping bricks to the town. The blockhouse was finished in the following January. From these records it appears that shipping physical bricks from Tayouan to Wancan only occurred in these two months, and that thereafter brick was manufactured in Wancan once the brick-making skills had been mastered.

In the second case, the experience of another fort built in Tamsui provides more detailed information about the process. Following in the steps of the VOC, the Spanish built their short-lived colony in the Tamsui area in northern Taiwan from 1626, but left in 1642 after they were defeated by the VOC. After the Spanishdeparted , the VOC in Taiwan wanted to build a new fort in Fort Antonio. The fort was built quickly with bamboo in 1642, and rebuilt with bricks between 1644 and 1646. [10] Owing to great demand, the VOC sent 50 brick-makers to Tamsui and even changed the production site once to produce better qualitymaterials. [11] Up to May 1645, the storage of bricks in Tamsui increased dramatically in one year, reaching 300,000. [12] Considering the intense frequency of import records between 1644 and 1646 (see Table 1), it is possible that Chinese imported bricks were also shipped to Tamsui to support the Fort Antonio project, but locally produced bricks from Tamsui also played a great part.

The brick transfers by the VOC in Taiwan before 1650 were mainly to address defence concerns. After 1651, the brick transfersextended to other construction projects, such as housing or wells. The brick company houses were built for local officers or priests. The first brick transfer record without a strong defence objective was for two company houses in Soulang in 1651, and the local priest and officer even argued over who should take the better one. [13] Then brick company houses spread to other VOC selected places, such as Keelung, Favorolang, Mattau, and Tirosen (see Table 1 and Figure 1). Among them, Tirosen is a special case. The VOC sent a Chinese craftsman there, who built a kiln to support the local production for a priest's house and a well in the central square in the village. [14] In this case, a Chinese brick-maker was transferred across the network of the VOC instead of the material - brick - itself.

Brick export (1637)

Apart from exporting to different places within Taiwan island, Tayouan-made bricks were probably shipped to Firando (Hirado/ Nagasaki) in Japan. According to the record of 15 March 1637, the VOC in Firando asked the company in Taiwan to send 50,000 bricks, no matter whether made in China or Tayouan. [15] This record means that 13 years after 1624, the production of Taiwan locally made bricks was not only supporting the demand of the VOC in Taiwan, but was also producing enough for export. Moreover, it shows that bricks were now the goods instead of the ballast.

According to the archives, the spread of brick and its fabrication techniques, the production facilities (kilns) and the products (brickbuildings) appeared in several places in Taiwan. However, it is difficult to know precisely what kind of techniques they applied because there is a lack of relevant references in the archives and no kilns built in the Dutch period survive in the present day, only a few remains of brick construction. The study of bricks in the remains of Fort Zeelandia may indicate the different sources of brick, of imported and local production.

Bricks in the remains of Fort Zeelandia

Very little of brick construction from the Dutch period in Taiwan is left today. Among them, the remains of Fort Zeelandia may provide the best physical evidence. The brickwork shows the Dutch architectural knowledge of the period, and different brick sizes may indicate different production sources.

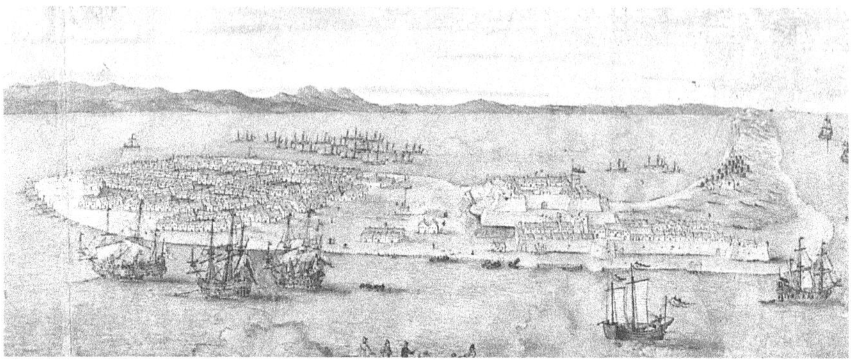

Fig. 2. Fort Zeelandia (about 1650), painted by Johannes Vingboon, commissioned by the VOC. (A collection of National Taiwan Museum) [16]

Fort Zeelandia was the first major project of the VOC in Taiwan, and it was also the first recorded brick building in Taiwan's architectural history. . Construction started in 1624, the interior fort was finished in 1632, with additional construction possibly continuing to the late 1650s. It is a typical Dutch Fort of the 17th Century, that used to consist of three concentric layers of walls, including residences for the officials, churches, barracks and jails (Fig. 2).

It is highly possible that the three layer construction of Fort Zeelandia was not planned from the beginning, but resulted in response to circumstances in different periods. Through archival research, the process of construction could be placed in four stages: timber wall, brick wall, external brick wall and brick reinforcement (see Fig. 3). The first stage lasted only a few months, between August and December 1624, when the VOC built the shape of the fort with timbers, for a temporary usage, as happens often when a new force arrives in a foreign land. In the meantime, the Taiwan VOC-governor Martinus Sonck planned to reinforce the construction with bricks and searched for a supply of this material. [17] The second stage lasted seven years, from 1625 to 1632, and scarps and bastions held with timber and clay were generally rebuilt with bricks, and the central fort with four bastions was all covered with a brick wall at the end of 1632. [18] The third stage ended in 1643. At that time, the central fort added an outside layer with four roundels, and two bastions were added to the external fort. [19] The final form was basically defined at this stage (see Fig 2, a drawing from 1650). Then at the fourth stage, the fort needed continuous repair because of the severe weather conditions and foundation problems, and the company reinforced it in the late 1650s in preparing the war with Koxinga. [20]

The continuous construction process caused a great demand for bricks but the number of bricks used to build Fort Zeelandia is a mystery. Considering the thickness of the walls mentioned in the archives and the size of the surviving bricks in the remains, the number of bricks used for the whole construction could be more than 5,000,000. [21] This is an enormous quantity, considering that the second layer of the inner fort and the outer fort were built after 1632, but only a few mentions of Chinese imported bricks exist in the archives. This is indirect evidence to prove that the local-made brick industry became the major supplier of the bricks of Fort Zeelandia's construction after 1632.

Today, only fractions of the original establishment are left. (Fig. 3) Remains of Fort Zeelandia are in three parts: two sections of walls of the outer fort, and part of the roundel of the outside walls of the inner fort. The condition of each part varies. The remains of the southern wall of the outer fort (Part B in Fig. 3 and left side of Fig. 4) are preserved in the finest condition among the whole ruin, about sixty five metres in length. Otherwise, the remains of the semi-circular bulwark of the inner fort (right of Fig. 4) is worthy of more attention because of its thickness – about 2.2 m.

The composition of the remains reveals the attitude to Fort Zeelandia by the VOC in Taiwan. Instead of earth or rubble in the middle core of the body of the wall, as happens often in walls of this size, the VOC chose a hard and excessive (intensive?) way to build the wall with a solid wall, up to 2.2 m thick. This is a noticeable feature since it is contrary to the traditional fort walls of the 17th century that usually had to be soft to absorb the impact of cannon balls in that period of gun powder. [22] It reflects the chronic insecurity of the VOC in Taiwan at that time, and proves that brick was a reliable material for the Dutch.. The fact that the brick was rendered with lime ash mortar demonstrates the need for the building to adjust to the local climate – hot, rainy, and windy.

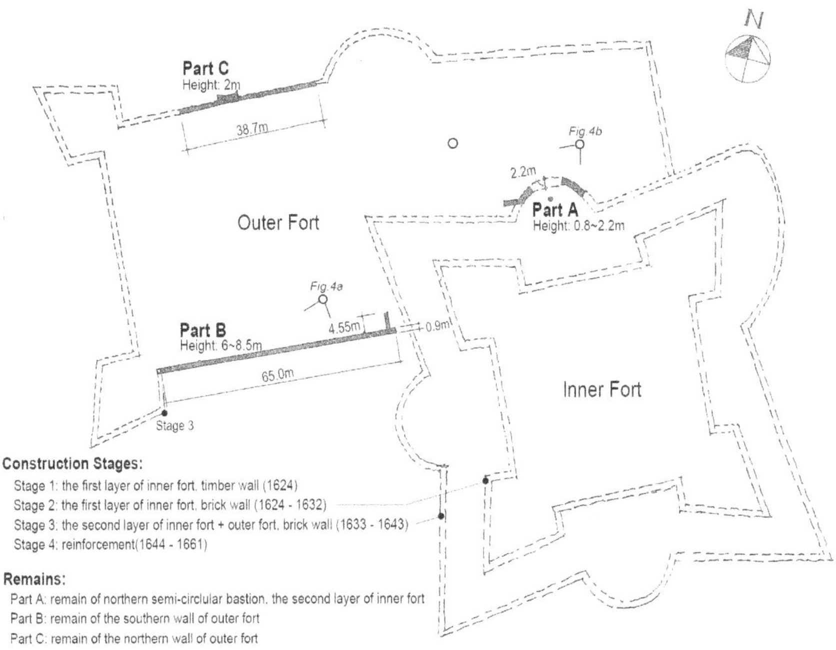

Fig. 3. Map of construction stages and remains of Fort Zeelandia (Author's drawing. Background diagram: [23])

The Dutch used a simple but functional brickwork bond to build this fort. From study of the remains, each layer use headers as the major element of the wall body, and a row of stretchers would be put on one side of the wall face. The brickwork pattern of the upper layer is reversed and horizontally shifts the distance of half a brick (Fig. 5 & Fig. 6). This maximizes the strength of the joints, suitable for its defence function. The level of each layer demonstrates the brickwork knowledge and craftsmanship, which can be clearly seen from the remains of the semi-circular bulwark of the inner fort. The average thickness of the joints of layers (?) is about 15 to 20 mm.

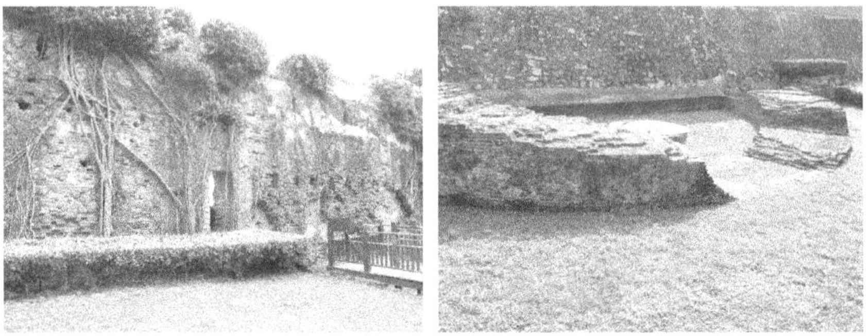

Fig. 4. A) north elevation remains of southern wall of outer fort; B) remains of the semi-circular bulwark of the inner fort (author's photos, 2011)

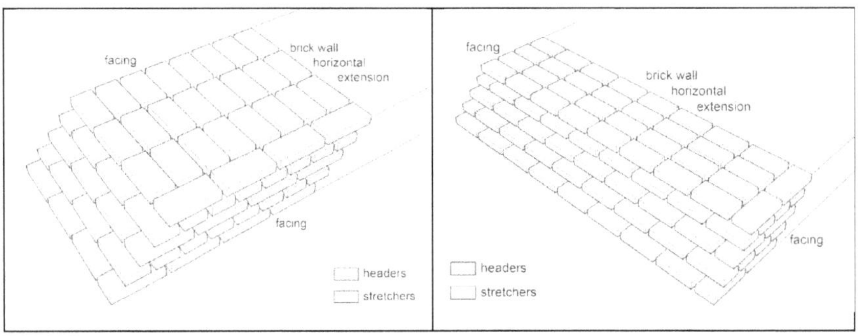

Fig. 5. Two brickwork patterns of remains: A) southern wall of outer fort; B) semi-circular bulwark of inner fort. (author's drawings)

In general, the colour of the bricks of Fort Zeelandia is orange red, and the texture of its appearance is slightly rough, without a glazed finish. Observing some surface frayed brick, it could be said the brick may have been fired for dozens of days (Fig. 6). The size of bricks varies, there are generally two types of thickness : One is longer and thinner, about 230 mm long, 115mm wide, and 35mm thick (Fig. 6a); the other is compact and thicker, about 210 mm long, 100mm wide, and 45mm thick (Fig. 6b). Considering the brick size could change by 5%-10% during firing, the major difference between these two bricks is thickness. The remains of the semi-circular bulwark of the inner fort were built with the former brick. But the remains of southern wall of the inner fort are much complicated, mixing these two types (see Fig. 7). The best example is its north side that shows clearly the two different types used, the facing one (45mm) being the thicker and the inner one (35mm) being the thinner.

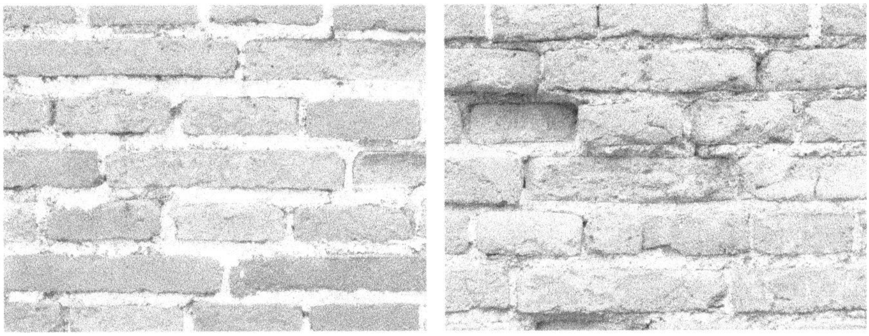

*Fig. 6. Two sizes of bricks on the north elevation remains of the southern wall of the outer fort: A) 230*115*35mm; B) 210*100*45mm. (author's photo, 2011)*

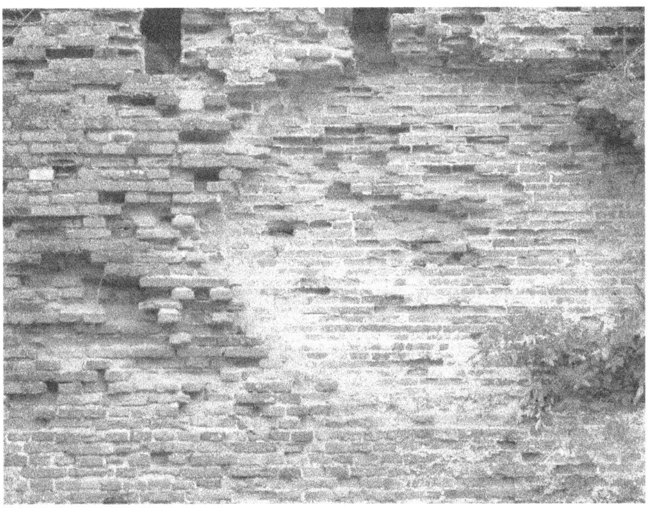

Fig. 7. Mixed sizes of bricks, north (inner) elevation, southern remains of the outer fort. (Part B in Fig. 3): thicker types facing a thinner-type inner core (author's photo, 2011)

The explanation for the two sizes of bricks is still a puzzle. One possibility is that they may represent the different sources. Comparing with common historical brick sizes in Fujian of (in?)China [24] and Amsterdam [25], the size of thinner one (230 x115 x35mm) is relatively close to the Amsterdam size; in addition, this dimension is rare in Fujian. Therefore, this type may have been locally fabricated under the VOC's supervision; some of them may even have come from Batavia or even Amsterdam as ballast. Alternatively, the thicker one (210 x100 x45mm) could have been imported from China; however, a similar brick size also appeared in Amsterdam. [26]. The exact sources of these two bricks still need more archaeological research to confirm their origins.

Conclusions

The research gave an outline of brick imported and localized in Taiwan under the Dutch in the 17th century. Unlike the company's usual habit of transferring bricks as ballast, the VOC in Taiwan introduced brick-making techniques to the local area. . However, the method of manufacture may not come from the Dutch. To acquire this reliable and familiar building material in East Asia, they decided to import bricks from China, using Han Chinese craftsmen for local fabrication at the same time. Through archive and fieldwork research, it appears that these two methods of obtaining bricks existed during the whole period of Dutch rule, a reliable building material for the VOC. This technology transfer included end products or expert fabrication skills, and the way they spread depended highly on support of authority, VOC.

Further research is needed to understand the process in more detail. For instance, further archaeological investigation may help to determine the source of the different sizes of bricks; the brick making techniques, repairs and design changes to support further discussion of the development of the bricks.

References

[1] R. Weston, *Materials, Form and Architecture.* London: Laurence King Publishing, 2008. P.21.
[2] S. Chiang, (Trans.), *De Missiven van de VOC-Governeur in Taiwan Aan de Gouverneur-Generaal Te Batavia 1: 1622-1626 (Letters of VOC-governor in Taiwan sent to Governor-General of Batavia 1: 1622-1626).* Tainan: National Museum of Taiwan History and Taiwan Historica, 2010. p.170.
[3] ibid., p.195; ibid., p.297; S. Chiang, (Trans.), *De Missiven van de VOC-Governeur in Taiwan Aan de Gouverneur-Generaal Te Batavia 2: 1627-1629 (Letters of VOC-governor in Taiwan sent to Governor-General of Batavia 1: 1627-1629).* Tainan: National Museum of Taiwan History and Taiwan Historica, 2010. p.191.
[4] Chiang, *Letters*, (Note 2) p.170.
[5] ibid., p.195.
[6] ibid., p.233.
[7] S. Chiang, (Trans.), *De dagregisters van het kasteel Zeelandia, Taiwan 1629-1662.* (4 vols), Tainan: Tainan City Government, 1999. Vol.1, p.156, p.161.
[8] Chiang, *Letters*, (Note 2) p.170.
[9] Chiang, *Journal*, (Note 7) 1999. Vol.1, p.247.
[10] Chiang, *Journal*, (Note 7) 2002. Vol. 2, p.270.
[11] ibid., p.312.
[12] ibid., p.407.
[13] Chiang, *Journal*, (Note 7) 2003. Vol. 3, p.259.
[14] Chiang, *Journal*, (Note 7) 2010. Vol. 4, p.162, p.329.
[15] Chiang, *Journal*, (Note 7) 1999. Vol. 1, p.302.
[16] National Taiwan Museum website: <http://www.ntm.gov.tw> [accessed on 20 December 2013]
[17] Chiang, *Letters*, (Note 2) p.170.
[18] Chiang, *Journal*, (Note 7) 1999. Vol.1, p.80.
[19] ibid., p.474; Chiang, *Journal*, (Note 7) 2002. Vol. 2, p.84
[20] Chiang, *Journal*, (Note 7) 2010. Vol. 4, p.379.
[21] The author's estimation, based on the research investigation and measure of remains.
[22] S. Pepper and N. Adams, *Firearms & Fortifications: Military Architecture and Siege Warfare in Sixteenth-Century Siena.* Chicago: University of Chicago Press, 1986.
[23] K. Tseng, *National Historic Monument Fort Zeelandia Investigation and Restoration Proposal.* Tainan: Tainan City: 1993. p.161.
[24] J. Gawronski and J. Veerkamp, "Bakstenen. Bouwstenen van Amsterdam", *Amsterdam. Monumenten & Archeologie*, no.3, 2003.
[25] C. Tsao. *Min-nan Traditional Architecture.* Xiamen: Xiamen University Press: 2006. p.86.
[26] J. Gawronski & J. Veerkamp, *Amsterdam* (Note 24).

Building by Direct Labour: the significance of Direct Labour Organisations (DLOs) in the provision of public housing in the UK 1890-1980

Linda Clarke
University of Westminster, UK

Introduction

> "When the honourable member says that the number of completed houses is not so great as he had hoped it would be, he must enquire for the reason into the organisation of the building industry, and not in the government's plan for housing" Aneurin Bevan 1946[1]

This statement by Aneurin Bevan well expresses some of the innate limitations of the private construction industry under capitalism to meet housing need and hence the reason why, at different periods of time in Britain over the past 120 years since 1894, local authorities have taken the responsibility for building with their own direct labour forces – or Direct Labour Organisations, DLOs, as they came to be known. These periods have tended to coincide with major social housebuilding programmes, including the first social housing under the Housing of the Working Classes Act in the 1890s, and then again after the first and second world wars, and – to an extent – with the Labour government in the 1960s. The role of the local authorities and the building workers they employed has thus been critical to the successful realization of public sector investment in social housing and public infrastructure construction, including the repair and maintenance of the public building stock.

The need for this public investment has been particularly acute in times of crisis when there has been a major need in particular for social housing for the working population, which the private sector has been unable or unwilling to fulfill. The local authorities have responsibilities for housing those in housing need, as well as for ensuring good training, employment and working conditions for the local population. It is therefore not surprising that their building programmes have often been exemplary with respect to providing comprehensive vocational education and training schemes, including the training and subsequent employment of women and those from the Black and Asian minority ethnic (BAME) population, setting health and safety and environmental standards, and maintaining good terms and conditions of employment negotiated with trade union involvement. Many of the social failings of the private construction sector – weak employer responsibility for training, adversity to trade unions, blacklisting, casual and insecure employment – that so often meant poor quality construction, extensive subcontracting chains and lack of integration of women and those from BAME groups have fallen on the DLOs to rectify, so showing that there are alternative ways of organizing the industry.

The DLOs raise many questions, which have been debated at length since their inception. Should they operate as a service, accountable on the basis of cost, or as an enterprise, trading at a profit? Can high quality, environmentally sustainable housing and public works be produced at reasonable cost? Are good employment and working conditions directly reflected in the standard of housing? Why did the DLOs largely disappear following the persistent attacks on them in the 1980s and 1990s? It is the purpose of this paper, written at a time of financial crisis and similar acute housing shortage, when local authorities such as Islington are again setting up building departments in house rather than outsourcing, to begin to address these questions, to celebrate and illustrate the importance of DLOs in Britain, and to consider how far they represent a historical phenomenon or can still play an important role, in particular in the provision of social housing.

The setting up of DLOs

Direct Labour Organisations were first set up in the 1890s in London soon after the passing of the Housing of the Working Classes Act of 1890 and at a time when the main function of local authorities was to act as rate-collecting agencies for paying the accounts of private contractors who were employed for all work. Underlying their establishment lay a number of important conditions. The first was the existence of extremely poor housing conditions, exposed by Charles Booth in his survey of the lives and occupations of the working classes in late nineteenth century London, the first volume of which appeared in 1889 and the second in 1891[2]. The second condition was proven collusion, 'ringing of contracts' and scamped work by contractors, in particular in relation to public building work for the Metropolitan Board of Works, the predecessor of the London County Council (LCC), clearly exposed with the building of Blackwall Tunnel and with the London school building programme. A third condition was the changing political environment, as local authorities were reorganized in order to provide better services and the 'Progressives' gained control of the new LCC, defeating the 'Moderates' who were more representative of the interests of local contractors and private companies. A fourth condition was the emerging strength of the trade unions, following the mass strikes of 1888 and 1889 and the organisation of those classed as 'labourers', and of the labour movement, especially through the success of the 'Fair Wages Movement' in fighting for improved trade union conditions and for the payment of 'fair' wages agreed within any particular locality. And a final condition was the growing unemployment of the 1880s and 1890s and the ineffective and temporary nature of public relief works intended to alleviate this. [3]

The campaign for direct labour became the focus for 'Progressives' in the LCC, for the 'Fair Wages Movement', for those agitating for better housing to remedy slum conditions and homelessness, and for the municipal socialists seeking to extend local authority services.[4] Indeed, one of the first acts of the Progressives was to bring in a 'fair wages' clause enforcing trade union conditions and pay on all building contracts given out by the council. In retaliation, contractors submitted very high tenders, in particular for a sewerage works at York Road, which attracted two identical tenders of £11,000. This, combined with proven collusive tendering for building cottages displaced by the construction of Blackwall Tunnel, led the LCC to set up its own Works Department, tasked with breaking the monopolistic position of contractors by building the sewer, which was completed for £6,854.[5] As John Burns, Progressive MP for Battersea, claimed:

> The establishment of a Works Department by the Council was inevitable. It was forced on us by the contractors themselves... their withdrawing of tenders and their systematic cornering of the Council. The new Department has completely revolutionized the old corrupt order of things. It has made the Council independent in its public works.[6]

The responsibilities of the local authorities of the time for a large range of services, including electricity, gas and transport, was reflected in the early activities of the LCC Works Department, which made all school furniture and carried out building work from a power station to new housing – including Chertsey House on the Boundary Street Estate, the first council house scheme in Britain. To support this activity, the LCC set up a variety of workshops and even set up its own brickworks.

The LCC's lead was soon followed by Battersea in 1894 and West Ham in 1896.[7] Battersea, with its large Progressive majority and with John Burns as its local MP, had perhaps the most widely extended role for its Direct Works, which had its own well-equipped joinery works, wheelwrights', blacksmiths', and fitters' shops. The department provided an entire building service, from having its own architects' department to new construction and repair and maintenance, and carried out all its own building work. Most notable was the building of the Latchmere Estate (Fig 1), opened in 1903, the first council estate in Britain to be built by the Council's own direct labour force, which followed a persistent and successful campaign, spurred on by Battersea Trades and Labour Councils, to obtain Local Government Acts of Parliament in 1899 and 1900 granting the right to build.[8] The Estate was composed of 315 dwellings, each with its own entrance and back garden, with baths and electricity supplied from the Council's own generating station and water supply from the artesian well sunk by the Council and also serving the adjacent Latchmere Baths. As Burns explained:

> The land has a communal origin, the streets bear democratic names; the whole plan, history and achievement is redolent of the common victory of the common people.[9]

It is perhaps not surprising that John Archer, a black man from Liverpool, was elected as councilor for Latchmere ward in 1906, becoming in 1913 the first black London mayor.

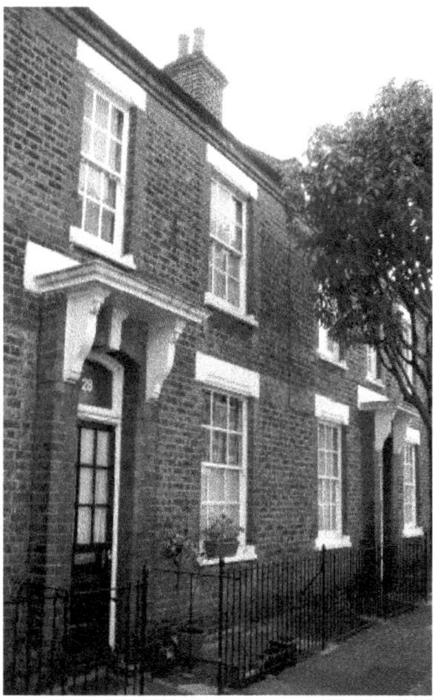

Fig. 1. Latchmere Estate early 1900s.

Central to the progressive vision was to provide good quality and affordably priced housing, with the DLOs as a service rather than operating as an enterprise, being seen as acting as a check on contractors and as non-profit making. Factors such as long term maintenance cost savings were also taken into account when justifying costs and, if outside tenders were lower than direct labour estimates, this might be taken as an indication of scamped work and poor wages and conditions.[10] As Mr Williams, architect, explained:

> The profit and loss fallacy is not indulged in at Battersea. I gathered that it was quite understood in Battersea that good materials, good and expeditious workmanship and proper conditions of labour were the points to be aimed at, and whether the eventual cost came out above or below the estimate the community benefited thereby in the end.[11]

The DLOs were thus not expected to tender in competition with contractors, which would have meant the council tendering for its own work, but were regarded as a means for the council to check on contractors rather than as a separate organization. Any department of the authority requiring work thus estimated the cost of this and, if the DLO accepted this estimate, then it carried out the work; if not, the work was put out to tender. Unlike other municipal activities, such as electricity and gas, the local authority was not actually selling a product so that building enterprise principles were not seen as applicable. Another important point was that up until the 1940s the building industry remained what G.D.H. Cole described as a 'great example of a 'time-work' industry'[12], so the DLOs were not beset by the problems with the application of bonus systems that dogged their post second-world-war development.

Withstanding attack

The success of the DLOs before the first world war can be gauged from the amount of work they undertook, with the LCC alone executing over 300 works to the value of £3million compared with the 170 works over the same period executed by contractors. As output in other sectors declined and wages and unemployment rose, contractors sought to re-gain control over public sector construction, emboldened by the weakened position of building workers. The Moderates had always sought to hamper the activities of the departments; indeed, they became so obstructive to the LCC Works Committee that the Progressive majority excluded them in 1898. However, when they appointed an outside auditor, Mr. Waterhouse, to report on the management of the Department, he certified that the 'present position of the Department from a commercial point of view indicates sound and careful management'[13]. Notwithstanding, the Moderates made direct works and 'municipal reform' one of their main issues in the 1908 LCC elections, claiming that the department was too costly and took away local employment. When they subsequently gained control of the council, they proceeded to ruthlessly destroy the Works Department and to sack its 3,000 workers, scrap all its undertakings and sell off the plant and equipment at far below its price.

The controversy this aroused was evident in the 1910 elections, when George Lansbury pledged to:

> ...work for the immediate re-establishment of the LCC Works Department to corruption. We shall employ our people direct, pay them proper wages, get a reasonable day's work for reasonable wages.[14]

However, the attacks were repeated in the 1913 election (Fig. 2), as evident from the campaign in Islington where the demands put forward by Jephson and Mills included:

- trade union wages and conditions of labour;
- direct employment to prevent contractors' rings;
- one rate for all London so that the rich shall pay their fair share;
- the taxation of site values so that ground landlords shall relieve the occupier;
- protecting ratepayers from trusts and monopolies.

Despite the attacks and with growing unrest over housing conditions and a stronger labour movement, after 1918 the number of DLOs steadily rose, with 70 new departments set up between 1919 and 1920 alone.[15] The building unions were reorganized and in part amalgamated along craft lines, coming under the federal umbrella of the National Federation of Building Trade Operatives (NFBTO). Their efforts enabled the DLOs to introduce better working conditions, including a 44 hour week and payment for 'wet' time, not introduced in the private sector till after the second world war. At the same time the legislative framework set up for state housing in 1919 meant that DLOs could play a large part in carrying out post-war housing schemes. Under the 1919 Housing Act, 70 direct labour schemes were sanctioned and by 1924 under the Wheatley Act DLOs were building ten per cent of all new council housing. The Wheatley Act provided higher subsidies and envisaged a fifteen-year programme of housing built by local authorities at rents affordable by the working classes. The success of the Act was evident for instance in Poplar where seven housing schemes containing 330 dwellings were developed. In the country overall by 1928 104,000 of post-war council houses had been built by direct labour, with some DLOs such as Glasgow set up in 1921 having extensive facilities and undertaking a wide range of work (Fig. 3).[16] By itself the private sector, still dominated by a myriad of small firms, would have been ill-equipped to cope with the scale of building required. Nevertheless, the attacks on the departments persisted, including the imposition of narrow costing criteria forcing them to compete with private contractors and their restriction to housing work only.

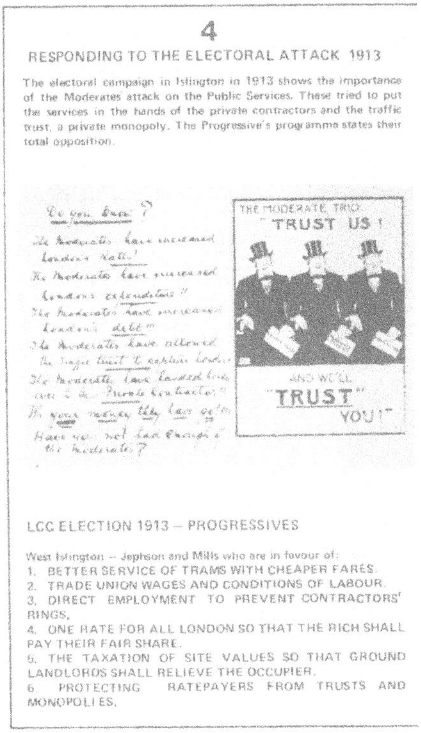
4

RESPONDING TO THE ELECTORAL ATTACK 1913

The electoral campaign in Islington in 1913 shows the importance of the Moderates attack on the Public Services. These tried to put the services in the hands of the private contractors and the traffic trust, a private monopoly. The Progressive's programme states their total opposition.

LCC ELECTION 1913 – PROGRESSIVES

West Islington – Jephson and Mills who are in favour of:

1. BETTER SERVICE OF TRAMS WITH CHEAPER FARES.
2. TRADE UNION WAGES AND CONDITIONS OF LABOUR.
3. DIRECT EMPLOYMENT TO PREVENT CONTRACTORS' RINGS.
4. ONE RATE FOR ALL LONDON SO THAT THE RICH SHALL PAY THEIR FAIR SHARE.
5. THE TAXATION OF SITE VALUES SO THAT GROUND LANDLORDS SHALL RELIEVE THE OCCUPIER.
6. PROTECTING RATEPAYERS FROM TRUSTS AND MONOPOLIES.

Fig. 2. 1913 Attacks on DLOs.

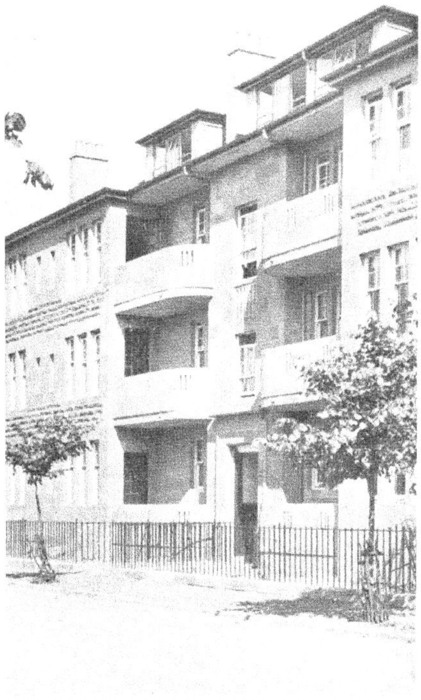

A fine example of stone constructed balcony flats built by Glasgow Corporation in 1939

Fig. 3. Glasgow DLO.

Post-war expansion

Just as after the first world-war, so too after the second world-war local authorities were confronted with excessive tenders and high building costs, at a time when 100,000 building workers were unemployed and many of the unemployed were available to carry out the post-war public housing programme.[17] Compared with 1939, the number of DLOs doubled to employ 70,000 by 1955 and 200,000 by 1967, with 41 local authorities building by direct labour alone. Many continued as before the war to carry out a wide range of activities (Fig 4) and to build extensive housing schemes (Fig 5). In labour strongholds such as St Pancras in London, for example, the new department, which included an advisory committee of trade unions, undertook all the housing maintenance and at times all the new council housing 'from start to finish' (Fig 6.).[18]

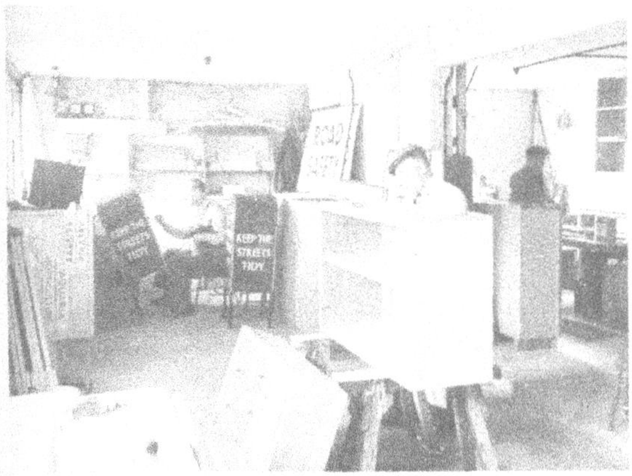

French polishing and sign-writing shop at the Direct Labour Section, Willesden

Fig. 4. Willesden Direct Labour workshop.

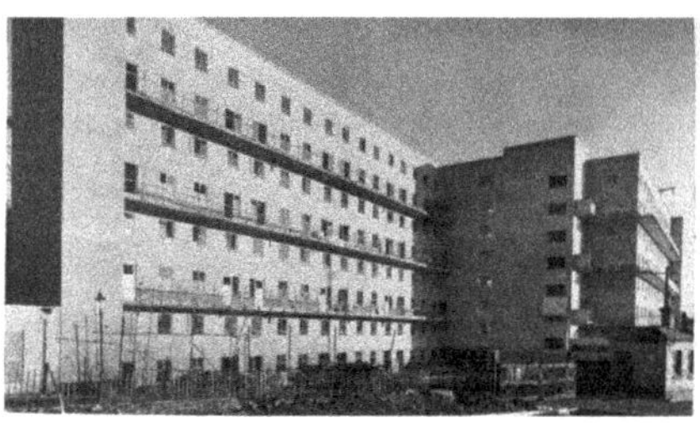

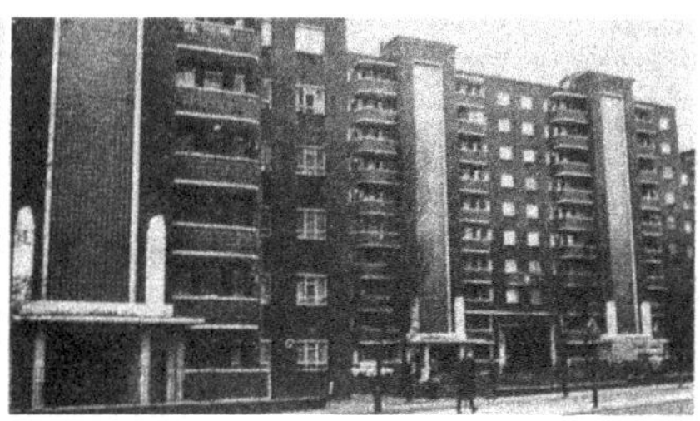

Fig. 5.(left) Maisonettes by Willesden DLO 1950s; Fig. 6. (right) St Pancras DLO, post WW2.

The departments were, however, confronted by particular obstacles. In the first place, payment by results was formally accepted by the building trade unions in 1947, so ending the long struggle against piecework of any kind.[19] At the same time, unlike before the war, most of the DLOs, with exceptions such as Dagenham, came under the Local Authority Agreement rather than the Working Rule Agreement, which meant that pay and conditions could be negotiated directly with the councils. With the expansion of the private construction sector during the 1950s and the Conservative return to power in 1951, the supply of labour available to DLOs was reduced and at the same time local authorities were forced to pay only trade union rates at the lowest local level, with bonus schemes limited to 20 per cent of this. Private contractors were thus able to draw labour away from DLOs.

This situation was compounded by a further obstacle introduced in 1959 when the Government imposed the requirement that DLOs must win every third contract in competition with local and national firms before they would be given loan sanction. Not only did this jeopardize continuity of employment in the departments but it flew in the face of the selective tendering, negotiated contracts and even 'package deals' which were increasingly demanded by the private sector. On the one hand the industry was arguing, and continues to argue, that direct labour could only be efficient in competition; on the other, that private contractors could operate more efficiently if they were not tendering in open competition. In 1965 the government temporarily revoked the one in three tendering requirement, so that by 1967 the DLOs reached their peak, employing nearly 200,000 in 1967.[20]

A further obstacle to the expansion of the DLOs was the growing use of industrialised systems in the 1950s and 1960s, which many departments were either too small or lacked the resources to deploy, though there were important exceptions. Manchester direct works did, however, take part in the initiative to form a consortium of local authorities to produce schools using what was termed the CLASP system, probably the only relatively successful example of systems building. Another example was provided by the newly constituted Wandsworth Council in 1965-6, where efforts were made to emulate the example of Edmonton (Fig 7), in transforming the DLO into a systems-building operation.[21] Here an arrangement was drawn up with the contractor Bovis Ltd to help the department to set up a site factory for concrete components for tower blocks.

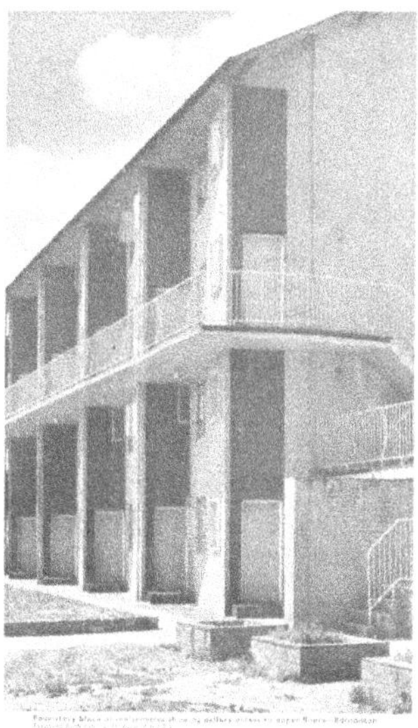
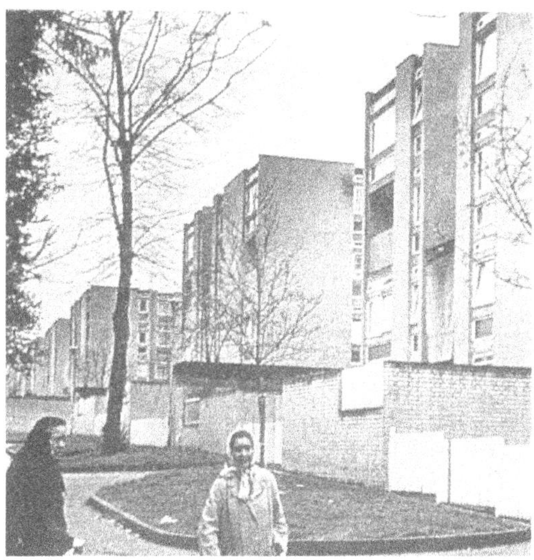

Fig 7. (left) Edmonton DLO 1950s; Fig. 8. (right) GLC DLO 1960s.

And a final obstacle was the restrictions on local authorities themselves. In 1965 the Manchester Corporation Act was passed in the face of opposition from the National Federation of Building Trade Employers (NFBTE) allowing the DLO to manufacture building components for other authorities. Added to this problem was local authority reorganization in 1972 when functions such as Transport,

Police, Fire and Highways came under different authorities. Under the Local Authority Act, the DLOs lost their right to construct new premises for other authorities.

Advantages of Direct Labour Organisations

By 1978 out of a total 717 local authorities, 537 had DLOs, employing 172,581 operatives and constituting 12 per cent of the Britain's construction workforce. Some departments, such as Manchester, Lincoln, Newcastle and Sheffield were very large, on a par with the big contractors, whilst others were small and confined their work mainly to maintenance. The GLC, for instance, did 90 per cent of all the maintenance of its 205,000 dwellings, saving £3.5m on the estimated cost of the work in 1976.[22] Lambeth employed over 1,000 operatives and carried out nearly £19m worth of work in 1976, saving nearly £1.5 m on tenders.[23] And Manchester employed over 5,000 workers and 630 apprentices, of whom 5 per cent were disabled, undertaking a variety of work from schools to manufactured components. By 1976 over a period of fifteen years the DLO had built 17,500 dwellings, 56 schools and other educational buildings, and 17 other major buildings, as well as having its own large workshop in Bessemer Street, which produced bespoke joinery and alone employed 419 workers.[24]

The very nature of private contractors' activities meant that DLOs became indispensable for local authorities if standards and building programmes were to be sustained. Contractors' profitability depended on the flexible use of capital and labour, enabling them to switch easily from one sector to another, and this was increasingly achieved by maintaining low overheads and only a core of employees through the use of plant-hire, casual employment and subcontracting. With heightened private industrial and commercial building activity and rising inflation in the early 1970s, it was often difficult for local authorities to find contractors willing to tender for public housing work, in itself leading to new DLOs being set up. In addition, the DLOs were indispensable for completing schemes left unfinished by bankrupt contractors, with the result that the Greater London Council doubled its workload in 1974 by this means alone (Fig 8).

A further advantage of the DLOs, well illustrated in the case of Manchester, was their superior record on training compared with the private sector despite not being in scope to the Construction Industry Training Board. Lambeth for example employed 1,000 operatives plus a further 100 apprentices. John Dowling described how St Pancras DLO set up its own training scheme and encouraged other London DLOs to do the same:

> We sent a circular to all London authorities asking them if they could give us the figures in relation to the number of apprentices they employed. And if not, were they prepared to take on the commitment... And I made sure that our department came in on that particular question ...and the next thing we had a fairly elaborate training establishment set up by the Council... The point was apprenticeships had been sadly lacking in London authorities at that particular time. We started the ball rolling and it's now a common feature on all local authorities.[25]

Another very significant advantage of the DLOs was the superior employment and working conditions they offered compared with the private sector, above all direct and secure employment. Indeed the Phelps Brown report of 1968[26] found that 81% of workers had one year's service or more. These superior employment conditions are epitomized again by Manchester where workers received superannuation, holiday and sick pay, and long service bonus, adding another 12.5 per cent to on-cost rates and 6 per cent to tender price on top of the 4 per cent added to on-cost rates for apprentice training covering fourteen different crafts.[27] Accident rates too were lower than those in the private construction industry, with the worst safety record of any industry. Wandsworth direct works, for instance, had a full-time safety officer and apprentice supervisor, as well as a depot containing all modern welfare facilities, all adding to the costs of running the department.[28]

Notable too has been the DLOs training and employment of women, particularly following the 1975 Sex Discrimination Act which made it illegal in employment or education to discriminate against anyone on the grounds of their sex. This provided the lever for women to access training in government training centres or Skillcentres and for many DLOs to take on women as trainees and subsequently as employees, including the London Boroughs of Camden, Haringey, Hackney and Southwark.[29] As a result, the service provided to tenants was improved and the exclusive practices of the private sector challenged.

Conclusions

This paper has concentrated on the emergence and development of the DLOs over the eighty year period from the 1890s to the 1970s, when they were depicted in great detail and to great effect in the pamphlet *Building by Direct Labour*[30]. Since that time, and following the severe attacks on both council housing and on direct labour in the Thatcher era, many DLOs have disappeared and with them the understanding that there are alternatives to the current organisation of the construction sector. In hindsight, we can see that the departments were undermined by a number of factors, not just directly related to government enthusiasm for speculative building. The bonus system, used as a means to attract labour into the DLOs, became a noose, hampering improvements in productivity and bearing no relation to collectively agreed rates. The strictly trade-based division of labour, evident too in apprenticeships, was a severe restraint when so many new occupations were becoming important and far broader occupational profiles were often required to meet the needs of tenants. And finally the restrictions placed on the local authorities themselves were a handicap when so much would have been gained through working and sharing with other authorities.

None of these undermining factors mean that DLOs cannot take on a new life and provide the means to alleviate today's severe housing shortages, the lack of training for young people today particularly in low energy construction, and the exclusive practices of the private sector with regard to the employment of women and those from black and Asian minority ethnic groups, and the blacklisting of trade unionists.

References

[1] From a speech by Aneurin Bevan, Ministry of Hiealth, October 21st, 1946.

[2] See Booth, C. *Life and Labour of the People,* Vol. I, 1989 and Vol. II, 1991.

[3] For further details, see: Clarke L., 'Direct Labour: 'goodbye to contractors'', chapter 9 in eds. Craig G., Mayo M., and Sharman N., *Jobs and Community Action,* Henley-on-Thames, Routledge and Kegan Paul, 1979; Clarke L, 'Die Bedeutung der Gemeindebetriebe für die sozialistische Bewegung in Groβbritannien, paper presented to the 5th International Bauhaus Colloquium, 27-30 June 1989, Weimar; and Direct Labour Collective, *Building with Direct Labour: local authority building and the crisis in the construction industry,* London: Conference of Socialist Economists, 1978.

[4] See Brown K.D., *John Burns,* Royal Historical Society, Studies in History Series, London, 1977.

[5] See Labour Research Department (LRD) *Direct Building,* London: LRD, 1929.

[6] *London* 6 January 1898, p. 3.

[7] Bradford Trades Council (1904) *London, Battersea and West Ham Municipal Works Departments,* official trade unionist report, Battersea Local History Library.

[8] Colin Thom, *Survey of London Vol. 50: Battersea Part 2; houses and housing,* English Heritage, Yale Books, 2013.

[9] Municipal Dreams, *The Latchmere Estate, Battersea: 'happy healthy homes for sober and industrious workmen'*, Municipal Dreams in Housing, 1 January 2013, p. 8.

[10] Clarke, 1979, op.cit.

[11] *London* (1898), 27 January.

[12] G.D.H. Col, *The payment of wages: a study in payment by results under the wage-system*, London: Allen and Unwin, 1918, p. 113.

[13] Cited in South West Bethnal Green Election address, 1910.

[14] George Lansbury 1910 election address.

[15] Labour Research Department (LRD) (1920), *Local Government notes*, Monthly Circular, November.

[16] Labour Research Department (LRD) (1929) *Direct Building*, London: LRD.

[17] See W.S. Hilton, *Building by Direct Labour: a national survey*, AUBTW, London, 1954.

[18] Interview by the author with John Dowling, shop steward and painter for St Pancras Direct Works, 1978.

[19] See L. Clarke, C. McGuire, and C. Wall C., 'The changing and distinct character of construction labour in Britain' in eds Dainty A. and Loosemore M., *HRM in Construction: Critical Perspectives*, Taylor and Francis (Oxon), 2012.

[20] See Direct Labour Collective and Clarke 1979, op.cit.

[21] See Thom op.cit.

[22] H. Graham, 'Getting the Best out of your Directly Employed Labour', Institute of Management Report on the most appropriate methods of accomplishing the expansion of municipal direct labour departments for new construction, building and civil engineering, Press Release, November 1976.

[23] Lambeth Construction Services (1976) Chairman's Report 1975-6.

[24] Manchester Direct Works, *Value for Money and Service*, Institute of Municipal Building Management Press Release 1976.

[25] Interview with John Dowling, op.cit.

[26] *Report of the Committee of Inquiry under Professor E.H. Phelps Brown into Certain Matters concerning Labour in Building and Civil Engineering*, HMSO, 1974.

[27] Manchester Direct Works, op.cit.

[28] John Tilley, *Building with Direct Labour: a Fresh Assessment*, London Borough of Wandsworth, November 1975

[29] See L. Clarke and C. Wall, 'Now you're in, now you're out: women's changing participation in the building trades in Britain' and C. Wall 'Any women can': 20 years of campaigning for access to training and employment in construction' in *Women in Construction*, CLR Studies, 2004; E. Michielsens, C. Wall, and L. Clarke, *A Fair Day's Work: Women in Direct Labour Organisations*, WAMT and ADLO, 1997.

[30] op.cit.

Knowledge transfer of vernacular finishing techniques in the interwar period: on the history of cimorné render

L. Dekeyser and A. Verdonck
Department of Architectural Engineering, Vrije Universiteit Brussel, Belgium

Summary

This paper sheds light on *cimorné* render and technique, the *cimorné* company and its founder. Moreover, the geographical spread of this technique and accompanying transfer of know how amongst contractors is questioned and investigated.

'Cimorné' is a combination of French words *ciment* and *orné* which refers to a 'decorated cement' and is quite similar to pebbledash. Instead of pebbles, opalescent *Marbrite* glass fragments were projected onto a wet cement mortar. Both the cement and opalescent glass were manufactured locally in the Hainaut province (South-West of Belgium) and blended together to create a 'vernacular finish'. In 1930 Pierre Pétroons presented 'Cimorné', a decorative façade finish and at the same time a company, at the International Exposition (*Exposition Internationale de la grande industrie, sciences et applications*). One year later, he patented the application technique in Belgium as abroad (France and the United Kingdom).

Since literature and previous research are lacking, other sources were explored. The testimonies of three former *cimorneurs* and fieldwork proved to be crucial for this study. *Cimorné* façades were listed and on-site investigations (including damage assessment) were combined with a study of patents, interwar period manuals and architectural periodicals. Since Belgian modernist architects like Louis-Herman De Koninck and Renaat Braem opted for *cimorné*, their archival records contained some fragmentary documents in which this decorative render and its technique is shortly mentioned. But despite the lacking of advertisements or a well-organised training center for contractors, the *cimorné* technique was spread throughout Belgium in a short period of time and it became a popular finish in the mid 1930s.

Introduction

Cimorné render consists of regional raw materials such as Portland cement and opalescent glass and became quite popular in the 1930s throughout Belgium. Pierre Pétroons (1897-1969) developed this decorative façade render and vernacular finish around 1928, by projecting glass cullet from the *marbrite* production onto a coloured cement render. The story goes that he visited the *Verreries de Fauquez* glass factory and noticed crushed *marbrite* glass, which was stored by colour in wooden trays, waiting to be added later on to a new glass batch and therewith facilitating the remelting process. [1] Pétroons called both the finishing technique and his company *'Cimorné'* and presented them at the *Exposition Internationale* in 1930 (Liège, Belgium). One year later, Pétroons patented his *'procédé de revêtement de construction'* in Belgium and later on he applied for patent specifications abroad. [2] After launching his small business, cimorné render became quite popular in Belgium. The economic advantage suited well in times of crises such as the 1930s, since the price of *cimorné* was about 35 to 50 % cheaper than traditional finishes. [3] Both the use of raw materials and particularly the reuse of *marbrite* waste elucidate the low cost. Due to this advantage, *cimorné* was applied all over Belgium: from the Walloon provinces in the south to the coastline in the northwest of Belgium. *Cimorné* was used either to renovate local labourer's cottages or to coat façades of new built villas. Despite this spread and wide application from about 1931 onwards, the use of *cimorné* as a façade finish for new buildings decreased and already stopped around 1939. On the other hand renovations and transformations of façades with *cimorné* render were carried out until the early 1960s, when the *S.A. Verreries de Fauquez* stopped manufacturing *marbrite* glass. The spread of the *cimorné* technique and transfer of know how will be discussed in this paper. Moreover, the transfer of the technique's characteristics and working details between contractors and spread throughout Belgium in such a short period will be illustrated.

The *cimorné* craftsmanship developed very locally since technological experience and details were generally communicated from contractor to contractor. Therefore a huge lack of written sources about this peculiar and typical Belgian cement render was noticed. Hardly any written document or advertisement was found, which confirms the presumption of a predominantly oral transmission of the *cimorné* technique: *'les informations dont nous disposons au sujet de l'inventeur du cimorné, Pierre Pétroons, sont lacunaires. Elles reposent essentiellement sur des témoignages oraux recueillis dans le cadre de cette étude auprès de personnes qui l'ont connu ou côtoyé.'* [4] Other sources were explored, in order to reconstruct this vernacular finish's history and the accompanying knowledge transfer. This paper also expands on dealing with this vernacular finish and lack of sources.

Besides contractors, other actors like architects and proprietors exchanged information, but it is not clear how they perceived this innovative finish. Moreover, their motives to choose *cimorné* are unknown. Nevertheless, Belgian leading modernist architects as Louis Herman De Koninck, Renaat Braem, Gaston Eysselinck, Geo Henderick, Emile De Nil, etc. opted for *cimorné* render. Presumably these progressive architects proposed the innovative façade finish to their clients, ambitioning a plain rendered and thus modernist building. Although a variety of colours and decorations was available – thanks to the manufacturing of almost 40 hues of *marbrite* glass – most new built villas were rendered with a white, beige of grey *cimorné*. When renovating a house, more decorated variants occurred. But who were the decision makers with regard to the used colours and decorative patterns? Since *cimorné* render consisted of a characteristic blend of coloured *marbrite* glass fragments (Fig. 1), the preparation process and used glass fragments determined the final result and thus the contractor was of major importance. Therefore, the contribution of the glass' manufacturer, the architect, the contractor and the builder was investigated.

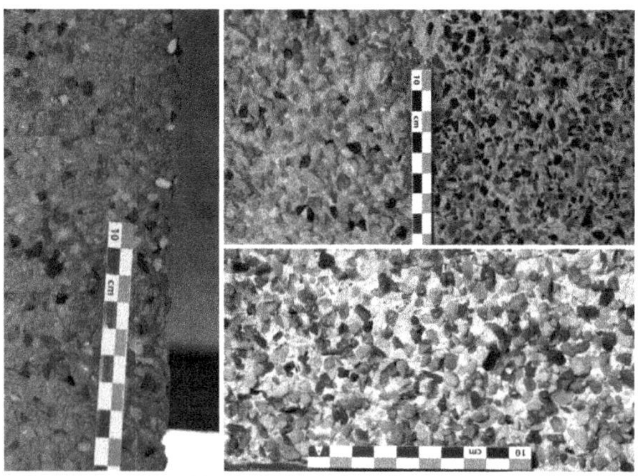

Fig. 1. Cimorné is characterized by a blend of coloured marbrite glass fragments. Each contractor mixed his own blends and applied a style of cimorné decorations. (Photo L. Dekeyser)

Cimorné, both a vernacular finish and a company name

Pierre Pétroons first worked as a plasterer on behalf of contractor Victor Pays in Braine-l'Alleud (about 20 kilometers South of Brussels, Belgium). Around 1928 he started his experiments with crushed bottle glass and waste from the *marbrite* glass production. Likely inspired by Italian terrazzo techniques, he wanted to create a textured and unique coloured finish. Therefore he used two regionally available raw materials: a Portland cement render, produced in the Hainaut Province, and millions of *Marbrite* glass fragments produced by the *S.A. Verreries de Fauquez*, which was located at a distance of about 20 kilometres from Pétroons' home. He transformed these ordinary raw materials into an innovative façade finish by applying a pigmented cement render with a blend of manually projected coloured opalescent glass fragments.

Pétroons became a businessman and promoted himself as a contractor. *Cimorné* was launched and shown to the public for the first time at the *Exposition Internationale* in Liège in 1930, in the *Fauquez* glass pavilion. Presumably Pétroons and Jos-Laurent Pays (the son of Pétroons' former employer Victor Pays) cooperated and even developed the *cimorné* render together. [5] However, Pétroons decided to patent the rendering technique singly, and later on he improved this original patent (1932) and protected his invention abroad (1932 French patent, 1933 and 1938 British patents). [6] With their co-operation having already stopped around 1935, Pays became an independent *cimorneur* and founded his own company in the *Rue de La Station* in Braine-l'Alleud, not far from Pétroons business in the *Chaussée de Tubize*. [7] Also in the *Chaussée de Tubize*, Pétroons built a villa demonstrating a variety of colourful inside and outside applications of *cimorné* renders (Fig. 2). The baseboard was carried out in red-orange *cimorné* in order to mimic stone. Even the fence and driveway were covered with *cimorné*. Pétroons' house, designed by architect Castiaux (nephew of Jos-Laurent Pays), was depicted in journals published by the Fauquez factory in order to promote their glass products (Fig. 3).

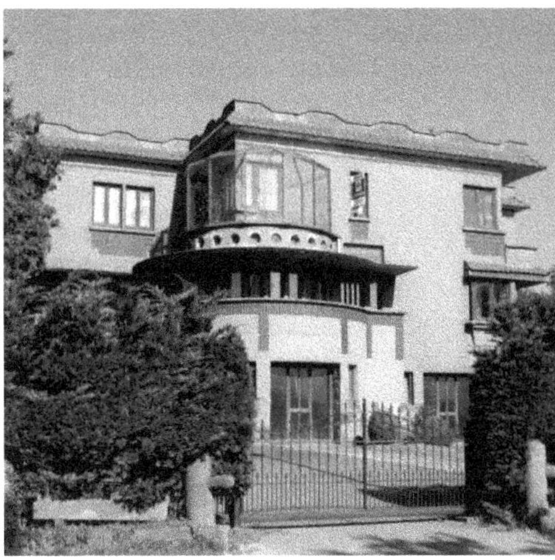

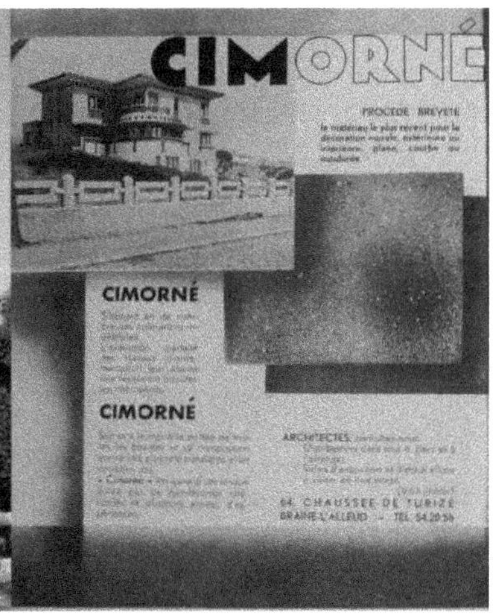

Fig. 2. (left) Pétroons villa was built in 1935, in order to demonstrate a variety of inside and outside applications and colours of cimorné render. (Photo L. Dekeyser) Fig. 3. (right) A rare advertisement for cimorné render, at the first page of L'Art Mural, showing the Pétroons' villa. (L'Art Mural, n°1, 1937, Archives d'Architecture Moderne)

Although Pétroons invented and patented the *cimorné* technique, he did not sell it as a ready-mix product. Instead he trained workmen to master the rendering technique and sent his staff to architects and entrepreneurs at the building site in order to demonstrate and learn the technique of projecting the glass fragments, therewith facilitating the choice for *cimorné* render. An information letter from the *Cimorné* company addressed to contractors and architects (AAM c.1930s) explained the *cimorné* technique and states: '*... once the cement render is executed, a technician is at your disposal to teach the projection technique during one or two days...*'. [8] Because of the increasing number of trained artisans, *cimorné* was widely and rapid diffused. Although the story started in the region of Braine-l'Alleud (hometown of both Pétroons and Pays), *cimorné* spread quite rapidly to other Belgian regions and even to the north of France. Despite foreign patents, no *cimorné* façades were found in Great Britain up until now.

The *marbrite* glass waste was crushed in the *Fauquez* glass factory by an industrial crushing machine, as described by patent GB481752: '*(...) Next the glass is crushed between inclined pivoted jaws which have fluted surfaces. Straightaway the coarsely broken glass is subjected to a second crushing and grinding treatment between adjustable rollers the surfaces of which have rounded ribs.*' [9] Likewise, particles of different sizes were obtained. Laurent Pays mostly applied four to six millimetres grains because the smaller ones (of about three millimetres) were more expensive. Pétroons also crushed *marbrite* glass by means of a homemade machine, thus creating his own little stock of *marbrite* glass fragments. [10] Meanwhile he visited the *Fauquez* factory several times a week to collect glass cullet and crushed *marbrite* glass. It is however unknown if Pétroons and the Fauquez company signed a contract to confirm their cooperation, but both parties benefited from this win-win game: the *Fauquez* factory sold their glass cullet, and Pétroons found a cheap raw material for his low cost decorative render.

Later on a network of distributors of the *Fauquez* company arose and sold crushed *marbrite* fragments by size and colour in bags of 50 kilograms in their stores. These sales points, where glass products of the *Fauquez* factory were demonstrated in a showroom, were spread over Belgium and listed in *L'art Mural*, a journal published from 1937 until 1939 by *l'Union Belge des marbritiers*. A *cimorné* advertisement in the first issue demands architects to visit the showrooms: '*Architectes, consultez-nous, distributeurs dans tout le pays et à l'étranger. Salles d'exposition et d'échantillons à visiter en tout temps.*' [11] Architects, builders and entrepreneurs got to know *cimorné* mainly by word of mouth. [12] Next they had the opportunity to chose colours and buy raw materials at one of the *Marbritier* sales points, of which Van Overstraeten (Brussels), Panné (Aalst) and *S.A. des Glaces de Gand* (Ghent) were repeatedly mentioned by former plasterers. These *marbritiers* also installed and tooled *marbrite* glass panels. [13] It is however not clear if the *cimorné* technique was also passed on via the retailers. According to oral testimonies, the contractor showed samples to his clients and together they decided which colours and decorative patterns were applied. Laurent Pays was one of the interviewed plasterers, his father Jos-Laurent Pays cooperated with Pétroons in the early 1930s. Laurent Pays bought *marbrite* glass both at the *Cimorné* company and the *Fauquez* factory. From the 1960s onwards, when the *Verreries de Fauquez* stopped producing *marbrite* glass, he searched for other opalescent glasses in Binche (*S.A. des verres séciaux de Bonne-Espérance*) and ground the glass himself by means of a grinder based on an ordinary mill.

Besides establishing a network of distributors and publishing their own journal, the *Verreries de Fauquez* promoted their glass products by attending various expositions. Remarkably, they had their own pavilion at some of these prestigious exhibitions, where new materials like *marbrite* glass were presented to the public. The *Exposition Internationale des Arts Décoratifs* (1925, Paris) was the take-off of *marbrite*'s success story. In the highlight of the 1930s, the *Fauquez* factory attended exhibitions like *Exposition Internationale de la grande industrie, sciences et applications, art wallon ancien* (1930,

Liège), *Exposition Coloniale Internationale de Paris* (1931, Paris), *Exposition Universelle et Internationale* (1935, Brussels) and *Exposition internationale du bâtiment, des travaux publics et des arts décoratifs* (1939, Brussels). On the latter exposition, architect J. Storck designed a pavilion for the *Union Belge des marbritiers* and the *Cimorné* company had its own stand (Fig. 4). However, not much later the application of *cimorné* reduced enormously. Perhaps, this decrease was caused by changing trends, problems with earlier *cimorné* façades or the dawning of the Second World War.

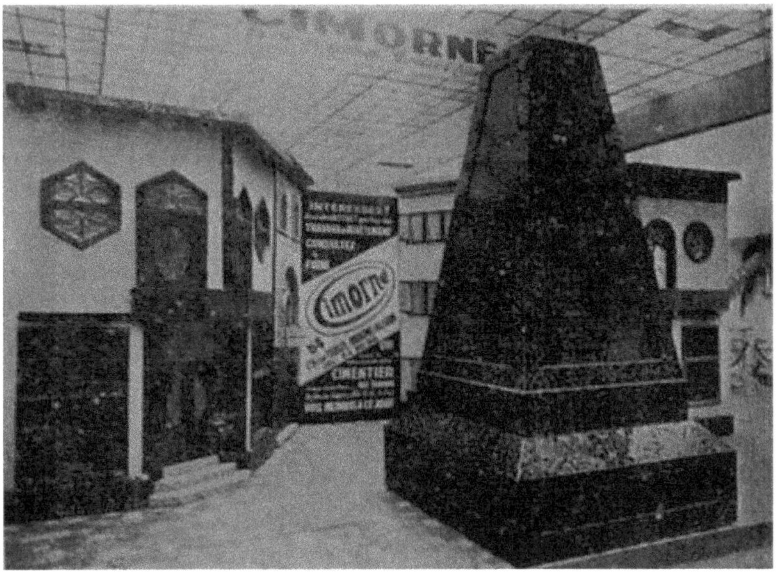

Fig. 4. The cimorné stand at the Exposition Internationale du Bâtiment. (Photo: l'art mural, n°6, 1939, p.22)

How to reconstruct the history of this vernacular finishing?

The *cimorné* technique was mainly orally communicated between Pierre Pétroons and other artisans, therefore hardly any written or iconographical documents remained Moreover, the desire to create a *cimorné* rendered façade resulted from word of mouth. Only some local publications about houses in the Walloon province [14] and an unpublished dissertation about tombstones covered up in *cimorné* [15] treat this subject. However, they all focus on Walloon applications (in the southern part of Belgium) and on renovations of labourer's cottages and rural heritage. Our research work however recently revealed the original formula, application method and working details of *cimorné* render, but the spreading of this technique and know how remained vague. [16]

As a result of this significant lack of knowledge and sources, new information had to be explored. But how can the history of a vernacular, regional developed and orally transferred technique be reconstructed without any written sources? No archives of Pétroons' company were preserved and those of the *Fauquez* glass factory were partly lost after its closure in the 1980s. Therefore, other archival documents like interwar periodicals, patents and some rare building specifications were searched for. Most artisans who possessed an extensive knowledge of *cimorné*, the technique and spread, are retired or have already died. However, former plasterers like Laurent Pays (Braine-l'Alleud), Gilbert Schoonheydt (Halle) and Romano Tondat (Eeklo) testified about their know how. Also Lucien Brancart, grandson of the founder of the *Fauquez* factory, was interviewed. Being the former director of the hollow container glass department, he was not directly involved with the *marbrite* production. Nevertheless, he remembers crucial information about the development and export and he elucidates the cooperation with Pétroons and the *Cimorné* company.

In addition to archival records and oral testimonies, patents dating from the period 1880-1940 were the main starting point to obtain additional technical information and to determine whether *cimorné* was known and successful abroad. Yet the absence of terms like '*cimorné*' or '*marbrite*' (or other commercial product names) in the patent's title rendered the search of this peculiar patent difficult. Two Belgian patents dating from March and November 1931 dealt with *cimorné*, its original formula and application method. The first invention patent by Pétroons described the application of a cement mortar and projection of glass fragments on top. After a couple of months, Pétroons also protected a machine that threw the glass fragments onto the rendered wall and he called it *cimorneuse.* Because he invested time and money in an improvement patent, Pétroons probably expected financial profit and the first patent was maybe even used in practice. International patents were found via the online database Espacenet, proving that Pétroons exported his ideas to France (FR730628, 18 august 1932) and Britain (GB390913 en GB481752, 20 April 1933 and 17 March 1938). Pétroons, together with the *Verreries de Fauquez* and Fernand Rene Lang, applied the latter. This shows the cooperation between the *Fauquez* factory and Pétroons. Very often, an inventor was financially supported by a company that automatically became their co-patentee.

Besides written and oral sources, the buildings themselves acted as primary sources. As yet, only tombstones, rural heritage and labourer's cottages with *cimorné* render were described in a local publication and some unpublished dissertations. [17] However, the Belgian street view suggested many other *cimorné* applications. After extensive fieldwork, a list was set up in order to discover more *cimorné* façades, to collect samples (of both renovated and new built buildings) and to shed light on the spread of the *cimorné* technique. First, existing databases of listed buildings and heritage were consulted: *Inventaris Onroerend Erfgoed* (inventaris.onroerenderfgoed.be) for Flanders Region, *Inventaris van het Bouwkundig Erfgoed* (www.irismonument.be) for the Brussels Capital Region and *l'Inventaire du Patrimoine Wallon* (not yet available online) for the Walloon Region. During this search other terms or product names came forward, because *cimorné* was sometimes mistakenly called '*granitobepleistering*' (granito finish, terrazzo), '*glaskorrelbezetting*' (glass granules coating), '*sierbeton*' (decorative concrete), '*granito stuc*' (granito stucco). All of these refer to a (cement) render in which decorative granulates are incorporated. In contrast with other decorative renders, *cimorné* is characterized by opalescent glass granules, which were thrown onto the cemented surface and of which the surface was not washed or polished afterwards. The confusing naming made it hard to track information about *cimorné*. Moreover, these databases only contain listed and inventoried buildings and are thus incomplete for *cimorné* applications.

Next literature and dissertations about the interwar period, '*modern bouwen*', '*nieuwe bouwen'* (modern movement and functionalism movement buildings) and 'art deco' (art deco buildings) in Belgian cities were consulted. [18] Hence different modernist *cimorné* façades were traced. This automatically led to the discovery of many others, since the preference for *cimorné* was transferred between builders by word of mouth and *cimorné* applications are thus grouped in streets or quarters. Next to contemporary texts, architectural periodicals of the time described lots of innovative building materials and techniques. Belgian interwar period journals such as *Bâtir, L'émulation, Tekhné, La Cité, Clarté, Le Document, L'époque, L'équerre, Opbouwen, Techniques des Travaux* were searched for *cimorné* and similar decorative renders. Again no *cimorné* advertisements were found and hardly any article introduced new information.

Heritage experts and amateurs were addressed via a call for information in heritage related journals [19] and onto the website and a facebook page for '*Expertisecentrum voor Technisch, Wetenschappelijk en Industrieel Erfgoed*' (ETWIE, an organisation with expertise in technical, scientific and industrial heritage). This resulted in the discovery of some former plasterers and a variety of *cimorné* buildings. Letters were sent to local heritage agencies in Flanders [20], city guides and regional history associations. Thanks to particular information obtained by Christian Reumont and Pieter De Raedt and the bookkeeping of Laurent Pays, a list of about 500 buildings was composed during the last three years. Continuously, the database will keep growing as new *cimorné* façades are still found every week.

Results of the database: spread of the *cimorné* technique

The abovementioned methodology led to a database of *cimorné* buildings, including basic information like addresses, date of construction, architect and typology. The façade view is described by subdivisions like 'building components' and 'colours'. In some cases the contractor and origin of the raw materials (namely the origin of opalescent glass) are known. Other data such as the current state, the availability of archival records and whether the building is listed, complete each construction's coordinates. Based on this unfinished list, limited conclusions can already be formulated.

The clear distinction between renovated façades and new built constructions is striking: each of them is characterized by a particular geographical area and time period. The first *cimorné* buildings date from about 1932, only one year after the *cimorné* technique was patented by Pétroons. Modernist buildings were covered up with *cimorné* in the period from 1932 until 1939, with a peak at around 1935. On the other hand, the renovation and transformation of façades by means of *cimorné* render was carried out from the 1930s until the 1960s, although their number strongly decreased after 1940. Pays however executed *cimorné* works after the Second World War in Blankenberge (1946-1947). Apparently this municipality prohibited the use of this innovative render, because 'it was not in accordance with the local building style'. [21] This myth returns several times in testimonies, but was not confirmed by municipal building regulations. [22] Even rumours about the power of *cimorné* renders to inflict injury on people were widespread.

Next to their extensive period of application, renovation projects are also more widespread. Mainly labourer's cottages, rural and modest buildings were carried out in colourful and richly decorated *cimorné*. Either entire façades or components like lintels, door and window frames were ornamented. Most of the new built constructions were sober and had a grey, white or beige *cimorné* rendered front view. Renovated façades were mostly located at the countryside and in Wallonia, whereas in general new built constructions were present in the Brussels region and Flanders (cities and nearby the coastline). It is not clear whether the spread of *cimorné* façades can be linked to the presence of a *Verreries de Fauquez* distributor. On the other hand, the proximity of a contractor directly influenced the number and presence of *cimorné* applications because most *cimorneurs* plastered in their own neighbourhood and did not need to advertise because their work was spread by word of mouth. Laurent Pays namely plastered façades in the Walloon provinces (especially in Walloon Brabant and Liège), but sometimes he even worked in the Brussels region and Flanders.

The database contains almost exclusively residential buildings, most of them being single or two-family housing. Rarely, an apartment or commercial building was finished in cimorné render. Because of the labour-intensive character of the projection of the glass fragments, only limited areas of such large buildings were carried out in *cimorné* render. *Cimorné* letterings of pubs, shops and groceries have mostly disappeared, such as many tombstones.

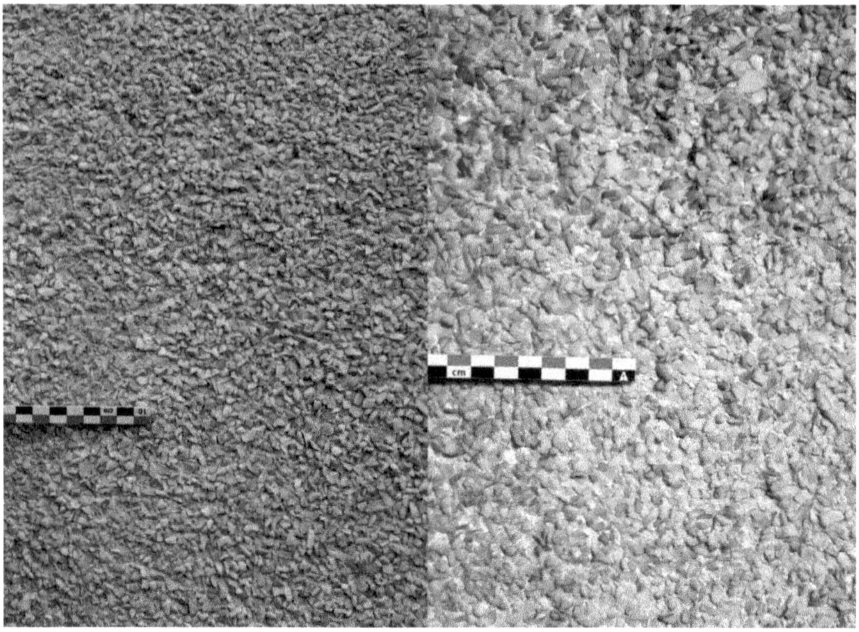

Fig. 5. Cimorné render at the Berteaux House of architect Louis Herman De Koninck (left), in comparison with 'Dura Glaskorrelbezetting' used by architect Renaat Braem in Deurne (right). (photo L. Dekeyser)

Some architect's names came forward in this preliminary database. Well-known Belgian modernist architects like Louis Herman De Koninck, Renaat Braem and Gaston Eysselinck experimented with completely grey, white or beige rendered *cimorné* façades. Archival records of Braem revealed the use of 'Dura Glaskorrelbezetting', a decorative render, which was quite similar to *cimorné* and contains white opalescent glass granules (Fig. 5). Progressive architects like J. Van Kriekinge, A. Malcorps, V. Coppe, E. De Nil, G. Henderick designed several *cimorné* constructions in Ghent and Brussels. Some regional architects such as Louis Legein (De Panne, Koksijde) and Gentiel Van Eeckhoutte (Waregem) favoured *cimorné* and frequently used it to ornament and emphasize horizontal or vertical building parts. The decision maker concerning the choice for *cimorné*, its colour and decorative patterns remains however obscure. In the case of renovating façades with *cimorné*, probably the artisans ruled, since each of them developed their own style. On the other hand, builders wanted to personalize their houses and choose their favourite colours based on samples that were shown to them by the contractor. In the case of new built constructions, no decorations and multicoloured façades were found during the fieldwork. This could point to decision-making by the modernist architect, who wanted to create a sober monolithic façade or to emphasize vertical and horizontal lines by the use of *cimorné*.

Conclusion

The setup of this database in order to shed light on the spread of *cimorné* render is a continuous work and still ongoing. For the first time, this study attempts to reconstruct the use and transfer of *cimorné*, a vernacular finishing technique that was solely spread by word of mouth. Information about the original composition, application method and characteristics of the *cimorné* technique, is gathered in order to resolve damage phenomena and enable future restorations. On the other hand, it is crucial to track the origin of this vernacular finish, and reconstruct how it was transferred. In a very limited time period, *cimorné* became popular in use across Belgium. Without any building specifications, advertisements or written documents, the trace of *cimorné* render was reconstructed in an innovative and experimental way. Therefore, retired plasterers testified about the transfer of knowledge and spread of this peculiar technique. With Pétroons' original patents and fieldwork on *cimorné* buildings as main sources, the existing knowledge was expanded to a great extent. Nevertheless, one has to be aware of the subjective stories told by both contractors and patent descriptions. Each of them considers their invention, story or execution method as highly relevant and correct. The truth remains vague and unknown, due to lacking facts, evidence and written source material. Remarkably this locally-developed decorative render, based on waste material, was spread and applied in a limited time period. Next this curiosity was commercialized and even adopted by well-known modernist architects. Despite the short but intense success and lack of knowledge, *cimorné* render belongs to our current heritage.

References

[1] L. Dekeyser, A. Verdonck and H. De Clercq, 'Cimorné cement render with opalescent glass granules: a decorative façade finish developed by innovative craftsmanship in the interwar period', *Journal of Architectural Conservation*, Volume 19, Issue 2, 2013, Taylor & Francis, pp.86-102.

[2] P. Pétroons, *'Procédé de revêtement en construction'*, Patent 383626, General State Archives, Brussels, Section Contemporary Archives, 30 November 1931.

[3] Brochure Cimorné, Archives d'Architecture Moderne, Brussels, January 1932: *'Il ne coûte que 35 à 50% du revêtement dit "lave" et parce qu'il supprime tout entretien, son coût déviant plus avantageux que le prix de revient du crépissage ordinaire'.*

[4] L. Gaiardo and C. Billen, *'les mausons en marbrite et cimorné en Wallonie'*, Carnets du Patrimoine, n°27, Ministère de la région Wallonne, Namur, 1999 , p.9.

[5] Interview with Laurent Pays, son of Jos-Laurent Pays, on 7 June 2011 and 20 December 2012.

[6] P. Pétroons, *'Improvements in or relating to apparatus for roughcasting building structures'*, Patent GB390913, Espacenet, 2 February 1933. Pétroons & Lang, *'Improvements in or relating to roughcast surfaces for building purposes*', Patent GB481752 , Espacenet, 24 September 1936.

[7] L.H. De Koninck, *1934 Villa Dr. Ley,* Archives d'Architecture Moderne, Brussels, box V.20. In the heading of a tender of February 20th 1935 Jos-Laurent Pays scored out his address and changed it from *Chaussée de Tubize* to *Rue de la Station*. Pétroons lived and worked at the Chaussée de Tubize 65.

[8] Information letter from the cimorné company to contractors and architects, Box 'cimorné', Archives d'Architecture Moderne, Brussels, 193 ?: *'...après cimentagé executé, un téchnicien est à sa disposition pour l'apprentissage de la projection, pendant un ou deux jours...'.*

[9] Pétroons & Lang, *'Improvements in or relating to roughcast surfaces for building purposes*', Patent GB481752 , Espacenet, 24 September 1936.

[10] L. Gaiardo, op. cit., p.11.

[11] *L'art Mural*, n°1, 1937, p.1. Author's translation: *'Architects and distributors troughout the country and abroad, consult us. Showrooms and samples to visit and see at any time.'*

[12] Interviews with Romano Tondat (30 August 2012), Laurent Pays (7 June 2011), Lucien Brancart (8 July 2011).

[13] *L'art Mural*, n°1, 1937, p.9.

[14] L. Gaiardo, C. Billen, *'les mausons en marbrite et cimorné en Wallonie'*, Carnets du Patrimoine, n°27, Ministère de la région Wallonne, Namur, 1999.

[15] L. Detry, *'Conservation / restauration des monuments funéraires en cimorné et mosaïverre'*, unpublished dissertation, Ecole Supérieure des Arts, Saint-Luc Liège, 2006.

[16] L. Dekeyser, L. Fontaine and A. Verdonck, 'Composition and application of cimorné finish: an interwar cement render decorated with coloured opalescent glass granules', in *3rd historic mortars conference,* University of the West of Scotland, Glasgow, 2013. And L. Dekeyser, A. Verdonck and H. De Clercq, 'Marbrite Fauquez opalescent and mass coloured glass: history, production and composition', in *Proceedings of the International Conference on Structural Analyses of Historical Constructions*, volume 2/3, Wroclaw, 2012, p.1106.

[17] L. Gaiardo, op. cit. and L. Détry, op. cit.

[18] Some of these are:

A. Demey, '*Art deco in Sint-Niklaas'*, Kleine Cultuurgidsen Provincie Oost-Vlaanderen, Gent, 1998. A. Demey, '*Interbellumarchitectuur en monumentenzorg'*, Kleine Cultuurgidsen Provincie Oost-Vlaanderen, Gent, 1997. A. Demey, '*Interbellumarchitectuur in Oost-Vlaanderen'*, Kleine Cultuurgidsen Provincie Oost-Vlaanderen, Gent, 1990. P. Devos, '*Interbellumarchitectuur in Ronse'*, Gent, 2005. P. e.a. Boekstal, '*Monumenten van de toekomst? Modern bouwen in Sint-Truiden 1900-1975'*, Open Monumentendag, Stadsbestuur Sint-Truiden, 2008. L. Meganck, '*Bouwen te Gent tijdens het interbellum (1919-1939)'*, unpublished dissertation, Kunstwetenschappen, Universiteit Gent, 2002. J. Vandenbreeden, F. Vanlaethem, '*Art deco en modernisme in België, architectuur in het interbellum'*, Tielt, 1996.

[19] *Interbellum* January-February 2012, *Industrie Erfgoed en Techniek* n° 2-3 2012 and *BRK-APRO Bulletin* 3rd trimester 2012.

[20] <www.erfgoedcellen.be>[accessed on 29 December 2013]

[21] Interview with Lucient Brancart, grandson of the founder of the Fauquez Factory, on 8 July 2011.

[22] The municipal building regulations of Blankenberge from 1933 onwards were checked, nothing was found about the prohibition of cimorné render.

Hang the roof: building the aerial site

Christopher Domin
Architecture Degree Program (Chair), CAPLA, University of Arizona, USA

A technologically innovative culture of architecture developed along the West Coast of Florida, USA immediately after World War II. Sarasota, FL became the epicenter for a regionally specific and technologically astute strain of post-war modernism in the United States. [1] Paul Rudolph (1918-97) and Victor Lundy (1923-) both established architectural practices in Sarasota in the 1950s after graduating from Harvard University Graduate School of Design under the direction of Walter Gropius. Each separately began to produce a body of work that was fed by healthy competition and critical friction supplied by a robust cohort of young architects in the area. As a result, Lundy and Rudolph forged a unique trajectory for their generation and continue to inspire innovative work.

Lundy and Rudolph experimented with materials and construction processes as they confronted the reality of inhabiting a sub-tropical environment before the prevalent use of air-conditioning. Defining characteristics of the work include clarity of construction, structural honesty and material economy, but spatial and formal concerns intertwined with a structural systems agenda in the evolution of projects over time. A unique strain of material development can be traced through roofing strategies as Rudolph and Lundy's work matured through the 1950s. The suspended roof structure, for instance, became a device that both architects advanced to a high level of specificity and sophistication in their Florida work. The nuanced articulation of the lightweight hung roof plane provided an antidote to prevalent post and beam flat roof strategies of the period and created the opportunity for unique interior spatial definition. This robust delineation of the sheltering canopy focused attention on the aerial project site and reinvigorated a relationship between earth and sky for inhabitants.

Within the larger context of constructive innovation in architecture and structural engineering during the era, Paul Rudolph's residential experiments in tensile roof design was positioned at the beginning of a fertile trajectory. As precedent setting work, these developments necessitated implementation of original material assembly systems employed within the building due to structural concerns related to the thin plane of the roof. This simultaneous focus on structural, spatial, and material conditions provides fruitful primary source material for building science research.

During WWII Paul Rudolph was stationed at the Brooklyn Navy Yard where he became aware of the ruthless efficiency of shipbuilding techniques for the war effort. [2] Rudolph's post also provided direct access to the architecture culture of Manhattan and the active programming schedule at the Museum of Modern Art (MoMA). For Rudolph, museum exhibits developed in the Department of Architecture and Design, which was founded by his Harvard colleague Philip Johnson, were of particular interest. During the 1940s, MoMA curator Elizabeth B. Mock, organized a series of influential installations and catalogues including Built in the USA (1944), Tomorrow's Small House: Models and Plans (1945) and The Architecture of Bridges (1949) that attempted to frame the current state of architectural discourse and set a clear trajectory for the postwar future. In the MoMA publication, *If You Want to Build a House*, the curator attempts to define the distinctive character of the architecture on display in the museum.

> "Modern architecture isn't just another imitative style. It is an attitude towards life, an approach which starts with living people and their needs, physical and emotional, and tries to meet them as directly as possible, with the best procurable means." [3]

As Rudolph and Lundy entered into private practice in the 1950s, they would become part of this public process, through building and publishing, of defining an attitude toward space and materials for this generation. Philip Johnson returned to MoMA's Department of Architecture and Design from 1945-49 as Director and Chief Curator. During this period, Rudolph often visited the museum and would certainly have been aware of Elizabeth Mock's influential book, The *Architecture of Bridges*, published by the Museum in 1949 in which she outlines the importance of the efficient structural logic employed by John Roebling in his suspension bridges from the late 1800's, including the Brooklyn Bridge (Fig. 1).

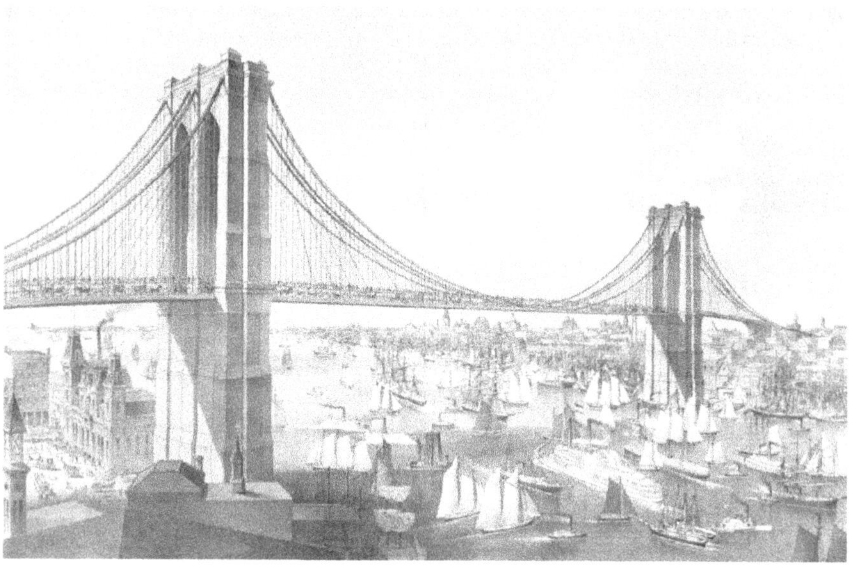

Fig. 1. The Great East River Suspension Bridge, Currier & Ives c1874 (Library of Congress, Prints and Photographs Division)

During Rudolph's work at the Brooklyn Navy Yard, the bridge was in prominent view to the south, spanning the East River as the major southern connection from Manhattan to the outer boroughs. In contrast to the hectic pace of the yard, this epitome of structural grace exhibited in the tensile steel construction was persistently looming in the distance. As work continued for the war effort, Rudolph was offered ample opportunity to marvel at the streamlined beauty of the navy destroyers in the harbor and consider future implications for the suspension bridge as a simple force diagram in service of material economy and lightness. [4] Elizabeth Mock asserts that:

> "a bridge is at once the most tangible and most abstract of architectural problems. As such, it is capable of extraordinary purity, thought it may perhaps never achieve the richness and depth of expression that are possible in buildings of more complex human motivation... Since the reality of a bridge lies in its structure, the art of bridge building lies in the recognition and development of the beauty latent in those structural forms that most effectively exploit the strength and special properties of a given material." [5]

Understanding the structural diagram was only one piece of the puzzle, it is in the application to a need in the culture, whether as shelter or transportation, in a unique time and place that transforms pure structural logic into built work invested with meaning. In the case of Roebling, the apparent thinness of the tensile steel members is understood in the context of its massive masonry piers that are punctuated with gothic-inspired pointed arches. This complex construction can be seen as a transitional composition at a juncture in engineering practice between two ways of working—one of lightness, and the other, mass. Elizabeth Mock continues,

> "Beauty is not automatic; technical perfection alone is not enough. A great engineer is not a slave to his formulas. He is an artist who uses his calculations as tools to create working shapes as inevitable and harmonious in their appearance as the natural laws behind them. He handles his material with poetic insight, revealing its inmost nature while extracting its ultimate strength through structure appropriate to its unique powers... Today, we boast the most powerful materials of all time: steel and steel-reinforced concrete. But there is a curious reluctance to explore their ultimate possibilities and accept their full esthetic implications— a reluctance based on the idea that massiveness is itself a virtue, as it was in the days when stone was the only strong, permanent, therefore honorable material." [6]

After Rudolph returned to civilian life, his work in Florida resumed with Ralph Twitchell. As the national economy began to expand again, architects such as Rudolph facilitated technology transfer from the war effort into domestic production with projects such as the Healy Guest House. Here the ideal of structural continuity is explored using standard steel components and other building materials that were generally available in the region. In a project description, the architects share an affinity for Mock's work on Bridges: *"For many years architects and engineers have realized the beautiful efficiency of tension structures of all kinds."* [7]

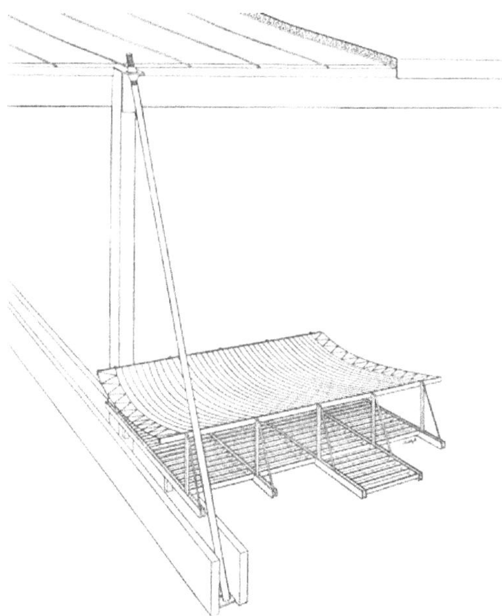

Fig. 2. Composite Drawing by Paul Rudolph, 1950 Healy Guest House, Twitchell + Rudolph Architects (Library of Congress, Prints and Photographs Division)

As seen in an early Rudolph hybrid drawing for the Healy Guest House (Fig. 2), steel, timber and concrete are all working in unison and can no longer be considered discrete building components. The hybrid drawing juxtaposes two scales: a tensile catenary and edge frame detail surrounding a small-scale perspective of the entire composition. Thin steel cables or wire are evident along with attenuated wood piers and threaded steel rod counter bracing. The entire structural idea is displayed at this stage including flat trusses placed at the terminal bays in order to reduce lateral movement in the roof and allow for significant amounts of glazing on the north and south building elevations. A highly reduced roof profile is evident in the drawing and apparently includes two elements, an insulating roof panel and waterproofing membrane. Rudolph also deftly connects the house to Bayou Louise via a cantilevered deck that leverages the already projecting floor beams and tensile braces as an inhabitable platform thrust into the canal. In the Healy Guest House, mass is reduced to minimal concrete piers hidden from sight under the floor deck. Dematerialization of the roof and wall, in conjunction with strategic site planning, were employed by Rudolph and Twitchell to subtly tie this work to its immediate context.

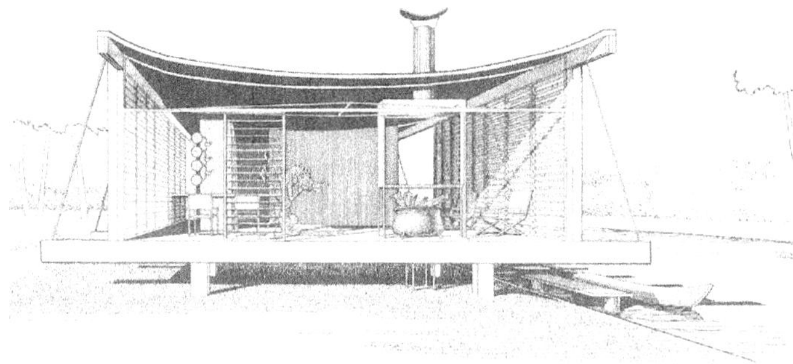

Fig. 3. Healy Guest House (Cocoon House) Perspective Drawing looking north Twitchell + Rudolph Architects, 1950 (Library of Congress, Prints and Photographs Division)

The economy of the catenary roof form can be viewed as a vestige of material scarcity and a postwar necessity, but for Rudolph and his Harvard cohort this thinness was both an ethical and an aesthetic concern. When the Healy Guest House was first published in the architectural press, Rudolph articulated the problem of defining an integrated edge condition and resolving tensile forces at the cast-in-place concrete foundation, but this small scale experiment in catenary roof construction came with several material technology challenges, including providing a waterproof roof membrane that was flexible enough to respond to dynamic movement in the roof plane and meet the prerequisite challenge of tenuity and elegance. (Fig. 3)

> " The form of this cottage in Sarasota Fla., is the result of using steel in tension for the roof structure. Architects and engineers have long recognized the beautiful efficiency and expressiveness of steel in pure tension, but no one, to our knowledge, has solved the problem of making the resultant structure stable. But why should a roof structure be stable if a way could be found to keep it watertight? One of us (Paul Rudolph, who spent the war in the Navy) had seen the moth-balling process used to protect warships when they were put in storage in 1946. The process in involves covering gun turrets, etc., with a wire frame, spraying the frame first with a mixture of Saran and vinyl plastics, and then giving the resulting "cocoon" a final coat of clear vinyl." [8]

Dow Chemical Company found a vast application for Saran, a polyvinylidene chloride (PVDC) compound that was developed in the 1930's in service to the postwar decommissioning or moth-balling of navy ships and equipment. PVDC was praised by the United States Navy for its relative impermeability to water, mold resistance and elastic qualities. An article in Life Magazine from 1946 presented the decommissioning effort of its 1,750 ship fleet in heroic terms. A Navy Destroyer vessel wrapped in a 'cocoonlike shroud' illustrates the process. In another photograph, a team of soldiers is creating a weather-proof shell over a gun turret by spraying a web of plastic over a taut fabric matrix. [9] These assets were stored at a variety of coastal facilities including Green Cove Springs, Florida according to the article. Rudolph continued the optimistic outlook for Saran in domestic applications for light waterproof roof membranes in published results for projects such as the Healy Guest House or 'Cocoon' House.

> "the advantages of this 'cocoon' were impressive: It can be stretched to almost three times its normal state , and still remain watertight. Navy tests have indicated that it will last for 30 years. It proved to be the answer to our flexible roofing problem, especially when 'cocoon' turned out to be only 15% more expensive than our usual tar and gravel finish." [10]

As the project went into construction under the direction of Jack Twitchell, Ralph Twitchell's cousin, some details were altered, but the idea remained constant. Flat steel barstock, which interfaced logically with flexible modular fiber insulation panels in cross-section replaced the round stock shown in the early hybrid rendering. Threaded rod was replaced with square stock in the tension-loaded exterior guy-assembly— at the timber frame edge, tensile force is directed toward the projecting beams and resolved at the concrete foundation piers. Water was directed off of the roof plan by gradually increasing the length of the catenary support by two inches from the center outward to the north and south. In order to resist uplift and potential flutter, the suspended roof plane was directly attached to fixed transverse wall partitions and the architects provided an option for vertical tie-backs to be installed in the open dining and living area at the threat of hurricane force winds. [11] The flat steel truss inserted into the hung assembly in order to laterally stiffen the lightweight roof plane is clearly evident in the construction photos taken by Joseph Steinmetz (Fig. 4a, 4b).

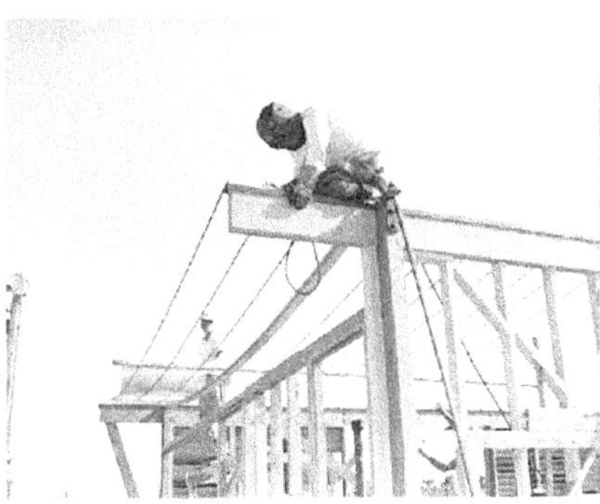

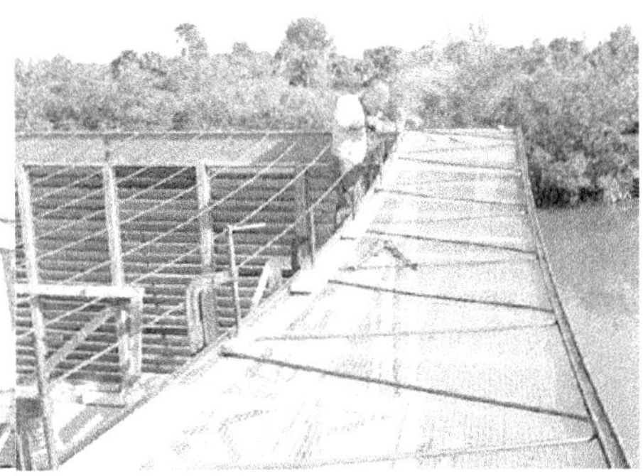

Fig. 4a, 4b (right). Healy Guest House, Twitchell + Rudolph Architects, 1950 (Joseph Steinmetz, Florida Photographic Collection: State Archives of Florida)

After the completion of the project, Rudolph articulated his discomfort with this structural application at the scale of a guest cottage. In his mind, the spanning potential seemed more suitable to larger buildings, but he recognized that architectural experimentation often took place in form of humble investigations.

The connection of this construction to its surroundings was more successful from the architect's perspective: *"we wanted to demonstrate that harmony between the work of nature and the work of man can be brought about by clearly differentiating the two."* [12]

Rudolph and Twitchell further developed the suspended-roof strategy in the Coward Residence (1951) and the Wheelan Cottages (1951). For these projects, a ridge beam is introduced to the tensile assembly, allowing for a tent-like interior and upward spatial condition in the living areas and carport. These projects continued the kit-of-parts organized for the Healy Guest House and the architects arranged them in response to more demanding programmatic requirements. Paul Rudolph's suspended roof investigations were widely published and had a lasting influence on the careers of structural engineers and architects alike. [13] Soon after these projects were finished, Rudolph split from the Twitchell partnership and entered into private practice.

Following service in the infantry and completing his Master of Architecture Degree at Harvard Graduate School of Design under the direction of Walter Gropius and Marcel Breuer, Victor Lundy relocated to Sarasota, FL. He opened his Sarasota Office in 1954 and quickly established himself as an artist/architect due to his prodigious drawing talents. Because of Lundy's connection to the Sarasota Arts Community, Sarasota Civic Booster and former President of United Press Association, Karl L. Bickel recruited him to design a building for the Sarasota Chamber of Commerce, which was completed in 1956. [14] This pagoda shaped pavilion with glazed Japanese roof tiles began a series of spatial and structural investigations in Lundy's office based on the potential for upwardly soaring wood laminated roof beams utilized in consort with operable walls of glass at ground level— connecting earth and sky in a single composition. For Lundy, wood technology was at the center of these early roof structure investigations:

> " In an age of complete uncertainty and tension, it is reassuring and universally appealing to all Americans to be reminded of their great heritage—the beautiful old wood structures of New England, still standing and beautifully preserved... traditions of beautiful wood buildings in the US like Howard Green's work in California—the interest of some of its Japanese derivations. When you talk about fundamentals and timeless beauty, you reach everyone, I think. Wood is basic and is our heritage." [15]

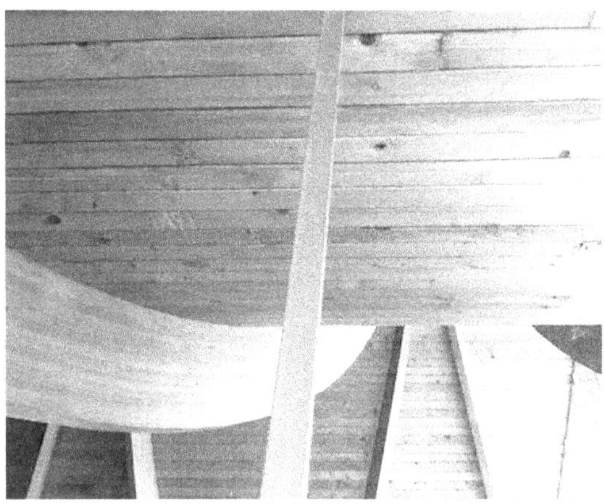

Fig. 5. Roof Structure at St. Paul's Lutheran Fellowship Hall, Victor Lundy Architect(photo by Christopher Domin, 2011)

After a secular detour in the Chamber of Commerce Building, Victor Lundy returned the wood pagoda roof form, in an array of variants, to spaces of worship for congregations in and around Sarasota (Fig. 5). During the late 1950s, St. Paul's Lutheran Evangelical Church contracted with Lundy to design a building complex, which included a Fellowship Hall, an Education Building and a Sanctuary. This productive working relationship would extend for more than ten years, until the completion of a main sanctuary in 1969.

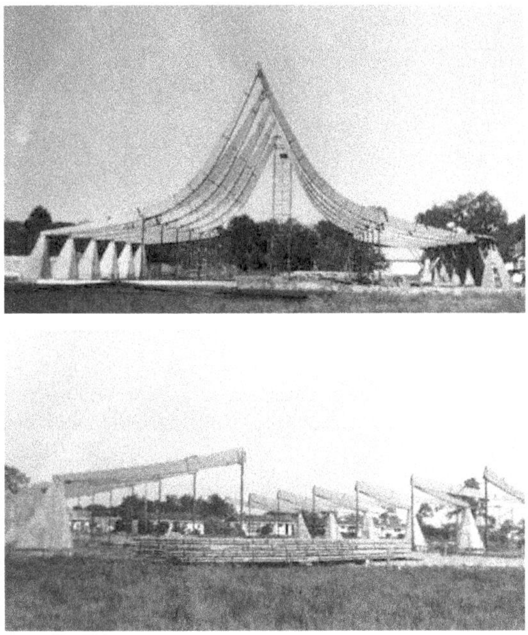

Fig. 6a (top), 6b. St. Paul's Lutheran Fellowship Hall, 1958-59 Victor Lundy Architect (St. Paul's Lutheran Church Archive)

Upon acceptance of the masterplan by the congregation, construction began on the Fellowship Hall, which would also serve as the worship space until the introduction of a larger Sanctuary building. In this bilaterally symmetrical plan and section layout, Lundy associates God with the vertical axis, which is defined by upward projecting wood laminated beams. Tongue and groove wood roof decking add to the autumnal warmth of the interior environment, which is reinforced by the stone clad piers and western end wall. In this space, the congregants rest on terra firma corresponding with the horizontal axis of the earth. Pairs of splayed beams shelter porches to the north and south and above the Fellowship Hall. The linked beams rise to a height of 11.3 meters where they are mechanically fastened at the ridge (Fig. 6a, 6b). The interconnected assembly of scissor-beams displays an affinity for the structural elegance achieved in Gothic Cathedrals and development of pointed arch, ribbed vault and expansive implementation of stained glass in the Middle Ages. These soaring masonry cathedrals, at their best, efficiently directed forces down through the vault and into the columns. This structural synergy freed the enclosure walls of structural performance allowing for expansive translucency. Lundy's Fellowship Hall effectively directs structural forces through the combined effort of beams, columns and piers, allowing for ample glazing at the seating area and main entry; all protected from the sun by the great sheltering hat of the roof.

In Lundy's Fellowship hall, the wood structural elements appear to effortlessly rise uninterrupted from the masonry piers. The interior space is reminiscent of the inverted hull of a wooden ship, graciously sheltering the fellowship. Uninterrupted gazing connects the interior to the north and south porticos. Ample illumination at the east entry logically contrasts the minimal intrusion of light to the west chancel area. As necessary, narrow steel columns assist the lofted ascending beams at the point of building enclosure. This interrelated wood, steel and masonry/concrete structural system creates the unique impression of the roof rising and suspending simultaneously, or perfectly balancing tension and compression in a closed structural loop.

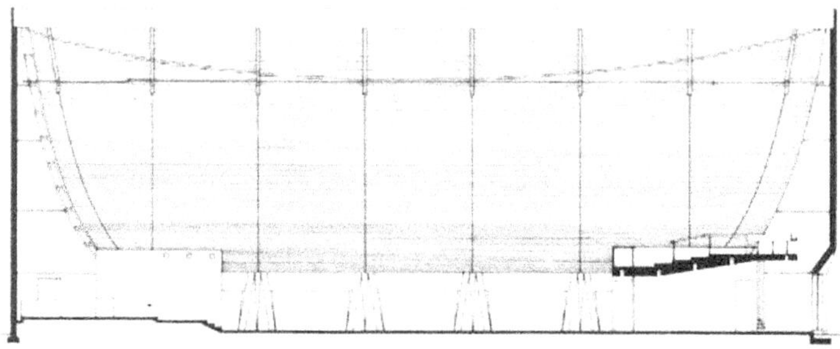

Fig. 7. St Paul's Lutheran Church Sanctuary Longitudinal Section facing east: 1966-69, Victor Lundy Architect with Fred N. Severud, Structural Engineer: Severud-Perrone-Sturm-Conlin-Bandel. (St. Paul's Lutheran Church Archive)

For the larger and more ambitious structural agenda of the Sanctuary Building (Fig. 7), Victor Lundy consulted with Structural Engineer Fred Severud (Severud-Perrone-Sturm-Conlin-Bandel) in order to achieve his aspiration for an elemental and highly principled space of worship.

> "I wanted a form that developed logically out of engineering, proportions, dimensions, and purpose. The result is a tent form, one of the most ancient of enclosures. And the promise of the form outside is fulfilled inside. The building is essentially one clear-spanned space, 90ft.wide by 139 ft. long by 50 high at the center... I tried to make the space an uninterrupted volume sheltered by a great curtain of wood— without an insistent pattern of structural framing to break it up." [16]

The defining feature of the Sanctuary is a hung wooden roof. In order to achieve this single curvature suspension structure, the architect and engineer designed an ingenious edge condition at the apex. A steel ridge truss frame allows the wood membrane to graciously drape the worship space from above and mitigate dynamic forces placed on the roof. In order to resist overturning, an integrated system of cast-in-place piers and walls support the tensile loaded roof system at the ground plane. Hannskarl Bandel, a structural engineer, and later partner with Fred Severud in his New York based consultancy firm, outlined the unique engineering problems associated with suspended roof design in a published article:

> "While cables are certainly at present the most efficient devices for carrying loads in pure tension, without bending stress or buckling limitations, their anchorage frequently presents a difficult problem. Unfortunately, the anchor forces are a multiple of the vertical cable loads, due to the fact that from the esthetic point of view the cable sag in midspan should be the smallest possible. The flexibility of cables creates a second problem, as this flexibility, if not restricted, causes dangerous movements and flutter phenomena in cable suspended structures." [17]

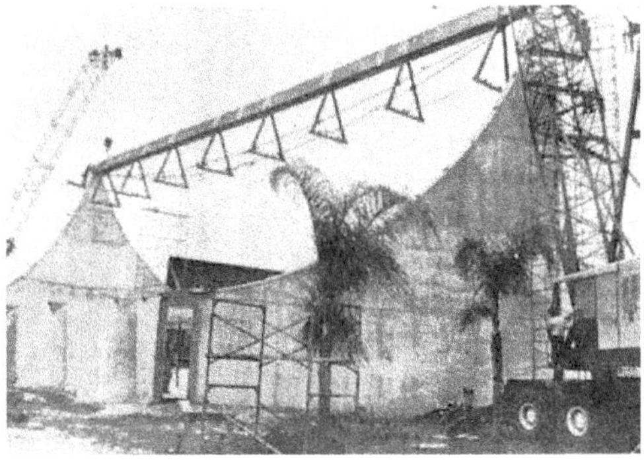

Fig. 8. St Paul's Lutheran Church: Sanctuary Construction Photo, 1969 (St. Paul's Lutheran Church Archive)

In order to support the roof on freely suspended cables, a hybrid truss was proposed by the Severud Consultency that is composed of a continuous, and relatively slender, triangular steel ridge beam and eight triangular frames (18ft spacing). The frames are linked with pairs of steel roof support cables (Fig. 8). Within the open frames, two steel cables in tension are attached to the base of the triangular frame in order to brace against horizontal wind loads. [18] Also, steel cables in tension are connected to the sides of the open frames along a parabolic curve in order to reinforce the ridge beam at midspan (Fig. 9). In order to further resist wind uplift forces, taut steel cables span the worship space and connect the two halves of the roof. The cable supported wood membrane ceiling is protected from the elements by a neoprene roofing system.

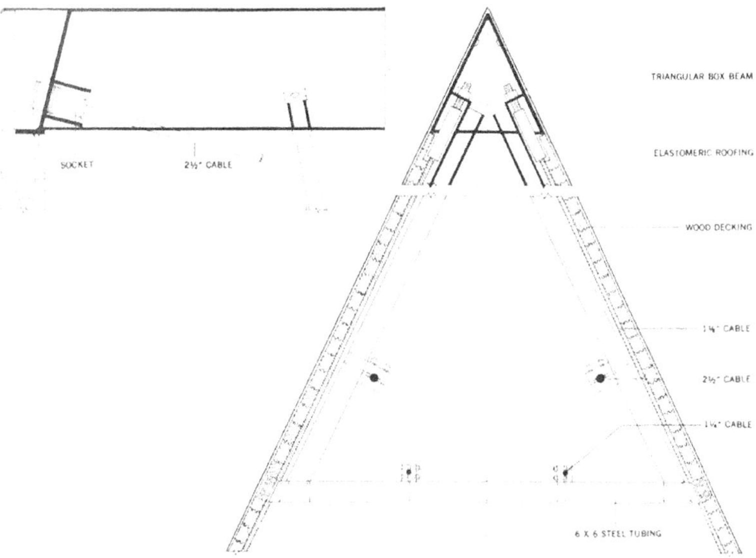

Fig. 9. St Paul's Lutheran Church: Sanctuary Ridge Beam Section Drawing (St. Paul's Lutheran Church Archive)

Cast-in-place concrete walls to the north and south provide privacy for the interior and support the steel ridge beam above. The end walls are angled in plan for stability and trace the catenary curve of the steel cable roof structure in profile as the wall interfaces with the roof system. Plastic-coated plywood panels measuring 4-foot x 12-foot are utilized in the concrete formwork assembly along with a systematic form-tie pattern. As concrete is formed by wood, and its impression defines the outward appearance and module of the walls, wood technology, and its application, begins to imbue this building with a distinctly human scale.

Direct axial entry into the building is interrupted by a service core with balcony above. The more intricate program of the Sanctuary allowed Lundy to slow the pace of procession into the worship space and divide the circulation into two side aisles that allow congregants to enter the main space of the sanctuary at a low-point in the roof profile. The adjacent structural piers are isolated from sidewalls by a slice of light. In fact, the first view of the interior is diagonally across the space with the roof soaring upward toward the ridge truss with pulpit below. Light is introduced from above the chancel through a thin slot of glazing between the northern roof edge and concrete end wall (Fig. 10). Lundy is keen to allow natural illumination strategically into the worship space, creating a veritable temple of light.

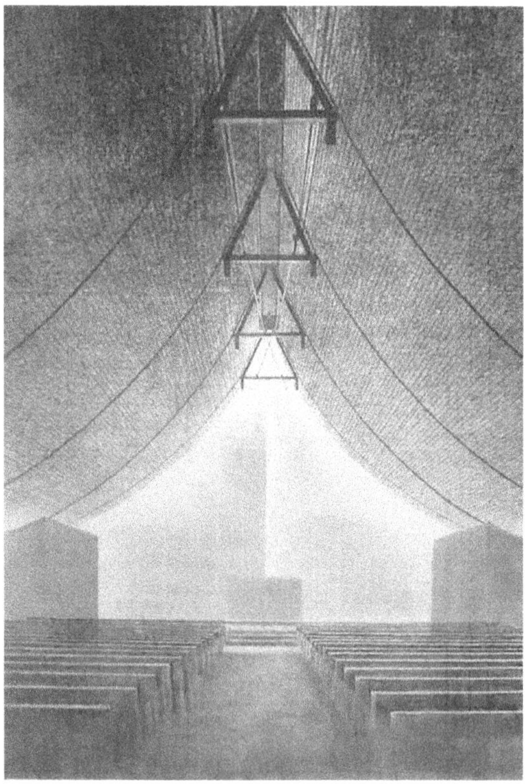

Fig. 10. St Paul's Lutheran Church: Sanctuary Perspective Drawing looking north, 2006, Victor Lundy (Library of Congress, Prints and Photographs Division) Note: steel cables connecting the east and west roof planes are not shown.

> " I am very concerned these days with the control of light. In the St. Paul's Sanctuary, I wanted to make a great, cool, dark space—a place of shade in contract to the persistent brightness outside... So there is a burst of light at the beginning, where one enters, and light at the end of the sanctuary, where the word is spoken from the pulpit and where the monolithic granite altar, polished on top and flame treated at the sides, stands." [19]

During the design process for the Sanctuary Building, Lundy and Severud were dedicated to creating a clear and irreducible spatial condition, while meaningfully confronting the complex structural and material condition that this building strategy necessitated. Technological and engineering solutions facilitated the architectural ideal and expanded the impact of the overall spatial experience. Through the metaphor and physical presence of the archaic tent-form, the architect was able to serve the clients well and produce an architecture that continues to awaken his profession.

Victor Lundy recently described the cloud patterns near his house in Marfa, Texas as "Roofs, over the myriad rock, hill and mountain forms," [20] one can imagine the cast-in-place concrete walls at St. Paul's Sanctuary as geology rising out of the earth, gathering space, and a benevolent wood cloud descending over the congregation. Of course, the spaces created by Rudolph and Lundy under suspended roofs are technologically precise and defined by measurable forces in the world, but the resultant spatial synergy is certainly more than a sum of its constituent material parts. In this work, architecture and engineering are strategically aligned in purpose and nurtured by a unique place in the world— a synthesis of architecture, engineering and landscape to achieve a logically situated way of building.

References

[1] This research expands upon work with Joseph King found in *Paul Rudolph: The Florida Houses*, published by Princeton Architectural Press. Primary source material for this article is gathered from the Victor Lundy and Paul Rudolph Archives at the United States Library of Congress: Prints and Photographs Division, the St. Paul Lutheran Church (Sarasota, FL USA) Archive, the Florida Department of State Joseph Steinmetz Archive, and original photography by Christopher Domin.

[2] C. Domin & J. King, *Paul Rudolph: The Florida Houses.* New York: Princeton Architectural Press, 2001. p.30. (from essay by Joseph King)

[3] E. Mock, *If You Want to Build a House.* New York: Museum of Modern Art, January 1946.

[4] Domin, Christopher and J. King, (Note 2) p.30.

[5] E. Mock, *The Architecture of Bridges.* New York: MoMA/Simon and Schuster, 1949. p.7.

[6] E. Mock, (Note 5) p.7.

[7] P. Rudolph & R. Twitchell, "Notes Concerning the Healy Cottage, Sarasota, Florida" undated entry in Ralph Twitchell papers from the Collection of Patti Jo Rice.

[8] P. Rudolph & R. Twitchell, "Cocoon House", *Architectural Forum*, June 1951, p.157.

[9] "U.S. Zippers Up Its Fleet", *Life*, vol.21, Sept. 9, 1946. p.32.
Note: Also see G. Rothstein, "Sprayed-on Vinyl-plastic Sheeting", *Progressive Architecture*, July 1952, pp. 99-102.

[10] P. Rudolph & R. Twitchell, (Note 8), p.157.

[11] P. Rudolph & R. Twitchell, (Note 8), p.158.
Note: A hook is clearly visible in the Healy Guest House photos taken by Ezra Stoller soon after construction, located directly above dining table (top image).

[12] P. Rudolph & R. Twitchell, (Note 8), p.158.

[13] F. Otto, *Tensile Structures, Volume 2.* Cambridge: The MIT Press, 1967, p.17.

Note: This entry appears in the English language edition of the publication under the heading: "Frei Otto: Basic Concepts and Survey of Tensile Structures".
[14] Victor Lundy interview by C. Domin. March 1, 2006
[15] V. Lundy, "Address to West Coast Lumberman's Association", Portland, Oregon. March 1959. Note: cited in 'Victor A. Lundy and the development of the American Architectural Tradition". Architecture: Form + Functions, 1964-65, p.26.
[16] V. Lundy, "Church Under a Great Tent", *Architectural Forum*, vol. 133, July/August 1970, p.77.
[17] H. Bandel, "Hung Roofs and their Structural Solutions", in N. Esquillan & Y. Saillard (eds.), *Hanging Roofs: Proceedings of the IASS Colloquium on Hanging Roofs, Paris 9-11, July 1962*. New York: John Wiley and Sons, 1963, p.33.
[18] F. Severud & R. Corbelletti, "Hung Roofs", *Progressive Architecture*, vol. XXXVII, No. 3, March 1956, p.100.
[19] V. Lundy, (Note 16) p.77.
[20] V. Lundy, from letter to Christopher Domin, April 25, 2006.

Construction history and alteration of the oak roof structure of the Sidney Sussex College Hall

Patrick Fleming, Sofia Singler and Michael Ramage
Department of Architecture, University of Cambridge, UK

Summary

In the following paper we present a study on the roof structure of the Sidney Sussex College Hall in Cambridge. The tie-and-collar-beam roof was originally constructed in the last decade of the sixteenth century as an exposed oak structure. Since its construction, however, the roof has undergone significant alterations, in part to address structural concerns, but also to attract more students by remodeling the College to make it appear more 'fashionable' in the eighteenth century. The original Elizabethan timber structure was at that time hidden behind a Rococo ceiling that still defines the space today. To better understand the roof's original construction and later conversion, we have performed on-site measurements, including the location and dimensions of key structural components and joints. Archival drawings also provide valuable insight into the roof's history and complement observations made on-site. The documentation and analysis of this roof presents a case study for the transitional period in English carpentry in the seventeenth century and brings to light a little-known but wonderfully wrought timer roof.

Introduction

The seventeenth century marked an important transition in the history of English architecture and construction. Before this time, timber roofs in England formed a significant architectural feature in the spaces of halls, churches, houses, and barns, and were often left as exposed structures. 'Cruck' roofs with large curved elements, sawn from crooked trees, were common, as were purlin roofs with relatively simple king or queen posts and curved or diagonal braces. The latter type has appeared in studies by Hewett [1], alongside more elaborate roof structures formed with timber arches and hammer-beams. For example, the hammer-beam roof of Westminster Hall, built in the late fourteenth century, is often cited as an exemplar, and a pinnacle of medieval carpentry in England. [2,3] Even though all of these structures used typical mortise and tenon joints with limited strength, spans of around 20 m, like that of Westminster Hall, could be achieved due to their steep pitches of roughly 45-55 degrees. The steep pitch and height of these roofs created their own space with dominant structure that could be readily seen from below.

With the introduction of architectural principles from the Italian Renaissance, a new kind of truss-roof type began to emerge in England in the seventeenth century. In contrast to earlier roof types, the truss roof was usually completed as a concealed structure behind ornamental ceilings and panels. More importantly, however, the truss roof also offered the possibility of even larger spans and greater structural efficiency. This accomplishment came about primarily due to new ways of detailing the joints of the king post connections, thereby providing increased structural action over the whole height of the truss. Later in the seventeenth century, comparable spans of about 20 m could be achieved with significantly shallower pitched roofs, with trusses having relatively less structural depth than in the previous centuries, as can be seen in Christopher Wren's Sheldonian Theatre in Oxford. Due to their improved structural efficiency, these trussed roofs also used significantly less timber. With these advantages in structure and construction, and association with and necessary use for achieving new architectural form and space, English carpentry began a transition from medieval exposed roofs, often with high pitches, to low-pitched roofs with trusses hidden above ceilings.

In an effort to further study this time of discontinuity in English architecture and carpentry, we present a case study of the timber roof above the Sidney Sussex College Hall in Cambridge. With its original construction in the late 1590s, the roof presents an example of a timber roof originally built as an exposed structure, just before the introduction of the Italian truss roof. The roof's simple design with a minimal number of arched elements and lack of diagonal bracing make it unlike those discussed in the literature on medieval carpentry and timber roofs. Moreover, the roof underwent subsequent alterations in the eighteenth and nineteenth centuries to convert the dining hall into a Rococo-style room, with a ceiling that would completely conceal the structure above. The original roof construction provides a unique example of late-medieval carpentry, while the subsequent alterations show how new techniques and designs from abroad could not only replace these older types, but also adapt them for changing architectural fashions and trends. This study primarily involves measurements of the roof in its current state, including the dimensions of the principal and secondary members and details of their typical joints. We use these measurements to create a detailed 3D digital model (Fig. 1) as a representation of the roof structure. We have also collected and studied archival material alongside the measuring and modelling process to better understand changes in the roof after its original construction. Together these measurements and archival drawings provide a detailed account of the Sidney Sussex College Hall roof, and a case study for the literature on seventeenth-century timber roofs in England.

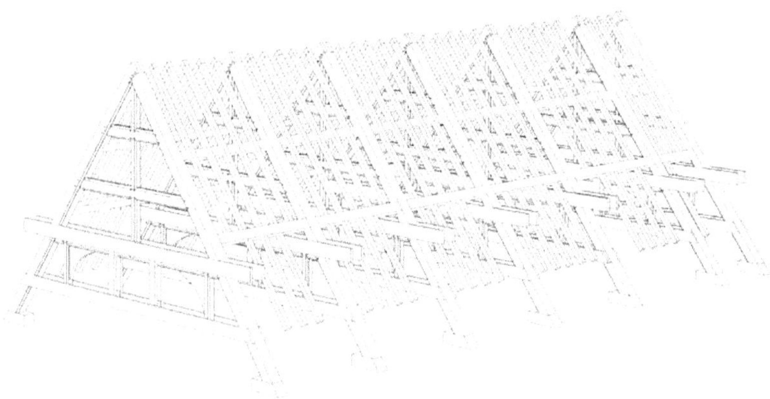

Fig. 1. Overview of the timber roof structure modeled from on-site measurements (Authors).

On-site observations and measurements

Preliminary on-site measurements of the roof started in June 2013. The authors' first visit to the now enclosed attic space began by examining the general layout of the structure and measuring the roof's overall ceiling area of 9.23 m x17.83 m. The roof itself consists of six bays with seven principal frames (Fig. 2), and several of the roof's secondary rafters have had their lower sections removed. Instead of relying on arched elements or diagonal bracing like most medieval roofs, the basic form of the principal frame in the Sidney Sussex College Hall roof is a tie-and-collar beam structure with a king post segmented between the collar beams. Apart from the rafters, the only arched or diagonal members in the roof structure are the small planed and profiled brace plates immediately above the ceiling level. The freedom to build without additional diagonal braces or more structurally-efficient elements probably came about due to the roof's relatively short span and steep pitch of 52 degrees. As the roof was originally designed and built as a prominent exposed structure, most of the timber is smoothly planed oak, with very little ornamentation except rounded and profiled edges on the members of the principal frames (Fig. 3). Much rougher, unplaned elements are also present in the roof today, and were added in modification and repair efforts after the roof's original construction.

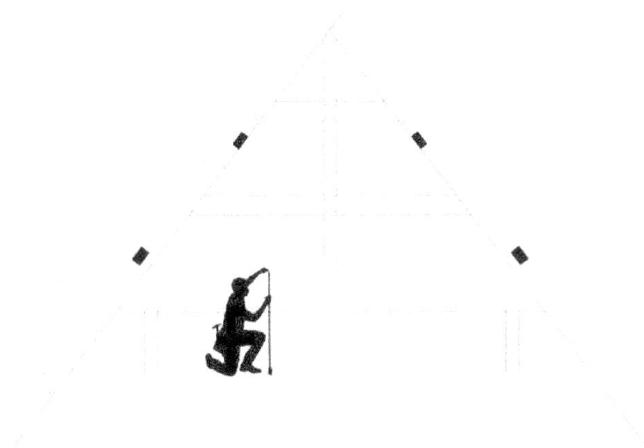

Fig. 2. The principal frame of the Sidney Sussex College Hall roof with hidden lines showing joint details and purlin locations (Authors).

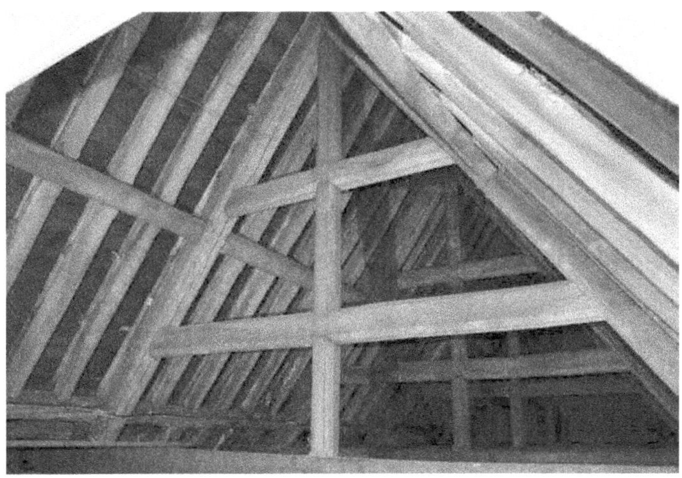

Fig. 3. The principal frames with smoothly planed oak with rounded and profiled edges (Authors).

More detailed measurements and analysis of the roof provided the basis for the 3D model and drawings already shown in Figures 1 and 2. While the original sixteenth-century construction appears to follow a simple but unusual arrangement of the main elements in the principal frames, the structure has typical pegged mortise and tenon joints like those seen in most medieval roofs. According to Yeomans [4], the key distinction between a medieval roof with basic frame elements and the later truss roofs lies in the joints of the king post: medieval roofs, with mortise and tenon joints like those seen in the present study could not achieve full structural action due to the limited capacity of the king post's mortise and tenon joints; instead of mortise and tenon joints, the truss roof used metal strapping around its tie beam and featured a king post with an enlarged top head for bearing against the principal rafters. [4] With several discontinuous king post segments between multiple collar beams, however, the roof may have been able to achieve adequate structural action over the majority of its depth. On the other hand, the presence of some iron bolts and straps in the roof structure today suggest that the roof's original design may have been inadequate. These issues are discussed in greater detail later on, as the iron connection details are found with the roof's rough timber elements and were added after the original construction, probably in the nineteenth century.

The internal dimensions of mortise and tenon joints can often pose challenges for obtaining accurate measurements of a structure's key details. While every effort was made to perform a systematic and complete survey with an additional site visit in November 2013, some details of the roof structure could not be physically accessed or even seen due to the repairs and additions associated with the Rococo ceiling. Some elements in the 3D model were therefore estimated or simply left incomplete, as in the case of the joints between the four short vertical posts and the tie beam. The dimensions of the king post's mortise and tenon joints (Fig. 2), however, were measured from an exposed mortise found on the bottom side of the lowest collar beam in the roof today (Fig. 4). The large broken fragments of wood still left in some of these mortises, along with the distinct colour contrast around these joints suggest that they were once used as a regular structural connection, or possibly to hold a decorative feature in the roof. These open mortises, their modification and disuse raise several questions regarding the subsequent alterations to the roof after its original construction.

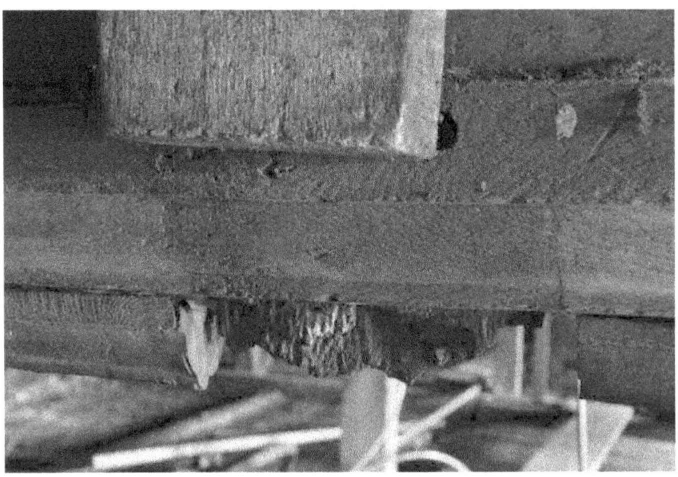

Fig. 4. An exposed mortise reveals some broken tenon wood still inside the joint and a darkened area of wood on the beam directly above (Authors).

Alterations

Following the initial on-site visit, archival records and drawings were examined to better understand the details and extent of the alterations to the roof. Reference material from the Sidney Sussex College archive confirmed that the oak for the roof members was sourced from the royal forests at Rockingham in Northamptonshire, and that the members were originally planed and profiled. [5] Records from the College archive also explain that in the mid-eighteenth century, the College took on the project to re-model the Hall entirely. The Elizabethan space was deemed inadequately impressive, and Gonville and Cauis Fellow and amateur Palladian architect James Burrough was commissioned to redesign the space. The wainscoting and seating were replaced and the whole space was refashioned in the Rococo style in 1752. [5] Burrough's plans included a flat ceiling along with new mouldings and cornices that would conceal the existing exposed roof structure. However, reasons for re-modelling the hall were not only limited to aesthetic considerations: another underlying aim was to repair some decaying parts of the building, most likely at the roof's base and rafter feet. The original pendant posts were removed, and new pairs of beams were also attached to the base of the posts to create a new tie beam for installing the ceiling. These alterations are illustrated with coloured overlays on a drawing from the College archive (Fig. 5). With the addition of this new tie beam, the roof's structural issues would most likely have been considered resolved without the need for any significant future alterations.

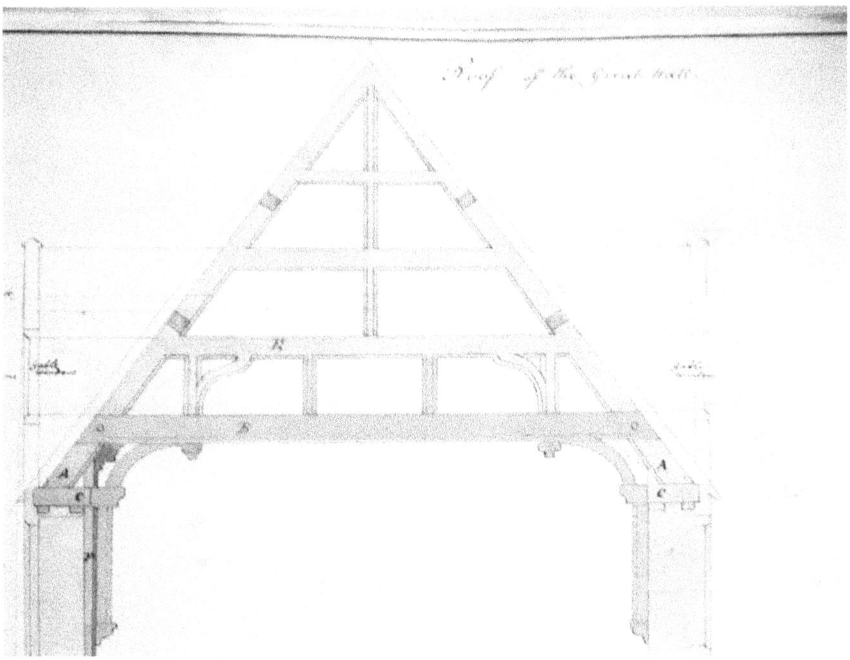

Fig. 5. A Sidney Sussex College archive drawing showing parts of the roof that were added (green overlay) and removed (blue overlay) by Burrough in 1752.

Less than a century later, in the early nineteenth century, the College had to consider the Hall's construction yet again, but this time for purely structural reasons. Archival records note that the hall's eastern wall was giving way, possibly from the forces of the timber roof. To address these structural concerns, the College commissioned an architect, Jeffrey Wyatt, to develop a design in 1821 (Fig. 6). Compared to our own measurements and drawings, Wyatt's drawings (Figs 5 and 6) show some obvious errors, such as the detail and shape of the roof's planed and profiled brace plates, along with the purlin dimensions and locations. Nevertheless, Wyatt proposed and executed a design involving a pair of beams bolted at lowest segment of the king post (Fig. 6), where the new beams extended out of the roof frames to sit on the brick walls built up from below. Buttresses were also erected to provide further support for the outward movement of the eastern brick wall. Based on these records and the observations and measurements made on-site, the sequence of these major modifications (Fig. 7) is shown for clarity and to complement the archival drawings. Since the 1821 alterations, the College has also added a relatively short steel element to support an isolated area of rot in one of the primary rafters at the east side of the roof. This latest alteration (Fig. 8) was probably completed sometime in the late twentieth century and demonstrates that the roof's structure is a relevant issue not only for scholarship purposes, but also for informing future alterations.

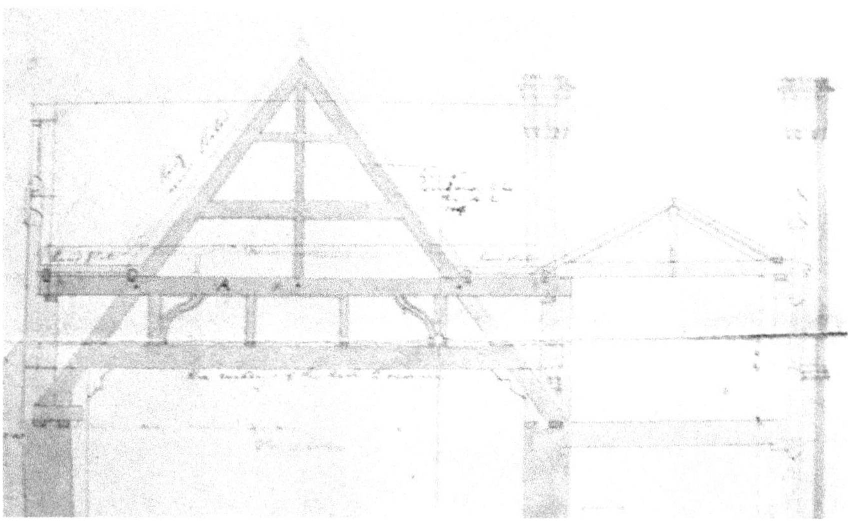

Fig. 6. A Sidney Sussex College archive drawing showing parts of the roof that were added (green overlay) by Wyatt in 1821.

Fig. 7. Summary diagram showing the roof's (a) original construction in the 1590s, (b) transformation with a Rococo ceiling by James Burrough in 1752, and (c) structural repairs by Jeffery Wyatt in 1821 (Authors).

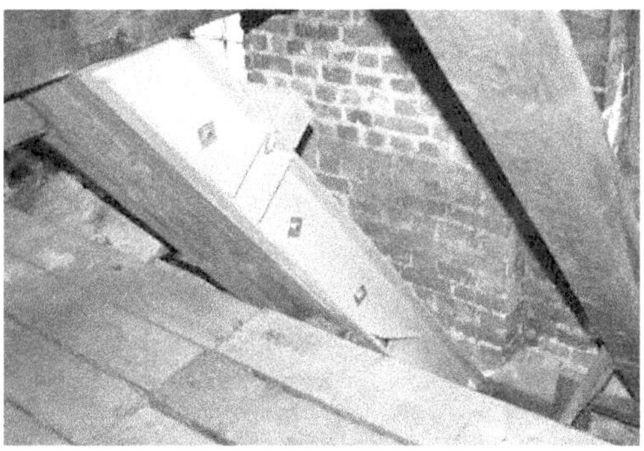

Fig. 8. The most recent alteration to the roof is a steel element added in the late twentieth century (Authors).

Discussion of the Transition from Open Roof to Concealed Structure

Drawings from the College archive have helped to better understand the sequence and layers of alterations seen in the roof today. While some of these alterations were performed for stylistic purposes, those that were structural suggest an inadequacy in the roof's original design that caused structural problems later on. Despite taking on-site measurements and checking archival material, aspects of this roof still remain unclear and open to discussion, including the origins of these structural problems and especially the movement in the eastern wall. For example, the wall itself, its foundations, or the ground below may have been the origin of the problems, causing the roof members to dislocate and also shift. Alternatively, the roof itself may have been unstable and thrusting outward and causing the eastern wall to shift. In the latter scenario, the effective structural action of the roof comes into question, and its ability to work as a truss despite only having basic mortise and tenon joints. The present understanding of the roof's original construction (Fig. 5) also raises further questions regarding the structural effectiveness of the original design. Without any horizontal tie beam at the current ceiling level of the hall, and with just short horizontal ties to make a hammer-beam like arrangement, it is possible that the pendants and arched brace plates were structurally insufficient and may have been the weak point in the design. If short horizontal beams were not even used in the original construction, however, these pendants and arches may have only served a decorative purpose. Their removal in the eighteenth century and the transition from an exposed to hidden structure should not have caused any further structural problems, especially considering the addition of the complete tie beam. The structural problems of the hall must have therefore started with the original design of the roof itself, or in the eastern wall, or perhaps both. Closer examination of the eastern wall and its history will remain as work for the future.

Conclusion

On-site measurements, modelling, and archival drawings and records have provided a detailed account of the Sidney Sussex College Hall timber roof. With its original construction dating back to the 1590s, the roof represents an interesting example of late medieval timber carpentry. Like most medieval timber roofs, the Sidney Sussex College Hall roof was originally constructed as an exposed structure with simple mortise and tenon joints, but featured very little bracing or diagonal elements apart from the main rafters. The roof's transition to a concealed structure for stylistic purposes also involved major changes in terms of removing pendant elements, and adding a new tie beam for installing a Rococo ceiling. Further changes were also made in the nineteenth century with the addition of more structural elements to stabilize the roof and hall structure below. The hall's eastern wall or the original roof design itself were mostly likely the cause of these ongoing structural issues, rather than the changes and alterations performed during the transition from open to concealed structure. The Sidney Sussex College Hall roof and its subsequent alterations show how the structures and joining techniques of the later Italian truss roof type could be used not only to supersede medieval structures in England, but also re-appropriate existing structures to work with new architectural principles.

Acknowledgements

The authors kindly acknowledge the Sidney Sussex College Archivist, Mr. Nicholas Rogers, for providing drawings and reference material for this study. The authors also thank Dr. James Campbell for providing valuable guidance and advice in the early stages of the study.

References

[1] C. A. Hewett, *The Development of Carpentry, 1200-1700: An Essex Study*. Newton Abbot: David and Charles Publishers Limited, 1969.

[2] C. A. Hewett, *English Historic Carpentry*. Chichester: Pillimore & Co. Ltd., 1980.

[3] E.W. Tarn, *The Construction of Roofs of Wood and Iron*. London: Crosby Lockwood and Co., 1883.

[4] D. T. Yeomans, *The Trussed Roof: Its History and Development*. Hants: Scolar Press, 1992.

[5] R. Humphreys, *Sidney Sussex: A History*. Cambridge: Sidney Sussex College, 2009.

The traditional wooden shipyards on the northwest coast of Spain

Oscar Fuertes Dopico and Joaquin Fernandez Madrid
University of Coruña, Spain

Resume

Wooden boat building (known in Portugal and Spain as "Carpinteria de Ribera") is today a traditional activity with a craft characteristic carried out by small companies dedicated to the construction, repair and maintenance of small boats where wood is the major element of construction.

The traditional shipyards are located in strategic places along the northwest of Spain and in the Atlantic seaboard. All countries of the Atlantic seaboard have a similar historical evolution of boatbuilding, but only in Galicia (Spanish autonomous community in the northwest) are there shipyards that still build wooden boats for the fishing industry.

The shipwright built their shipyard with the same tools and technology used to build a boat. Originally the shipyard started as a small facility to store tools and materials, but finally become a building which provided shelter during the process of building the craft. In this way the traditional shipyards are directly linked to the conservation of the local culture and landscape, and are part of the ethnographic heritage of the Galician coast.

It was in the late nineteenth century when they started to settle down and build their own shipyards: wooden structures, primarily large airy decks like upside down ships, perfectly prepared to withstand strong gusts of wind and to shelter both the carpenter and the boat from the rain in winter and the sun in summer. This paper presents a study of the shipyards still present in Galicia and focuses on the analysis of the structural similarities of both constructions: the boat and the shipyard.

Introduction

Wooden boat building is known in Spain and Portugal as *Carpintería de Ribera* and is nowadays a traditional artisanal trade practised by small companies dedicated to the construction, repair, and maintenance of boats in which wood is the principal building material.

Spain's north-western coastline, in particular that of Galicia, totals over 1,000km, which has brought about the predominance of sea-related activities, trades and livelihoods amongst its inhabitants, shaping Galicia as a fishing country.

During the Middle Ages fishing practices involved entrapping fish with nets near the shore side, but from the eighteenth century the fishing method known as *Xeito* became more prevalent. This method consisted of two or three fishermen sailing a boat inshore to fish with nets [1], which gave rise to the widespread use of boats as a means to access different fishing grounds.

Over time different types of wooden boats began to appear - *Bote*, *Dorna*, *Gamela*, *Lancha*, *Trincado*... - each with an idiosyncratic morphology stemming from the particular cultural context of the coastal areas in which they were developed. [2] Thus, the figure of the *carpintero de ribera* (wooden boat builder) emerged as the artisan, builder and repairer of these wooden craft.

At first the *carpinteros de ribera* did not have a permanent workshop in which to build their boats. Rather, they would travel to the coastal area from which an order had been made and there erected temporary sheds. These were low quality structures made of wood and used only during the time taken to build the boat. Once the boat was completed the shed was taken down, leaving no traces of it on the landscape. From 1940 onward the obtaining of licences for the construction of shipyards and carpentry workshops became more accessible [3] and a number of such permanent facilities began to appear along the coast, erected by the shipwrights themselves with the very same tools and techniques that they used for building boats.

The present paper is a study of these so-called *carpinterías de ribera*, traditional shipyards of the Galician coast. Three traditional shipyards located at different points of the Galician coastline are analyzed in terms of their construction features: *Roseva,* in Cabana de Bergantiños; *Cerson* in Outes, and, finally, *O Charango*, in Muros.

Traditional shipyards on the Galician coastline

Traditional shipyards appear as landmarks along the entire Atlantic coastline, but it is in Galicia where a larger number of these structures have survived. Galicia has important forest resources and a maritime vocation, and came to have hundreds of shipyards scattered along its *rías* (coastal inlets).

In 1930, according to figures from the "Anuario Estadístico de España" ("Statistical Yearbook of Spain"), there were 62 shipyards in Galicia, a figure that had increased to almost 100 by 1942, the year in which 96 facilities were reported. In 1996 the number of facilities had fallen to 73, and would continue to fall to the 25 active shipyards reported in the sectorial study of 2006. The dramatic decrease in traditional shipyards became more evident in 2010 with their inventorying and cataloguing. [4] Out of the 110 facilities which had been inventoried, 78 have disappeared, although their existence is recorded. Of the remaining 32 facilities, there are 22 which are still active. Out of these, only ten can be classified as traditional shipyards (Fig. 1).

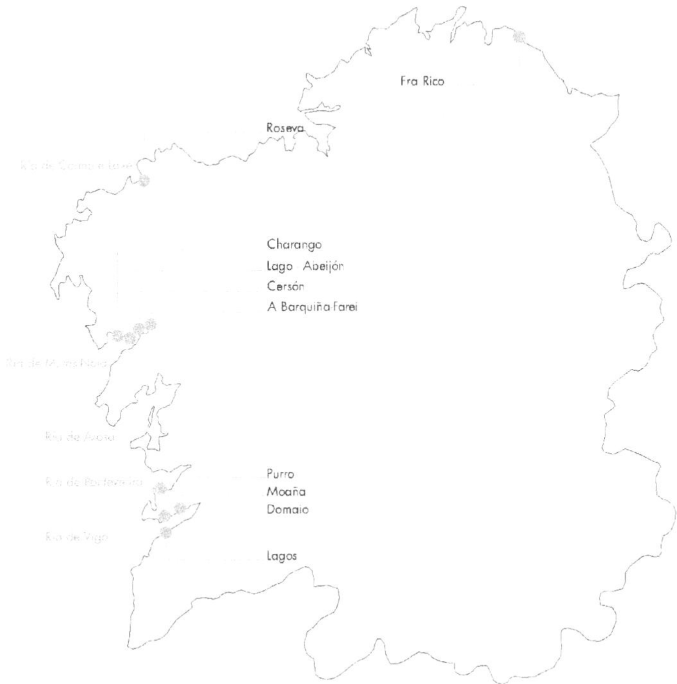

Fig. 1. Map of the Traditional Shipyards on the Galician coastline (O. Fuertes).

Architectural structure

Having analyzed the location of the *carpinterías de ribera*, it is significant that all of them are situated by coves inside the *rías*, which, as is well known, are river valleys drowned by the sea, forming numerous natural sheltered areas appropriate for boats to drop anchor and thus constituting well situated sites for a shipyard.

These are constructions which are embedded in a specific landscape, responding admirably to the topographical, climatic and environmental features of the site they occupy, and which are perfectly integrated into their surroundings.

Traditional shipyards comprise the following elements: work platform - a large area of land or beach where the boats are built; the slipway - a slope leading down into the water; the storehouse - a closed structure for the storage of tools; the sawmill - the place where the timber was squared up and sawn; and, finally, the shed - a covered area used as a workshop.

Several aspects relevant to the understanding of how the shipyards were conceptualized are highlighted below:

Use of raw materials from the area, usually wood and stone, which due to their colour and texture attain a high degree of fusion into the landscape.

Rigorous formal economy, with a perfect synthesis between form and function.

Direct and functional architectural language, which favours integration with its immediate surroundings. In general, buildings consisting of light wooden structures which adapt to the topography.

Continuity between inside and outside. The work space flows freely between the covered shed and workshop and the open yard.

Spacious interior areas, due to the need of bright ample spaces for the construction of the craft.

Construction features

The traditional shipyard is a classic wooden frame construction, the width and height of which depends on the size of the boat to be built. A great roof supported by wooden posts (columns) is mounted to give shelter to an ample work space, with the most exposed vertical faces closed with timber boards to adverse weather conditions. The roof is similar to an inverted boat and is covered with timber planks finished off with battens (Fig. 2). The eaves drain the water directly onto the ground or may incorporate gutters made of eucalyptus logs which have been halved and emptied.

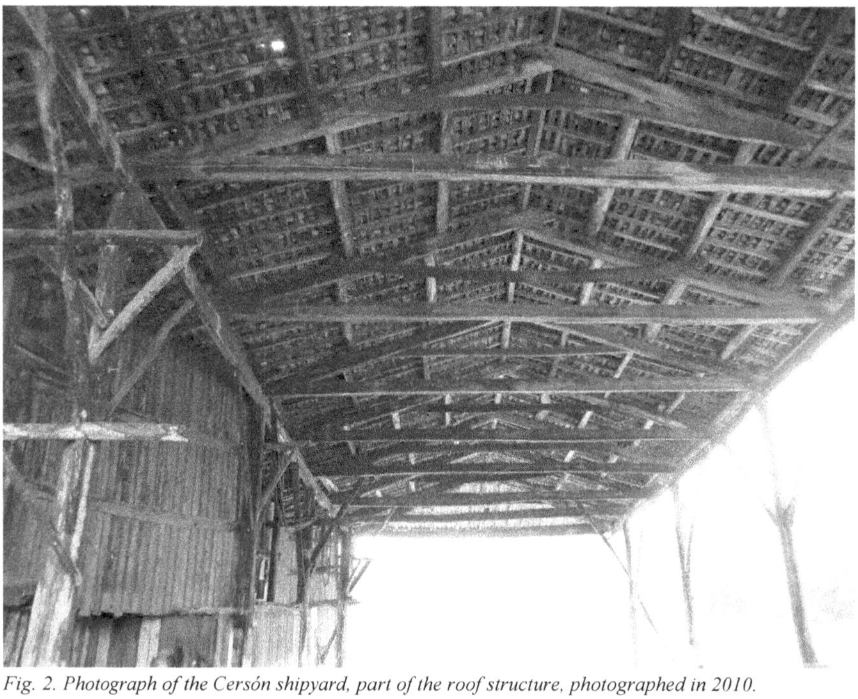

Fig. 2. Photograph of the Cersón shipyard, part of the roof structure, photographed in 2010.

In the construction of the shipyards, traditional carpentry techniques are applied, such as collar roof and trussed beam structures, wood board assemblies, or sheathing the frame with planks to provide stability and rigidity. In any case, the building methods are basic but stand out for the coherence achieved between the form of the construction and the materials used. It remains to be said that this is a clear instance of technology transfer in wooden constructions: symbiosis between vernacular architecture and the building of traditional boats.

Out of the whole structure of the shipyard, it is the roof layout which presents the higher parallelism with the frame structure of the boats. In addition, another element of parallelism is the outer wooden cover of both constructions, given that in many cases it is made of horizontal planks fastened edge to edge, similar to the hull planks of the boats, which were fixed to the ribs. [6] The hull of a wooden boat usually consists of planking fastened to frames and a keel. Keel and frames are traditionally made of hardwoods such as oak while planking can be oak but is more often softwood such as pine, larch or cedar.

Roseva shipyard:

The *Roseva* shipyard (Fig. 3) is a big building with dimensions of 24 x 9.5 metres, which has survived on a stretch of coastline profoundly modified by the construction of a seafront. Despite this, it has been possible to maintain Roseva's coastal wall, made of granite masonry, which clearly delimits the slipway and the exterior work platform. The building is bright and its structure divides it into two differentiated areas, an ample work area, rising 5 metres in height, which can be identified as the workshop shed, and a lower building at 3 metres which comprises the storehouse and the sawmill.

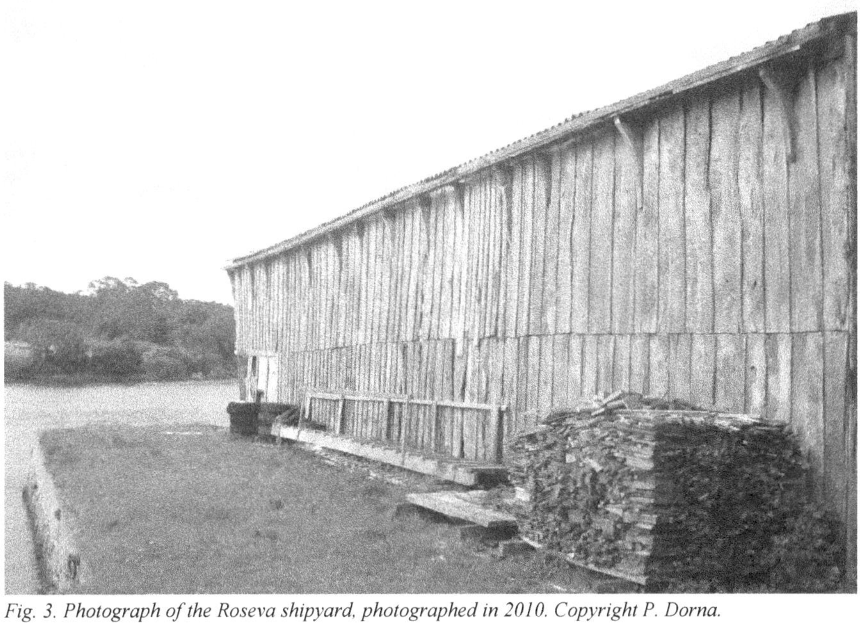

Fig. 3. Photograph of the Roseva shipyard, photographed in 2010. Copyright P. Dorna.

The main structure bridges a 9.5 metre long span through seven open frames, each with two 4-metre-high pine wood posts (columns) squared to 10x20cm, on which the rafters are seated. This is similar to the close couple roof except that the horizontal ties are now placed higher up the roof, and they are called collars. At the joint between the rafter and the collars, there is an inclined strut, with the function of exerting the load directly to the posts. The structure of timber posts is seated on different stone elements, most of them granite.

The roof surfaces rest on the rafters via seven pine wood purlins that are variably squared, three on each side, and one purlin on top over the central part, as a ridge piece. The inclined planes of the roof direct the rainwater to roof gutters made of halved timber logs and fixed through curved brackets.

Cersón shipyard:

The *Cersón* shipyard (Fig. 4) is a 37 x 9.5 metre building which stands out as a slender wooden structure. Above the work platform, delimited by a coastal wall and the slipway, both in stonework, there is one single building which comprises the storehouse, the sawmill and the workshop shed (Fig. 5). The storehouse and the sawmill serve as protection against the wind, being placed on the side most exposed to the dominant winds.

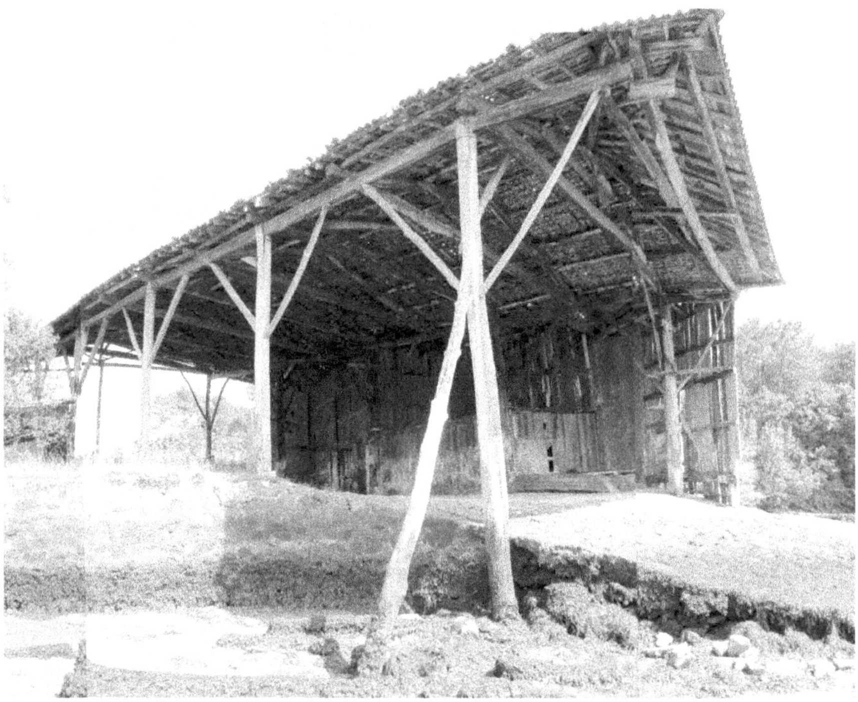

Fig. 4. Photograph of the Cersón shipyard, photographed in 2010. Copyright P.Dorna.

The structure is formed of two open frames, with six and five posts respectively. It is a slender and bright structure due to the posts having a 25cm diameter, rising 8 metres high and being placed 6.4 metres apart from each other. In each frame there is a big wooden beam joining the posts and on which there rest ten wooden trusses formed by two rafters (top chords), a collar and a bottom chord bridging a 9 metre span. This structure is remarkable because every other trussed beam is supported by inclined struts rising from the two adjacent posts, allowing the omission of the vertical supports at these points. The timber posts rise from grey granite block which insulate the timber from the damp of the ground.

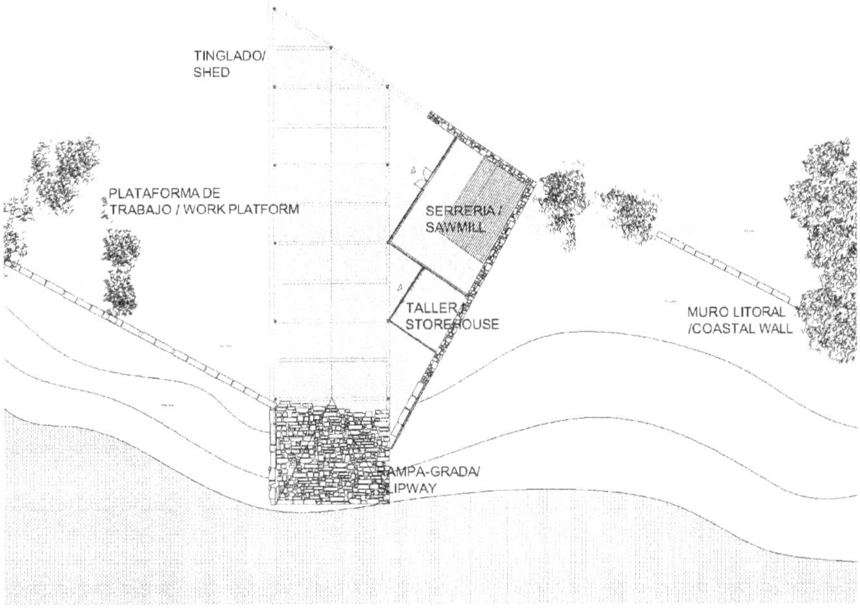

Fig. 5. Ground floor : Cersón Shipyard. (O. Fuertes)

The trussed roof structure extends into a third plane that covers the slipway. The roof sheathing rest on the trusses by means of eight pine wood purlins that are variably squared, with the side purlins functioning as eaves and the central one performing as a ridge. Over these, plain tiles are laid over timber battens and lattice panels to form the covering. The inclined planes of the roof drain the rainwater directly to the ground.

The workshop (shed) was built attached to two constructions, the storehouse and the sawmill, which have stone walls on their faces most exposed to the rain and vertical timbered sheathing on the other faces.

O Charango shipyard:

The *O Charango* shipyard (Fig. 6) comprises several modules, each with a specific function. Above the work platform, delimited by a coastal wall and the slipway, there are three buildings: the storehouse, the sawmill and the workshop shed.

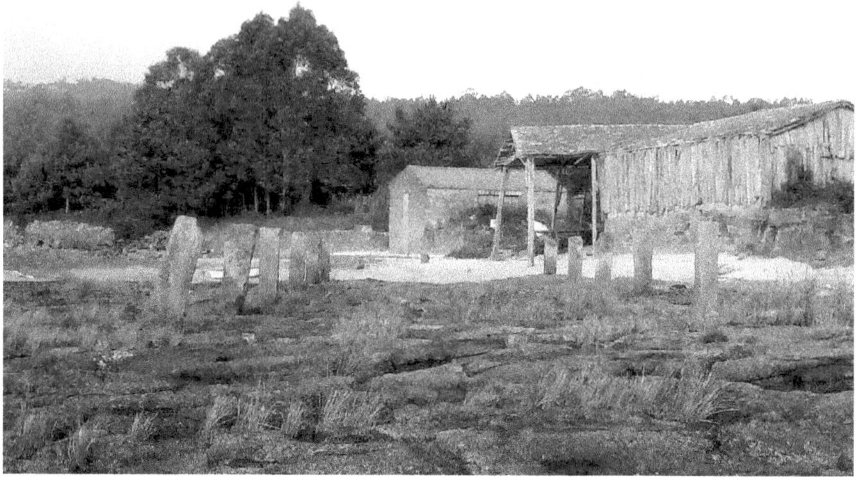

Fig. 6. Photograph of the O Charango shipyard, photographed in 2000. Copyright FGCMF.

The storehouse (Fig. 7) is a sturdy construction erected with masonry load-bearing walls which support the trussed roof structure. Inside, it keeps the most valuable possessions of the boat builders: the tools and the plans. It was also the place where the drawings and details were crafted. The building is perfectly protected against any wind direction and force.

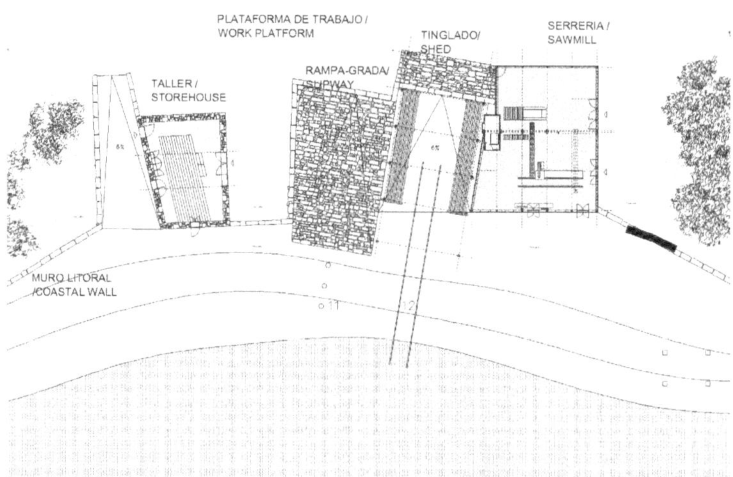

Fig. 7. Ground floor: O Charango Shipyard. (O. Fuertes)

All the modules have tamped earth floors. This type of earthen paving is an excellent solution as it allows the thrusting of timber posts into the floor, which is of great advantage when building boats. There is a small wooden dais in the storehouse like a small office, on which the shipwrights could be placed away from the ground, and in the sawmill, where small boats are sometimes assembled, there are two pits, one metre deep by 70cm wide, which were used to work under the hull.

The structure of the sawmill bridges a 7 metre span, achieved through eight open frames, each with two posts standing 2.5 metres high, squared to 15x15cm and made of eucalyptus timber, over which rest the timber roof trusses. The trusses are formed by two rafters and a collar (bottom chord), both held by struts. The timber post structure rests on a series bases of grey granite blocks on one side, and on the coastal wall on the other side.

All of the facilities have wooden roof trusses (Fig. 8). The cover roof of the sawmill is formed by five 10x5cm pine wood purlins, which support a sheathing of 20x2.5cm pine boards, creating two inclined planes with battens spaced at 28cm, over which plain tiles are laid as a covering. The inclined planes of the roof drain the rainwater to timber gutters, and these in turn allow discharge to the ground.

The external enclosure (facade system) is made of timber lattice panels, 12 to 15 cm wide and 2.5cm thick, laid vertically with 2 to 5cm separation in between, and in some cases closed by flashing. This lattice is anchored to a pine wood substructure formed by vertical and horizontal upright panels - like wall studs.

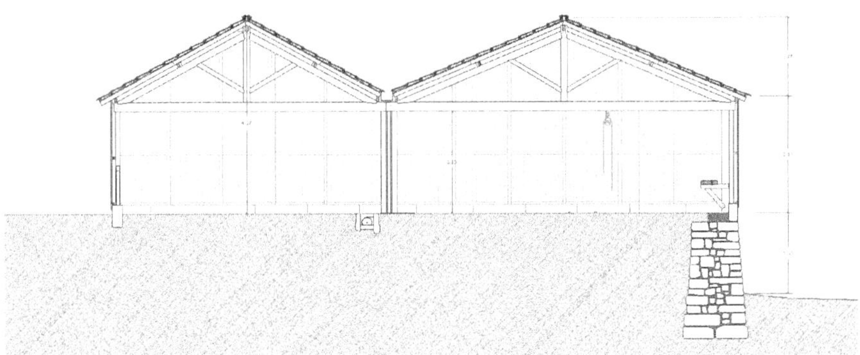

Fig. 8. Constructive section for the sawmill: O Charango Shipyard. (P. Dorna)

Conclusion

The traditional shipyards are the result of the collective experience of a trade, that of the wooden boat builders. From the analysis undertaken above it can be concluded that traditional shipyards are direct expressions of the confluence of naval and architectural technology and constitute a unique example of ethnographic heritage which is at high risk of disappearing.

Acknowledgments

Part of this research has been possible thanks to Dorna project. DORNA promotes the conservation and recovery of the traditional nautical heritage of the European Atlantic coastal regions.

References

[1] J. Díaz de Rábago, *La industria Pesquera en Galicia*. Fundación Barrie, 1884, p18.
[2] S. Mörling, *Lanchas e Dornas: a estabilidade cultural e a morfoloxía das embarcacions na costa occidental de Galicia*. Xunta de Galicia, 2005, p.227.
[3] J.M. Aguado, *La carpintería de ribera en Galicia (1940-2000)*. Universidad de Coruña, 2000, p.24.
[4] P. Taboada and O. Fuertes, *Dorna Astilleros*. COAG. 2010, pp.76-77.
[5] M.Verney, *Complete amateur boat building*. Tutor S.A., 1995, p.93.

Some instances of arch and vaulting construction at the New York Public Library

José Antonio García-Ares and Fabio Tellia
Foster + Partners, UK

Introduction

On May 24th, 1911, the main branch of the New York Public Library opened its doors to the public. Its architects John Merven Carrère and Thomas Hastings had both studied architecture in Paris at the *École Nationale Supérieure des Beaux-Arts* and met later in New York when working for McKim Mead & White. From there they moved on to set up their own practice and become well established architects. In 1897, with a bold design in their own fresh *Beaux-Arts* style, they won the competition to design the New York Public Library building beating their former employers. This was to become their masterpiece and, as their best works, coupled monumentality with practicality.

The study of stereotomy in the 19th century was deeply influenced by the invention of descriptive geometry by Gapard Monge (1799) and was a discipline of great importance at the École des Beaux-Arts. It not only provided a powerful tool to solve practical construction questions but as well as an educational one to develop spatial analytical abilities. It suited very well the composition method of design taught at the *École*. [1] The New York Public Library project provided an opportunity like no other to put all those skills into practice to serve the creation of spaces of great monumentality. For us it offers, as well, a good example to analyze the transfer of knowledge and skills in a time in which it was common for American young aspiring architects to travel to Paris to receive an Academic education. This flow of knowledge America-Europe continued after their return through travels, books and as well by means of new generations of architects working at Carrère and Hastings office which were encouraged to travel to Paris to study at the *École*.

Sadly, Carrère and Hastings had not started to receive, until recently, the attention their architectural achievements deserve, and there are yet only few publications devoted to their architectural production. The most relevant of those being the two volumes with the firms complete works by Hewitt et al. (2006), the monograph on the New York Public Library flagship building by Reed and Morrone ([1986] 2011) and the beautifully illustrated volume on Carrere and Hastings masterworks by Ossman and Ewing (2011). In any case, given the scope and aim of these studies, they cannot be concerned with specific construction aspects or details. The present article, by analyzing some representative examples of arch and vaulting construction in the New York Public Library, aims to contribute to deepen the understanding of these spaces, their design and the way in which they were built. [2]

The 42nd Street Staircase; a Palladian stair

The staircase that serves visitors accessing through the 42nd street entrance on the North façade presents a very particular configuration. It consists of pairs of parallel vaulted flights rising in opposite directions (East-West) and meeting in the centre on a shared landing between the floors. This allows the user to switch direction if required depending on the intended destination.

Reed ([1986] 2011, 235) thinks they are inspired by Pierre Lescot's stairs in the wing he built for Henry II at the Louvre and this idea is supported by Lemos (Hewitt et al. 2006, 321). It may definitely have been some influence from Lescot's stairs but we think, however, that the model for this staircase is to be found in Palladio's famous *Quattro Libri dell'Architettura*. At the end of the *Libro Primo* he includes a type of staircase with the very same arrangement in plan and section as this by Carrère and Hastings (Palladio, 1570, I: 66). He explains that "those [stairs] which are in the church of Sancto Apostol [Rome], towards Monte Cavallo, are very fine; they were double and many have since taken Models thereof: they led to a Temple seated on the top of the Mountain" (Palladio, 1570, I:64,66) later on in the *Libro Quarto* when describing this temple adds more information; "There were very commodious stairs, that went up to the Temple(…)which, going one over another, led to the courts on the sides of the Temple" (Palladio 1570, IV:41-42). If we look at the plate he provides to illustrate this type of stairs (Fig. 1) we see that the flights are supported by sloping vaults —as in the Library— and groin vaults occur at the landings which as well are connected by arched voids.

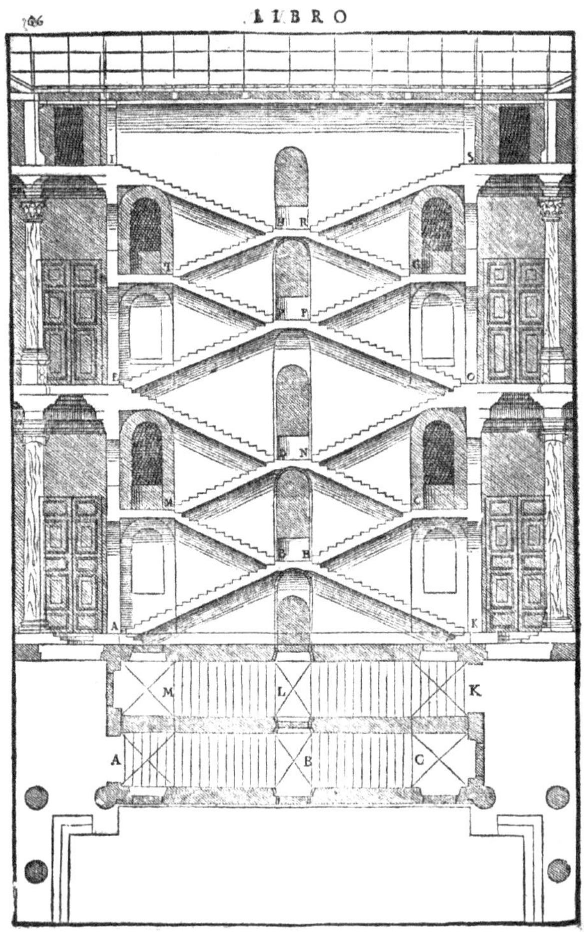

Fig. 1. Stairs from Palladio's Quattro Libri dell'Architettura ([1571] 1581, 66)

At the library, while the landings between the first and second floors are topped with sail vaults and connected by an opening with a semicircular arch, the landings between ground and first floors have straight barrel vaults and are connected through an opening with a voussoired lintel below the impost line. [3]

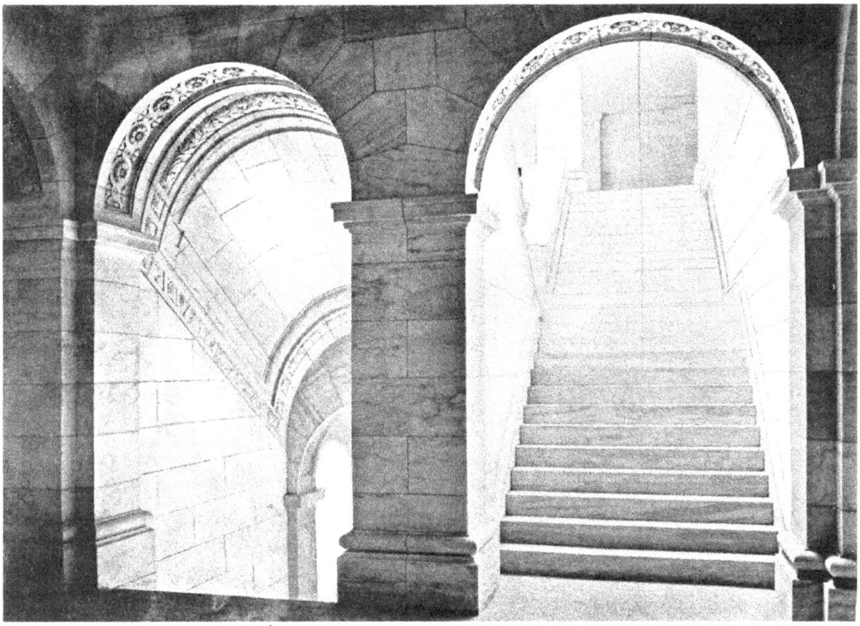

Fig. 2. Historic photo of the 42nd Street stairs (David 1910)

Geometric and constructive configuration of the vaults

The sloping vaults above the flights consist of an inclined cylindrical surface whose normal cross section is semicircular. This surface when cut by a vertical plane generates an ellipse as can be seen when the vault meets the arches that frame the landings. These arches are semicircular with the keystones lower than the crown of the sloping vaults to avoid the elliptical profile to manifest in the archway. These sloping vaults were a typical study case in the stone cutting treatises. In this specific case, the vault's circular cross section simplifies the cutting of most of the voussoirs but the stereotomy of the ones at both ends is still complex and laborious. Carrère and Hastings drawing shows the marble vaults being 6 in. (15 cm) thick with brickwork on top and a photo taken during the construction shows the extrados of a brick sloping barrel vault above one of the stair's flights. The two sail vaults at the upper landing are spherical surfaces. The particularity here is that the coursing is not concentric as usually happens. Each of the seven courses is a vertical arch spanning in the direction of the flights and the faces of the voussoirs in contact with the walls or adjacent courses are vertical.

The Spherical Niches on the Fifth Avenue Façade and Portico

Semi-dome niches of 8ft 8 in. (2.64 m.) diameter can be seen to either side of the Fifth Avenue portico backing MacMonnies' statues of Beauty and Truth. Their stereotomy is characterized by round horizontal courses. For this type of coursing, the method to obtain the templates that are needed to cut the vousssoirs out of stone blocks assumes each of the spherical horizontal courses as being a conic surface and, since cones are developable, the intrados of the voussoirs can be drawn in two dimensions. Although this method is the most commonly used there are instances in which a different approach may be more appropriate to develop the spherical surfaces of the intrados.

One of those cases happens at the coffered niches in the end piers of the portico. Here the diagrid pattern that generates the coffering defines the geometry of the voussoirs. The resulting shape of the typical voussoir is a stone block conformed as a diamond whose diagonals are orthogonal and aligned with the horizontal and vertical directions. The method, we think, has been used here approaches the developing of the spherical intrados by diving the surface according to meridians. Subsequently it considers those lunes as being cylindrical so can be developed and decoration drawn in two dimensions.

This coffering pattern motive was used already in Ancient Roman architecture, as is possible to observe in the apse of the Temple of Venus in Rome. Carrere and Hasting may have been influenced by Philibert de l'Orme's decoration for the dome of Anet Chapel or most probably the niche in the vestibule of Peruzzi's Pietro Massimi palace in Rome as drawn by Letaroully ([1840] 1920, 3: pl. 285, 286) which is very similar. [4]

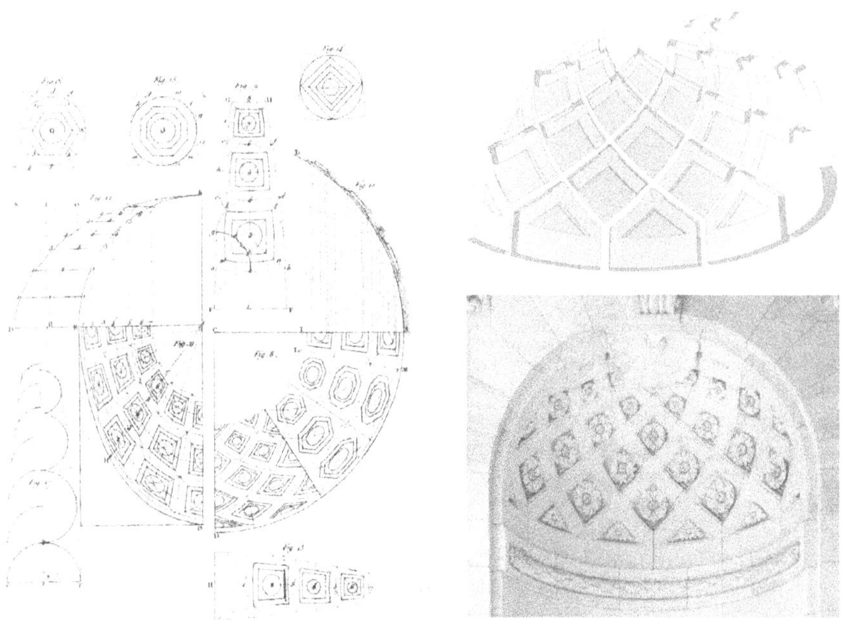

Fig. 3. Development of a spherical surface as shown by Rondelet ([1812] 1828, pl 50) Axonometric diagram and photo of semi dome niches in the 5th Avenue Portico (right)

Less ambitious but precisely executed as well are the hemispherical niches in the interior of the building, which embedded in the thick stone walls are placed above some of the doorways to frame marble busts.

Astor Hall vault

There are many precedents of oval barrel vaults which could have inspired the design of Astor Hall and its vaulted ceiling. We are going to mention two that believe may have had a more direct influence on the architects. One of them is a built example, *La Salle des Caryatides* located on the ground floor of Pierre Lescot's 16th century Renaissance wing at the Louvre. As in the library, it is vaulted with an oval barrel with lunettes on both sides and the articulation of the walls with the piers framed by columns is very similar to what we can see in Astor Hall. The other is again from an engraving by Letaroully ([1840] 1920, 3: pl. 326) depicting a "house once inhabited by Michelangelo" in Rome. Comparing this with Astor Hall the similarities are striking. The lay-out of the vault's decoration, the two arches on the end side and even the stairs resemble to a great extent those of Astor Hall.

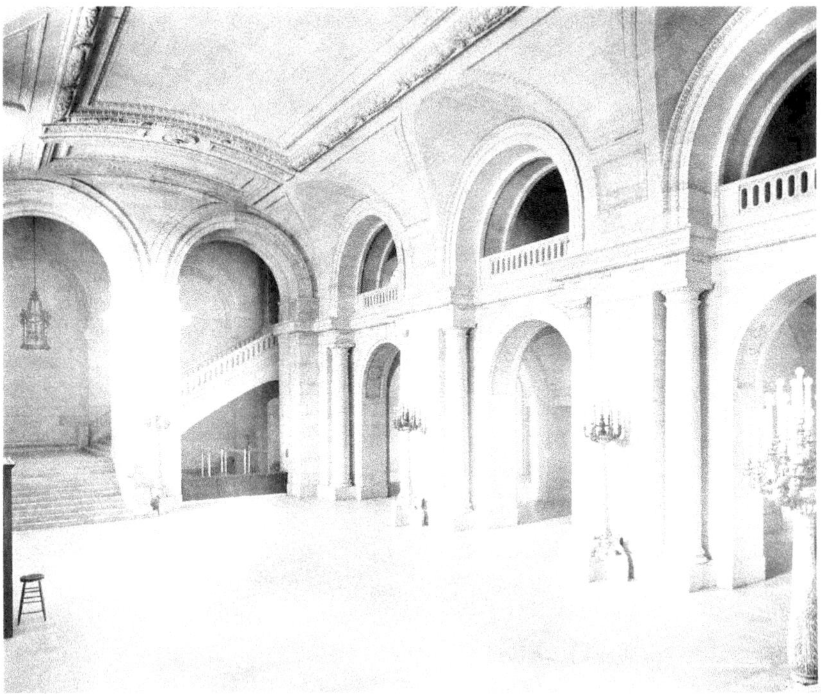

Fig. 4. Main Entrance Hall, the New York Public Library, 1910. Library of the Congress LC-DIG-det-4a24343 (digital file from original) LC-D4-72708 (b&w glass neg.)

Geometric configuration and some stereotomic questions

The main vault runs in the N-S direction with a length of 76ft 8in. (23.37 m.). Its profile is a stilted half oval with the larger arch having a radius of 38ft 7 ½ in. (11.77m.), the smaller ones of 10ft 2in. (3.1 m.) and spanning 41ft 1in. (12.52 m.). These give a 1/3 rise/span ratio and in plan each of its three bays is a rectangle of 41ft 1in. (12.52 m.) by 25ft 7in. (7.8 m.) which is very close to the golden ratio. The imposts are raised 1ft 3 43/64in. (39.8 cm) above the top of the cornice. The vaults over the staircases on both sides have a semicircular profile with a radius of 9 ft 6 in. (2.89 m.) and run in the E-W direction with a length of 41ft 1in. (12.52 m.). [5] Both the main vaults and the side ones have lunettes with curved ridges. [6]

The lunettes of the main vault present a very particular geometry. The front elevation of the groin curve resulting of the intersection of the lunette with the main barrel vault as shown in Carrère and Hastings drawings (Fig. 5, a) does not match the curve that result of intersections with cones, spheres, ellipsoids or cylinders, and it is neither in plan nor in elevation a recognizable bidimensional curve. Apart from the plan, the front and the side elevations this curve appears represented in a third drawing by Carrère and Hastings. It is a "Developed Plan of Ceiling" that shows the setting-out of the joints and the decoration in detail (Fig.5 c). Here this curve can be matched by using circular arcs as an oval with several centres. We could then make the hypothesis that this intersection curve was first drawn unrolled in two dimensions. We have to consider, as well, that in the entire design process physical models could have played an important role; with sheets of paper or other flexible material wrapped around a plaster cylinder to quickly test design options. It is plausible that once they were satisfied with the curve obtained it could have been drawn unrolled as a composite curve made up of tangent circular arcs. Then by means of descriptive geometry the drawings for the front and side projections could have been produced and thus the groin curve defined in three dimensions.

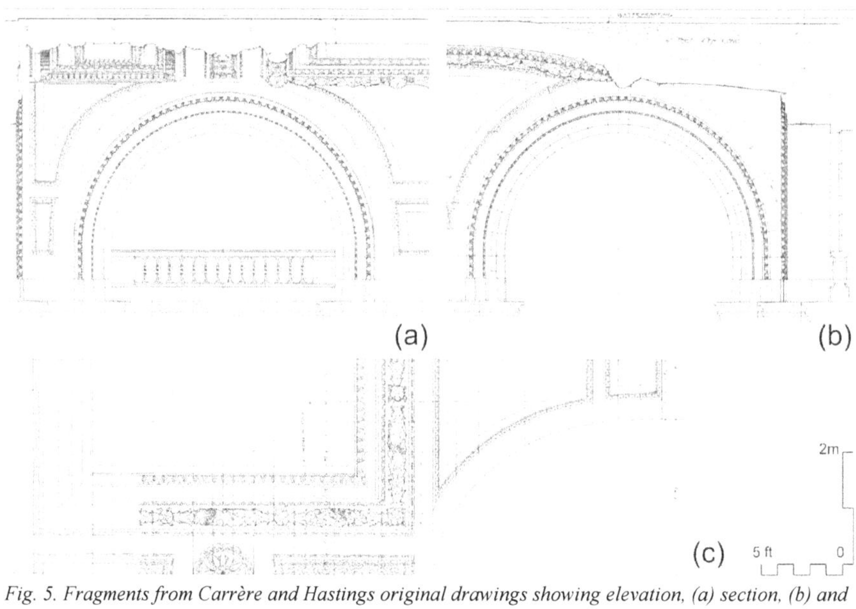

Fig. 5. Fragments from Carrère and Hastings original drawings showing elevation, (a) section, (b) and developed intrados (c). Images courtesy of the New York public Library.

Looking at the lunette front and side elevation drawings (Fig.5 a, b) it can be observed that the staggered joints resulting of the intersection of the lunette's intrados by vertical planes parallel to the longitudinal axis of the main barrel vault are circular arcs. Therefore the surface of the lunettes intrados would have been generated by a series of frontal arcs with the springs on the groin previously obtained and the crown on the ridge line. Based on the architects' side elevation (Fig. 5 b) the ridge has been considered as an arc with horizontal tangent at the point it meets the barrel vault. The first of the generating frontal arcs is the semicircular arch at the edge of the vault and from it the radiuses grow as the lunette advances into the space. [7] (Fig.6)

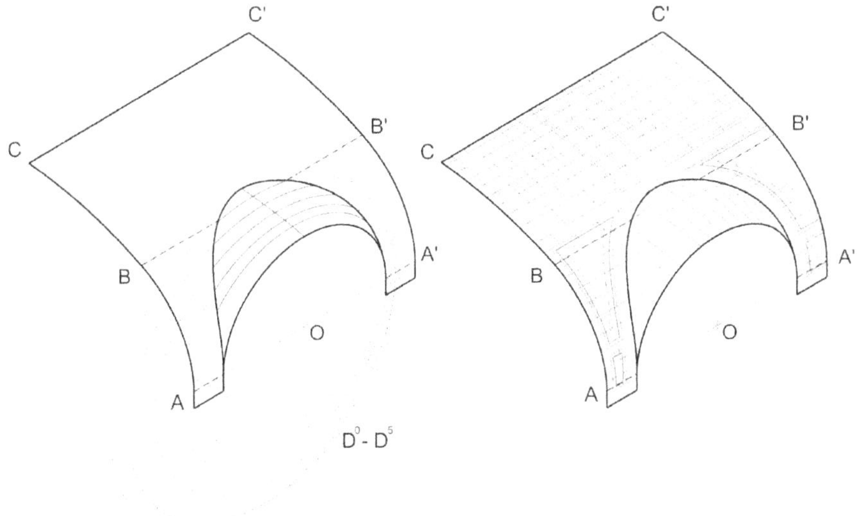

Fig. 6. Geometrical configuration of lunettes (left) with voussoirs's joints (right). AA' impost line. BB' tangency line where vault's curvature changes. CC' vault's crown. D^0-D^5 circular arcs generating lunette's geometry.

The setting-out of the courses both in the lunette and the barrel vault seeks usually to generate courses of the same width to produce regularity throughout the intrados. [8] When the geometries of both surfaces are unlike the direction of those joints rarely meet the groin at the same point. Nevertheless the canonical solution *–penetration filée–* (Pérouse de Montclos 1982, 112) asks for the joints to meet at the intersection groin and this need will determine the stereotomy of the voussoirs at the intersection. The typical procedure to achieve this convergence is letting the radial planes that generate the lunette coursing continue beyond the groin until meeting the joints of the courses in the main vault. [9] In the lunettes of Astor Hall the procedure followed is the opposite and are the radial planes that generate the barrel vault coursing the ones that continue beyond the groin until meeting the joints of the lunette. This is quite rare and they could have seen it in Monduit's stone cutting treatise published in 1889. In any case the cutting of the stone blocks at this intersection between the lunette and the vault is certainly complex and it is not strange that plaster models such as the one shown in Fig. 7 were produced to clarify it to the stone masons (Hewitt et al. 2006, 296).

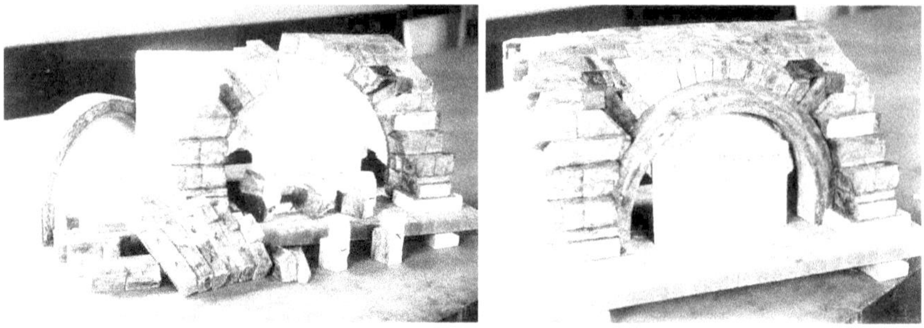

Fig. 7. Plaster Models showing coursing of lunettes and barrel vault. December 1904. Images Courtesy of the New York Public Library. Record IDs: 311065 and 311069

The decoration

A reason that would explain the importance of the drawing showing the developed plan of the vaulted ceiling (Fig. 5 c) is the presence of decoration on the vault´s extrados. The best way to ensure alignment and coordination of the joints with the decorative motives is to draw the decoration flat beforehand on this developed plan where all the surfaces of the voussoirs' intrados are shown in true magnitude. This is how the setting out of the courses can be designed and adjusted if needed to suit the decorative program and vice versa. Other tool extensively used to test and communicate design ideas, anticipate potential problems and find the appropriate solutions were the plaster models. All the significant ornamentation details of the library were mocked-up at full scale before its installation in the building. There is a photo showing a scaled plaster model of two bays of Astor Hall´s vault including all the decoration and the stone joints.

From the photos of the construction is possible to observe that the stone blocks at the moment of being installed were shaped only with the gross profile of the decoration. The carver, once the voussoirs were set in place, had to imprint with the help of flexible templates outlines that could provide guidance in the subsequent process of carving the bass-relieves.

Some structural and construction considerations

After having analyzed its formal and geometrical configuration it is necessary, we think, to devote some time to study its structural behavior. Our aim is to clarify some inaccurate ideas present in existing publications. It is surprising to reed in Reed ([1986] 2011, 83) that "While this stereotomic vault is designed to be self-sustaining, it is, in truth, affixed to the floor above by a layer of concrete. The architects wanted to avoid spalling…which might have occurred had the stones alone supported the pressure". Lemos in Hewitt et al. (2006) following Reed´s idea writes "The barrel-vaulted ceiling is constructed of marble voussoirs and designed to be self supporting, though as a precaution a layer of concrete supports the ceiling from above".

There are many evidences that confirm this is a true vault not only designed but as well working as such. First it is the actual thickness of the vault; the thinnest voussoirs at the crown are 1ft 6 ½ in. (46 cm) thick and they grow up to about 2 ft 1½ in. (64.8 cm) at the springs. If it was to be affixed to whatever structure above it, the reasonable would have been to reduce its thickness in order to limit its weight. Then, "affixing" the different voussoirs would have required some sort of ties or other elements to mechanically attach the stones to the supporting structure. We have not found evidence of the use of these elements either in construction photos or any of Carrère and Hastings drawings. These drawings show only 3 in. (7.2 cm) between the extrados and the floor level, just the necessary to install the marble flooring and its bedding. The supposedly "layer of concrete" required to support itself and the considerable weight of the marble vault spanning 41ft 1in. (12.52 m.) would, not only have required to be reinforced but as well to have considerable downstand beams to viably deliver the necessary structural capacity. Here there is just no room for any of these. One characteristic of the vault is the very high quality of the construction –this is in fact a constant in all the marble construction in the library– with very thin joints. As Huerta (2004) indicates, since tests carried out in the late 19th century by Tourtay it is proved that thick mortar joints reduce the resistance of masonry and Delbecq (1983) has explained why thin mortar joints are a good measure to minimize the risk of spalling. Other factor that could cause concentration of tension and the subsequent risk of chipping and fracture of the stone blocks is the deformation in the geometry of the vault. It can be caused by insufficient abutment or soil giving way under action of the vault's thrust. In this case there is no wanting in abutment; on the East provided by the mass of the Fifth Avenue portico and on the West by the doubled arched gallery. On both sides solid brickwork walls rise more than 20 ft (6 m.) above the vault's crown. Furthermore the soil underneath was well compacted as the Croton Distributing Reservoir stood on the site for more than 50 years prior to the construction of the library and rock is quite superficial. Lastly, deterioration of the mortar such as disaggregation could lead as well to localized concentrations of pressure at the joints. Here Carrère and Hastings took the precaution of protecting the extrados of the vault against any water infiltration that could affect the mortar with a waterproof membrane as seen in their drawings.

There is a source from 1916 –just five years after the library opened to the public– which refers to the structural nature of the main vault and those supporting the stairs. This is the official guide of the New York Public Library published by the library itself where we can read: "On a first visit, however, the sightseer should use the main entrance on Fifth Avenue, in order to see the lobby, which rises through two stories, with broad staircases to the right and left. The flying arches of these staircases are of seventeen feet span, and are all of marble without any brick or metal work whatever. The marble used in the lobby is from Vermont. The ceiling is a true marble vault of forty feet span, supporting itself and the floor over it, with no metal whatever, except some reinforcing rods buried in the concrete filling in the floor above." Photos taken during the construction and the original drawings by Carrère and Hastings confirm both that the side staircases are built entirely in marble and the presence of "some reinforcing rods". By the drawings and the construction photos these rods seem to be located diagonally only at the two corners formed by the façade and the two end walls. They penetrate into the portico almost to its outer face next to the capitals. There may be intended to help to withstand the thrust of the lateral vaults above the stairs –which run in the opposite direction– since the abutment on those end flanks is limited. In any case being located that high it is arguable they can efficiently perform this action. Note that the extrados of these lateral vaults is 8 in. (20 cm) lower that the main vault allowing for these rods to run between the vault and the paving of the floor above. [10]

In order to assess the structural behaviour of this vault we carried out an initial [10] stability analysis within the theoretical framework of the lower bound limit analysis of masonry structures as developed by Heyman (1995). [12] To calculate the loads a specific weight of 169.2 lb/ft^3 (27.1 kN/m^3) has been considered for the Danby marble vossoirs, 150 lb/ft^3 (24.0 kN/m^3) for the Portland cement concrete filing, 120 lb/ft^3 (19.2 kN/m^3) for the brickwork and a uniformly distributed load of 100 lb/ft^2 (4.9 kN/m^2). First, looking at the vault in isolation, it is possible to draw a line of thrust within a hypothetical arch of 1/7 the thickness of the actual vault. This indicates that the vault has a geometrical safety factor of, at least, 7 or in other words that the arch is, at least, 7 times thicker than the thinnest arch required to support the loads considered. (Heyman 1969, Huerta, 1992). This is assuming the worst case scenario of the loads being carried only by the marble vault. In our case the valleys of the vault are filled with unreinforced Portland cement concrete up to the level of the crown. This adds to its stability as allows pathways for the transmission of thrust outside the boundaries of the vault's masonry making the vault immensely resistant.

This high geometrical safety factor is nothing unusual, given the geometry of Astor Hall vault, since low-rise vaults are very stable and can be built with thin voussoirs (Heyman 1977). On the other hand low-rise vaults and arches, as well due to its geometry, produce substantial outward thrust requiring important abutment. Here is where we should look at the vault and adjacent structures as a system in equilibrium. Figure 8 shows half of the cross section of the vault and the arched galleries to the West being E the Gottesman Hall wall. Lines of thrust haven been drawn. We see that the horizontal component of the thrust produced by the vault is considerable and the pier at B, if on its own, would have given way with the subsequent collapse of the vault. Nevertheless the architects built the double arched gallery to configure a buttressing system that could provide convenient abutment. This function is, in this way, shared by the stabilizing weights of the walls at B, D, and E connected by arches for the three to work as a system. On the portico side there is the same need for abutment. Here, the stabilizing weight is provided by the substantial piers of solid masonry –pierced by arched passageways to visually lighten them– which, in order to provide the necessary weight, need to rise behind the portico's pediment and can be appreciated in the Solomon Room on the third floor. The solidity of the portico's architectural composition is therefore determined as well by its structural function. An open portico solution, without the piers, would have been unable to provide the required abutment.

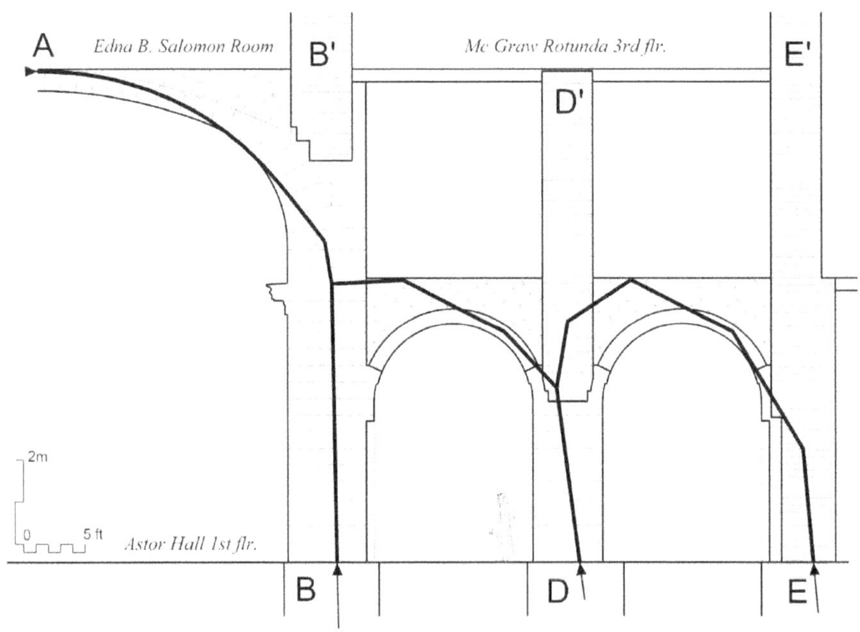

Fig. 8. Astor Hall E-W cross section with lines of thrust (for the vault the minimum thrust line). Walls at B and E have been considered 23 ft and 50 ft high above 3rd floor level.

As we have seen it would have been possible to build this vault much thinner, as a false hung vault or even as a tile vault as the ones built by Guastavino at the main entrance hall of McKim, Mead and White's Boston Public Library. Carrère and Hastings intent to create a truly monumental edifice not only on the appearance or effect but as well in its internal constitution made them to adopt a solution close in the materials and in the way of working to the ancient classical monuments they so much admired, a solution that could last for centuries and welcome the visitors to one of New York's most loved institutions. This monumental vault in Astor Hall remains as a testimony of their ambitious and determination. [13]

Conclusions and future research

Apart from the three main examples of vaults and arches we have mentioned there are many other that would deserve attention. [14] All of them show the great attention to detail and exquisite workmanship present in the building. This display of technical vaulting construction and the involved structural and stereotomic expertise cannot be understood, and would have not been possible, without Carrère and Hastings education in France. Even the excessive geometrical complication of some stereotomic solutions such as the lunettes of Astor Hall's vault are a typical characteristic of the XIX century stereotomy they learned in Paris. Further studies of the specific stereotomic methods in light of the stone cutting treatises they could have known are pending. Gaining more information on the actual building is required to confirm the geometrical hypothesis exposed and know better its material configuration. Additional studies would be required to analyze the abundant information that exists on the construction and design stages. Documents such as construction photographs, original contracts, specifications and project correspondence could all provide valuable data regarding the contractors, the architects and the way in which the New York Public Library Building was designed and built.

Our aim trying to trace the origin of some of Carrère and Hastings aesthetic and technical references has been to bring insights to better value and enjoy the achievements of their architecture. At the same time we would like to stress the importance of the studies on the field of Construction History to deepen our understanding of the built environment.

Acknowledgments

The New York Public Library kindly granted permission to reproduce the original documents included in the figures and its Digital Gallery has been an important resource during the research stage. We would like to give thanks as well to professor Enrique Rabasa for his valuable comments, to Jose Manuel Berg Aldoney for his assistance in developing the illustrations and to professors Santiago Huerta and Jacques Hayman for their comments on Figure 8.

References

[1] See Schlichting (2001, 130-139) on the French Beaux Art education of American architects.

[2] The present study is part of an ongoing research on historic construction aspects of the New York Public Library. The intention here is not deal with the specific stereotomy procedures which would require a much longer discourse.

[3] There are as well drawings by Leonardo Da Vinci (Manuscrit B (Ms 2173), 68 v) and Serlio (1540, 87) showing a similar stair although in these cases without intermediate landings.

[4] Hitchcock (1944) explains the important influence that Letarouilly's drawings had on 19th century American architects.

[5] All the dimensions are for the intrados.

[6] Some initial photogrammetric studies carried our during this research indicate a great level of concurrence between the geometry of the build vault and Carrere and Hastings drawings. A more detailed survey, however, would be required to be conclusive.

[7] Centering shown in one of the construction photos takes the form of these frontal arcs. The geometrical procedure could resemble Frezier's ([1737-1739]1754,2:,Pl 51] version of the *arrière voussure de Marseille* where the frontal sections are circular. Although Frezier's has a straight ridge later versions of this arrière voussure are shown with a curved ridge as in Astor Hall. (Coulon 1885)

[8] The coursing of the lunette is the result of cutting it by a sheaf of planes whose horizontal axis passes through the centre of the semicircular arch at the edge of the lunette.

[9] In fact, in this typical way is how the joints are arranged in a photo of a two bay scaled plaster model of the vault. New York Puclic Library Visual Materials. Record ID: 310125

[10] 10. Regarding these rods Carrère and Hastings drawings refer to drawing "813". This drawing may bring light to the actual purpose of the rods. However, we have not yet been able to see it or confirm it exists.

[11] At this stage it has been studied the typical oval profile of the vault without considering the lunettes.

[12] Heyman (1995), in order to be able to apply the basic theorems of limit analysis to the masonry structures characterizes the material with three basic assumptions: Masonry has unlimited compressive strength, no tensile strength, and sliding failure does not occur. Consequently, the lower bound [safe] theorem states that for a specific structure and load distribution, as long as it is possible to find an equilibrium state in compression, however unlikely that state seems to be, the structure is stable and the collapse cannot occur for the given loading (Heyman 1995). Graphically this equals to the possibility of drawing a line of thrust within the masonry. All the above implies that the analysis of masonry structures is not mainly a question of strength but of stability.

[13] There was even an original scheme with marble ceilings throughout the public spaces. Lemos in (Hewitt et al. 2006).

[14] Other example that deserves mention are the descending arches above the staircases in the McGraw Rotunda where the coffered ruled surface of the intrados transitions in-between two arches of the same radius, one of them stilted. Interesting coving jambs and arches can be seen in the external segmental arches of McGraw Rotunda´s upper windows and in doorways of the first and second floors.

Bibliography

1. James W. P. Campbell and Will Pryce, *The Library: a World History*, 2013.
2. A. G. Coulon, *Nouveau vignole des menuisiers: ouvrage théorique et pratique, utile aux ouvriers, maîtres et entrepreneurs...* Paris: Dalmont [u.a.], 1855.
3. Leonardo Da Vinci, "Manuscrit B, Ms2173." Bibliothèque de l'Institut de France.
4. A. C. David, "The New York Public Library." *The Architectural Record* 28 (3), 1910, pp.144–72.
5. Jean-Michel Delbecq, "Analyse de La Stabilité Des Ponts En Maçonnerie Par La Théorie Du Calcul à La Rupture". Ecole Nationale des Ponts et Chaussées, 1983.
6. Amédée François Frézier, *La théorie et la pratique de la coupe des pierres et des bois... Traité de stéréotomie ...* 3 vols. A Paris : Chez Charles-Antoine Jombert, imprimeur du roy en son artillerie, à l'image Notre-Dame, 1754.
7. Mark Alan Hewitt, Kate Lemos, William Morrison, and Charles D. Warren. *Carrère & Hastings Architects*. New York: Acanthus Press, 2006.
8. Jacques Heyman, *The Stone Skeleton: Structural Engineering of Masonry Architecture*. Cambridge; New York: Cambridge University Press, 1995.

9. Henry-Russell Hitchcock, "Frank Lloyd Wright and the 'Academic Tradition' of the Early Eighteen-Nineties." *Journal of the Warburg and Courtauld Institutes* 7, 1944, p.46.
10. Santiago Huerta Fernández, *Arcos, Bóvedas y Cúpulas : Geometría y Equilibrio En El Cálculo Tradicional de Estructuras de Fábrica*. Madrid: Instituto Juan de Herrera, 2004.
11. Paul Marie Letarouilly, *Édifices de Rome moderne; ou, Recueil des palais, maisons, églises, couvents et autres monuments publics et particuliers les plus remarquables de la ville de Rome*. 3 vols. Washington, The Reprint Co., 1920.
12. Louis Monduit, Alexandre Denis, and Charles Juliot. *Traité théorique et pratique de la stéréotomie au point de vue de la coupe des pierres*. Paris: Juliot, 1889.
13. "NYPL Digital Gallery." <http://digitalgallery.nypl.org/>
14. Laurie Ossman, Heather P Ewing, and Steven Brooke. *Carrère & Hastings: The Masterworks*. New York: Rizzoli, 2011.
15. José Carlos Palacios, *Trazas y cortes de cantería en el renacimiento español*. Madrid: Ediciones Munillaleria, 2003.
16. Andrea Palladio, *I qvattro libri dell'architettvra di Andrea Palladio : ne' quali, dopo un breue trattato de' cinque ordini, & di quelli auertimenti, che sono piu necessarij nel fabricare;* In Venetia : Appresso Dominico de' Franceschi, 1570.
17. Jean-Marie Pérouse de Montclos, *L'Architecture à la française: XVIe, XVIIe, XVIIIe siècles*. Paris: Picard, 1982.
18. Enrique Rabasa Díaz, *Forma y construcción en piedra: de la cantería medieval a la estereotomía del siglo XIX*. Madrid: Akal Ed, 2000.
19. Henry Hope Reed, and Francis Morrone. *The New York Public Library: The Architecture and Decoration of the Stephen A. Schwarzman Building*. New York; London: W.W. Norton & Co., 2011.
20. Jean Baptiste Rondelet, *Traité Théorique et Pratique de L'art de Bâtir*. Traité Théorique et Pratique de L'art de Bâtir. Paris, 1828.
21. Kurt C. Schlichting, *Grand Central Terminal: Railroads, Engineering, and Architecture in New York City*. Baltimore, Md: Johns Hopkins University Press, 2001.
22. Sebastiano Serlio, *Il terzo libro di Sabastiano Serlio Bolognese...* Venice: Francesco Marcolini da Forli (Italian, Forli ca. 1500–after 1559 Venice), 1540.

Two Egyptian curves revisited

John Gelder
RIBA Enterprises, UK

Summary

Construction drawings of two vault curves survive from ancient Egypt. One has been described in the literature as an arc of a circle, with most authorities agreeing on a 5 cubit radius, and the other as an arc of an ellipse constructed using a pair of 3:4:5 right-angled triangles. Again, this has been accepted by most authorities. The contention here is that both curves actually describe arcs of a circle of 6 cubit radius. This would mean that there is no hard evidence that the ancient Egyptians understood the construction of true ellipses.

Introduction

This article revisits the derivation of two ancient Egyptian sketches for vault curves. Documentation of curves in architecture has always been problematic (until the advent of CAD), both for designers and for builders. For example, in the 1980s the author was responsible for documenting the set-out of a façade in London (1 St Pauls Churchyard) based on a pair of large tangentially intersecting arcs. They were designed first as arc segments based on their centre points, radii and angles. The two arcs were then converted into straight line segments of equal length, and the endpoints of each segment were documented as x and y offsets from the planning grid. The contractor set-out the segments as vectors.

This process used three different methods of setting out, one of which was the use of offsets. Use of offsets was understood in ancient Egypt, as the two drawings of vaults discussed here indicate. For the Saqqara vault curve, the offset dimensions are given – they could not after all be measured from the small not-to-scale ostracon sketch. For the Rameses VI vault curve, the offset dimensions are not given – instead they can be taken directly from the full-size 1:1 in-situ sketch.

In spite of these differences, and the 1550 years or so between them, it is suggested that both curves were derived in the same manner, contrary to previous analyses. The principle followed here is to find the simplest curves that fit all the points in the two originals. Circles, then, are preferred to ellipses. Circles that fit all the points are preferred to those that don't.

Saqqara vault curve on ostracon

The first sketch curve examined was found in 1925 at Djoser's Step Pyramid (3rd Dynasty, c. 2650 BC) at Saqqara, making it perhaps the oldest extant Egyptian working drawing (Fig. 1).[1] It is now at the Cairo Museum (CM 50036). Gunn suggested that the curve describes the extant vault of a solid 'saddle-back' construction erected near where the ostracon was found.[2]

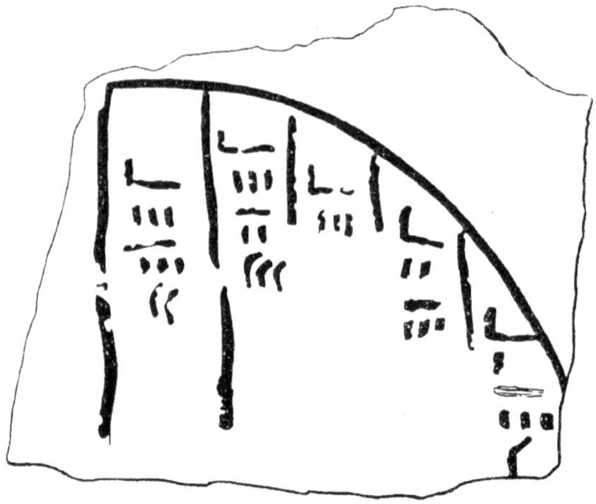

Fig.1. Sketch of ostracon (from Gunn, 1926, Fig. 1)

It is on a limestone flake or ostracon of 150 x 175 mm, and shows a curve and five offsets dimensioned in royal cubits, palms and digits, where 1 cubit = 7 palms, and 1 palm = 4 digits (shorter cubits of 6 or 5 palms do not generate a smooth curve, and besides, they were not used in construction). The baseline isn't shown. In interpreting the sketch, it is assumed that the offsets are 1 cubit apart, and that the sixth (right-hand) offset is zero (Tab.1).

Tab.1. Ostracon offsets

Cubits	Palms	Digits	Total digits	
3	3	2	3 x 28 + 3 x 4 + 2 =	**98**
3	2	3	3 x 28 + 2 x 4 + 3 =	**95**
3	0	0	3 x 28 =	**84**
2	3	0	2 x 28 + 3 x 4 =	**68**
1	3	1	1 x 28 + 3 x 4 + 1 =	**41**
0	0	0	assumed	**0**

Heisel suggested that a full-size plot of the curve was used to determine the coordinates recorded here – perhaps that document was on papyrus[3] and archived (unlikely given the size of the arc and the available sizes of papyrus), or on sand and lost – and that this is the on-site version.[4] Rossi agrees with the latter explanation.[5] But how was the curve plotted?

Gunn suggested that this curve is not part of an ellipse, but nor is it part of a circle (otherwise he considered that an offset sketch such as this would not be needed).[6] Daressy disagreed, suggesting that the curve is actually part of a circle of 5 cubit radius, with centre of origin 1½ cubits below the baseline and aligned with the first offset.[7] But this means that the assumed sixth offset of zero cannot be part of the curve (hence the very large error for this offset in Tab. 2). Rossi accepted this, also noting that 'little mistakes' had been made in converting the original full-scale drawing to the ostracon.[8] In terms of fitting all the points, Daressy's suggested curve fails, at offset 6 in particular.

Tab. 2. Tabular comparison of ostracon and 5 cubit arc

Offset (L to R)	CM 50036 above original baseline (digits) [+]	Daressy's 5 cubit arc [○]		Difference between original and Daressy's 5 cubit arc (digits)
		Above proposed origin (digits)	Above original baseline (digits)	
1	**98**	140	98	0
2	**95**	137.2	95.2	0.2
3	**84**	128.3	86.3	2.3
4	**68**	112	70	2
5	**41**	84	42	1
6	**0**	0	-42	-42

Moyer's online 'Circular solution to Saqqara ostracon' improves on Daressy, by demonstrating that a circle of radius 167 digits (5 cubits 6 palms 3 digits) provides a good fit to all the coordinates given in the sketch (including the assumed sixth offset of zero), but it is not clear where he locates the point of origin.[9] Also, the choice of 167 by the designer would be surprising given a general preference in Egyptian design for whole numbers of cubits and for simple fractions.

Proposals to date for the construction of the curve include the following:

- Gunn (1926): Not an ellipse, not circular either.
- Daressy (1927): 5 cubit (5:0:0) radius.
- Clarke & Engelbach (1930): None suggested; not an ellipse.
- Hoberman (1985): 5 cubit radius, after Daressy.[10]
- Badawy (1986): Elliptical curve.[11]
- Arnold (1991): None suggested.[12]
- Heisel (1993): None suggested.
- Adam (2001): None suggested.[13]
- Lumpkin (2002): None suggested.[14]
- Goyon et al. (2004): 5 cubit radius, after Daressy.[15]
- Rossi (2004): Supports Daressy, i.e. 5 cubit radius.
- Moyer (c. 2005): 167 digit (5:6:3) radius.

Fortunately a close match to all the coordinates can be achieved using a more rational (than Moyer) circular arc of 6 cubits (168 digits) radius, with a rational point of origin – half a cubit (14 digits) to the left of the first offset, and 2½ cubits (70 digits) below the baseline. The dimensions in Tab. 3 are in digits (about 18 mm). The equation $y = \sqrt{(r^2 - x^2)}$ is used to generate the values for the '6 cubit radius', as it is for the '5 cubit radius'. We still have some 'little mistakes' though, which will have arisen when setting out, measuring or transcribing the full-scale source plot. The 6 cubit radius would give the curve on this small sketch a scale of roughly 1:23 (Heisel has 1:20), though the offsets are not to scale.

Tab. 3. Tabular comparison of ostracon and 6 cubit arc

Offset (L to R)	x-axis (digits)		y-axis (digits)				Difference between original and 6 cubit arcs (digits)
			CM 50036 [+]		6 cubit radius [○]		
	To right of proposed origin	To right of first offset	Above proposed origin	Above original baseline	Above proposed origin	Above original baseline	
origin	0	-14	-	-	168.0	98	-
1	14	**0**	168	**98**	167.4	97.4	-0.6
2	42	**28**	165	**95**	162.7	92.7	-2.3
3	70	**56**	154	**84**	152.7	82.7	-1.3
4	98	**84**	138	**68**	136.5	66.5	-1.5
5	126	**112**	111	**41**	111.1	41.1	0.1
6	154	**140**	70	**0**	67.1	-2.9	-2.9
7	168	154	-	-	0.0	-70.0	-

A comparison of the three curves (ostracon, 5 cubit arc, 6 cubit arc – plotted in Fig. 2) shows the suggested 6 cubit arc is a slightly better fit than Daressy's proposed 5 cubit arc for 3 of the offsets, a very much better fit for the 6th, and a slightly worse fit for 2 of the offsets. It is picking up the 6th offset that's the clincher for the 6 cubit arc.

The extant ruined 'saddle-back' was about 11.1.1½ across (5775+75+30 = 5880 mm, where 1 cubit = 525 mm) at ground level, whereas the 6 cubit radius would suggest a designed width of 11 cubits – an error of about 2% overall.[16]

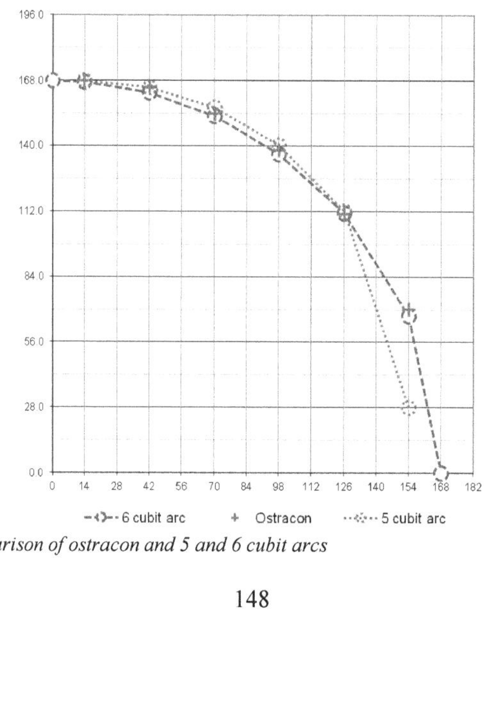

Fig. 2. Plotted comparison of ostracon and 5 and 6 cubit arcs

Rameses VI vault curve in-situ

The second sketch curve examined was first reported in 1907, and is (or was) in-situ, drawn on a level white-washed (or plastered) rock face in black ink, in the entry way to the underground tomb KV9 in the Valley of the Kings, on the left (south) side, high up near the overhang (Fig. 3).[17] This tomb is that of Rameses VI (20th Dynasty, c. 1100 BC). The curve represents the vault of the tomb's burial chamber ceiling.[18]

Fig. 3. Sketch of in-situ curve (after Daressy, 1907, Fig. 1)

This drawing shows an arc and a baseline with regular offset points corresponding to points on the arc. Daressy provided accurate dimensions, enabling others to check his conclusions (Tab. 4). Workers would have measured the offsets from this full-scale (1:1) drawing and used them in excavating the vault in the tomb. This curve is clearly intended as an aid to construction, and is likely to be the original plot – there is enough room on the rock face to generate the curve in-situ using a water level and a plumb bob.[19] However, Daressy suggests it was first drawn on the ground and then measured and transferred to the wall, for the convenience of the tomb workmen. Rossi concurs with this suggestion.

Tab. 4. Averaged coordinates of in-situ curve (after Daressy, 1907)

Offset (L to R)	x-coordinate	y-coordinate
22 (arc intersects baseline)	-3166.5	0
21	-3076.0	174.5
20	?	376
19	-2785.0	545
18	-2641.0	710
17	-2491.0	?
16	-2344.5	920
15	-2196.5	1022.5
14	-2048.5	1121
13	-1900.0	1204.5
12	-1757.0	1272.5
11	-1616.0	1336.5
10	-1469.0	1389
9	-1321.0	1420
8	-1172.5	1456.5
7	-1026.5	1486.5

Offset (L to R)	x-coordinate	y-coordinate
6	-881.0	1512.5
5	-731.0	1543.5
4	-584.5	1561.5
3	-440.5	1575.5
2	-294.5	1581.5
1	-147.0	1585
0 (apex of curve)	0.0	1586
1	146.5	1583.5
2	295.5	1582.5
3	441.0	1573.5
4	587.0	1561
5	735.0	1540
6	880.0	1511
7	1024.5	destroyed
8	1173.5	
9	1320.0	
10	1467.0	
11	1615.5	
12	1758.0	
13	1905.5	
14	2050.5	
15	2199.5	
16	2345.0	
17	2493.0	
18	2639.0	
19	2786.0	
20	2932.5	

Following Daressy no commentators have questioned his suggestion that this curve is an ellipse generated using a pair of 3:4:5 triangles (Tab. 5).[20] This ellipse would have been generated using two fixed foci, 6 cubits apart, and a cord loop of 16 cubits. Proposals to date for the construction of the curve include the following:

- Daressy (1907): Ellipse, based on two 3:4:5 triangles.
- Hoberman (1985): Ellipse, after Daressy.
- Badawy (1986): Ellipse, after Daressy.
- Arnold (1991): Ellipse, after Daressy.[21]
- Heisel (1993): None suggested.[22]
- Rossi (2004): Ellipse, after Daressy.[23]
- Moyer (c. 2005): Ellipse, after Daressy.[24]

Tab. 5. Tabular comparison of in-situ curve and Daressy ellipse

Offset (L to R)	x-coordinate		y-coordinate (offset)		Difference in offsets between original and Daressy ellipse
	Original curve	Daressy ellipse	Original curve [+]	Daressy ellipse [□]	
22	-3166.5	-3170.0	0	0	0
21	-3076.0	-3078.6	174.5	130	-45
20	?	-2932.0	376	320	-56
19	-2785.0	-2785.4	545	485	-60
18	-2641.0	-2638.8	710	625	-85
17	-2491.0	-2492.2	?	755	?
16	-2344.5	-2345.6	920	865	-55
15	-2196.5	-2199.0	1022.5	970	-53
14	-2048.5	-2052.4	1121	1055	-66
13	-1900.0	-1905.8	1204.5	1140	-65
12	-1757.0	-1759.2	1272.5	1210	-63
11	-1616.0	-1612.6	1336.5	1270	-67
10	-1469.0	-1466.0	1389	1330	-59
9	-1321.0	-1319.4	1420	1380	-40
8	-1172.5	-1172.8	1456.5	1420	-37
7	-1026.5	-1026.2	1486.5	1455	-32
6	-881.0	-879.6	1512.5	1490	-23
5	-731.0	-733.0	1543.5	1520	-24
4	-584.5	-586.4	1561.5	1545	-17
3	-440.5	-439.8	1575.5	1560	-16
2	-294.5	-293.2	1581.5	1570	-12
1	-147.0	-146.6	1585	1580	-5
0	0.0	0.0	1586	1583	-3
1	146.5	146.6	1583.5	1580	-4
2	295.5	293.2	1582.5	1570	-13
3	441.0	439.8	1573.5	1560	-14
4	587.0	586.4	1561	1545	-16
5	735.0	733.0	1540	1520	-20
6	880.0	879.6	1511	1490	-21

Apart from its complexity, there are several problems with Daressy's proposal. First, his tables showing how the ellipse matches the drawing actually show many substantial discrepancies (up to 70 mm on his own admission), which he puts down to stretching of the cord used to draw the ellipse (we should give them credit for being able to manage this), the curve transfer process (it was probably drawn in-situ), and the thickness of the line (over 2.5 mm, quite precise at this scale).[25] Rossi agrees with these explanations of the discrepancies.

Second, his 3:4:5 triangle is actually 12/3:16/3:20/3 cubits – this assumes that a royal cubit divided into 6 palms is being used.[26] But the divisions in the drawing's baseline are 2 palms of a royal cubit divided into 7 palms.[27] It is unlikely that the designer will have mixed royal/6 and royal/7 in the same project and, in any case, royal/7 is used elsewhere for this tomb (and, indeed, for all Egyptian royal construction projects as far as the author is aware).

Third, Daressy uses a cubit of 527.7 mm. The original drawing uses a cubit of 513 mm, if we assume that the baseline intervals are of exactly 2 palms each (146.6 mm on average), and provide for the 90 mm remainder on the ends of the baseline. Tab. 6 (row 4) shows that using Daressy's cubit results in no remainder at all.

Finally, the executed vault has a flattened top – refer to the section and photographs in the online Atlas – whereas Daressy's ellipse curves right across.[28]

Tab. 6. Daressy's survey interpreted using cubits of 527.7 and 513 mm

Dimension	Survey	Cubit		Comment
		527.7 mm	513 mm	
Height above baseline at apex of vault curve	1586 mm	3.0.0	3.0.2½	The basis for Daressy's construction
Height above baseline at four divisions to one side of apex	1561 mm	2.6.3	3.0.1	Corresponds to top of 6 cubit arc in author's construction
Baseline divisions (average)	147 mm	0.2.0	0.2.0	Daressy makes no attempt to convert this
Half total baseline length	3167 mm	6.0.0	6.1.1	Each half has 21 (6 x 7/2) equal divisions and a 90 mm remainder

A less complex fit (i.e. the 3:4:5 triangle is not needed, and nor is an ellipse) can be achieved using two simple circular arcs of 6 royal cubit radius (3077.6 mm), one for each end of the vault, with a flattened section in the middle (as in the executed vault). It is also a better fit – the discrepancies are smaller (a maximum of 48 mm, and only 23 mm if we ignore those at the flat section – Tab. 7). A royal cubit of 7 palms is used, of 513 mm, providing 90 mm baseline remainders (Tab. 6, row 4).

The centre point of each arc is located 3 cubits below the original drawing's baseline, and 1 cubit (3.5 baseline divisions) to each side of its vertical centre line. This generates a flattened top 2 cubits across (the flattened part of the executed vault is about twice this width).

Tab. 7. Tabular comparison of in-situ curve and 6 cubit arc

Offset (L to R)	x-coordinate		y-coordinate (offset)			Difference in offsets between original and 6 cubit radius with flat top
			Original curve [+]	6 cubit radius with flat top [○]		
	146.6 mm centres	Original curve		Above origin	Above baseline	
arc	2656	3166.5	0	1556	17	17
21	2566	3076.0	174.5	1701	162	-13
20	2419	?	376	1903	364	-12
19	2272	2785.0	545	2076	537	-8
18	2126	2641.0	710	2226	687	-23
17	1979	2491.0	?	2357	818	?
16	1833	2344.5	920	2473	934	14
15	1686	2196.5	1022.5	2575	1036	14
14	1539	2048.5	1121	2665	1126	5
13	1393	1900.0	1204.5	2745	1206	1
12	1246	1757.0	1272.5	2814	1275	3
11	1100	1616.0	1336.5	2875	1336	-1
10	953	1469.0	1389	2927	1388	-1
9	806	1321.0	1420	2971	1432	12
8	660	1172.5	1456.5	3006	1467	11
7	513	1026.5	1486.5	3035	1496	9
6	367	881.0	1512.5	3056	1517	5
5	220	731.0	1543.5	3070	1531	-12
4	73	584.5	1561.5	3077	1538	-23
3	-73	440.5	1575.5	3077	1538	-38
2	-220	294.5	1581.5	3077	1538	-44
1	-367	147.0	1585	3077	1538	-47
0	-513	0.0	1586	3077	1538	-48
1	-660	146.5	1583.5	3077	1538	-46
2	-806	295.5	1582.5	3077	1538	-45
3	-953	441.0	1573.5	3077	1538	-36
4	-1100	587.0	1561	3077	1538	-23
5	-1246	735.0	1540	3070	1531	-9
6	-1393	880.0	1511	3056	1517	6

A comparison of the three curves (in-situ, ellipse, 6 cubit arc – plotted in Fig. 4) shows the suggested 6 cubit arc is a much better fit than Daressy's proposed ellipse for all of the offsets along the arc. For simplicity, the rows below the heavy line in Tab. 7 are not plotted – that is, the flat section and the beginnings of the second arc are not shown – and the plot is mirrored left to right.

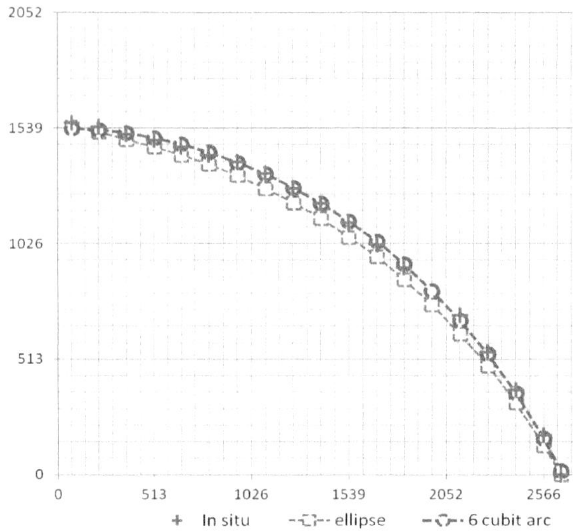

Fig. 4. Plotted comparison of in-situ curve, Daressy ellipse and 6 cubit arc

Unfortunately, much of Daressy's conclusion therefore falls:

> "This simple sketch gives us therefore much information: … we furnish a value of the cubit under Ramses VI, and … we understand that one thousand two hundred years before our era the ellipse was understood and used for works of art."[29]

We do have a value for the cubit used in this sketch, but it isn't Daressy's. The curve can be generated simply, using arcs of a circle – there is no need for ellipses, and no need to invoke 3:4:5 triangles.

Why six cubits?

Both vault curves have similar construction – both describe circular arcs of a 6 cubit radius.[30] But why 6 cubits? This is probably not a coincidence, but arises simply from the use on both projects of a 12 cubit loop pulled tight, with one end of the loop at the centre, and the other at the arc. Such a loop would have been available on construction sites because it could also have been used to set-out a right-angled 3:4:5 cubit triangle – the so-called 'rope stretcher's triangle', which in turn is very useful in setting out rectangular structures and spaces on site.[31]

The 12 cubit loop could also have been used to set-out a 4:4:4 cubit equilateral triangle. Interestingly, moving the loop from a 3:4:5 triangle with a 4 cubit base to this equilateral triangle might have revealed that that the loop could be used to generate a new class of curve: true ellipses, with foci at each end of the base of the triangle. It is hard to believe that they didn't discover this. However, any such ellipse based on a 3:4:5 triangle (and a 12 cubit loop) does not match the paired 3:4:5 construction proposed by Daressy (which requires a 16 cubit loop).

Conclusion

The principle followed in this analysis is to find the simplest curves that fit all the points in the two original sketches. Previous suggestions over the past century or so do not satisfy this principle for either curve, but, within acceptable tolerances, a circular arc of 6 royal cubit radius satisfies it for both.

Following from this, the case for the use of the 3:4:5 triangle in ancient Egypt is indirectly reinforced (the curves could have been made using a loop which also generates this triangle), and the direct case for the ancient Egyptians knowing about the construction of true ellipses fails, but a very indirect case for this might still be made (the same loop can generate ellipses).

References

[1] First reported in Battiscombe Gunn, "An architect's diagram of the Third Dynasty", *Annales du Service des Antiquites de l'Egypte* 26 (1926), pp. 197-202.

[2] Gunn, p. 200.

[3] Suggested by Somers Clarke & Reginald Engelbach, *Ancient Egyptian construction and architecture*, Dover Publications, New York, 1990 (reprint of: *Ancient Egyptian masonry: The building craft*, Oxford University Press, 1930), pp. 52-56.

[4] Joachim P. Heisel, *Antike Bauzeichnungen*, Wissenschaftliche Buchgesellschaft, Darmstadt, 1993, pp. 129-131, #Ä18.

[5] Corinna Rossi, *Architecture and mathematics in ancient Egypt*, Cambridge University Press, 2004, pp.115-116.

[6] Gunn, p. 200.

[7] Georges Daressy, "Tracé d'une voûte datant de la IIIe dynastie", *Annales du Service des Antiquités de l'Egypte* 27 (1927), pp. 157-160.

[8] Rossi, p. 115.

[9] Ernest Moyer, "Circular solution to Saqqara ostracon", *Egypt origins*. <www.egyptorigins.org/saqost3.htm> [accessed on 1 January 2014]

[10] Max Hoberman, "Two architect's sketches", *Journal of the Society of Architectural Historians* 44 (1985), pp. 380-383.

[11] Alexander Badawy, "Ancient constructional diagrams in Egyptian architecture", *Gazette des Beaux-Arts* 107 (1986), pp. 51-56.

[12] Dieter Arnold, *Building in Egypt: Pharaonic stone masonry*. Oxford University Press, 1991, pp. 8-9.

[13] Jean-Pierre Adam, "Maquettes et dessins d'Égypte: le projet et sa presentation", in Béatrice Muller (ed.), *"Maquettes architecturales" de l'Antiquité*, Actes du Colloque de Strasbourg 1998, Travaux du Centre de 147echerché sur le proche-orient et la grèce antiques 17, de Boccard, Paris, 2001, pp. 211-225.

[14] Beatrice Lumpkin, "Mathematics used in Egyptian construction and bookkeeping", *The Mathematical Intelligencer* 24/2 (2002), pp. 20-25.

[15] Jean-Claude Goyon et al., *La construction Pharaonique du Moyen Empire a l'époque gréco-romaine*, A&J Picard, Paris, 2004, p. 79.

[16] Though other values for a royal cubit are possible – they tend to vary a little from project to project.

[17] First reported in Georges Daressy, "Un 149race égyptien d'une voûte elliptique", *Annales du Service des Antiquites de l'Egypte* 8 (1907), pp. 237-241.

[18] Kent Weeks, 'KV9 – Rameses V and Rameses VI', *Atlas of the Valley of the Kings, Theban Mapping Project* <www.thebanmappingproject.com> [accessed on 1 January 2014]

[19] These tools would also have been used in executing the vault in the tomb. Setting out the curve on the ground would have been very much harder.

[20] This is, as far as the author is aware, the only hard evidence that the ancient Egyptians understood the construction of true ellipses. Extraordinary claims require extraordinary evidence.

[21] Arnold, p. 22.

[22] Heisel, pp. 135-136, #Ä21.

[23] Rossi, pp. 115, 117. She discusses the idea at some length, and is surprisingly uncritical.

[24] Ernest Moyer, "The Ramesses VI ellipse", *Egypt origins*. See <www.egyptorigins.org/ramesellipse.htm> [accessed on 1 January 2014]

[25] A recalculation of Daressy's ellipse, not included here, agrees with his figures within 7 mm, and is generally within 3 mm. The formula is $y = \sqrt{(b^2 - b^2x^2/a^2)}$, where a and b are the semi-axes.

[26] Actually, Daressy (1907, p. 238) has 6 1/3 = 19/3 rather than 6 2/3 = 20/3, an error in his halving of 13 1/3 cubits.

[27] Daressy does not observe that the divisions are 2 palms.

[28] Weeks, *Atlas* online.

[29] Daressy, 1907, p. 240.

[30] It would be very surprising if this applied to all, or even most, extant ancient Egyptian vault curves. This analysis has not been done, but an issue here is the likely discrepancy between the designed vault and the executed vault.

[31] We might speculate that this loop would have also appealed to the astronomically-minded ancient Egyptians because, laid out as a circle (radius of about 2 cubits), its 12 facets can be mapped to the 36 decans, the 12 months of the year (each of 30 days) and the 24 hours in a day – all Egyptian inventions (Wells, in Walker, C. (ed.) (1996) *Astronomy before the telescope*, British Museum Press, London, pp. 28-41). In other words, it would have religious significance.

A proposal for cataloguing the building techniques in Roman-Byzantine and Islamic Jordanian Hawrān

Stratigraphy and Chrono-typology: the Umm as-Surab village

Piero Gilento
Department of Historical Sciences and Cultural Heritage, University of Siena, Italy

Introduction

The state of the art of the research in the Hawrān region

The southern Hawrān (part of Roman Provincia Arabia), currently divided between Southern Syria and Northern Jordan (Fig. 1), is a highly homogeneous geographic region and, at the same time, a culturally rich area, characterized in ancient times by numerous interactions between sedentary and nomadic populations. It is also a strategic transition zone of caravan and pilgrimage routes linking the Mediterranean coast to the Arabian Peninsula, passing through major urban centres, such as Bosra and Damascus. Historical and social events, together with the use of an extremely durable building material like basalt, created the conditions for good preservation of the ancient habitat in the region, at least until the end of the 1970's. From that point, the situation changed abruptly due to a strong demographic growth, linked to wider processes of urbanization in the area and intensive land-use. Since 1974, the Southern Syrian Hawrān has been the object of intensive research by the French Archaeological Mission, which has significantly increased our understanding of the ancient settlement dynamics and building systems, from the Bronze Age to modern times. [1]

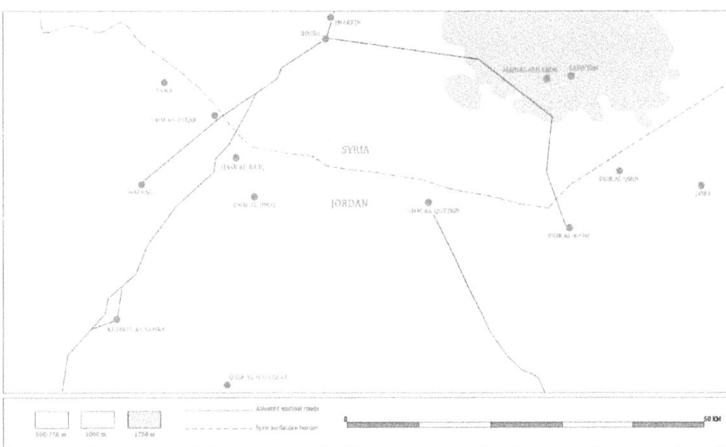

Fig. 1. The geographical localisation of the Umm as-Sura site near the Syro-Jordanian border.

Despite this large project in the Syrian area, we can point to an almost total lack of intensive archaeological research on the built heritage and ancient landscape of the Northern Jordanian territories. There are some exceptions to this: the survey of Byzantine and Islamic architecture of the Jordanian Hawrān in the early 1980's, the study of the Jawa Bronze Age site, the on-going research in the village of Umm al-Jimāl, a specific project on the ancient landscape of the southern Hawrān, and the research project in the Khirbet al-Samra village. [2] Thus, the local situation of ancient settlements and construction history in the Jordanian Hawrān needs to be more deeply investigated.

The Building Archaeology in Jordan project

The *Building Archaeology in Jordan* project is focused on the study of Roman-Byzantine and Islamic architecture in Jordan. [3] Directed by Professor Roberto Parenti, this multidisciplinary project is funded by the University of Siena and the Italian Ministry of Foreign Affairs and supported by the Department of Antiquities of Jordan. One of the project's topics is the analysis of building techniques and materials of the *Provincia Arabia* from the Nabataean period to Umayyad times. One of the goals is to compare the large 'local' eastern building traditions with Roman building techniques. In three seasons (2009, 2011, 2012) we focused our research on the Umm as-Surab site (Fig. 2) and on the northern area of Jordan (Mafraq governorate), studying building techniques and materials with archaeological and archaeometric tools (mortar analysis) and developing a new approach to the building survey of complex sites with photogrammetric tools to obtain high-definition 3D models of buildings and surfaces.

Work methodology: studying building techniques

Construction techniques are part of material (the buildings) and immaterial heritage (ideas and knowledge at the basis of their execution). A key role for the development of a building technique is the environment and its resources. The functions, durability and aesthetics of a structure are strictly linked to the environment. These features, however, depend on *human know-how*: a corpus of knowledge developed through the transmission of techniques and rules, elaborated and codified only after long periods of testing (*empirical knowledge*) and influenced by tastes and skills dictated by the times (*trends*) as well as by specific demands of patrons and the ability of the artisans. A better understanding of the traditional building methods in a territory means entering into a background that is as historical-anthropological as it is technical-structural, which may open up points of view for the analysis of social and cultural processes, which are otherwise difficult to analyse. The methodological approaches employed in the research project are stratigraphy and chrono-typology. [4] Architectural stratigraphy is the sum total of construction acts, reconsideration and changes to the initial plan of a building, as well as any collapses and reconstructions due to human factors (demolition) or natural ones (such as earthquakes). All of this information is recorded in the structure of the building itself and must be meticulously identified and decoded. Decoding signs left on the horizontal and vertical surfaces of buildings reveal the passage of time for these structures, which is one of the main questions in archaeological research. Stratigraphy thus becomes an analytical methodology that can give reliable information and help delineate the outlines of construction techniques, the architectural typologies and, in particular, the history of a building. The basic concept of chrono-typology is to identify particular constructive elements associating them with the period when they were used. For any single element, distinctive features may include materials, construction techniques, shapes or dimensions. It has to be noted that chrono-typology applies to element types, not to the whole building. It was for this reason we took into account the different typologies of building techniques (walls, arches, windows, doors, etc.). Data becomes meaningful when a large production of the same common-use element is identified. The core of the method is a table (the so-called chrono-typological table) that records the meaningful instances of constructive elements, associated with an appropriate date period, obtained by other dating methods: surveys, excavations, epigraphs, inscriptions and building typologies (for example, mosques and minarets, which did not exist prior to a certain date); the period boundary is marked by the oldest and most recent occurrences of the element in an homogeneous territory: in our case the Hawrān region.

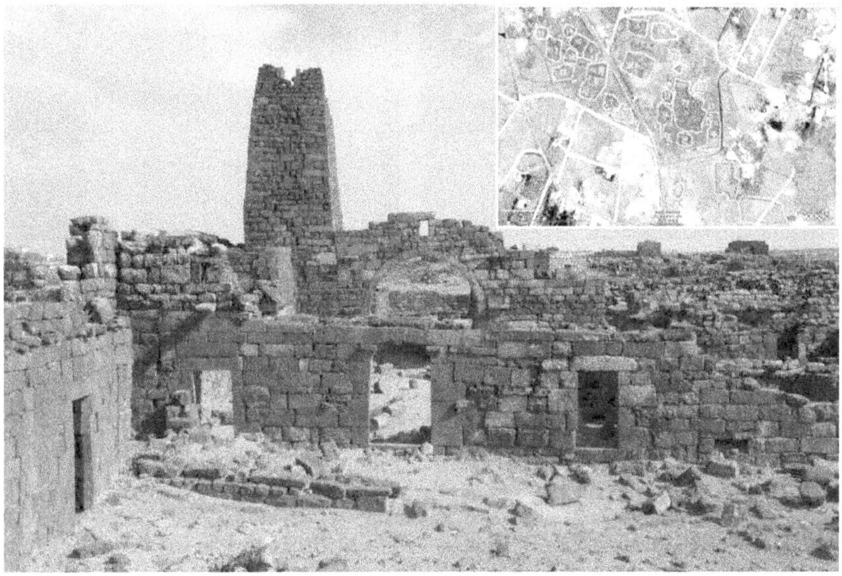

Fig. 2. The front of the Church of the Saints Sergius and Bacchus (TU28). Top-right the plan of the Umm as-Surab village divided into 28 Topographical Units (TU). Plan realized by F. Saliola and S. Anastasio (source: Google).

The buildings survey

In the research project the attention was focused on the representation and characterization of building techniques recorded during the field survey activities. A good digital survey was essential to study the typologies of the different structures recorded in the field work in order to create chrono-typologies to compare the building techniques and the architectural typologies with other areas. (Fig. 3) Practical operations consisted of systematic surveying of remaining standing structures of the walls and buildings under consideration, and the subsequent recording of their technical-formal characteristics and relative stratigraphic analysis. Later development of the study of these buildings will seek to provide concrete answers at least as far as relative chronology is concerned. Field work will be based fundamentally on topographic and photographic surveying with a method of creating three-dimensional photogrammetric models by utilizing triple photos taken with a calibrated camera and a tripod, which is well suited, for its practicality, to an extensive survey. [5] The rapid application of this methodology will allow for the recording of a large amount of data during a limited period in the field.

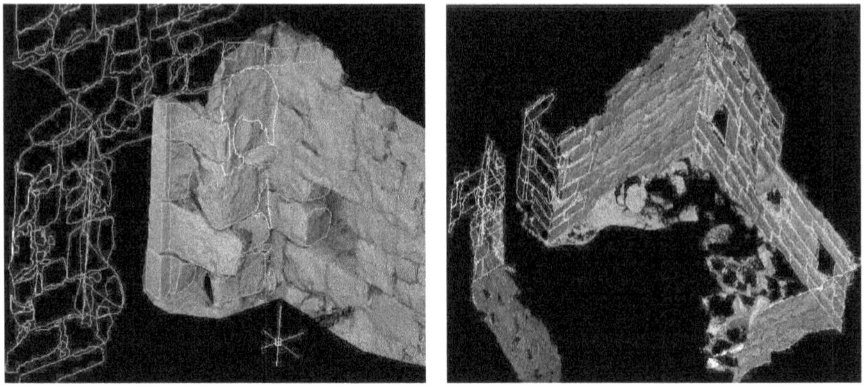

Fig. 3. Post-processing of the photogrammetric survey. The building techniques were recorded directly on the 3D models at high photographic resolution.

The building techniques of the Umm as-Surab village

Main issues of the research project

In the analysis of a structural system, one of the main elements to be taken into consideration is the construction material. The only widely available material in the southern Hawrān is basalt. This material, along with the near-total absence of wood for construction, generated an absolutely unique local situation in terms of the building techniques developed in the territories today comprised within southern Syria and northern Jordan. The relevant literature indicates that it is particularly difficult to obtain chronological information from the study of construction techniques for various reasons, which we agree with in part: the re-utilization of materials, continuity of techniques, the physical properties of basalt (its hardness and thus the difficulty in identifying signs of handling or manipulation on the surface), and lack of reliable chronological contexts. [6] Temporal indications may be provided by decorative and sculptural elements which, however, are not always found in the area's villages, or are limited to particular contexts like churches or large architectural complexes.

In the case of the Jordanian Hawrān, and specifically Umm as-Surab, we find a mid-sized village (11 ha), with still relatively well-conserved buildings, rather limited (in some cases non-existent) decorative elements and, in contrast, a rich stratigraphy of standing structures. Stratigraphic analysis, along with the individuation and description of construction techniques, can enhance our knowledge of a few aspects of the architectural culture of the Jordanian Hawrān. Despite the homogeneity of the material and its continual re-use, actions impacting architecture are in fact well identifiable because they generally correspond to a precise building phase (reconstruction or vertical expansion of a wall, demolition or insertion of an aperture, etc.). Our work focused on identifying indicators that could help us to determine such differences, especially through the study of wall-building techniques.

The work was organized to begin with a precise analysis of the material structure of individual buildings, which was then expanded to the broader scale of observation through targeted comparisons with other complexes in the region.

The conceptual path can be summarized as follows:

1. Stratigraphic analysis of standing structures;

2. Contextualization of data we obtain with previously available data for each building from digs and epigraphs;

3. Cross-referencing of information on different complexes: initial typological comparisons;

4. Precise comparisons with other buildings from the area: the scale of observation is broadened;

5. Proposal of a chrono-typology of wall construction techniques.

Walls

The point of departure for our analysis was the study of wall structures, as they are initially better analyzable than other construction techniques. Likewise, analysis of walls allows us to carry out targeted comparisons with all of the structural elements linked to them: apertures, arches, roofs and coverings.

With stratigraphic analysis, we thus identified the typological differences between one wall building technique and the adjacent one, recording physical relationships. These comparisons led to an initial chronological seriation within each individual building analyzed. The next step was to contextualize the information gleaned, still in a raw state, refining the analysis through an initial cross-referencing of available data for each individual complex. From this point of view, initial associations were made between various buildings, leading to a precise seriation, summarized in a chrono-typological table. Springing from a general description of wall structures, which are usually made up of two facings, we then took into consideration the presence, frequency and arrangement of headers in the section. In fact, headers can contribute to characterizing a wall structure because in some cases they are completely absent, while in others they are present but with an absolutely irregular arrangement, and in still others are arranged in more or less orderly rows. Based on stratigraphic analyses of standing structures at Umm as-Surab and Umm al-Jimāl villages, the irregular presence of headers can be attributed to the Early Byzantine period building phases, while the wall building technique using regular rows of headers alternated with four or five rows of stretchers to the Late Byzantine. Although it is not to be excluded that some building techniques could be used during different periods with other techniques.

Particular attention was given to construction details and the characteristics linking various elements, taking into consideration:

- Type of wall structure and construction characteristics (regular or irregular structure texture; with or without headers);

- Quality of linkage between vertical walls (cantonal structure anchorings);

- Quality of linkage between horizontal elements (suspended floors, roofs and coverings) and walls;

- Type of orientations.

It should be noted that one problem faced in reading – and thus understanding – these structures concerns the continued occupation of some sites in the southern Hawrān into the modern era, and the replication of ancient construction techniques by the local Druse population. [7] In any case, Druse techniques present characteristics that make them distinguishable from ancient ones, such as the types of apertures. Help in this regard is provided by period photos present in reports made by earlier scholars. A direct comparison with these materials aided in the reconstruction of a few ancient assets that had been completely lost (Fig. 4), and also alerted us to the possibility of replications – in some cases very well done – of traditional construction systems employed not only by the Druse people but also during more recent restoration work.[8]

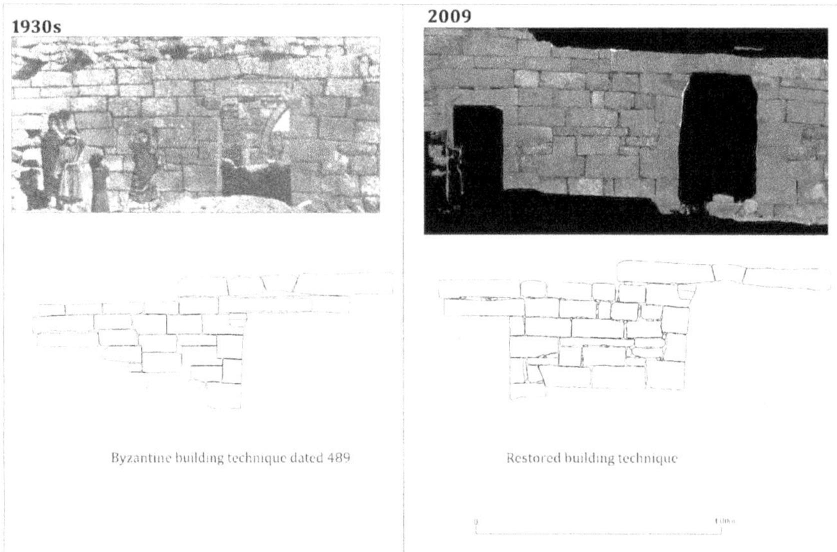

Fig. 4. The Church of Saints Sergius and Bacchus. Right, the building technique of the façade dated by the lintel inscription of the main door at 489 A.D. Right, the façade restored during last years.

The study of the tools utilized in construction and the textures on the faces of the building blocks, although complex, may provide equally interesting clues. (Fig. 5) For example, at Umm as-Surab we noted utilization of a chisel with a 45-degree inclination in the oldest phases of the Church of Saints Sergius and Bacchus and in the so-called Western Church. (Fig. 5d) This same technique is present on the blocks of the Commodus Gate at Umm al-Jimāl, dating to 176-180 A.D. (Fig. 5c) The 'economic abossed face' technique, widespread in the Nabatean and Roman period in other towns such as Bosra, is almost completely absent at Umm as-Surab. (Fig. 5b) For the Byzantine period, a very widespread technique, especially for the finishing of faces on the door and window blocks, was the use of a pointed chisel, which made the surface particularly smooth. (Fig. 5e)

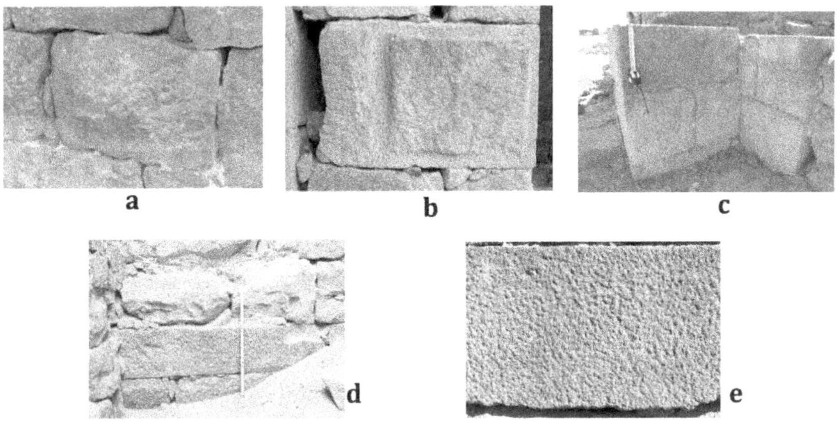

Fig. 5. Different tool trackings recorded in the Umm as-Surab village: a, b, d, e. The Gate of Commodus (c) where were used both chisel and pointed chisel.

Pilasters

In the Hawrān, the use of pilasters as weight-bearing elements for ceiling and roof support arches was quite widespread in the Roman-Byzantine era in both public and domestic buildings. This structural element is also present, however, in the architecture of the Negev, again linked to transversal ceiling and roof support arches. At the Mampsis site, pilasters are divided into two categories: *free-standing* and *engaged pilasters*. The former are pilasters independent of walls, in the middle of a room, or resting against a wall but without structural ties to it. The latter are pilasters linked to the wall structure by horizontal elements that join with the wall structure forming a single weight-bearing element. Digs carried out at Mampsis have revealed that these elements were already in use in Nabatean architecture, and comparisons have been made with the Umm al-Walid site. [9]

The system of construction of these pilasters is a regular header-and-stretcher alternation of perfectly-squared blocks, forming a sturdy interlocking structure. Their base is a stretcher block about 0.55 m wide, and they continue upwards with the alternation of two-block headers, up to a maximum of six rows. On the seventh rests the capital, from which the impost of the arch springs.

The height of the pilasters system is relatively standard, around 1.80 m. The same technological solutions have been found in the Hawrān, although in this case the material used is basalt, while in the Negev it is limestone. The Church of Saints Sergius and Bacchus at Umm as-Surab conserves some excellent examples of pilasters, which allow us to study the construction methods employed. These pilasters are structurally linked to the mesh of the wall – the church's counter-façade, and mark the alternation of the three entrances to the church. The main entrance conserved an epigraph that allows us to date part of the wall, especially the pilasters, with absolute certainty to 489 A.D. This date indicates that the technique was used over a lengthy period of time and in different geographic zones, a sign of cultural influences which, as we will see, can also be demonstrated by other construction elements. There are several well-documented examples from Umm al-Jimāl in both civic and religious buildings.

Doors and windows

In the villages of the Hawrān, the good state of conservation of the structures has allowed for observation and study of systems of apertures. Doors here are made up of separate elements assembled during wall construction: the threshold, door jambs – generally consisting of four or five blocks per side - and the architrave, which caps the upper part of the door and is overlaid by a system of support and relief, usually created through a separate aperture, of which there can be many variations. The threshold, usually in a single piece, has an external leaf higher than the level of the block, which allows the movement of the pane towards the exterior to be blocked. There may be one or two leaves, either in wood or basalt, set into the aperture system by means of hinges on the threshold and the architrave. Various solutions are used to complete the upper part of the door system above the architrave. In fact, in some cases we have found that one or two other beams are positioned in accordance with the same structure as the architrave. This solution is not one of the best adopted in the area, because in this case the weight of the wall is discharged directly onto the main architrave, as is demonstrated by numerous lesions found on architraves utilizing this system. A second method employs a relieving arch inserted into a tri-lithic or monolithic system, positioned atop the main architrave. The third solution calls for an aperture above the architrave. As we have already mentioned, there are many variations of this type. At the sites we visited (Umm as-Surab, Samā, Qasr al-Bajj, Umm al-Jimāl) we have found small, square or rectangular apertures, formed, like the doors themselves, of separate elements assembled during construction. At Umm al-Jimāl, there are also other types of apertures - especially in cases of prestigious private structures - including circular ones, which thus become *oculi*.

In other cases, found above all in the Batanea and in the Syrian Hawrān, apertures are richly decorated, presenting a precise stylistic scheme reflecting the wealth of the inhabitants. Windows are also composed of systems of blocks assembled during wall construction.

Roofs and suspended floors

In the Hawrān, roofs and suspended floors are made up of stone elements that serve to seal the top of the house or delimit levels. The system of construction of roofs and suspended floors is highly dependent on corbels on which basalt beams traversing the space rest. The corbels are single-block elements inserted into the wall section and jutting out from the wall profile far enough to hold the end of the beam (usually 0.25-0.30 m). The principle is the same as that of stairs – in fact, they are self-supporting individual elements. In other cases, these elements can be associated with other corbels, with a maximum of three overlaid rows. This device allowed the system to support a significantly greater load.

Buildings in the Hawrān usually have two or more levels, with some types, such as towers, reaching notable heights. Roofs are almost always flat and, in some cases, are still sound today, covered with a compact mixture of straw and earth that provides good insulation, not only from water but also from heat. However, the roofs we see today in the villages of the southern Hawrān, at least at the sites we have studied, are later reconstructions, difficult to date, which almost never correspond to the original roofs. The flat roofs we described can be divided into two broad categories, generally influenced by the size of the rooms. In cases where the rooms are of modest size and the space to cover between one bracket and another is no greater than around three meters, this distance allows for a roof of beams resting laterally on corbels. In cases where the larger size of rooms requires it, the roof beams are supported by a transversal arch or a group of parallel arches, usually three, which function as a weight-bearing structure for the transversally-arranged ones above. However, we have noted that room size is not a distinguishing element with regard to the presence of a transversal arch. Even in very small spaces, arches may be used to support roofs and suspended floors. Another technological device is the use of smaller beams linking the upper corners of the walls. In fact, they are beams passing transversally from one side to the other, in the upper corner zone of the two walls. This solution guarantees greater stability. We have no evidence of the existence of suspended floors or roofs in wood. H. C. Butler, however, describing the Church of Saints Sergius and Bacchus at Umm as-Surab, hypothesizes that the roof of the central nave had a wood double gable. [10]

Based on the study of arches and wall apparatuses, we found that at the Umm as-Surab site, most of the roofs have been rebuilt. Determining a precise chronology of events is still a complex matter, but we can nonetheless understand when an arch is original or has been partially or totally reconstructed. In fact, in an arch in its original position, the edges of each individual block are perfectly aligned with the adjacent ones. When an arch has been reconstructed, is it generally reassembled using the blocks from the original arch, but they no longer follow the order of the initial build. In this case, we can note inconsistencies in the edges, misalignments of the blocks, various types of reutilized materials (epigraphs, decorative elements) – all signs of reconstruction activities.

Stonemarks

We were able to discern a few examples of builders' and chiselers' marks in the column drums of the Church of Saints Sergius and Bacchus at Umm as-Surab, in the form of the letters 'A' (Fig. 6a-top) and a sort of upside-down 'P.' (Fig. 6d) Also on the column drums are markings for the assembly of elements during construction. (Fig. 6b) On one column drum in the courtyard of the *Praetorium* of Umm al-Jimāl, a mark directly comparable with the letter 'A' was found (Fig. 6a-bottom), almost surely a sign of activity by the same workmen. The function of these marks in the construction process is not clear, but it is important to study the matter further in order to better understand traditional construction dynamics.

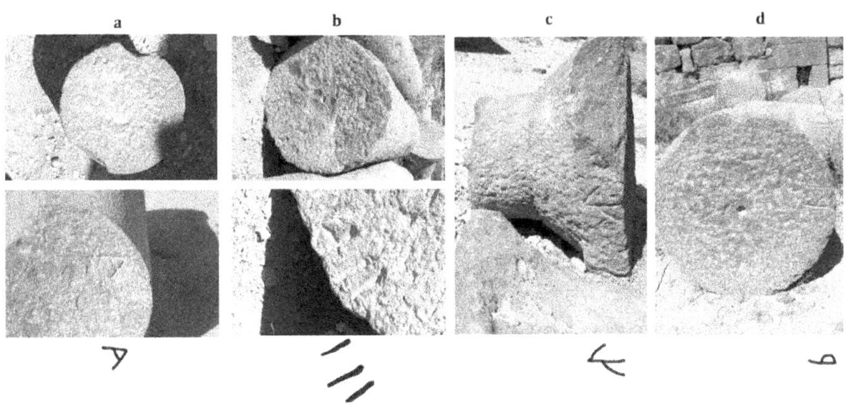

Fig. 6. Different stonemarks recorded in the Church of Saints Sergius and Bacchus at Umm as-Surab.

Conclusions

In conclusion, we can assert that an in-depth review of material remains necessary if we want to understand not only the material and technical history of a building, but above all the social and cultural history of an entire territory. In this case, construction techniques are a source of information that must simply be interpreted correctly. In this regard, stratigraphic reading of standing structures, along with instrumentation for adequate and prompt recording of material structures, can open new cognitive and interpretative horizons. At the same time, it can also serve as a means of documentation and conservation of a heritage rendered fragile and threatened by historical and political events in the region. The immense task of recording and describing the architecture of the Hawrān carried out by H.C. Butler in the early 20th century must be looked to as a prerequisite and a caveat for performing a thorough job of critical observation, description and recording of the material structure. Only in this manner can we be an active part in the process of understanding and conserving architectural and archaeological heritage, with the aim of producing, and later passing down, information that is not only quantitatively, but above all qualitatively relevant.

The results should be considered as a work in progress; the subsequent steps in the research are as follows:

1. to collect more information in the field. If we can collect more samples in the field, we will be able to refine the chrono-typological table;

2. to link masonry techniques with other building techniques: arches, doors, windows, horizontal elements;

3. to make targeted archaeological excavations to link vertical and horizontal stratigraphies, and thus provide a reliable overall stratigraphy and propose absolute chronologies.

References

[1] J. M. Dentzer (ed.), *HAURAN I. Recherches archéologiques sur la Syrie du Sud a l'èpoque hellénistique et romaine*, Première et Deuxième partie, Paris: Ifapo, 1983. P. Clauss-Balty (ed.), *HAURAN III. L'Habitat dans les campagnes de Syrie du Sud aux époques classique et médiévale*, Beyrouth: Ifpo, 2008. M. Al-Maqdissi, F. Braemer, J.M. Dentzer (eds.), *HAURAN V. La Syrie du Sud du néolithique à l'Antiquité tardive. Recherches récentes.* Damas: Ifpo, 2010.

[2] G.R.D. King, "Byzantine and Islamic sites in northern and eastern Jordan" pp.79-91, in *Proceedings of the Sixteenth Seminar for Arabian studies, Oxford 1982*. Oxford, 1983. S.W. Helms, *Jawa. Lost city of the black desert.* Methuen, 1981. B. De Vries (ed.), *Umm el-Jimal. A Frontier Town and its Landscape in Northern Jordan.* Portsmouth, 1988. D.L. Kennedy, P. Freeman, "Sourthern Hauran Survey 1992", *Levant*, XXVII, 1995, pp. 39-73. J.B. Humbert, A. Desreumaux, *Khirbet es-Samra, Jordanie. 1. La voie romaine, le cimetière, les documents épigraphiques.* Paris, 1988.

[3] In the framework of the *Building Archaeology in Jordan* project was developed my PhD research titled "*Stratigrafia e tipologie costruttive dello Hawrān giordano. Verifica delle potenzialità di un antico strumento archeologico: la registrazione degli edifci*" ("*Stratigraphy and Building Typologies of the Jordanian Hawrān. Checking the potentialities of an ancient archaeological tool: the building survey*") hold at the University of Siena on June 27, 2013.

[4] A. Boato, D. Pittaluga, "Building Archaeology: A Non-Destructive Archaeology" in *Proceedings of 15th World Conference on Nondestructive Testing Roma 2000.* It is possible to find the article in the website: <http://ndt.net/article/wcndt00/general/board.htm> [accessed on 23 December 2013]

[5] R. Parenti, P. Gilento, F. Ceccaroni, "Three dimensional data and the recording of material structure", in Proceedings of 3DARCH-2011. *International Archives of Photogrammetry, Remote Sensing and Spatial Information Sciences, Trento 2010.* Trento, 2010.

[6] J. C. Bessac, "Le basalte de Syrie du Sud: quelques repères techniques, économiques et chronologique", in M. Al-Maqdissi, F. Braemer, J.M. Dentzer (eds.), *HAURAN V. La Syrie du Sud du néolithique à l'Antiquité tardive. Recherches récentes*, Damas: Ifpo, 2010, pp. 413-423.

[7] R.M. Brown, "The Druze Experience at Umm al-Jimal: Remarks on the History and Archaeology of the Early 20th Century Settlement", *Amman, Studies in the history and archaeology of Jordan*, vol.10, 2009, pp. 377-388.

[8] I am referring, for example, to the façade of the Church of Umm as-Surab, recently restored. It was documented by R. Bartoccini with a photo taken between 1929 and 1938 -Fig. 4left- (S. Anastasio's personal communication who is working on the cataloguing of the R. Bartoccini's personal archives) and by G.R.D. King in the 1980s. The exterior wall has been completely transformed in comparison to the original one.

[9] A. Negev, *The Architecture of Mampsis. The Middle and Late Nabatean Period.* Volume I, Jerusalem, 1988.

[10] H.C. Butler, *Ancient Architecture in Syria, Division II, Section A, Part 2, The Southern Hauran*, Leiden, 1909.

The transformation of Drumlanrig Castle at the end of seventeenth-century

Cristina González-Longo
Department of Architecture, University of Strathclyde, UK

Summary

The transformation of Drumlanrig Castle between 1679-98 makes it one of the most original and interesting buildings of its time in Britain. It was carried out at almost the same time as the Royal Palace of Holyrood, whose design, construction and procurement influenced the making of Drumlanrig. James Smith, one of the mason-contractors at Holyrood, went to work at Drumlanrig as an independent architect for the first time, providing a unique design that was in continuity with local practices but also aware of contemporary Continental architectural developments. The careful selection of craftsmen, techniques and materials make this building one of the finest in Scotland. Although the original drawings and accounts of the project have now disappeared, it is possible to trace the history of its design and construction through a series of documents and drawings at Drumlanrig Castle and by looking at the building itself. This paper will unravel the transformation of the building at the end of seventeenth-century, identifying the people, skills, materials, technologies and practices involved and discussing how the design ideas were implemented during the construction.

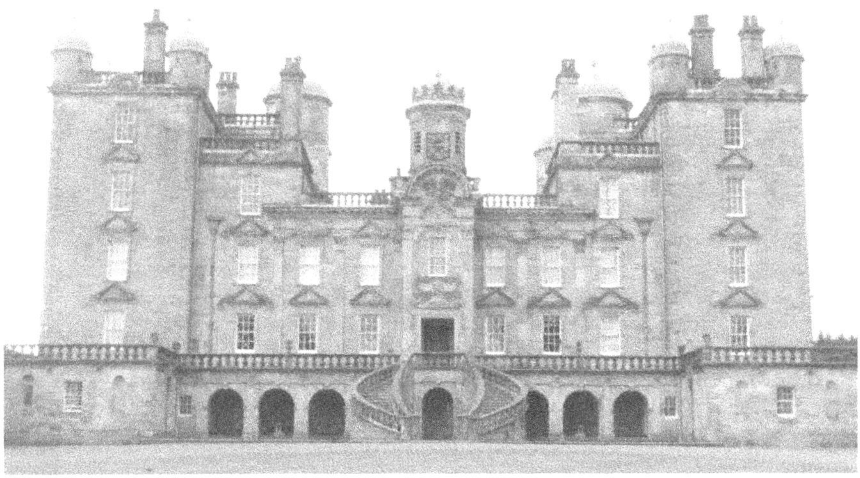

Fig. 1. North elevation of Drumlanrig Castle.

Introduction

John Summerson considered Drumlanrig Castle the most remarkable building emerging from the tradition initiated by the King Mastermason William Wallace (d.1631), *'the last great gesture of the Scottish castle style'* and *'obstinately Scottish'*.[1] He did not make any reference to the fact that the building incorporates an earlier building, which makes it even more remarkable. The design of Drumlanrig innovates but at the same time respects tradition and local identity, including the romantic idea of the Scottish castle (Fig. 1).

Perhaps for all these attributes, despite its apparent non-classical design, Colen Campbell featured it in his *Vitruvius Britannicus.* [2] He does not name the designer, maybe due to the fact the building is the result of various phases and design collaborations. There was also a great deal of the client's input, including a drawing of the front of the house done by the 2nd Duke.[3] However, archive documents clearly show a structure of designers and contractors, *'Mr James Smith'* performing the role of architect and the gardener *'Cornelius'* as landscape architect.

The transformation of Drumlanrig Castle occurred between 1679-98, almost at the same time as Holyrood Palace, which design, construction, procurement systems and professionals involved influenced it. C. T. Ramage talks about a *"plan made of wood, with proper elevations, which it is evident the master of works... had closely followed"*; which unfortunately has disappeared. [4] The original drawings and accounts of the project have also disappeared, but it is possible to trace the history of its design and construction through documents and drawings at Drumlanrig Castle archives and by looking at the building itself.

The first Castle

The seventeenth-century iron yett behind the oak front door and the North ends of the corridors along the East and West ranges are some of the few indications of the castellated nature of the building. The Douglas family owned land in this area from at least 1388. There is mention of a castle on the site in an indenture dated 29 May 1429. [5] From the remains still visible and a plan dated 1608, we can assume that this first castle was similar to others: a rectangular barmkin enclosure accommodating four ranges of courtyard buildings, with the north-west tower as keep.

An inventory made at the death of James Douglas, 7th Lord of Drumlanrig (1498-1578) mentions that he *'beildit the haill house and pallice of Drumlanrig'* [6]. The main building would have taken by that time the appearance of a tower-house (N and W ranges) with a turnpike stair in the corner and James would have built additional ranges around the same quadrangular courtyard we see today.

Visible remains from this period would be the North wall of the basement (under the terrace), the barrel vaults in the basement (fig. 2), the ribbed cross vault at the Chapter Room, to the North of the West quarter and the vault at the Duke's Study below. Other remains are visible all around the bottom of the building, where the new windows seem to be located at the top of the old enclosure wall. Remains are also visible at the lower part of the East external walls, including gun-holes and blocked windows, and at the courtyard, where the windows' surroundings are not built as a basement window (like the new window to the stair tower) but appear as partially buried (Fig. 3). There is an obvious difference in the colour, tooling and dimensions of the older stone, as well as in the window surroundings and wall thickness, being around 2.5m at the bottom of the East external wall.

Fig. 2. Plan of the basement showing the location of the vaults (RCAHMS).

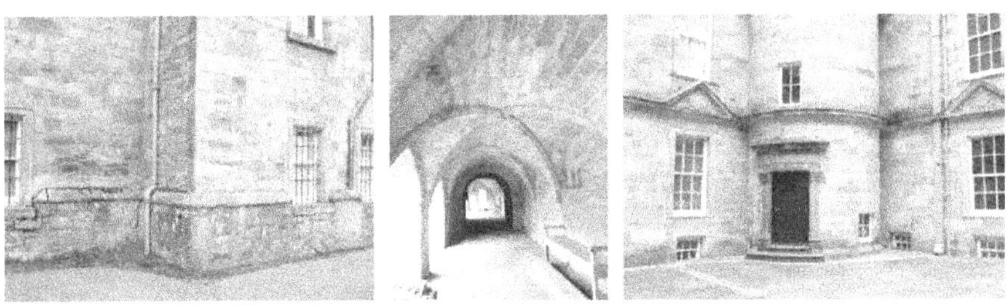

Fig. 3. Remains of the older castle.

William Douglas, 1st Earl of Queensberry and grandfather of the 1st Duke, wanted to transform the castle. Two plans dated 1608 and 1615 (Fig. 4) show the building around a square courtyard plan and two plans dated 1618 (Fig. 5) illustrate the proposal for the rebuilding of the South quarter and the retention of the existing north-west tower. There is also a document dated 1618 describing the works. The Earl asked for estimates *'for the reedifreing and building up of the South quarter'* and there is a 1622 letter about wood required from David Anderson, burgess of Aberdeen to his *'guid frend Thomas Ballantyne, wright'*, then at Drumlanrig. [7]

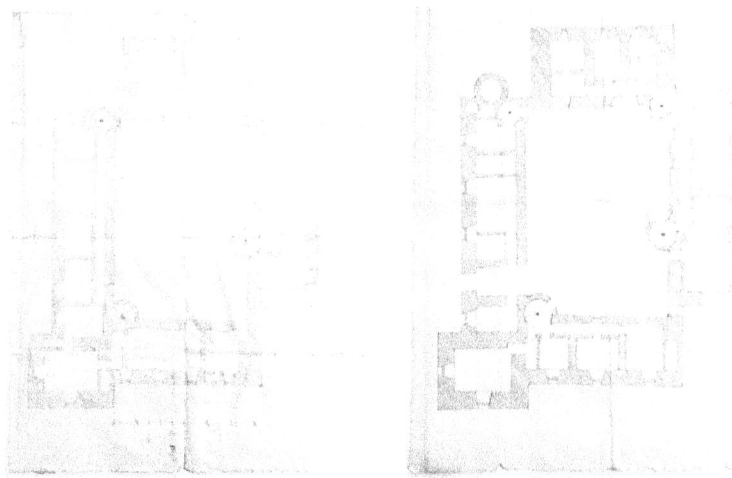

Fig. 4.Plans of Drumlanrig Castle 'unreformed' dated 1608 (left) and 1615 (right). North on the left. (Drumlanrig Castle Archive).

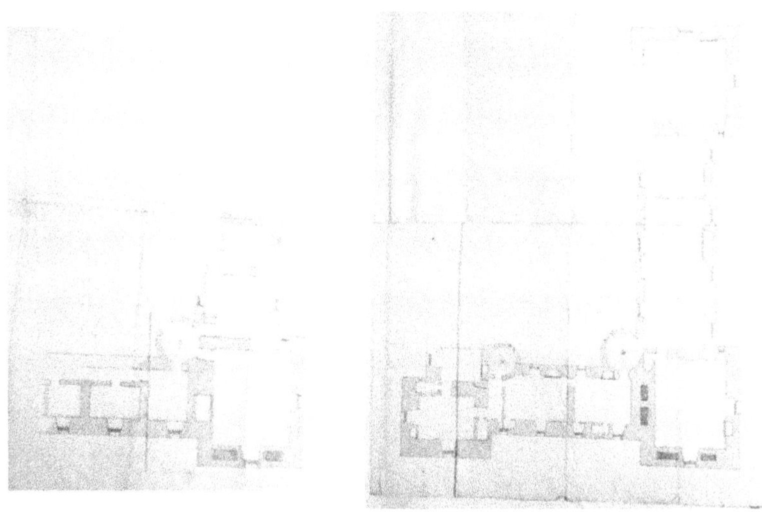

Fig. 5. Ground and Third floor Plan 'as reformed' dated 1618, showing the existing NW and the new SW and SE towers (Drumlanrig Castle Archive).

The 1st Duke of Queensberry's transformation (1679-95)

Sir William Douglas (1637-1695), 3rd Earl of Drumlanrig wanted to transform the Castle into one of the most impressive houses of Britain, probably in view of his rising political career. He became the 1st Duke of Queensberry in 1684 and King James VII's Lord High Commissioner to the Parliament in 1685. The starting point for the transformation of Drumlanrig Castle seems to have been the design produced for his grandfather (Fig. 5) as the Earl marked on the back of the 1618 plan: '*A memorandum of my grandfather anent repairing ye House of Drum to be looked over and advised be Sr Wm Bruise* [Sir William Bruce]'. An additional note, with different handwriting refers to '*Mr. Mills* [Robert Mylne's] *draughts both for the house and bridge and estimett for the bridge*'. Both Bruce, Royal Surveyor, and Mylne, Royal Master Mason, were working at the time at Holyrood Palace.

There is a contract dated 1686 which confirms Smith's role in the project, as the plumber William Waldhave had to carry out the works '*conforme to the synd draughts, and that at the sight of Mr James Smith or ony his Grace is pleased to apoynt for surveying the said work*'. [8] Most probably the imprisonment of William Bruce in 1678 facilitated the appointment of Smith as architect for the project, becoming his first important documented commission as independent architect. Smith had married Janet, the eldest daughter of Mylne in 1679, when he was working in Holyrood Palace. Smith's life and work have been discussed before, including his way up the profession, from *measson*, like his father, to become the leading architect in Scotland at the end of seventeenth-century. [9]

The basic configuration of the building – a square with one turreted tower in each angle, built around a courtyard with one turret staircase in each angle –is usually associated with George Heriot's Hospital in Edinburgh, founded in 1628 and completed over the next sixty years. Although it is not clear who was Heriot's designer, the building is the result of the directions of Dr Blacanquall, executor of Heriot's will, and the work of the leading master masons of the time: William Wallace, William Aytoun, James Murray, John and Robert Mylne. The latter is probably the most clear connection between the two buildings.

But Drumlanrig is different, as it included extensive reuse of the existing castle, extending the building in elevation and regularising its exterior. Only few of the great Scottish houses at the end of seventeenth-century were completely new constructions. At Drumlanrig, the traditional method of extending a tower-house horizontally, the tower-and-jamb with the staircase in the re-entrant angle, was employed. There is a tower staircase in each of the corners of the internal courtyard, which was quite out of fashion at the time, but appropriate for an ancient paternal seat, the house of the *noblesse d'épée*. We find however some innovations, like the vertical extension, the 'wrapping up' of the existing buildings, the creation of a new main entrance at first floor level and the location of the public spaces in the central sections and private ones to the sides.[10]

With the 1st Duke's intervention, Drumlanrig was converted in a regularised courtyard palace with tall square towers in each corner, a theatrical façade of magnificent polished pink stone and a magical skyline of ogee roofs. Large terraces with a bold balustrade, all around the front and the top of building, unified the design. By1691 there were 64 hearths at the Castle.[11]

The terrace and main entrance is reached from the forecourt by a divided horseshoe stair with a striking balustrade.[12] The mouldings of the balusters follow in parallel the curve of the flight of the stairs, resulting in a very complex geometry (Fig. 7), which together with the *Dorico Barbáro* at the portico of the lower level, show a clear reference to Juan Caramuel de Lobkowitz's treatise *Architectura Obliqua*.[13] The horseshoe stair led into an open loggia through the entrance porch. The porch is the base of a tower surmounted by a semicircular pediment and a clock within a *tempietto*-like element with Ducal coronet above. This tower marks the main entrance (Fig. 6), configuring the façade as a civic building – like the *Palazzo Senatorio* at the Capitoline Hill in Rome - rather than as a private residence.

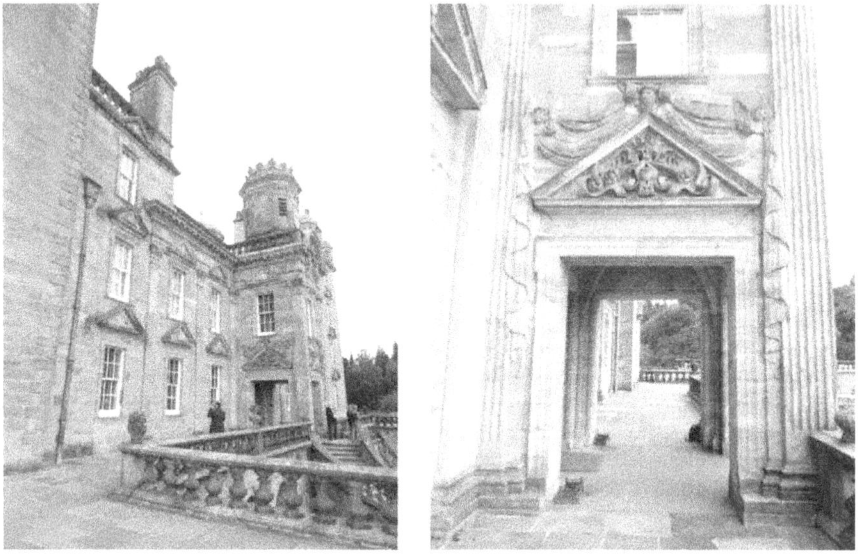

Fig. 6. Front terrace and stonework at the main entrance.

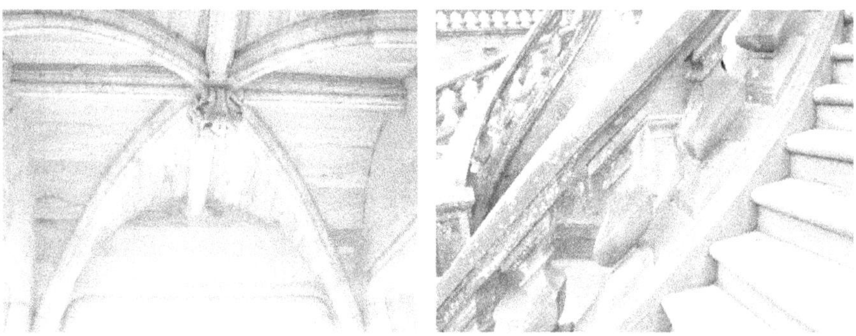

Fig. 7. Vault above main entrance and balusters of stair.

In overall composition, the building refers again to the *Palazzo Senatorio*, with a giant order of fluted composite pilasters in the North elevation, used for the first time in Scotland, [14] a balustrade topping the building and a tower in the middle. These pilasters, an element which appears in other of Smith's works, helped to unify and monumentalise existing fabrics, but was also a technique to better join the masonry with the existing one. By rotating the balusters 45 degrees it improves sunlight penetration in the North facade. There is an identical rotated balustrade above the main entrance in Holyrood, probably also Smith's design. The need to adjust the composition to the location of the existing windows in the

North elevation caused an unorthodox overlapping of the order with the windows' pediments, giving the elevation more movement.

The portico at the lower level of the North elevation is roofed with ribbed cross vaults springing from corbels in the manner of c.1500 (Fig. 3), [15] but they could be late seventeenth-century, like the cross vault covering the main entrance (Fig. 7).

The 2nd Duke of Queensberry's improvements (1695-98)

Although William was not able to travel to complete his education due to debts incurred during Cromwell's rule, his sons, James (1662 – 1711) - who became the 2nd Duke after his father's death in 1695 - and Henry went on a Grand Tour of two and a half years, visiting France and Italy and spending a considerable time in Rome. Smith designed William's magnificent mausoleum at the nearby church of Durisdeer and remained engaged in the Castle, as the 2nd Duke says in a letter dated 22 April 1696 that he '*is to go to Drumlanrig with Mr Smith*'. [16]

James Nasmith was responsible for providing materials, tools and resources, for the payments and keeping the books as well as making sure Lukup completed previously instructed works. The instructions refer to James Smith and the Duke's gardener, Cornelius, as those who would give the specific design and construction instructions to the workmen ('*advise and direct*') in both Drumlanrig and Queensberry House, in a collaborative way, with a great input from the Duke. He went into a great deal of detail, specifying design, materials and procedures, with a clear interest in finding the cheapest way to purchase the specified materials and in the '*handsomest and newest fashion*'. As well as providing drawings for the building and garden features, including a 'well of Love' in the park and pedestals for lead statues, Smith also produced architectural elements such as marble fireplaces.

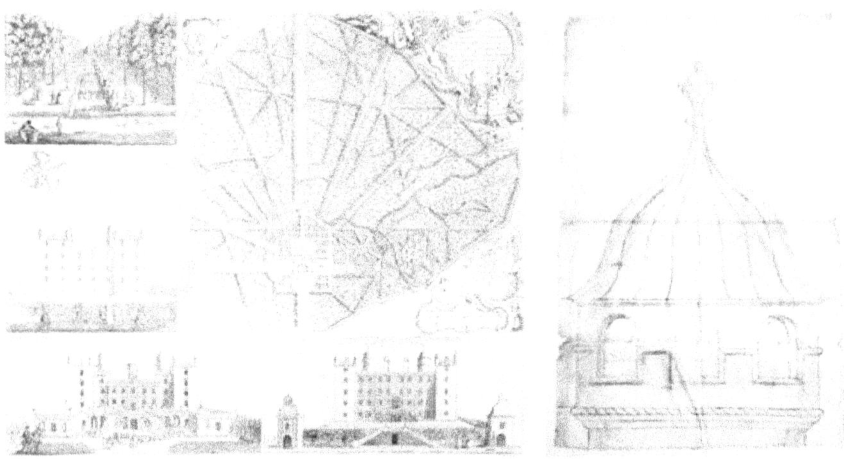

Fig. 8. John Rocque plan and Roof of stair turret (Drumlanrig Archive)

The 1739 engraving *Plan of ye Garden and Plantation of Drumlanrig* by John Rocque is the most important evidence of the appearance of the Castle and garden in the early 18th century (Fig. 8). The North front is shown without its later service wings, added in the early nineteenth-century. The plan also shows the gardens created by the first three Dukes, including the cascade on the hill to the South, served by an aqueduct two miles long (Fig. 8 top left), and later abandoned by the 4th Duke, 'Old Q'. The gardens were originally organised in terraces, but later converted into slopes, apart from the upper one, which is around 10m high. This was probably related to structural problems of the dykes due to water ingress, as they had already been collapsing in 1697.[17]

The procurement of the works

There is a series of letters dated 1687-90 from the 1st Duke of Queensberry to his cousin, the Laird of Dornock, who was his caretaker in the area. They make clear that Smith was the architect of the project, overseeing William Lukup.[18] Lukup is named as Master of Works of Drumlanrig on his tomb, employed between 1679-90.[19] He subcontracted work, as revealed in a contract of 1686 with the plumber Waldhave, who provided the water piping to the building, the drainage and leadwork to the roof. [20] The works were finished in 1697 by another plumber John Carnabie, with whom a maintenance contract to keep the lead roofs and water works in good order, with a yearly salary, was also agreed.[21] The ogee roofs of the two pavilions to the South were commissioned by James Smith to the wright, Alexander Acock.[22]

While Lukup seemed to have executed builders' work, instructions given by the 2nd Duke to James Nasmith, gardener, in September 1697, made clear that Nasmith had strictly the role of Master of Works. He was the administrative and financial officer, responsible for the organisation of the building operations, the engagement of craftsmen, purchase of materials and keeping and controlling the accounts. The project at Drumlanrig followed the direct labour system, in a similar way the Royal Works, local authorities, the Church and some noblemen did, with staged contracts that permitted adjustments as works progressed. [23]

The clock in the tower of the main entrance shows the date 1686, the same appearing in documents concerning a visit of two Dutch stone-carvers, Peter Paul Boyse and Cornelius van Nerven. They had worked before for three months at Kinross, Bruce's own house,[24] where Smith and his servitor John Hamilton were also working at the time.[25] They seem to have done most of the fine carving at the North elevation, [26] including the large swags of fruit in the front entrance and a great Queensberry coat of arms under the clock tower.

The Duke had to endure the rising cost of the building, probably affected also by the turbulent times in the country, in particular when James VII left for France in 1688. Works were however progressing and marble was sent from Edinburgh just when the first rumours began that the Prince of Orange was going to invade England.

The construction sequence

Dates carved on the window lintels of the stair-turrets to each of the four corners of the courtyard provide the sequence of the construction: the building started in 1679 at the Northeast corner and developed clockwise, about every two years, around the courtyard until 1689, finishing with the top of stair-towers and the north-east tower.

In July 1684 there were 66 men, including 31 masons and 8 wrights, employed in the works. [27] In 1686 Peter Paul Boyse and Cornelius van Nerven were working in the North elevation, and perhaps before, as one of the swagged curtains and seaweed scrolls, carved around the West entrance of the porch, bears a marquess's coronet, so is datable to 1682-4 while the other, at the East entrance, with a duke's coronet, should be after 1684.

From the contract with Waldhave, we know that by 1686 works in the North quarter (apart from the wrightwork to the Gallery), NW tower, including the three turrets above it, and the clock house were completed and ready to receive the leadwork. The other stair-turrets were already finished. A drawing of one of the ogee roofs of the four stair turrets existing in the building is still in the archive at Drumlanrig, showing a more elaborated design than the existing ones (Fig. 8). Waldhave also covered the garden pavilions with ogee roofs "*conform to ther proporsions*" and provided other leadwork to the house, including sash and case window weights. As mentioned before, the leadworks would be finished in 1697 by John Carnabie.

The fine decorative ironwork of the balustrade of the South balcony and stair was made by James Horn of Kirkaldy, who worked with Smith on other buildings in the 1680s. The lead sundial, dated 1692, is by Thomas Wynne of London. The Great Oak Staircase was added in 1697.

Materials and techniques

The characteristic pink sandstone of Drumlanrig, which becomes purple when wet, is an impure arkosic arenite sandstone.[28] It comes from the nearby King's Quarry, within the Buccleuch Estates, which at the time of the transformation of the Castle was largely available. There is also evidence of quarrying stone in 1698 to the westward of the house, in order to reduce transportation costs.[29] It has a bedded character and it is typically a fine to coarse grained, pink to lilac banded sandstone, with relatively abundant clay minerals. It has thin bands and localised patches of grey in areas where the iron oxides minerals are less oxidised. This crystaline local stone reflects light and it has a relatively high quality as building stone, with good cohesion, strength and durability. Although the bedding and texture of the stone make it in theory unsuitable for high quality ashlar and detailed carving work, its use in Drumlanrig Castle has demonstrated its durability and 'lively' quality, despite its vulnerability to damage from continued water penetration due to the presence of clay minerals.[30]

The North façade has finely carved ashlar stone with fine joints and rusticated quoin dressed stones (Fig. 6), with both good geometry and craftsmanship; this can also be seen in the internal walls of the South West stair tower. The chimneys and turrets are also in ashlar. The rest of the elevations are rough-hewn rubble, which were harled with pink lime-based mortar, with exposed quoin dressed stones. Most probably, as at Melville House and his own house, Newhailes, Smith covered these walls with pressed-back harl and regular ruling-out to imitate coursed ashlar.

Great care was taken in the selection of stone blocks for particular locations, as can be seen in the vault above the main entrance (Fig. 7). At this time, the quarriers were part of the building workforce. The consistent colour and careful selection of the stone at Drumlanrig were certainly the result of a painstaking process, perhaps trying to evocate the red marble (*diaspro*) which Smith could have seen in Rome.[31]

The display of flat lead roofs, with a total absence of the traditional slate roofs, is a clear 'showing-off' of the local Queensberry lead mines. As well as the main block, Smith added his characteristic ogee roofed pavilions to the front of the house and flanking the South front (Fig. 8). Nasmith was instructed in 1697 to search for the cheapest timber, at Whytheaven or Leith and in 1698 he met merchants at Dumfries to see if rates for timber from Norway would be cheaper than at Leith. In October 1698 white lead and oil to paint the windows was brought from Holland .[32] Smith also added the Great Oak

Staircase giving access to the formal rooms at the first floor, a freestanding stair displaying virtuoso turning of spirals.

The building shows details which would become some of Smith's characteristics: ogee roofs, regularly disposed fenestration, moulded lugged architraves, fluted giant Corinthian pilasters and modest marble or stone mouldings in fireplace surrounds between the fire and the timber panelling. He incorporates vernacular, gothic and 'oblique' elements at the same time. As Charles McKean put it: '*even when they* [Bruce and Smith] *were designing in the grand manner, their Scottish accent made itself heard'*; he considered that '*Smith created a contemporary palace that visitors allowed to be noble*'. For McKean, the influence of Drumlanrig is clear in William Adam's Duff House, and beyond [33] and it will certainly influence Smith's later work.[34]

Conclusions

Although the original drawings and accounts of the transformation of Drumlanrig Castle between 1679-98 have now disappeared, it has been possible to trace the history of its design and construction through the building itself and a series of documents and drawings at Drumlanrig Castle. This has allowed the identification of the people, skills, materials, technologies and practices involved in the project. The careful selection of craftsmen, techniques and materials make this building one of the finest in Scotland.

The retention of the older fabric and the reference to the older dynasty of the Douglas clan are two important reasons why the building is still called Castle and not Palace, as its appearance could suggest. It is one of the most original and interesting buildings of its time in Britain, where theory and practice, local and international, come together.

The existence at the end of seventeenth-century of larger projects with educated clients such as the Dukes of Queensberry, closely involved in the process, changed the nature of the projects, becoming more complex and collaborative. This allowed for the emergence of architects such as James Smith, who like his clients, with university education and European travel experience, was able to understand their design intentions and cultural aspirations.

There is a clear shift in the procurement of the works carried out by the 1st and the 2nd Dukes, most probably influenced by the works at Holyrood Palace. The first phase works at Drumlanrig were designed and controlled by Smith and executed by Lukup (subcontracting parts of the work such as the lead to the roofs) and in the second phase Nasmith is brought in to administer the works, and the roles of each individual become more specialised.

The integration of the house and landscape works is clear, with a continuous design process which allowed for changes as the project proceeded. The main works were however planned and contracted beforehand although some serious delays occurred. The 2nd Duke was also able to carry out the works in Drumlanrig at the same time as those at Queensberry House, the Duke's new residence in Edinburgh, and with the same designers.

The 1st and 2nd Dukes of Queensberry were Smith's greatest patrons and Drumlanrig is an exemplar of good patronage which nourished artistic freedom and an effective collaboration and exchange of ideas. It responded to the aspirations and limited resources of an emerging country and nobility. Drumlanrig has a world-class quality, with continental references but solidly rooted in local traditions. Its architecture and construction resulted from a mixture of traditional Scottish construction methods and architectural references from abroad. Despite the extensive inclusion of fragments of the earlier Castle, thanks to the skills and ability of the client and professionals involved, Drumlanrig achieved harmony, or, as Alberti would put it, *concinnitas*.

Acknowledgements

My gratitude to David Munro, Duke Richard, the 10th Duke of Buccleuch and Queensberry, Aonghus MacKechnie, the late Charles McKean, John Dunbar, Dimitris Theodossopoulos and Luis Albornoz-Parra, for their generosity, inspiration and scholarship and to all the staff at Drumlanrig Castle for their help during the research, in particular to Claire Oram for organising my first visit, Stuart St John for a great tour and Betty Cook for the delicious lunches.

References

[1] J. Summerson, *Architecture in Britain 1530-1830*, Yale, 1993 (1st ed. 1953), p. 509-10.

[2] C. Campbell, 1715-1725. *Vitruvius Britannicus*. Reprint by Harris, J. 1967. New York: Benjamin Blom.

[3] *Instructions to James Nasmith*, 13th December 1697, item 7. Bundle 1124, Drumlanrig Castle Archive.

[4] C.T. Ramage, *Drumlanrig and the Douglases*, 1876, p. 22-25.

[5] M. Girouard, "Dumlanrig Castle, Dumfriesshire-I", *Country Life*, 25 August 1960, p. 378.

[6] J. Montagu Douglas Scott, *Drumlanrig. The Castle, its People and its Paintings*. Caïque Publishing, The Buccleuch Estates, 2010, p. 9.

[7] The plans and document can be found in a volume compiled in 1908 by the historian The Hon. Hew Dalrymple. Cf. D. Munro, *Buccleuch and Queensberry Estate Historical Note No. 26*; Letter dated 19th April 1622. Bundle 1123, Drumlanrig Castle Archive.

[8] J.G. Dunbar and K. Davies, "Some Seventeenth-Century Building Contracts", in Scottish History Society, *Miscellany*, XI, 1990, p. 269-327.

[9] C. Gonzalez-Longo, 'James Smith and Rome', *Architectural Heritage*. Volume 23, 2013, pp. 75-96; C. González-Longo and D. Theodossopoulos, 'From Master Mason to Architect: James Smith's Construction Techniques at the End of 17th Century in Scotland' in R. Carvais, A. Guillerme, V. Negre and J. Sakarovitch (eds.) *Nuts & Bolts of Construction History. Culture, Technology and Society*. Proceedings of the 4th International Conference of Construction History. Picard, Paris, 2012.

[10] C. González-Longo, 'Using old stuff and thinking in a new way: material culture, conservation and fashion in architecture', in P. Emmons, J. Lomholt, J. Hendrix (eds.) *The Cultural Role of Architecture: Contemporary and Historical Perspectives*, London-New York, 2012.

[11] *List of hearths in the castle of Drumlangrige*. Papers of the Leslie family, Earls of Leven and Melville, NAS GD26/7/328.

[12] This staircase is believed to have been replaced c.1860 by Charles Howitt copying the original forms, but not evidence has been found so far.

[13] C. Gonzalez-Longo, 'James Smith and Rome', *Architectural Heritage*. Volume 23, 2013, pp. 75-96

[14] M. Glendinning, A. MacKechnie, R. Oran, *The architecture of Scottish Government*, Dundee, 2004, p.107. J. Macaulay (*The Classical Country House in Scotland*, 1987, pp. 27-9) gives the unbuilt hunting lodge of Charleval (began in 1570) as another possible reference.

[15] J. Gifford, Dumfries and Galloway, *The Buildings of Scotland*, Penguin Books, 1996, p. 223.

[16] Papers of the Montague-Douglas-Scott Family, Dukes of Buccleuch, NAS, GD224/1031/1.

[17] *Instructions*, cit. 11th September 1697, item 60.

[18] Bundle 1220, Drumlanrig Castle Archive.

[19] J.G. Dunbar,"The organisation of the building industry in Scotland during the 17th century" in SVBWG, *Building Construction in Scotland. Some Historical & Regional Aspects*, Edinburgh and Dundee, 1976, p. 8.
[20] Dunbar, *Some Seventeenth-Century Building Contracts*, cit.
[21] *Instructions*, cit., 11th September 1697, item 80; 22nd October 1698, item 5.
[22] *Instructions*, cit., 22nd October 1698, item 29.
[23] Dunbar, *The organisation of the building industry in Scotland*, cit.
[24] Kinross muniments, Cf. Girouard, "Dumlanrig Castle, Dumfriesshire-I", p. 381.
[25] D. MacGibbon and T. Ross, *The Castellated and domestic architecture of Scotland, from the twelfth to the eighteenth century*, v. 4, 1892, p. 566.
[26] Montagu Douglas Scott, *Drumlanrig*, cit., p. 9.
[27] Gifford, *Dumfries*, cit., p. 136.
[28] E.K. Hyslop, E. A. Tracey, L. J. Albornoz-Parra, *Assessing the Potential for the reopening of a building stone quarry: King's Quarry, Thornhill, Dumfriesshire*, British Geological Survey Open Report OR/10/040, 2010.
[29] *Instructions*, cit., 22nd October 1698, item 59.
[30] Hyslop *et al.*, *King's Quarry*, cit.
[31] Gonzalez-Longo, *James Smith and Rome*, cit.
[32] *Instructions*, cit., 1697-98.
[33] C. McKean, The Scottish Chateau, Sutton Publishing Ltd, 2001, pp. 257-8.
[34] T. Ruddock, 1981. "Dalkeith House Pediment: Its Form and Construction" in *The Scottish Georgian Society*, Bulletin No. 8.

The introduction of ready-mix rendering mortars for stone imitation in Belgium [1920-1940]

Y. Govaerts[1], A. Verdonck[2], M. de Bouw[3] and W. Meulebroeck[4]
1, 2. Department of Architectural Engineering, Vrije Universiteit Brussel, Belgium; 3. Lab of Renovation, Belgian Building Research Institute (BBRI) and Department of Architectural Science: Conservation of Monuments and Sites - Artesis University College Antwerp, Belgium; 4. Department of Applied Physics and Photonics, Brussels Photonics Team (B-PHOT), Vrije Universiteit Brussel, Belgium

Summary

In the late nineteenth century, the introduction of Portland cement led to innovative compositions towards decorative rendering mortars. These finishes were often developed as a substitute for other, more expensive materials, like sandstone. In order to create an almost perfect imitation of a natural stone masonry, imitation renders were frequently applied on façades. After roughening the texture, simulated joints were drawn into the wet surface. These artisanal imitation mortars were prepared on site by craftsmen, according to their own formula and skills. Generally such mortars contained lime, white cement, crushed stone and mica particles. Due to variations in composition and application technique, almost none of these façade finishes had an identical appearance.

According to Belgian architectural periodicals and plasterer's manuals (1900-1940), ready-mix mortars gained popularity after World War I. Adding water to these dry prefabricated mixtures resulted in a uniform stone-like surface, which met the standard requirements.

This paper sheds light on successfully imported foreign ready-mix mortars, such as *Terranova*, but also on innovative Belgian renders like *Dura* and *Supra*.

Knowledge of how these stone imitating renders have been used, manufactured, mixed and applied is essential to propose viable restoration strategies. By means of written historic sources such as journals, patents, manuals and advertisements, this paper will focus on the technical aspects of the mentioned ready-mix renders and their related socio-cultural background.

Towards premixed renders

With the invention of artificial Portland cement by the Englishman Joseph Aspdin in 1824 [1], the development of new materials moved on apace. Compared to the efforts and time taken for the production of traditional lime mortars, Portland cement was an attractive alternative as binder. [2] Moreover, because of its high quality, easy workability and a quick setting this cement was introduced in concrete mixtures and masonry mortars. Builders also appreciated the particular glaucous mortar colour, which was similar to real Portland stone. As a consequence, Portland cement became gradually an indispensable ingredient for façade rendering mortars. Around 1850 the product was imported into the Netherlands and Belgium, where the render properties were experienced in a positive way. This Portland stone render was presented as a waterproof and durable finish, which implied that the mortar surface would not crack or display staining patterns. Also from an economical point of view, this uniform render was recommended because it was no longer necessary to apply paint layers, and people could build façades using cheaper brickwork. From 1875 onwards, ashlar imitations by using the relative expensive Portland cement were no longer fashionable and were considered as boring and ugly. As a reaction to the less attractive grey cement plasters, people started to manipulate the cement substances by adding pigments or by replacing sand with brick powder or white sand. Unfortunately these interventions did not meet the expectations. Painting cement layers was not an option because the chemical components interacted with the paint whilst hardening. [3] Once again preference inclined towards real natural stones. This process was stimulated by Leopold II, Belgian King and shareholder of several national quarries [4]. In the same period, a German factory in Heidelberg developed white cement for renderings with identical properties as ordinary Portland cement. Due to a lack of interest, it had to cease its activities. [5] However, white cement was rediscovered in the early twentieth century, as a result of a new architectural vision in which 'smoothness' and 'whiteness' were synonyms for hygiene. White cements gradually gained attention, such as *C.B.R. blanc*, *Dyckerhoff Weisz*, *Harmiblanc* or *Snowcrete* [5]. *C.B.R. blanc* was produced by the *Cimenteries et briqueteries réunies* company, which had 6 factories in Belgium. Generally, white cements were crucial to the constituents of ready-mixed stone imitating renders since they were commonly indicated as main ingredient. Later advertisements show that even C.B.R. launched *Simili Euville C.B.R.* and *Simili Carrare C.B.R.*, both ready-mix plasters based on *C.B.R. extra blanc* [6]. New façade materials were imported around 1910, including stone dust 'steenmeel', an important ingredient of artificial sand stone. *Steenmeel* contained crushed natural stone, river sand, cement and colourful pigments. To obtain a convincing stone appearance, the surface was often treated by a professional stonecutter. [7] Probably this is one of the first premixed mortars in Belgium.

Leaving the dynamic curves of the Art Nouveau behind, the Art Deco movement was introduced in the 1920s. This period was characterized by rationality and optimism, which led to clean geometries with sober, but colourful finishing materials. During the interwar period, it was rather uncommon to render new brick masonry façades. [8] Influenced by the traditional Flemish regional architecture and the Amsterdam School Style, brick was again appreciated by prominent Belgian modernists, such as *Van de Velde*, *Hoste*, *Stynen* and *Van Steenbergen* [4]. External plasters were usually applied in case of a concrete support or to hide damage patterns as a result of atmospheric conditions or renovation campaigns. The economic crisis caused by the stock market crash of 1929 seemed to have an influence on the sale of building materials. There was an increased interest in cheap finishes. In order to comply with the needs of the construction industry, a growing prefabrication and standardization process was a logical outcome.

Advertisements in monthly architectural journals show a tendency towards less expensive alternatives for natural stone [9], like stone imitating renders and artificial stones. There were three frequently used techniques to create an imitation of natural stone. On the one hand a special concrete mixture reproduced the aspect, colour and texture of stone. Secondly, an imitative render could be cast into a mould, together with ordinary concrete as inner mass (the so-called *pierre artificielle*). On the other hand a stone-like render could be applied as a layer on wall surfaces. The latter method was economically most interesting. [10] Young monuments illustrate how imitating mortars have been implemented onto façades. Sometimes the whole outer surface was plastered, resembling a stone masonry through simulated joints, for example *De Roma*, a cinema complex from 1928 in Antwerp, Belgium (Fig. 1, left). Other cases, such as *Breivelde Castle* in Zottegem, show a multiple material usage by alternating brickwork with ornamental imitative strips, brackets or rendered bay-windows (Fig. 1, right).

Ready-mix mortars in the interwar period

To combine knowledge concerning the building materials at the time, Dutch professors like *Van der Kloes*, *Wattjes*, *Scharroo*, *Schorteldoek* and others published signifying manuals for practical construction. Their technical literature reached many architects and contractors in The Netherlands and Flanders, who considered it presumably as a standard reference.

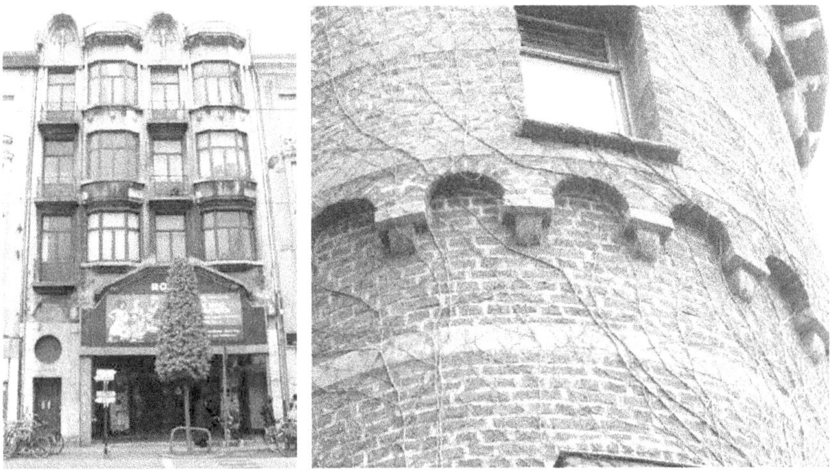

Fig. 1. The De Roma cinema building in Antwerp (left) and Breivelde Castle in Zottegem (right) (Pictures taken by the author, 2013)

In 1929, the manufacturers of Portland cement in Belgium published an overview of render typologies. [10] Due to the nature of crushed stone aggregates and various colour additives, no composition was identical. *Ordinary cement renders* were greyish but uniform mortars, obtained by mixing artificial Portland cement and sand (preferably from the Rhine river) with a ratio of 1:2. A *white stone imitation* was realised by adding white sand or crushed white limestone to bright Portland cement. This binder contained an additional 10% of fat slaked lime. On the other hand, *Sandstone imitation renders* consisted of fine crunched sandstone particles (greenish, pink or red sandstone from the city of Diest) with cement and appropriate colourful additives. *Coloured renders* varied in accordance with the desired shades and contained cement and mineral pigments. The amount of colouring additives could not exceed 10% of the mixture; otherwise it would harm the binding process. [10]

At the end of the interwar period, Poptie described why the building industry had lost faith in the plastering skills. The main purpose of exterior rendering was the protection of the underlying masonry against weather conditions. To achieve such durable mortar layer, plasterers had to prepare the mortar composition according to predefined proportions. However, this was not always the case. Many professionals used coarse lime as a rendering layer, with a finishing layer of a fat lime-cement mix. This mixture sometimes contained partially hardened cement particles. Too fat mortars were applied to create even smoother surfaces than was possible using the available materials. As a consequence cracks appeared in the surface, making it no longer waterproof and hence less insulating. During cold winters, the infiltrated water turned into ice, causing an expansion in the mass, with a poor adherent top layer as a result. [8] Negative experiences with these in situ composed cement mortars probably led towards an increased practice of ready-mix renders. Since the ingredients were already mixed in correct proportions at the factory, a uniform result of high quality was acquired. At least this was the case if a right amount of water was added to the prepacked dry mixture.

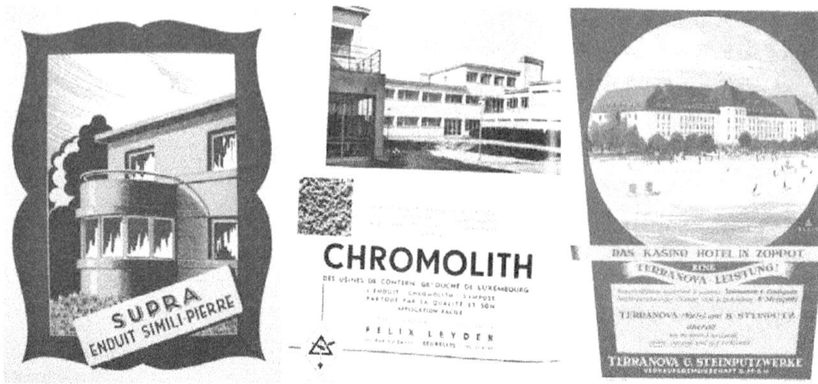

Fig. 2. Advertisements of Supra (AAM, Brussels), Chromolith (La Technique des Travaux, edition October 1936) and Terranova (Sokrates-digital.de)

Decorative plasters for both interior and exterior use were available under different brands, for example *Terranova, Terrasit, Granitine and Chromolith* (Fig. 2). [11, 13, 14] Most of these mixtures were imported from Germany, where an important artificial stone 'Kunststein' industry existed. As a counter-movement, Belgian mortar companies claimed their market shares, launching premixed renders such as *Dura, Simili Euville C.B.R.* and *Supra* (Fig. 2). Unfortunately, the composition of these commercial ready-mix renders was never mentioned due to their trade secret. [12]

Technical background of popular brand names

Analysis of seven Belgian architectural periodicals [12] between 1919 and 1940 highlights the use of *Terranova* in modern case studies of leading architects. [16] Although *Terranova* advertisements are rare, it is mentioned by its trade name in articles, unlike other brand names, which were often referred to as ordinary 'stone imitating renders'.

Moreover, *Terranova* was the only trademark, which was extensively described and discussed in the Dutch plasterer's manuals. [7, 14, 15] However this product entered the Belgian building industry just after World War I, and it existed already for almost 25 years. Carl August Kapferer developed the *Terranova* render in 1893 near Freihung, Germany. At the turn of the century, the company's success enabled it to expand and set up production units throughout Europe. [17] Exporting to the neighbouring countries, including their colonies, was common practise. Three different *Terranova* mixtures were available, varying from coarse to fine-grained aggregates. They were designed to simulate the appearances of bright sandstone. Depending on the type, one bag of 50 kg dry powder was needed to finish a surface of approximately 1.8 to 3.5 m^2. [14]

Given the emerging trend to use simple cheap materials, Belgian firms also started to manufacture alternatives for natural stone after World War I. The *Dura* trademark was certified in 1928 as general name for "powder for coverings", artificial stones, concrete elements and other building materials. [18] Five years earlier, Adriaan Martens had established a small family business *NV Kempische Betonwerken* in Kalmthout (Antwerp), which had a considerable impact on the local architecture. *Dura* renders and artificial stones mainly appeared in the Antwerp region, where they were used in different building typologies such as mansions, hotels, pubs, social housing, schools, sculptures and even castles. [19] The workshops were situated next to the Kalmthout train station. This strategic location enabled the company to import raw materials from thc Walloon region and to export products, using their own private track. Employees had to work outside beneath a wooden shelter where they composed the mixtures, cast artificial stones and texturized them according to the wishes of the customers (Fig. 3). The light limestone coloured *Dura* mortar also had an indoor application, which is illustrated by the interior of the public swimming pool in Antwerp (Veldstraat, 1932). During the 1930s, it was essential to finish pool halls in a hygienic, sustainable way. Because of the high humidity levels, it was no longer allowed to use wood or iron elements, as they would degrade prematurely. In this case, a 'water resistant French stone finish' was applied above a tiled plinth. [20]

Between 1920 and 1930 the Brussels Etablissements Cantillana company also produced a French stone imitating render *Supra*. [25] Cantillana was established in 1875 and the firm still exists today. Specializing in plasters for interior and exterior application and pointing mortars, Cantillana currently has outlets in the Benelux, Germany and France. [27] *Supra* was promoted as a waterproof plaster with a hardness of the best available concrete. The dry mixture contained crushed stones to obtain different colours, just like the other two renders. One 50 kg bag was sufficient to apply a mortar layer with a thickness of 6 mm on a surface of 4 m². [25]

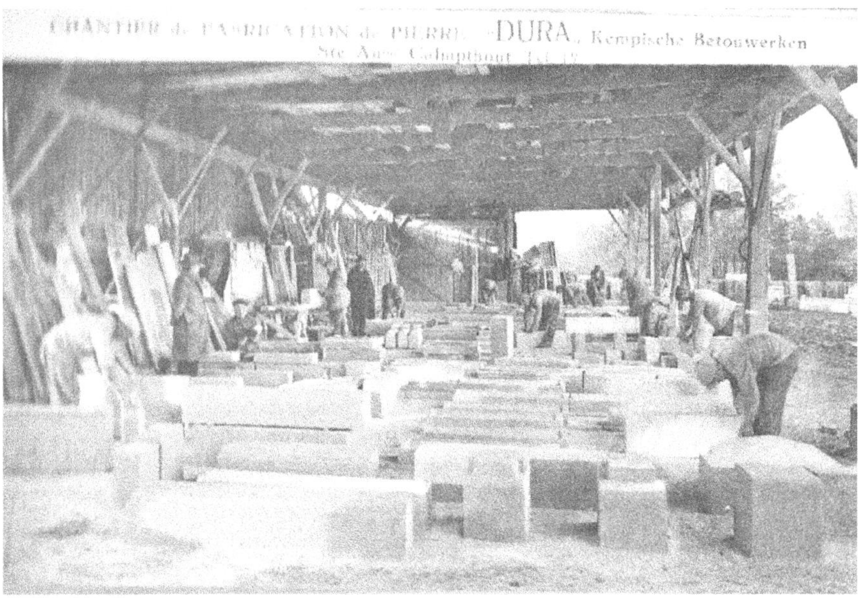

Fig. 3. Manufacturing artificial stones at the Dura factory anno 1936 (Private archive Gilbert Uitdenhouwen)

Rendering layers

Every exterior decorative plasterwork is composed of two or more layers. Base layers or renderings are intended to create a flat surface, which has to provide sufficient adhesion with the top layer at the same time. Advertisements and historic manuals recommend unanimously following ratios when applying ordinary *Terranova*: one part of artificial Portland cement, one part shell lime and five parts of coarse sand, mixed with an appropriate volume of clean water. These proportions are also valid for the *Supra* render, besides the fact the manufacturers recommended usage of slaked lime or hydraulic lime from the city of Tournai instead of shell lime. Since it was advisable to have a rather absorbing base layer, a lime cement mortar was a good solution. In case of the coarser *Terranova grana*, a rendering layer with one volume part of cement and three parts of sand was appropriate. [14]

Unlike these written instructions, optical analysis of such layers indicates the presence of four units of binder material to five units of aggregate. [21] The binder can be classified between strong hydraulic lime and early cement. The aggregate is mainly quartz sand with a smaller amount of feldspar, trass and stone fragments. This layer is applied with a thickness of about 5 mm, but has to be approximately 15 mm according to the regulations [7, 14]. Although the ingredients are correct, the proportions used are not in accordance with regulations. The fatter the mixture, the more shrinkage and stresses will occur in the surface. Finally, to improve the adhesion with the top layer, grooves are drawn into this wet preparation layer.

Determining historic compositions

Since authentic formulas of these prepacked plasters are never mentioned, it is not obvious how to reproduce them for restoration purposes. An extensive patent search and complementary laboratory analysis can provide useful information concerning original compositions.

After the trade name *Terranova Industrie* was certificated in 1896, all produced materials (plaster, imitation stone and imitation facing brick) were called after their manufacturer.

Developper C.A. Kapferer requested a few patents concerning mortar compositions without specifying his trademark, but we can assume he means *Terranova*.

The oldest accessible patent dates from September 4, 1899. The focus lies on improvements in the manufacture of cement. [22] The goal of the patent was to obtain a more attractive colour and to use this cement both for decorative as well as structural purposes. Clay (kaolin), feldspar (orthoclase), lime and pure quartz sand, the main components of the formula, were burned to produce cement clinker. On one hand feldspar was necessary for the production of soluble silicic acid, which was required for binding and on the other hand it improved the workability. In 1911 *Terranova* becomes a ready-mix dry mortar and it was no longer recommended to paint the rendering after application, because the mortar itself had improved its aesthetic qualities. "*... The new plastering mortars were employed where the plastering was not to be subsequently coated with paint. Very fine architectural effects were obtained by staining the mortar, by adding granular rocks and minerals to the slaked lime ...*". [23] Another patent suggested adding a mixture of oil, acetone and starch substances in order to distribute the additives uniformly into the lime paste. In 1926, an 'improvement patent' referred to mortars for architectural purposes. To create a sparkling, stone imitating effect, particles of mica, natural stone or stone-dust were added to the top layer. Apparently, the lime particles seemed to cover up these micas. To avoid such visual complications, organic additives like glycerines were added.

A reasonably intact *Terranova* finish can be found on the façades of *De Roma* (Fig. 1, left). Analysis of a sample taken from the west façade and examined in the laboratories of the Royal Institute for Cultural Heritage provides additional information. [21] Since calcium, silicon, and a small amount of magnesium, potassium and aluminium (feldspar) were discovered, the binder can presumably be identified as a combination of white cement and lime. Note that the composition of historic 'white cement' was different from ordinary white coloured cement. 'White cement' was known as heavy white lime mixed with about 10% of bright coloured cement. [28] Subsidiary companies in Italy used Duralbo cement (an extremely white high-strength Portland cement) and Cromocemento (a tinted high-strength Portland cement) to compose the binder [17]. The aggregate is composed of 50% quartz sand and 50% limestone fragments, which are derived from coarse crushed Euville stone. A catalogue from 1936 proclaimed that the stone fragments consisted of Muschelkalk, Travertine and Dolomite rock. [13] The sand particles are characterized as angular, average coarse units. As expressed in the patents, muscovite particles with a length of a few millimetres are also present in the sample. Only the ratio of binder to aggregate is a missing factor, due to the presence of limestone particles in the aggregate. [21]

An example of the *Dura* plaster can be found on the awning of the Sint-Lievens College in Antwerp (1929). A sample was lifted and again examined by the Royal Institute for Cultural Heritage. Three layers were detected through optical microscopy: one coarse granular layer and two upper fine-grained layers. The top layer has a thickness of about 3 mm and contains sharp sand particles (50 – 250 μm) with a small amount of potassium feldspars. Crunched lime fragments and cement traces are noticed, which indicates the use of a *bad mortar* (a mix of Portland cement and hydraulic lime). Over the years, the *Dura* render has been painted several times. An interview with a former employee revealed the ratio of binder to aggregate. Three volume units of aggregates were mixed with one volume unit of binder, which was white cement. [24]

Finishing techniques

Depending on the grain size of the finishing mortar, different layer thicknesses were suggested. It was recommended to apply a layer of 5-8 mm for a fine structured render, 10 mm for medium and 12 mm for coarse mortars. [7, 25] Choosing a thinner top layer would involve an exposure of the base layer, because the final treatment included some material losses. The top layer was always applied on a pre-wetted rough rendering layer using trowel and float. *"... The final stage is achieved by means of a sharp steel lath, making it possible to scratch the Terranova layer: it is clear that this action can only be carried out when the top layer is hardened and dry ..."* . [7, 26]

Through the scratching procedure, loose mortar particles were removed from the wall surface, creating a rough stone appearance (Fig. 4). Another source suggests the use of a commercial 'Terranova Scratcher', which was nothing but a wooden plank perforated with iron nails. [15] The application technique of the coarser version *Terranova grana* was slightly different. After roughening the preparation layer, the top mortar was pressed with a wooden float to compress the grains. As a consequence, binder sludge would come to the surface and had to be wiped off using a brush before the hardening process initiated. In order to obtain a regular surface, it was important to repeat this treatment. Three or four days later, the remaining binder sludge needed to be removed with diluted hydrochloric acid. [15] *Supra* could be finished as an ordinary cement render, with particular attention to the grain compression. Depending on the season, craftsmen had to wait 8 to 12 days before cutting or chiseling the hardened mortar. After scratching, brushing, combing or cutting the plasterwork, simulated joints could be drawn into the surface by means of a small wooden instrument. Finally these joints were filled with a greyish cement to represent a genuine natural stone masonry (Fig. 5).

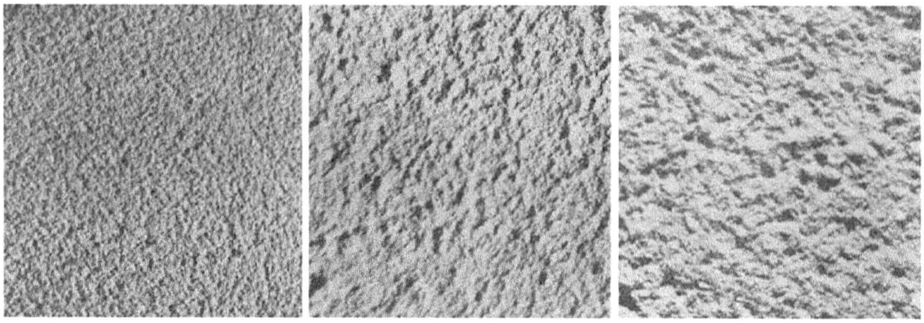

Fig. 4. Examples of different grainy mortars after scratching process (A. Poptie, Handboek voor den stucadoor, 1948)

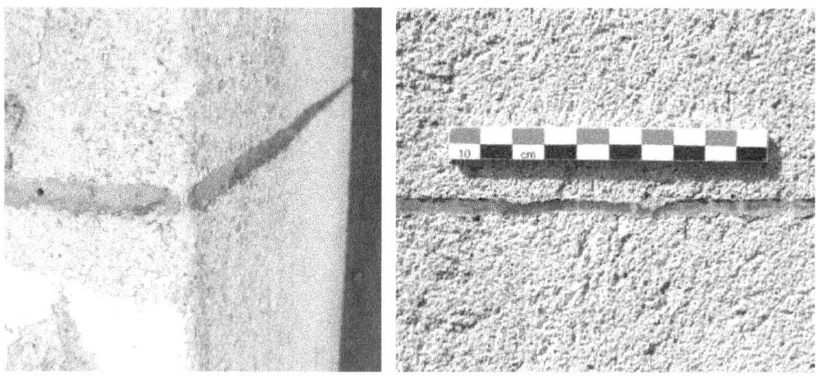

Fig. 5. Simulated joints to create a masonry aspect, case De Roma (Pictures taken by the author, 2012)

Conclusions

This paper gives an overview of frequently used premixed renders for stone imitation in Belgium during the interwar period. Starting with the development of Portland cement, a remarkable evolution towards the introduction of popular brand names was briefly discussed in a socio-economical context. Focussing on material-technical aspects of *Terranova*, *Dura* and *Supra* allowed us to understand their raw composition and texturing techniques. However, further research is essential to gain insight into other historic renders. Due to an evolving policy of listing valuable interwar heritage, architects will rediscover these finishes soon during restoration campaigns. Because of the limited knowledge in modern literature, it is not always clear how to deal with these materials. Current restoration procedures often result in a reproduction of the plaster by means of trial-and-error. To optimize this time-consuming procedure it is important to delve deeper into the historic formulas, verify them and put them into practice.

References

[1] A. Henry and J. Stewart (eds.), *English Heritage Practical Building conservation: Mortars, renders and plasters.* Surrey: Ashgate Publishing Limited, 2011, pp. 18-19.

[2] B. Bennett, *The development of Portland Cement*. Retrieved from: <http://www.buildingconservation.com/articles/prtlndcmnt/prtlndcmnt.htm> [accessed on 13 December 2013]

[3] P.C. van der Heiden, "Portlandcementpleister: geen kleur is ook een kleur", in *Stuc, Kunst en Techniek*. Zwolle: Waanders, 2010, pp. 392-405.

[4] S. Migom, "Steen voor steen, een proeve tot architectuurgeschiedenis van Vlaanderen vanuit materiaaltechnisch standpunt", in *Grondstof, stof tot nadenken: Steen*. Oostkamp: Stichting Kunstboek & Erfgoed Vlaanderen, 2003, pp. 28-29.

[5] P.W. Scharroo, *Cement en Beton: Oud en Nieuw*. Amsterdam: L.J. Veens Uitgeversmaatschappij N.V., 1946, pp. 196-198.

[6] Simili advertisement, La Technique des Travaux, Brussels: Archives d'Architecture Moderne, May 1936.

[7] A. Poptie, *Handboek voor den stucadoor (deel 2)*. Haarlem: N.V. De Technische Uitgeverij H. Stam, 1948, pp. 18-37.

[8] A. Poptie, *Handboek voor de stucadoor (deel 1)*. Haarlem-Antwerp-Djakarta: N.V. De Technische Uitgeverij H. Stam, 1950, pp. 119-124.

[9] *Les ciments blancs en architecture commerciale*. Bâtir, nr. 9, 1933, pp. 350.

[10] *Pierre artificielle et décoration élémentaire*. in *Béton de ciment Portland artificiel*, nr. 7, Brussels: Groupement professionnel des fabricants de ciment Portland artificial de Belgique, 1929, pp. 42-46.

[11] C.W.J. Schorteldoek, *Het uitvoeren van gebouwen*. Amsterdam: Kosmos, 1920, pp. 250-252.

[12] Architectural periodicals *Bâtir*, *La Cité*, *L'Emulation*, *Clarté*, *La Technique des Travaux*, *Le Document* et *L'Equerre* were consulted at the Archives d'Architecture Moderne in Brussels.

[13] R. Stenvert, "Mooier voor minder: cementlagen en betonafwerking", in *Stuc, Kunst en Techniek*, Zwolle: Waanders, 2010, pp. 412-425.

[14] A.P. Oosterhof, *Handboek voor den stucadoor*. Amsterdam: Van Mantgem & De Does, 1932. pp. 62-71.

[15] G. Arendzen & J.J. Vriend, *Constructie en uitvoering van Modern Stucadoorwerk*. Amsterdam: NV Uitgeversmaatschappij Kosmos, 1936, pp. 53-58.

[16] "L'architecture en Belgique: Les Travaux de Jan B. Lauwers", La Cité, Vol. VII, nr 7, Brussels: Archives d'Architecture Moderne, 1928. pp. 85-95.

[17] E. Garda, "Terranova, history of a modern plaster: Smooth, hard, clean, perfect", in S. Huerta and J. de Herrera (eds.), *Proceedings of the First International Congress on Construction History*, Madrid: Instituto Juan Ferrera, 2003, pp. 970-977.

[18] Marques de fabrique et de commerce: *DURA*, nr. 7047, Brussels : Belgian General State Archives, dated 02/03/1928.

[19] Private collection of Mr. Gilbert Uitdenhouwen, residing in Kalmthout.

[20] L. De Clerq, R. Mertens and E. Denissen, "Het zwembad aan de Veldstraat in Antwerpen", in Monumenten, Landschappen en Archeologie, year 31, nr. 2, 2012. pp. 6-33.

[21] *Decoratieve bepleisteringen, materiaalanalysen, Staal M8, pierre-simili bepleistering in Borgerhout*, Scientific report, Brussels : KIK-IRPA, Royal Institute for Cultural heritage, 2011. 11 pages.

[22] C.A. Kapferer and Schleuning, *Improvements in the Manufacture of Cement*. The Firm of Terranova Industrie, patent 17853, dated 04/09/1899, Munich: Online German patent database DEPATISnet. 3 pages.

[23] C.A. Kapferer and J. Weber, *Method of improving the permeability to air of dry plastering-mortar*, patent 1180775, dated 15/08/1911, United States: Online German patent database DEPATISnet. 2 pages. Based on C.A. Kapferer, *Verfahren zur Verbesserung der Luftdurchlässigkeit von Trockenputz Mortel*, patent 74852, dated 30/08/1910, German Empire: Online German patent database DEPATISnet. 3 pages.

[24] *Antwerpen, Sint-Lievenscollege Mortelanalyses: Staal 3,* Scientific report, Brussels : KIK-IRPA, Royal Institute for Cultural heritage, 2006.
[25] *'Supra' Enduit Simili-Pierre*, Advertisement, Brussels: Archives d'Architecture Moderne, not dated.
[26] Translated from Dutch by the author.
[27] Cantillana, The history. Retrieved from: <http://www.cantillana.com/Organisation/History/tabid/62/language/en-GB/Default.aspx> [accessed on 30 December 2013]
[28] E. Leduc, *Chaux et Ciments*. Paris: Baillière J.B. & Fils, 1902, pp. 115-116.

The evolution of building techniques in the Italian Alpine context from the 1950s-1960s

The case of the Church in Corte di Cadore by Edoardo Gellner and Carlo Scarpa (1956-1961)

Laura Greco
Department of Civil Engineering, University of Calabria, Italy

Introduction

Alpine architecture in Italy underwent a powerful development in the twentieth century consequent to the diffusion of winter sports. During this period, the repertoire of architectural types and construction techniques revealed a renewal that took into account the general evolution in the Italian context. The phenomenon was more intense in the Alps during the period from the 1950s to the mid-1960s, affecting both the residential sector and hotels, in concurrence with the advancement of tourism.

The progress of building construction dealt with the utilization of steel, the employment of cast in situ reinforced concrete elements and precast processes, the improvement of systems, and the development of exterior wall insulation methods. In this frame of reference the use of reinforced concrete techniques increased, whose effectiveness in renewing the aesthetic and technical paradigms of buildings was predominant over other issues, which for several reasons were less significant.

In fact, steel construction was uncommon at that time in the Italian context, due to the difficulties of production of metallic components and the unfamiliarity of designers and builders with this procedure. The use of steel construction remained limited to specific typological sectors (factories, offices) and geographic areas (Milan, Genoa, Turin) only affecting alpine areas episodically, taking into consideration the insufficient adaptability of this method to mountain weather conditions. The insulation of walls and roofs, and the improvement of systems, were specialized aspects of the project. They also could not modify the main architectural features of buildings. However, the use of reinforced concrete techniques allowed the rationalization of the construction process of buildings (hotels, residences) with unusual proportions in the mountain landscape and promoted the renewal of architectural characters. In fact, the traditional wood-masonry construction system was revised encouraging the combination of wood and reinforced concrete elements in the aesthetic and construction organization of the buildings.

The reasons of technical feasibility were reinforced by cultural motivations promoted by architects as Carlo Mollino (1905-1975) and Gio Ponti (1891-1979), who, in their works, transferred the reinforced concrete procedure developed in the urban areas of Northern Italy. They appointed the technique with the role of defining the new stylistic identity of alpine architecture.

Modern construction in the Italian Alps: the role of the reinforced concrete technique

The beginnings of reinforced concrete structures in the alpine areas date back to the mid-1920s, and particularly affected the regions of Valle d'Aosta and Piedmont. The first hydroelectric power plants, dams, and small bridges were developed in the mountains, while downstream, offices and factories were build around the cities.[1]

In the second half of the 1930s, Gio Ponti completed a housing and shopping complex as well as the *Palazzo Assistenziale Nazionale* in Cogne, near Aosta. In the new holiday centers of the Italian Western Alps, housing and hotels were built. Among these buildings we note the *Duca d'Aosta* and *Torre* hotels (1932-1934) designed by engineer Vittorio Bonadè Bottino (1889-1979), within the urban planning of Sestrière, and the colony *IX Maggio* (1938), designed in Bardonecchia by the architect Gino Levi Montalcini (1907-1975).

After the Second World War, the reinforced concrete technique was confirmed as the favored procedure of a new step in the evolution of construction in mountainous areas. In this background Gio Ponti developed a second series of hydroelectric power plants [2], while the architect Carlo Mollino designed a group of villas and housing such as the *Casa del Sole* in Cervinia (1947-1955), the station-hotel in Saulz d'Oux (1947-48), and the *Maternità* housing and shopping complex (1951-1953). These buildings stood out for their reinforced concrete structures.

Edoardo Gellner (1909-2004), one of the most enterprising designers in the Italian Eastern Alps, carried out a research path in the 1950s, which was articulated between applications and theoretical considerations, suggesting exposed concrete as an element of stylistic innovation in mountain architecture. [3]

In the holiday village for the employees of the *Ente Nazionale Idrocarburi* (Eni), which was built in Corte di Cadore near Cortina d'Ampezzo between 1955 and 1963, Gellner developed his most ambitious architectural program. The holiday village in Corte di Cadore consisted of 600 single-family houses, a colony for 600 children, a camp for 200 boys, a church, hotels and a community centre. [4]

In the Eni village, the designer attempted to manage the timing of the project process and to industrialize construction through the use of precast reinforced concrete structures. Gellner first utilized precast concrete panels for the prototypes of the houses, but later abandoned them in favor of lightweight prefabricated systems (i.e. steel frameworks and sandwich façade panels) combined with traditional techniques. Gellner also studied the employment of precast reinforced concrete frameworks in the village community centre, but he abandoned this choice during the design process. In fact, the architect had to use mainly cast *in situ* reinforced structures, thus the industrialization of the construction process was limited to the panel façade, wood frames and furniture, which was designed by Gellner with the collaboration of factories.

The church of *Nostra Signora di Cadore* was one of the most powerful buildings of the village. (Fig. 1) Gellner collaborated with the architect Carlo Scarpa (1906-1978) in the design process. Concrete was the chosen material for the church. This building now represents a singular approach to the use of exposed concrete in the Italian Alps from that period. The building did not employ prefabricated elements in the final project. The designers in this case focused their attempts on the conjugation of the reinforced concrete technique with the landscape of Dolomites. The construction method reproduced the rocks of Dolomites surrounding the building in the sculptural effects of the exposed concrete walls. The link between the almost intact natural alpine environment and the artificial signs of the concrete mark the identity of this building, designed to live in a unique landscape using a technique that was born to be universal.

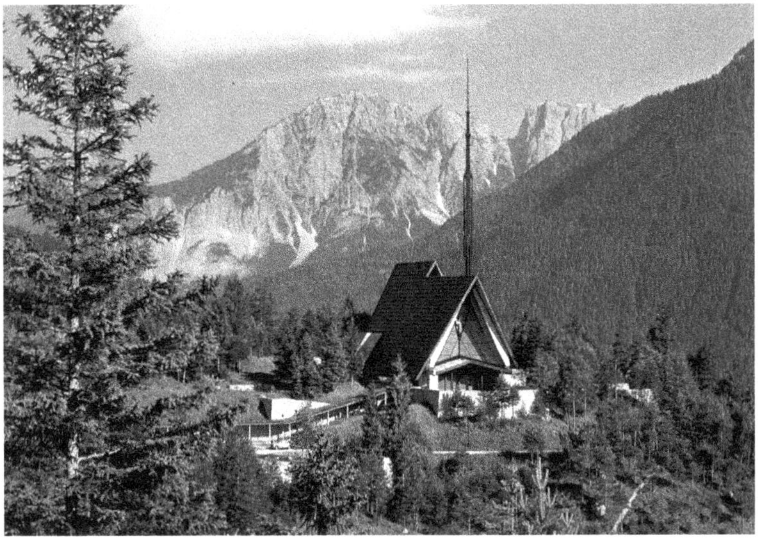

Fig. 1. The church in the landscape. In evidence the pitched roof (26 meters high) and the narrow bell tower (65 meters high). (Source: Eni Historical Archive, Rome)

It is worth mentioning that in the 1950s Brutalism took shape on the international scene, a movement that based its aesthetic program on the use of exposed concrete, continuing throughout the 1960s-1970s. In the Italian context, studies on this topic affected the research of engineers such as Pier Luigi Nervi (1891-1978) and only the works of a few architects such as Leonardo Ricci (1918-1994), Leonardo Savioli (1917-1982), and Vittoriano Viganò (1919-1996). Among the Italian architects was Carlo Scarpa, previously mentioned for designing the church in Corte di Cadore with Edoardo Gellner. Scarpa played a leading role in the exposed reinforced concrete architecture as evidenced by his projects culminating in the Tomba Brion (1970-1972) and the cemetery of Sanvito near Treviso (1973). He focused his work on the study of the mixture and the formwork to extend the aesthetic choices of the concrete technique.

The church of Nostra Signora di Cadore

The church in Corte di Cadore has a polygonal plan arranged in three naves. It can accommodate 350 worshipers. [5] The profile of the church is marked by a pitched roof (26 meters high at the ridge) and a narrow bell tower (65 meters high). The most widely used materials were concrete and steel for the structures. Wood was employed in the fronts and, with natural stone, in the interior design. The manse completed the project. The building is identified by a fruitful relationship between the architectural concept and the project of the reinforced concrete structures. This awareness was encouraged in early studies by the collaboration of Gellner with the Italian engineer Silvano Zorzi (1921-1994). In this case, the designers used concrete to connote the aesthetics of the church, enhancing the surface treatment of the exposed elements thanks to methods that had been carefully studied to create a connection between the building and surrounding landscape.

The church design process began in the summer of 1956, when Gellner received confirmation of the assignment. At this point, he involved Scarpa and they began to work on the preliminary project in early September. The drawings of this phase were dated 30 September 1956. [6]

Scarpa also worked on the project in 1957 and, for some periods between 1958-1961.

The first sketches showed a central plan scheme, which later evolved into the solution with three naves developed in the project dated December 1956. In the second scheme, the structure of the church was composed of two groups of columns on which lay longitudinal beams that supported the structure of the roof, based on ribs and horizontal rods. The plan was concluded by an asymmetrical transept. (Fig. 2) This was accomplished by a wall that broke the alignment with the nave and fold out, defining two asymmetric heads. On the opposite side, Scarpa studied the narthex and a side chapel. Here, the entrance wall was aligned parallel to the wall of the transept and ended with the side chapel. Two revolving doors marked the access to the church.

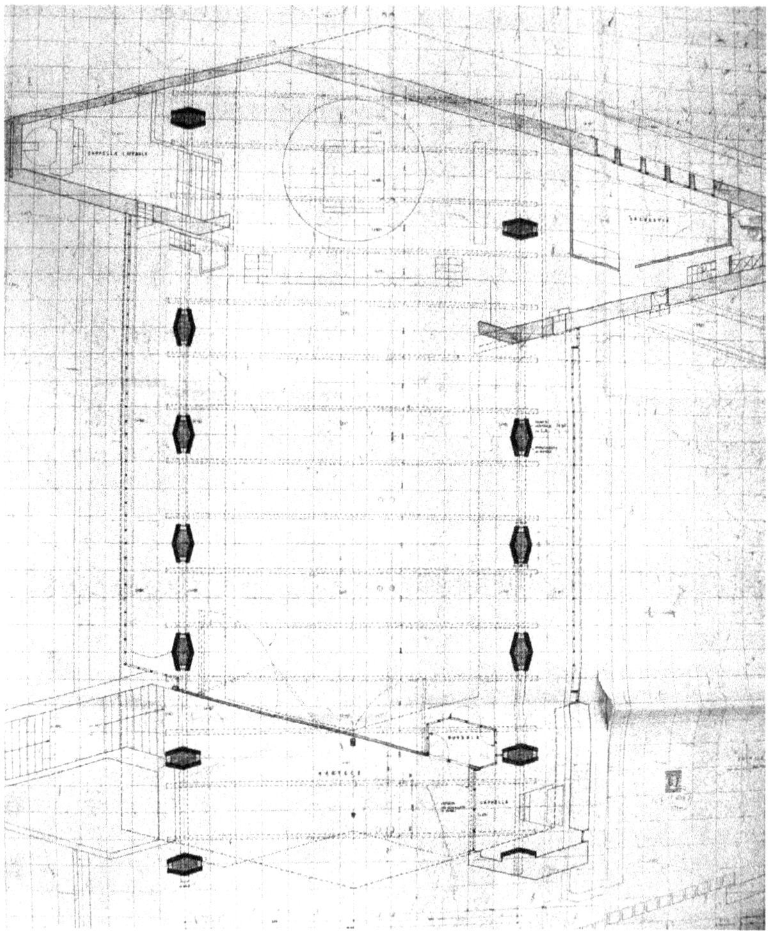

Fig. 2. The plan showing the scheme with three naves. In evidence the hexagonal reinforced concrete pillars. (Source: Eni Historical Archive, Rome)

During the design process the horizontal tie rods, were replaced by a pair of cross-diagonal rods, in accordance with the suggestion of Zorzi. In section 'A', dated 10 December 1956, the scheme of the roof consisted of prefabricated reinforced concrete trusses supported by longitudinal beams placed on top of reinforced concrete hexagonal pillars. Between the trusses were pairs of cross metal rods to brace the concrete structure. [7]

The 1956 project was revised. The new version of 1957 confirmed the plan, but reduced the modular grid of the building. In the first project, Gellner chose a module of 91. 5x91. 5 cm. In the 1957 solution he set a module of 122x122 cm, the distance between the pillars was 488 cm (122x4), and between the trusses 244 cm (122x2). Revisions of the solution developed during the spring of 1957 and continued up until the beginning of construction in the summer of 1958. The scheme for the roof was modified, in accordance with suggestions made by the engineer Carlo Cestelli Guidi, the new structural consultant. (Fig. 3) The rods were placed in the vertical plan and the section of the ribs and pillars were reduced. In reality, Gellner accepted only part of the proposals from Cestelli Guidi. While he reorganized the group of rods according to the instructions of the specialist, he also kept the crossed rods. [8] Gellner based his choices not only on structural advantage, but tried to combine the aesthetic value with engineering efficiency. He was determined to use the structural scheme and its parts as architectural key-figures of the church.

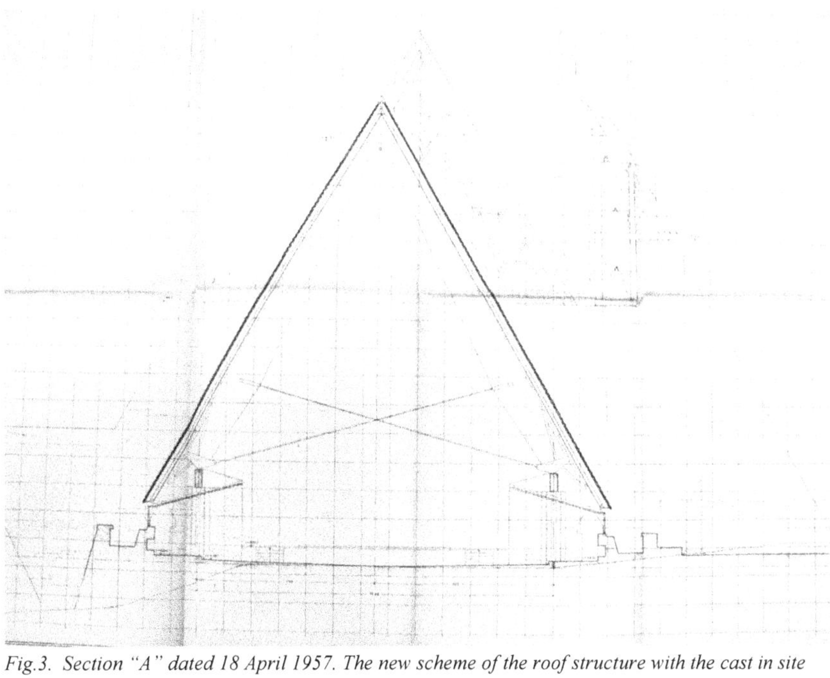

Fig.3. Section "A" dated 18 April 1957. The new scheme of the roof structure with the cast in site reinforced concrete ribs. (Source: Eni Historical Archive, Rome)

The designer employed the structural elements to order the interior space, articulating the depth and the height under the pitches, marking the perspective of the nave. The ribs were also modified. The use of prefabricated elements, planned in the project of 1956, was abandoned and instead they were replaced with cast in situ, developed by the general contractor S.I.C.C.I. [9] The section 'A' dated 18 April 1957 showed the new scheme of the roof structure.

Gellner and Scarpa studied pillar designs through a series of sketches concerning geometry and dimension, but at the same time, these elements were considered as part of the aesthetic properties of the building. They moved from different ideas on this topic, with particular reference to choices concerning the cladding of the hexagonal pillars. While Gellner preferred the use of large red stone slabs, Scarpa favored the use of stone plates with different heights, according to the solution developed by the architect in the Olivetti showroom in Venice (1957-1958). A drawing dated 1959 shows the final solution. The concrete core of the hexagonal pillar was clad on four sides with monolithic slabs of red stone, and on the other two finished with a ten centimeter layer of concrete cast in wood formwork constituted by planks of different sizes. (Fig. 4) Electric ducts were integrated in the cast. [10] Gellner and Scarpa also studied the cramps supporting the stone slabs. Gellner suggested thin metallic elements, and Scarpa proposed others with 20 millimeters thickness. Finally, Scarpa's idea was chosen and the red stone slabs were supported through cast iron plates bolted to bars inserted in the concrete layer. [11]

As mentioned, the designers used the exposed reinforced concrete to link the building to the landscape. The walls that indicated the volume of the transept and supported the terrace and the external ramp, assumed a leading role in the expressive connotation of the church. (Fig. 5) Gellner and Scarpa studied the cast in situ reinforced concrete process to enhance the aesthetic values of the exposed surfaces through the composition of the mixture and the design of the formwork. On the front side, near the chapel designed by Scarpa, there is another exposed reinforced concrete wall. Recent studies examined a portion of this wall. [12] The results demonstrated some details of the construction. The mixture was formed by split components of different diameters (sand 0-3 mm, chips 3-5 mm, fine and average crushed stone measuring respectively 5-20 mm and 20-40 mm). The wooden formwork consisted of horizontal staggered planks. The exposed surface now presents irregular and large porous areas. These technical choices suggest a link with another project by Scarpa carried out by the architect for the Venezuela Pavilion at the Biennale of Venice (1954-1956). In this case the designer used the same type of formwork and disarming. The exposed concrete surfaces have

a similar appearance. The correspondence between the two works allows you to hypothesize the influence of Scarpa in the church project. At the same time, the observation of the other buildings of the Eni village, and in particular of the exposed concrete walls which Gellner used to arrange the embankments, shows that the aesthetic choices of Scarpa were approved and also used by Gellner.

These rough surfaces of reinforced concrete, which are both similar to the walls used by Gellner in the other buildings of the village, and linked to the works of Scarpa, definitively marked the architecture of the church. They became the emblematic representation of the organic relationship between building and land, which was one of the key points of Gellner's architecture in alpine environments.

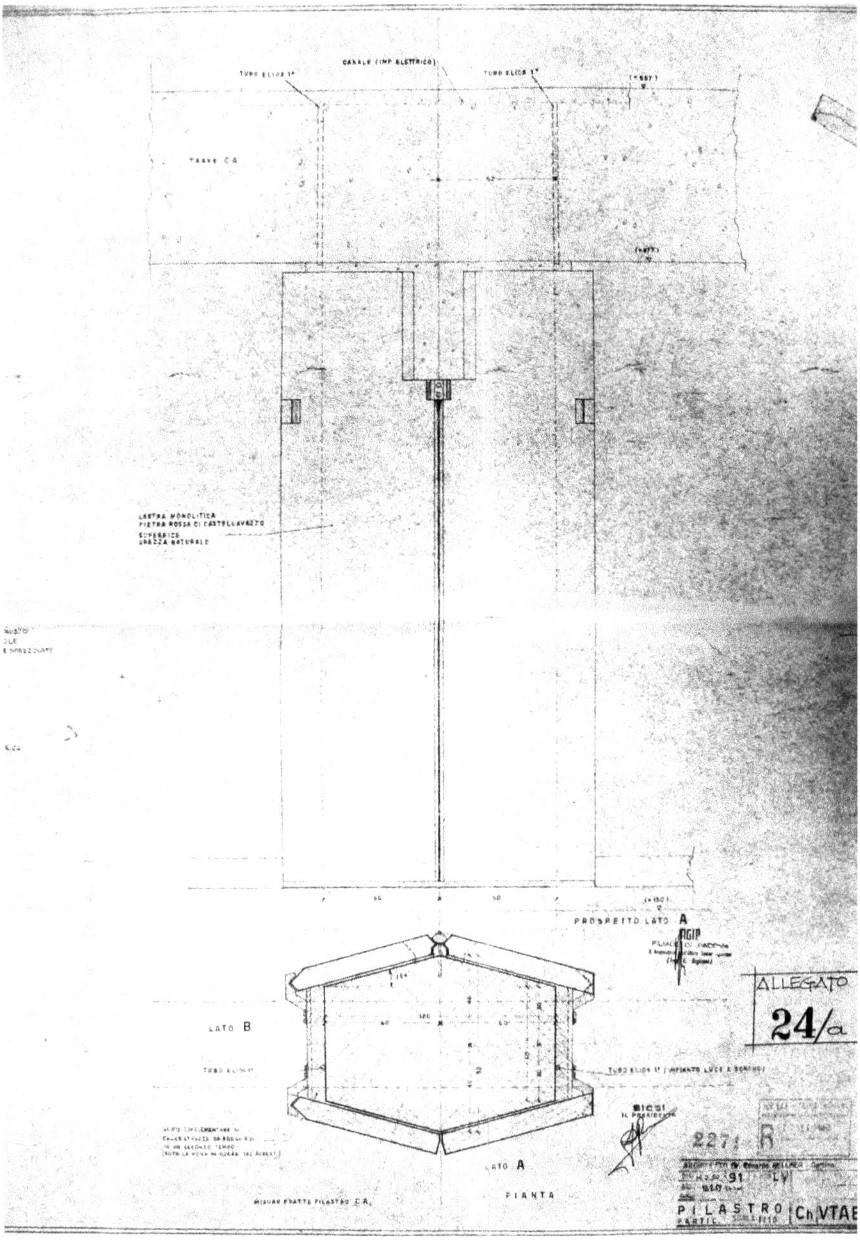

Fig. 4. A detail of the hexagonal pillar showing the cladding of the reinforced concrete core with red stone slabs. (Source: Eni Historical Archive, Rome)

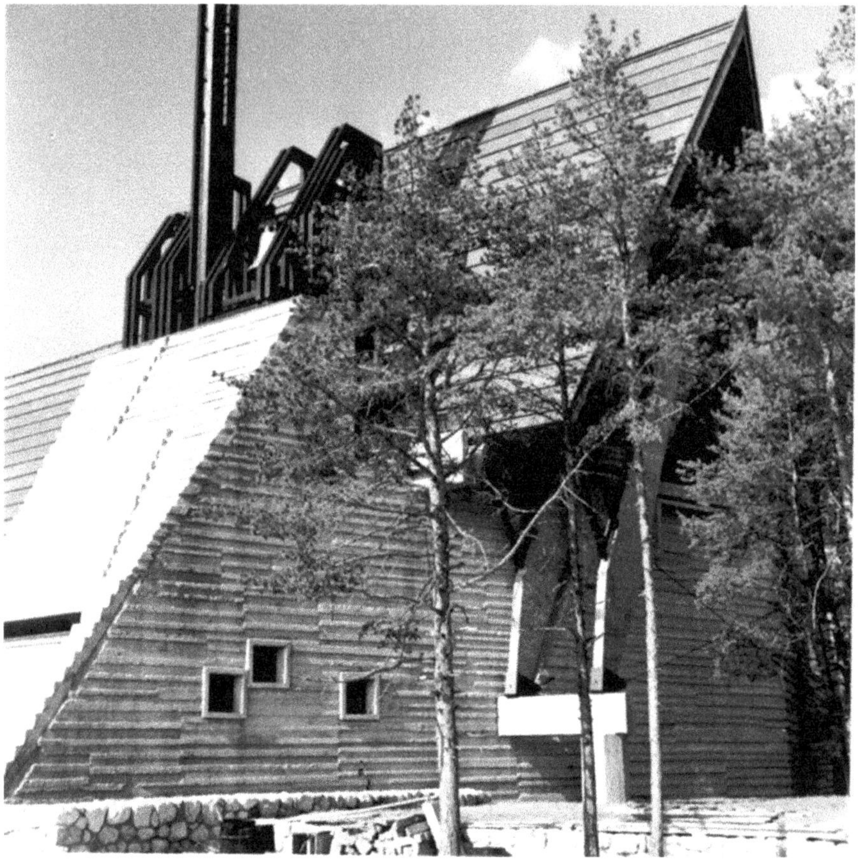

Fig.5. One of the exposed concrete walls of the church. In evidence are the horizontal signs of the wood formwork. (Source: Eni Historical Archive, Rome)

Conclusions

The case of the church in Corte di Cadore documents the evolution of building techniques after the Second World War in the Italian alpine geographical and cultural context. The example allows you to highlight points of affinity with the national framework of that time such as the prevalence of reinforced concrete on metal construction, the poor diffusion of prefabricated systems and the use of reinforced concrete to qualify the style of emblematic works, which marked the renewal of ways of life in Italian society. This correspondence between the construction technique innovation and the representation of the modern national economic and social program finds analogies with what happened in the rest of the country on the occasion of major events such as the Olympic Games (Rome 1960), the International Expositions (Turin 1961), the highways program (*Autostrada del Sole* 1956-1964) whose architectural and engineering achievements marked an advancement in the evolution of building construction in Italy in the twentieth century.

Likewise, the analysis of the alpine cultural context highlights some aspects of specificity The Eni village and the church in particular, evidence the desire of Gellner to protect and enhance the environmental features of a beautiful countryside. This regard for environmental features suggested to designers such as Gellner worked on the experimentation of the concrete mixture and the specialization of the construction process. This consequently led to a harmonization of the aesthetic program of buildings with the natural features of the contexts, limiting the use of standardized prefabricated systems

In a wider geographical and cultural framework, the Italian experience offers insight into the international situation to investigate the correlations and disparities between the diffusion and specialization of modern techniques in a particular landscape, raising questions about the delicate balance between the rationalization and universalization of the construction options and the identity properties of the places. This balance, investigated in the Italian cultural debate of the time, was considered to be at the base of the relationship between modern architecture and the history of contexts. In this sense, we consider the events of the church in Corte di Cadore as part of a more general and complex reading of the international building techniques carried out by Italian architectural scholarships and designers and documented in urban areas by the works of Franco Albini, BBPR, Gabetti and Isola and Carlo Scarpa.

References

[1] See L. Moretto (ed.), *Architettura moderna alpina in Valle d'Aosta*. Quart: Musumeci editore, 2003, pp. 13-28.

[2] See L. Bolzoni, *Architettura moderna nelle Alpi italiane dal 1900 alla fine degli anni Cinquanta*. Torino: Priuli & Verlucca, 2000.

[3] For a summary on the work of Edoardo Gellner see F. Mancuso, *Edoardo Gellner. Il mestiere di architetto*. Milan: Electa, 1996, pp. 111-126.

[4] For a summary on the project and the construction of the village, ibid., pp. 179-207.

[5] The drawings of the Church are preserved in the *Archivio del progetto* IUAV Venezia (IUAV from now on) and in the *Archivio storico* Eni in Rome (from now on ASE).

[6] For the chronology see E. Gellner, F. Mancuso, *Carlo Scarpa e Edoardo Gellner. La chiesa di Corte di Cadore*. Milan: Electa, 2000.

[7] See drawing n° 30 'Sezione A', dated 30 December 1956 (IUAV).

[8] Drawings n° 47 'Sezione A' and n° 46 'Sezione M' dated 18 April 1957 (ASE).

[9] For the longitudinal beams supporting the trusses of the roof see drawing no. 9830 '*Trave longitudinale portante costole tetto*' dated 29 January 1959 (ASE).

[10] Drawing n° 91 'Pilastro' 14 February 1959 (ASE).

[11] Drawing no. 167 '*Particolare rivestimento in pietra dei pilastri*' prog. Ch VTAB del 25 March 1960, 1:1, later updated between September and October of the same year (ASE).

[12] See Bruschi G., Faccio P., Pratali Maffei S., Scaramuzza P. *Il calcestruzzo nelle architetture di Carlo Scarpa. Forme, alterazioni, interventi*. Bologna: Editrice Compositori, 2005, pp. 103-109.

A history of structural defects and repairs, the Iron Bridge, Shropshire

Jackie Heath[1] and James Miller[2]
1. CEng, MEng, MSc, MICE, Associate, Ramboll, London, UK; 2. CEng, MA, FICE, FIStructE, Conservation Accredited Engineer, Technical Director, Ramboll, London, UK

Introduction

The Iron Bridge at Ironbridge in Shropshire was the first cast iron highway bridge in the world. The idea and form were conceived by the architect Thomas Farnolls Pritchard in 1773, inspired by the revolution that had been unfolding in Coalbrookdale, and encouraged by John Wilkinson, an ironmaster.[1] However, Pritchard died in 1777 just as construction of the bridge started. As the project proceeded, the responsibility of financing and for the delivery fell progressively upon the ironmaster and shareholder, Abraham Darby III, whose grandfather had first used coke instead of charcoal to fuel the production of pig iron in the valley. The iron structure of the bridge was raised in the summer of 1779, although delay in completing the very substantial stone abutments, constructed after the ironwork, meant that the crossing did not open until New Year's Day in 1781.[2]

The structural form of the bridge, with its concentric semi-circular skeleton, is widely accepted as an icon of the industrial revolution. However, as a prototype for cast iron it suffered from shortcomings in the original structural concept, features that were better resolved in the cast iron bridges that followed within twenty years. In particular, there was excessive redundancy and this, coupled with the inherently inflexible material, made it unable to easily accommodate movement. The result has been a structure that has accumulated extensive defects.

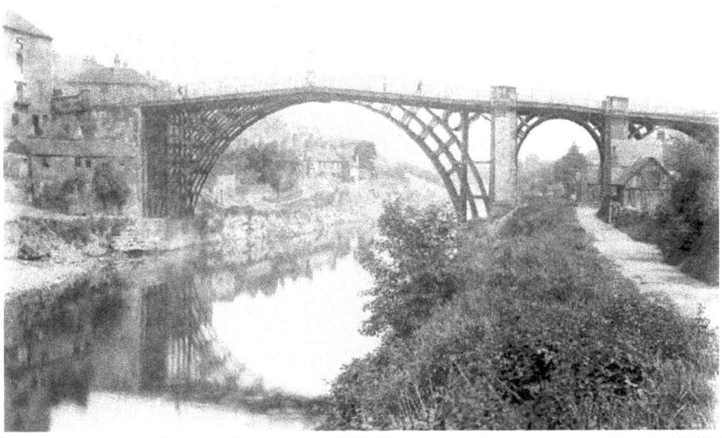

Fig. 1. General view of the Iron Bridge taken in 1897 (Ironbridge Gorge Museum Trust ref 1981.52). Note the change in slope of the deck over the inner verticals on the far side, indicative of cracked deck bearers.

The bridge was declared a Scheduled Monument in 1934 under the Ancient Monuments and Archaeological Areas Act; it was listed Grade I in 1983 under the Planning (Listed Buildings and Conservation Areas) Act No 1038659; and it was inscribed as part of the Ironbridge Gorge UNESCO World Heritage site in 1986.[3]

The crossing is now about 240 years old and has experienced alteration and repair throughout its life. This includes very early work to complete relatively local, unresolved elements of structure, two significant phases of rebuilding (1803 and 1821) to create the southern spans, a succession of ironwork repairs throughout the nineteenth and first half of the twentieth centuries, and the more recent installation of a concrete slab across the river bed to strut the arch springings apart (1973/4). The crossing today comprises the main cast iron span of around 30 metres and two southern cast iron spans each of around nine metres, with stone abutments and intermediate piers.

In 2013 the authors were asked to undertake an assessment of the bridge by custodians English Heritage. This paper discusses the history of defects and repairs at the Iron Bridge and reviews the geometry of the bridge in the light of recent archive research and laser scanning.[4]

History of defects, repairs and alterations

Information about the construction of the bridge relies on records in daybooks and some contemporary paintings and drawings. In particular a 1779 watercolour by Elias Martin, showing the partial erection of the main arches using timber falsework and ropes and provides a fascinating glimpse of the likely erection methods.[5] When the Iron Bridge was first completed, the main iron span was flanked by substantial stone abutments to both the north and south and, in particular, the south was not approached by the two iron spans seen today.

When the bridge was opened in 1779 it was missing the lowermost arcs of the upper ribs on the north bank. These were not inserted until 1791, twelve years after the original construction. Remarkably, the castings are understood to be hollow, instead of the original solid elements, this being confirmed by ultrasound testing in 1996.[6]

In 1798 tie rods inserted in both the north and south abutments, because cracks had appeared. They proved to be inadequate to save the south abutment, which was demolished soon afterwards.[7]

In 1801 Thomas Thomas, then engineer for the bridge, was asked to measure the main span and reported that both abutments had moved and that there were fractures in the ironwork on both sides of the river.[8] Neither his measurements nor the location of the damage were recorded (he possibly refers to cracked radials[9] or possibly the base plate on south side.[10] The advice was to reduce the weight of the abutments to avoid further damage to the main arch from ground movement. To that end in 1802 the whole southern abutment was demolished and replaced by a temporary timber deck supported off timber trestles, and the following year the temporary deck was replaced by two timber lattice beams supported on raking struts between new stone piers. John Simpson became the engineer in charge of these works in 1803 and, because he was a mason known to work with Thomas Telford and because they both subsequently received payment from the Trustees, it is thought that Telford may have provided advice for this work to the bridge.[11] When Simpson had the new stone piers built, the vertical members of main span were already noticeably out of plumb, reflecting the movement of the base towards the river; and so the upper sections of the vertical were built into the stonework of the pier.[12]

Between 1821 and 1823 the southern timber arches were replaced by two cast iron arches, each with five parallel frames comprising two bulb-ribbed quadrants in each, together with flanged deck plates that are bolted together. The ribs were unbraced which made them susceptible to further movement and compression. The ribs in the northern of the two spans, that closest to the river, evidently cracked. They were repaired using wrought iron gusset plates bolted in place in a number of successive phases: 1845-46, 1861 and 1879.[13] The implication is that horizontal ground movement continued throughout this period and that it was concentrated in the span closest to the main bridge, lowest in the valley. As a result this span became highly fractured and, although the fragments of the cast arches were retained in place, four additional cast iron asymmetric flange beams were added in 1880, one between each pair of ribs.

Little more is recorded about the inspection and repair of the main span ironwork until the late nineteenth century, perhaps because during this time horizontal ground movement was being absorbed in the adjacent southern span.

Sometime after 1880 the north abutment is reported to have bulged at mid-height, applying a horizontal force to the horizontal stays which in turn caused the inner cast iron verticals and some of the radials to crack. Furthermore, a photograph taken in 1890 shows the deck bearers to be cracked above the north quadrant, over the inner vertical: there is a clear change of angle.[14] This could be due to either settlement of the baseplates or a rising of the arches under horizontal compression or as a result of the cracking and 'kinking' of the verticals.[15] The laser point cloud had revealed this damage in three frames.[16]

By 1897 the two radials in the upstream frame on the north quadrant that rise to meet the deck at quarter-span had pulled away from the deck bearer beam. These were repaired by adding wrought iron rods to tie the elements together, to ensure continued location.[17]

On 17th December 1896 an earthquake, estimated to be 5.3 Richter scale with an epicentre in Hereford, appeared to cause a dipping of the baseplates inwards, towards the river. It is also likely to have allowed stresses to redistribute around the structure.

In 1902 Sir Benjamin Baker, who was then the engineer, together with John Fowler, of the Firth of Forth Railway Bridge and the Metropolitan underground railway, undertook a survey of the Iron Bridge, although this has not been found in archive searches. It is known that as a result of his report the Coalbrookdale Company cast replacement parts to be bolted in place where the edges of the deck plates were found to have either cracked or completely fallen off in a number of places. In addition, they supplied and fitted steel straps and cast iron spacer blocks to be inserted between the inner verticals, parallel to the bank, just above baseplate level, which will have enhanced the locational fixity of the arch springings (figure 2 image 4). They also supplied and fitted a steel sleeve repair to one of the diagonal braces.[18]

In 1923, a report by Basil Mott of Mott Hay and Anderson concluded that *'the tension stress is dangerously high and there is some risk in using the bridge for vehicular traffic'*.[19] He recommended that that the road width should be reduced by inserting two 5ft wide footpaths and that the thickness of the road construction should be decreased, in order to reduce the possible loads. Following discussion with Shropshire County Council, the bridge was closed altogether to vehicles in 1934. Basil Mott's 1923 inspection was principally responsible for reducing load on the bridge.

Shortly afterwards, in 1926, a scaffolding was erected for detailed inspection and a substantial phase of repair undertaken in that year and the following. The inspecting engineer Luther Griffiths recommended that saddles be installed to clamp the top of the inner verticals to the deck bearer beams, as these connections had failed comprehensively. These were iron castings, bolted around each of the ten failed joints.[20] Significantly also at this time, horizontal steel straps and spacers were introduced between the five frames at 'quarter points' (i.e. at roughly 22 ½ degree separation) around the two rib quadrants, presumably to address the lack of bracing or connectivity between the ribs, which are compression arches and therefore susceptible to buckling. A new strut was strapped to the diagonal on the south side; the circles were clamped to the uprights. A rather poor intervention was made in the choice of thin steel plate ogees to replace the original, cracked, thicker cast iron ones; these have flexed and stand at odds with the other, more substantial repair work.

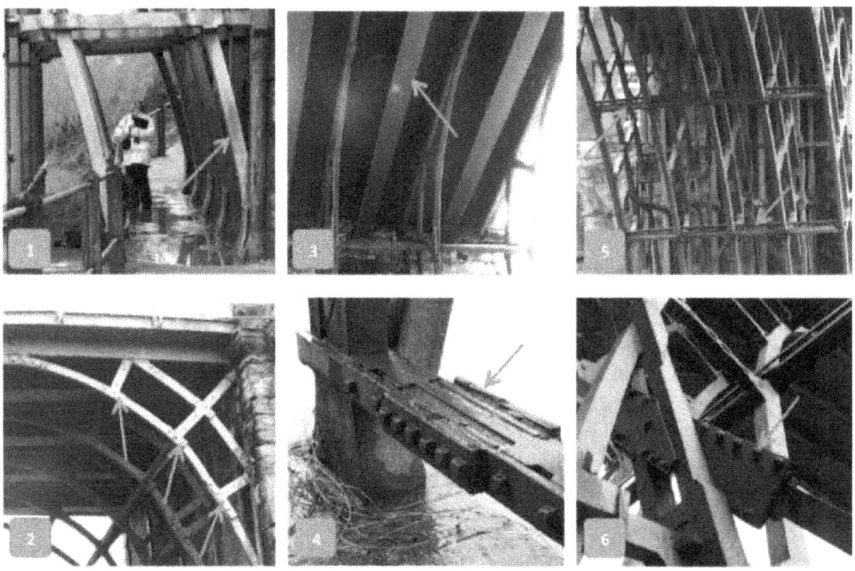

Fig. 2. (1) Lowermost arcs here not inserted until 1791; (2) Wrought iron gussets on southern span 1845-79; (3) Cast iron beams on southern span 1880; (4) Straps just above baseplate 1902; (5) Straps between ribs 1926; (6) Clamped repair to diagonal 1902.

Little work was undertaken on the bridge immediately after the Second World War, although a survey of cracking was carried out in 1948 and drawings marked-up to show the position of cracks in the five frames. In 1950, when Shropshire County Council (SCC) took on the ownership and repair of the bridge, it was neglected and poorly understood so they began a programme of measuring the span to determine whether the springings were moving together as was suspected.[21] A further inspection in 1960, showed an increase in the number and extent of cracking, in particular cracks at the ends of the short radial members, nearly all of which were in the northern quadrant of the bridge. From the mid-1960s a series of engineering reports were issued in quick succession: in 1966 (by H C Adams for Sandford Fawcett Wilton and Bell) on the condition of the bridge; in 1968 boreholes were drilled by Cementation Co to determine the condition of abutments and bedrock as part of investigation for a scheme to anchor the abutments with stressed anchors. Another report that year recommended the lightening of load in the north abutments (Anthony Blackwell for SCC). In 1969 J A Williams for Sandford Fawcett Wilton and Bell confirmed the recommendation for lightening of load in the abutments but also considered other proposals including that the abutments be anchored back into rock or strutting apart the abutments across the river bed, the latter was eventually agreed to be the preferred solution.[22] In 1970, a report by Morton and Moseley for Wolverhampton Polytechnic included details of a survey of fractures and contains the only recorded material tests on samples of iron taken from a fallen, original bracket.[23]

This sequence of surveys, investigations and reports built to most major intervention in the recent history of the bridge: the removal in 1972 of fill in the north abutment and replacement with various reinforced concrete works to hold stonework and roadway in place, and the installation in the summers of 1973 and 1974 of the concrete strutting slab across the river bed, to hold apart the two baseplate springings. This was a major civil engineering project requiring sheet piles to be driven, with some difficulty, to create a coffer dam before the strut could be cast in two sections. It was a significant achievement and appears to have at least slowed the movement in the gorge.[24]

Intervention since the 1970s has comprised major painting, repairs to the north abutment arch, and repairs above the arch to the deck and parapet. A hung scaffold was installed on the bridge in 1980, the ironwork cleaned to bare metal, inspected and painted for corrosion protection. Since 2000, monitoring works have been instigated, deck resurfacing carried, the replacement of deck wedges that locate and hold the main frames in position below the deck, and carbon fibre wrapping repairs to cracks in the structure of the balustrades.[25]

Geometries from laser scan data

A laser scan survey of the bridge was commissioned by English Heritage in 2012 to assist the commissioned bridge assessment.[26] Terrestrial laser scanning has been used with increasing accuracy and frequency as a tool for both data capture and, progressively, for archaeological recording.[27] The scan data provides a unique opportunity to examine the geometry of the structure in fine detail; it showed considerable variation between elements, for example in the cross section of the main arch members.

A key indicator of ground movement in the river gorge and consequential movement of the arch springings and deformation of the cast iron structure has been the dimensions of the main arch span measured between the springing points at the base of the ironwork. The original span was recorded as 100' 6" and measurements of the span have been taken fairly regularly since 1927. Unfortunately the physical origin for all the measurements is not precisely known and the measurements have been taken between different points at different times, making comparison with current measurements difficult. The accuracy of laser scan data permitted the arch geometry to be interrogated. The scan data for each quadrant of the main arch inner castings was found, as expected, to be a near-perfect arc of a circle. Assuming the bridge abutments to have been set-out correctly at the time of construction, thus forming an almost complete semi-circle, the scan data verifies an original span of 100' 6" (30.632m), a figure which accords with other references. This contrasts with the current span, extracted from the scan data, of 30.093m, which is an inward movement of about 540mm.[28]

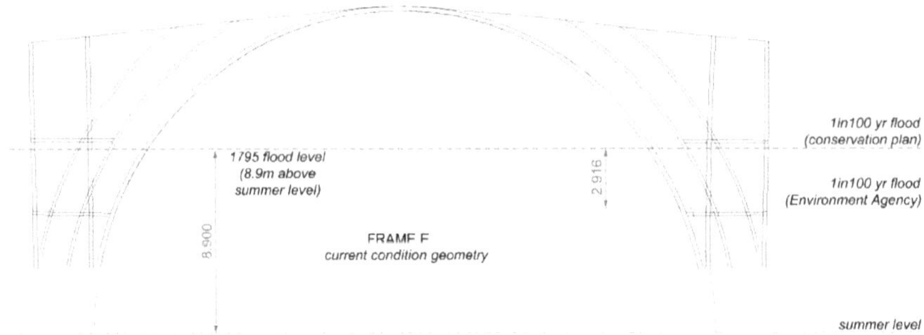

Fig. 3. Theoretical original span (arc in red) and current span taken from scan data, postulating an inward movement of around 540mm.

The laser scan data not only provides a starting point for analysis and an archaeological record, it also provides a complete geometric record of the element sizes, hitherto unavailable except by detailed, exhaustive measurement.[29] The new data is important because, whilst the Iron Bridge looks like a kit of interchangeable parts, in reality many apparently similar castings are in fact of different size, e.g. the north quadrant of the lower rib of the fourth frame was cast with a larger radius and smaller members. The parts were cast in open sand, this is noted on the 1782 drawing by J Phillips and is evidenced by the rough faces, pitting and concave surface of the members.[30] Further evidence from a 1999 site inspection suggested that no patterns were used for the large castings.[31] So it seems likely that the castings were adjusted at the time of construction, with components being made to fit the partially-erected structure. Opportunity was taken to compare the dimensions of apparently similar elements (Figure 4). The results show that, in particular, the sizes of radials connecting the three concentric arch ribs varied, with even more types identified than previous noted. This implies that fit was not easily achieved during erection, something inherent in a more highly redundant structure and itself pointing to the vulnerability of the structure to movement, such movement being in effect an 'imposed' lack of fit. This points to an interactive casting and erection process, where the secondary, smaller elements of the main structural frame were re-measured and cast to fit the overall geometry as the larger ribs and verticals were lifted and held in place.

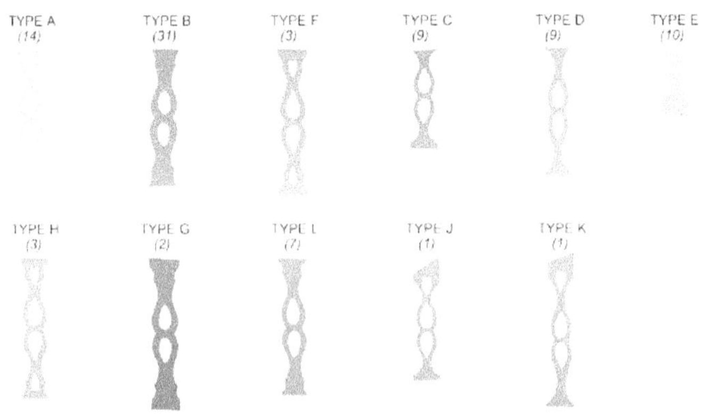

Fig. 4. Part types for radial members, taken from the point cloud data, showing that what might seem to be an interchangeable kit of parts is in fact made up of elements that were probably cast to fit.

Concluding remarks

The Iron Bridge has a unique form of construction that has become a recognisable international symbol. This may well have been borrowed from conventional timber arch construction, but the semi-circle is also familiar from contemporary masonry bridges and eighteenth century classical architecture. The shape of the bridge, reflected in the low summer water, creates a full circle that is a delight to the tourist.

However, it is perhaps best regarded as a show-case of what cast iron could be used for rather than necessarily the best way of using it. The design is problematic. The high degree of redundancy has led to the development of very many defects, as the bridge has responded in particular to large horizontal movements of the unstable ground found in this part of the Severn gorge. It is likely that the abutments have moved in by over half a metre since construction in 1779. Fortunately, the redundancy in the structure has also been ample to maintain its integrity, even with so many defects.

Two major interventions have taken place in the life the bridge: the reconstruction of the southern approach in both 1802 and 1823; and the reinforced concrete strengthening and strutting in 1972-74. Neither of these altered the essential appearance of the main span, both have been correct in their response to diagnosis and both have been proportionate in addressing major risks to the fabric.

Some significant names in the history of bridge engineering have advised on the condition and repair of the bridge. It is likely that Thomas Telford was consulted in 1803. Sir Benjamin Baker advised in 1902 and Basil Mott in 1923. This underlines the value placed on the heritage asset even before it was scheduled or listed.

There were concerns about the capacity of the structure from the early twentieth century onwards. There is no record and photograph of anything heavier than a car on the bridge - no traction engine or steam lorry - and the well-worn poster image of a small circus elephant standing on the structure seems only to confirm that even this was regarded as a heavy load. It appears to have never been used for heavy motorized vehicles.

There is an extensive and authentic patina of alterations and repairs to the ironwork of the bridge, in the form of early reconstruction, local rebuilding of stonework, and the insertion of straps, clamps, gussets, brackets and bolted castings. Each appears to have been preceded by survey and report and each is typical of its age. In this sense, the Iron Bridge is, with one or two small exceptions, an excellent example of modern conservation philosophy in action: using the best experience of the day to diagnose and repair the historic structure in a way that was appropriate to its continued use and is legible to future generations. Although repairs to the structure predate the modern conservation movement, it essentially adheres to contemporary conservation principles.

Laser scan data acquired for the bridge has provided the opportunity to examine the geometry of the overall span and individual components in detail. Assuming the bridge abutments to have been set-out correctly at the time of construction, thus forming an almost complete semi-circle, the scan data verifies an original span of 100' 6" (30.632m). This contrasts with the current span, extracted from the scan data, of 30.093m, which is an inward movement of about 540mm. It seems likely that a significant number of the smaller castings of the bridge were adjusted at the time of construction, with components being made to fit the partially-erected structure, which points to an interactive casting and erection process rather than a kit of interchangeable parts.

Acknowledgements

The authors are grateful for permission of their client and bridge's guardian, English Heritage, to publish this paper. Any views or opinions expressed are the authors' and not necessarily those of their client.

References

[1]N. Cossons and B. Trinder, *The Iron Bridge: Symbol of the Industrial Revolution*. 2nd ed. Chichester: Phillimore & Co Ltd., 2002.

[2] Ibid.

[3]D. de Hann and others, *Conservation Plan for the Iron Bridge, Ironbridge, Shropshire*. London: English Heritage, 2011.

[4]C. Brookes, J. Heath and J. Miller, *The Iron Bridge, Desk Study, Modelling Analysis and Assessment*. London: Ramboll, 2013.

[5]M. Elias, *The Iron Bridge*. [watercolour] (Stockholm, Skandia Company collection), 1779.

[6]D. de Hann, *Abraham Darby's Iron Bridge of 1779 Construction and Restoration*; (unpublished working paper) 1999.

[7] de Hann, *op. cit.*

[8] Cossons and Trinder, *op.cit*, p.47

[9] *ibid.*

[10] de Haan, *op.cit.*

[11] Cossons and Trinder, *op.cit.*, p.49.

[12] de Haan, *op.cit.*

[13] de Haan, *op.cit.*

[14] de Haan, *op.cit.*, p.149.

[15] de Haan, *op.cit.*

[16]C. Brookes, J. Heath and J. Miller, *op.cit.*

[17] de Haan, *op.cit.*

[18] de Haan, *op.cit.*

[19]B. Mott, *The Iron Bridge Shropshire*. London: Mott Hay and Anderson, 1923.

[20] de Haan, *op.cit.*

[21]A. Blackwell, *Historic Bridges of Shropshire.* Shrewsbury: Shropshire Libraries, 1985.

[22] Ibid.

[23]G. R. Morton, and A. F. Moseley, An Examination of Fractures in the First Iron Bridge at Coalbrookdale. *Journal of West Midlands Studies,* 1970 Special Publication no 2. Wolverhampton: Wolverhampton Polytechnic, 1970.

[24] Blackwell, *op.cit.*

[25] de Haan, *op.cit.*

[26] APR Services Limited, *Report on Laser Scanning of the Iron Bridge.* London: APR Services Limited, 2012.

[27]J. Miller, Laser Scanning: Surveying, Recording and Monitoring Historic Buildings. *The Building Conservation Directory, Cathedral Communications*. 2012.

[28] Brookes, Heath and Miller, *op.cit.*

[29] e.g. Ironbridge Archaeology, *The Iron Bridge Building Survey Record and Analysis.* Ironbridge: The Ironbridge Gorge Trust Museum, 2002.

[30]J. Phillips, *The Iron Bridge.* [Engineering Drawing] (IGMT.SS/MT.43, Ironbridge, Ironbridge Gorge Museum Trust),1782.

[31]D. de Hann, *The Iron Bridge Historic Building Survey Record and Analysis.* Birmingham: Ironbridge International Institute for Cultural Heritage, 2001.

Wren's domes

Jacques Heyman
Department of Engineering, University of Cambridge, UK

Summary

Wren introduced domes into England, and three different features of these may be distinguished in his designs. First, masonry domes could be used to roof very much larger areas than was possible with Romanesque or Gothic vaults, and Wren (with Hooke) was concerned to find the correct structural shape for such domes. A prime example is of course to be found in St Paul's Cathedral. Second, smaller spaces (but still relatively large) could be given the appearance of having a domed roof, when in reality the dome was non-structural and depended from a timber framework above. Two examples are to be found among the 51 parish churches in London rebuilt by the Wren practice after the Great Fire. Finally, in several of the rebuilt churches, the dome has no structural function, and is used as a decorative element.

A brief chronology

1625	Accession of Charles I
1632	Christopher Wren born
1649	Commonwealth
1651	Christopher Wren graduates from Oxford
1657	Christopher Wren Professor of Astronomy at Gresham College, London
1660	Restoration. Accession of Charles II
1661	Royal Society founded
	Christopher Wren Savilian Professor of Astronomy, Oxford
1664	Sheldonian Theatre, Oxford
	Pembroke College Chapel, Cambridge
1665	Plague year
	Christopher Wren visits Paris
1666	Christopher Wren's proposals for Old St Paul's
	Great Fire of London
1672-73	Great Model of second design for new St Paul's
1675	Warrant Design for new St Paul's
1685	Accession of James II

Ely Cathedral

The visitor to Ely may not at first quite understand the great open space at the centre of the Cathedral. The space is in fact unique in English Romanesque and Gothic; in all other such churches the four crossing piers constrict the area to the width of the nave, choir and transepts – typically 12 m square in England (cf 15 m in the cathedrals of the Ile de France). Standing at such a narrow crossing, the visitor will see the vault, in timber or stone, marking the level of the first storey of the tower above. At Ely, however, the vault covers not only the width of the nave and transepts, but also the side aisles; the space measures not 12 but 24 m. Figure 1 shows Bentham's engraving of 1771, and the start of the timber octagonal lantern supported by the vault may be seen.[1]

This configuration is a fourteenth century alteration of the original Norman design. The new church was started in 1071, shortly after the Conquest; the Romanesque cathedral was completed about a century later, and included a central (12 m square) crossing tower. This tower collapsed on 12 February 1322, after standing for two centuries – the precise cause will probably never be known, but is likely to have been geotechnical failure of the foundations. The original four Norman bays of the choir were either destroyed or were taken down and rebuilt in the Gothic style in the 1320s after the collapse, and at the same time an imaginative reconstruction was made of the central crossing.

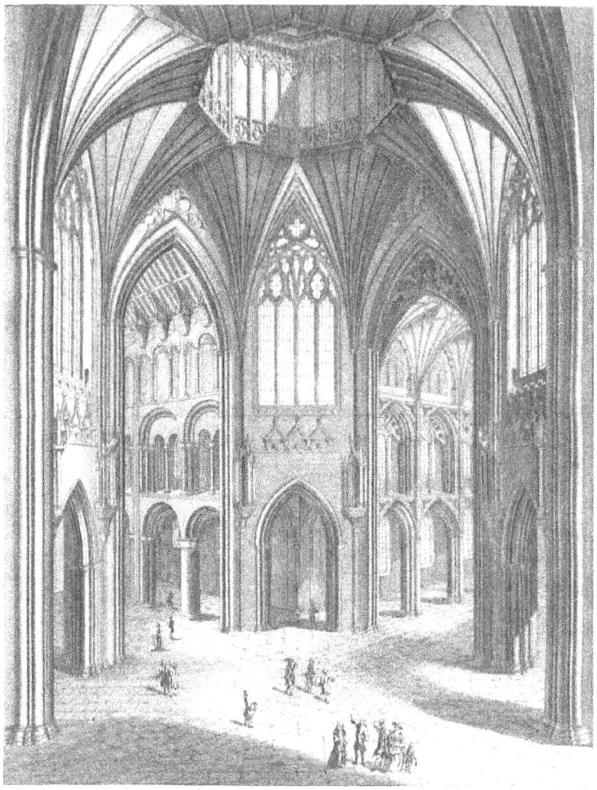

Fig. 1. Ely Cathedral, the crossing (Bentham 1771).

The redesign is associated with the name of Alan de Walsingham, who was then the Sacrist (and later Prior) of the Cathedral, but it is clear that expert advice was sought.[2] The design of the wooden octagonal lantern was in the hands of William Hurley, King's Carpenter – and a local carpenter, Geoffrey de Middleton, carried out the work. However, it may have been Walsingham who had the idea of suppressing the original crossing piers; Figure 2 shows on the left the arrangement before the collapse, and on the right the present positions of the piers. The 12 m square crossing has been opened up to an octagon of about double that size. (The foundations of the eight Norman piers were enlarged, and the piers themselves were encased with new masonry.)

It may be noted that the new octagon is not equilateral; there are four sides of 12 m and four of about $8^1/_2$ m. The eight piers carry masonry arches which in turn carry the octagonal masonry shell which forms the present "crossing tower" of the Cathedral. However, the vault is of timber (Fig. 1), and the wooden ribs spring from the same capitals on the piers as the masonry arches. These capitals are at about 19 m above the floor, and the vault rises to an octangular timber ring, the lower sill, at about 29 m. This lower sill is a regular octagon of "diameter" about 12 m, and the ingenious geometry of the construction has achieved this regularity from the irregular base below.

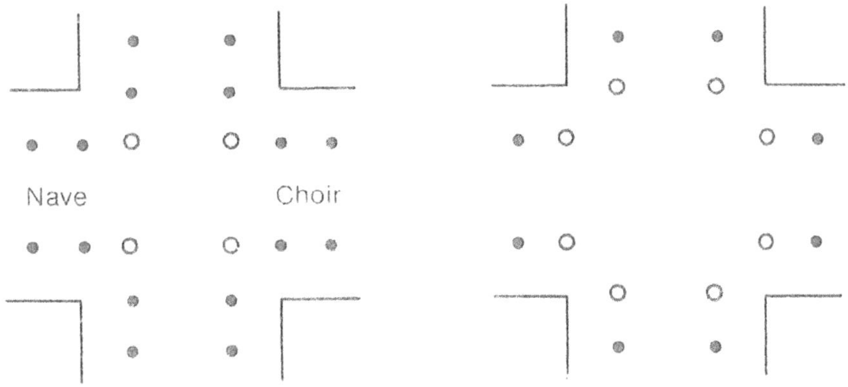

Fig. 2. Ely Cathedral, the crossing; left, before 1322, right, after reconstruction.

Matthew Wren and his nephew Christopher

Matthew Wren was appointed Bishop of Ely in 1638, and he was Bishop of Ely when he died in 1667; he was imprisoned in the Tower of London from 1642-60. He had graduated from Pembroke Hall, Cambridge (Pembroke College) in 1605, and he held various appointments in the University (Oxford and Cambridge) and the Church, including the Mastership of Peterhouse, Cambridge from 1625-34; it was in his Mastership that the College chapel was built, and consecrated in 1632. He knew the Royal family; he had been Dean of Windsor in 1628, and Chaplain to Prince Charles (Charles I) in 1622.

Matthew's nephew Christopher became an MA of Wadham College, Oxford, in 1653 and was appointed, at the age of 25, as Professor of Astronomy at Gresham College, London; four years later he returned to Oxford as Savilian Professor of Astronomy. He retained his interest in science throughout his life, and was a founder member of the Royal Society, becoming President 1680-82.

It was to a meeting of the Royal Society in 1663 that he exhibited his design of the Sheldonian Theatre in Oxford. It would seem that this was Wren's first architectural commission, and he was probably consulted initially on the design of the timber roof, required to span more than 21 m. Wren devised a triangulated truss system, of a form not far removed from those used for great medieval churches. Those trusses, however, spanned the "standard" 12 m, and Wren's 21 m tie beam was composed of smaller lengths, ingeniously locked together with scarfed joints. Building started in 1664, and Oxford University was closed a year later because of the Plague. Before that, however, Matthew Wren commissioned Christopher to design a new chapel for his old College, Pembroke (Matthew is interred in the crypt of the chapel).

In fact, as Grimstone has shown,[3] a preliminary design for the chapel had already been made, by the "architect" Edward Pearce. Pearce was a "medieval" master builder, of the kind who undertook design/build contracts, often using stone from their own quarries. Such men had deep knowledge of both the art and practice of their trade, and their expertise in executing important commissions could be recognised by the conferral of a knighthood. But Bishop Matthew, as donor and as a former Fellow of Pembroke, evidently imposed a young and inexperienced architect on the chapel project.

In fact, the days of men such as Pearce were, by accident, numbered. Within two years the Great Fire had destroyed so much of London that the medieval master builders could not manage the required enormous volume of work. Wren, who had not gone through the hoops of apprentice, journeyman, master, emerged as the first "modern" architect, prepared to design but not trained to build major projects. Medieval masters, with their gangs of craftsmen, were content to be hired as contractors for the rebuilding programme after the Fire, and Wren certainly used Edward Pearce as one of his (several) contractors at St Paul's Cathedral. Pearce was evidently happy to assume a similar role at Pembroke in 1644. (It might be truer to say[4] that it was not Wren, but his pupil Hawksmoor who was the first English architect, as someone trained in an established architectural practice.)

The architectural treatment of both the Sheldonian Theatre and Pembroke College Chapel show some stylistic infelicities, as might be expected in the first works of a young architect. Wren had indeed knowledge of the published Italian masters – Palladio, Alberti and Serlio – and his later work shows how this knowledge was assimilated in his designs.

Wren's domes

Both Palladio and Serlio have lengthy descriptions of the Roman dome of the Pantheon - *La Ritonda*. In 1655, before Wren visited Paris for the better part of a year, he had never seen a dome – none existed in England. He was armed with introductions from the Royal Society, and was able to meet eminent architects (for example Bernini), and to see domes recently completed or under construction (Lemercier's Sorbonne, and Mansart's Val-de-Grâce). Seventeenth century domes owed much to that of St Peter's, Rome, both to Bramante's original concept (illustrated in Serlio) and to Michelangelo's final design, executed by Giacomo della Porta, so that the forms of the domes he saw were not strange to Wren. He returned to England early in 1666, and an opportunity arose immediately for him to design his own first dome.

The Old Cathedral of St Paul's had fallen into a bad state of repair, and in 1663 a Royal Commission was set up to determine what action should be taken. Building at Pembroke Chapel and the Sheldonian had only just started, but despite this, the inexperienced Christopher was asked to report to the Commission. He did so, and he made a thorough survey of the existing fabric. The work of the Commission languished, but by the Spring of 1666 the Plague was over, and Wren made a further survey and report for the Commissioners. All were agreed that the crossing region was in a defective state; in Wren's words (*Parentalia 1750*):

> "... the Tower from Top to Bottom, and the next adjacent Part, are such a Heap of Deformities, that no judicious Architect will think it corrigible, by any Expence that can be laid out upon new dressing it, but that it will still remain unworthy of the rest of the Work, infirm and tottering; and for these Reasons, as I conjecture, was formally resolv'd to be taken down.

I cannot propose a better Remedy, than by cutting off the inner Corners of the Cross, to reduce this middle Part into a spacious *Dome* or *Rotunda*, with a *Cupola*, or hemispherical Roof, and upon the *Cupola*, (for the outward Ornament) a *Lantern* with a spiring Top . . ."[5]

The "cutting off of the inner corners" is exactly illustrated in Fig. 2 – Wren was creating an Ely space at the centre of Old St Paul's. The cross-section of his design is shown in Fig. 3; a cylindrical drum supports a masonry dome carrying the cupola, and the whole is surmounted by a lead-covered timber dome. It was decided on 27 August 1666 to recommend Wren's scheme to the Commissioners – on 2 September the Great Fire destroyed much of the City of London, and damaged irremediably Old St Paul's. It was of course the Fire that established Wren as England's greatest architect.

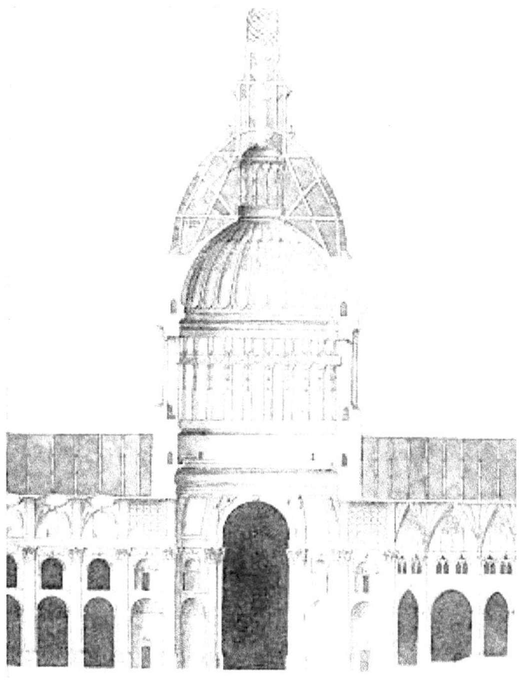

Fig. 3. Wren's proposal for a dome for Old St Paul's (Campbell 2008).

St Paul's Cathedral

Strangely, Wren's first design for the new St Paul's was a smallish conventional church, without a dome. A model was made, but the project was soon rejected, partly on the grounds of being insufficiently impressive for the cathedral church of a great city. Some five years later, in 1672, Wren had completed a design for a much more magnificent building, in the form of a Greek Cross and surmounted by a great dome, Fig. 4. This design was approved by Charles II, who commanded a model to be made; this Great Model, which may be seen in St Paul's Cathedral, was at a scale of 1 inch to 2 feet, and cost the enormous sum of over £500.

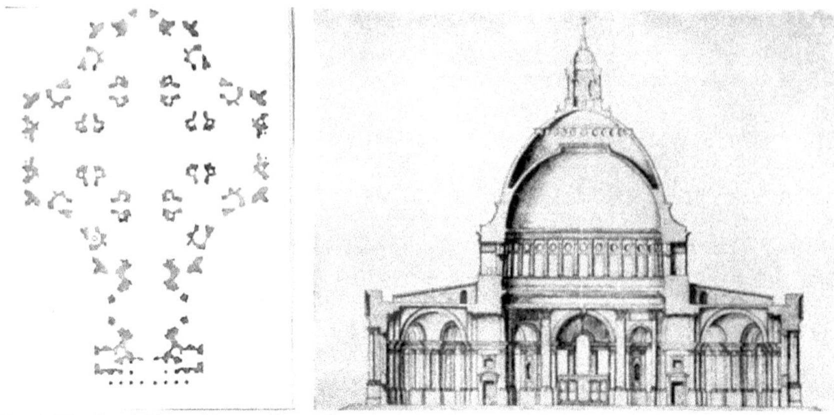

Fig. 4. The Greek Cross design, plan and section.

Quite apart from the "architectural" innovations in this design, there is a significant technical difference between the design of Wren's dome and the Renaissance domes he had seen in illustrations or in France. Those domes have profiles which become vertical as they come to rest on their supports; the dome of Fig. 4 is still spreading at its base. Robert Hooke had stated that the perfect shape for a (two-dimensional) (masonry) arch was that of an inverted hanging chain – such a chain will always slope at its points of support. In 1671 (just before the Great Model) he had declared that the corresponding ideal shape for a dome was that of a cubico–parabolical conoid, that is, the curve $y = ax^3$ rotated about the y axis.[6] One of Wren's drawings (Fig. 5), probably executed also in 1671, shows a study for this design, with the numbers clearly indicating the cubic parabola;[7] the profile of the dome does not become vertical as it descends to its supports.

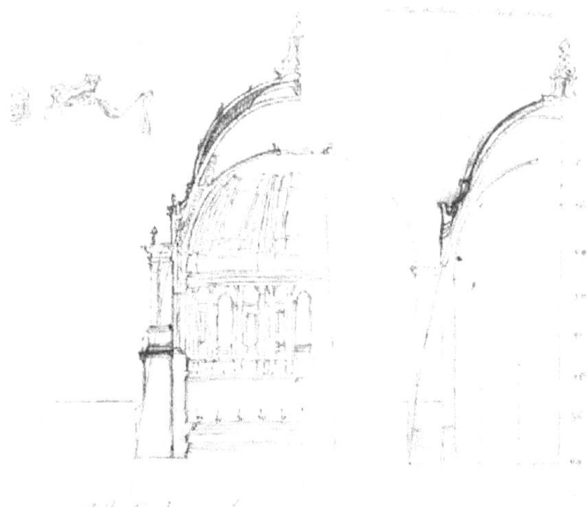

Fig. 5. Wren's sketch of the cubic parabola (Higgott 2009).

Although royal approval had been given, the Greek Cross design had not been agreed by the Dean and Chapter; indeed, they objected strongly to the absence of a traditional choir, and demanded a design with the usual eastern altar rather than the innovation of worship in the round. Thus a third and final stage was reached in the design of the present St Paul's, and the "Warrant Design" was approved in 1675, albeit with provision for Wren to make changes in detail. (It is perhaps ironic that the Dean and Chapter have, very recently, after three hundred years, reordered the Cathedral so that the main altar now stands at the crossing beneath the dome.)

Among the many detailed changes that Wren made in the thirty years of construction, some of the most significant were to the dome. The final design, completed in 1705, is shown in Fig. 6. There is an inner dome constructed with an open eye from two brick courses of total thickness 460 mm;[8] it is this dome which is seen from the ground, and defines the central space of the cathedral, about 34 m in diameter. Above lies a conical shell, again of brick but with some stone, which supports the 700 tonne ball, cross and lantern. The whole is surmounted by a timber construction with a lead covered outer surface; this is the dome seen from outside the cathedral.

It has been estimated that the total weight of masonry above the level of the supporting arches in Fig. 6, is something over 20 000 tons; this is probably about twice the weight of the conventional crossing tower of Old St Paul's.

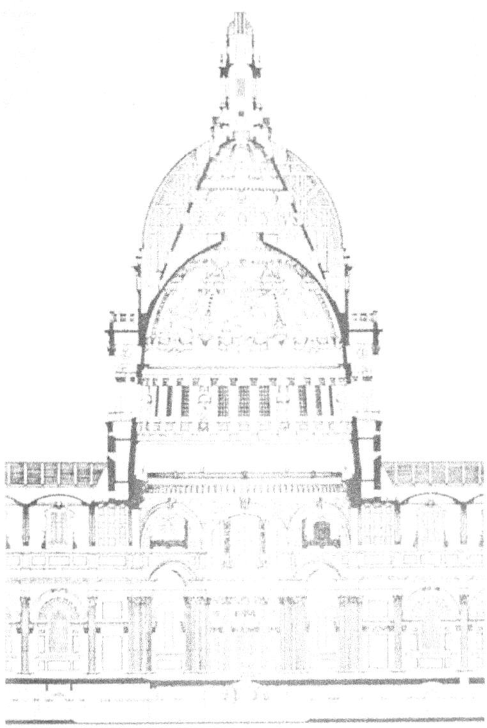

Fig. 6. St Paul's Cathedral, section.

It will be seen from Fig. 4 that the dome for the abandoned Greek Cross design (the Great Model) was to be supported on eight pairs of piers arranged as a regular octagon. The final design for St Paul's Cathedral has similar supports, but the octagon is not regular, and is of the kind illustrated in Fig. 2; the design has sometimes been criticized for this.

The mechanics of masonry domes

Hooke's cubico-parabolic conoid was in fact a rational guess at the "perfect" shape for a masonry dome. With certain assumptions, it may be shown that the true profile is vastly more complex mathematically, but is actually extraordinarily close to the cubic parabola. [9] In any case, Robert Hooke's diary entry in 1675 explicitly states that Wren based his design of the dome for St Paul's on his, Hooke's, "principle". The exact shape is irrelevant; the essential point of the design, grasped by Hooke and Wren, was that the profile should follow the "line of thrust", that it, the shape of the (equivalent three-dimensional) inverted hanging chain dictated by the equations of statics. The masonry never descends vertically.

(In more recent times the Spanish architect, Gaudi, also grasped totally the concept of non-verticality implied by his models based on complex arrangements of hanging chains. The design for the *Sagrada Familia* in Barcelona, for example, seems to use piers in the conventional way to support the high vaults, but these piers are in fact inclined inwards as they rise from the ground.)

Wren's parish churches

The rebuilding of the parish churches of London after the Fire enabled Wren to produce a great variety of designs, and in particular he appeared to delight in creating fanciful spires. Sometimes these rise directly from towers, and sometimes, as at St Peter's Cornhill, they surmount domes. In fact domes feature in many of the churches designed by the Wren practice; the churches are conventionally attributed to Wren himself, but St Edmund the King, for example, was created by Robert Hooke (as was the Monument), and Hawksmoor came to play an increasing part in the designs which left the offices of "Wren, Hooke and Partners".[10] St Edmund the King has a domed entrance lobby, and Wren added similar lobbies to the west end of St Clement Danes in the Strand.

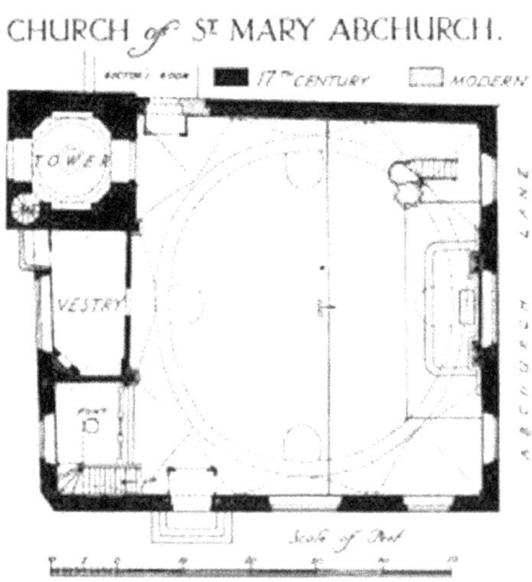

Fig. 7. St Mary Abchurch, plan.

Two domes, however, are on a larger scale (a third, at St Mary-at-Hill was largely destroyed by fire in 1988, and has been reconstructed). The larger of the two is that of St Mary Abchurch; the dome is contained within a (slightly irregular) rectangle (Fig. 7), and spans some 19 m. This dome is not structural; it is constructed from lath and plaster, and is suspended from a "space frame" of vertical and horizontal timbers, which is in turn covered by a wooden hipped roof.

The dome of St Stephen Walbrook is more spectacular, although of diameter only 14 m; it spans both the nave piers and an aisle either side, thus opening out an "Ely" space at the west end (Fig. 8). This dome is again of lath and plaster, and its form is reflected externally by a dome surmounting the "space frame".

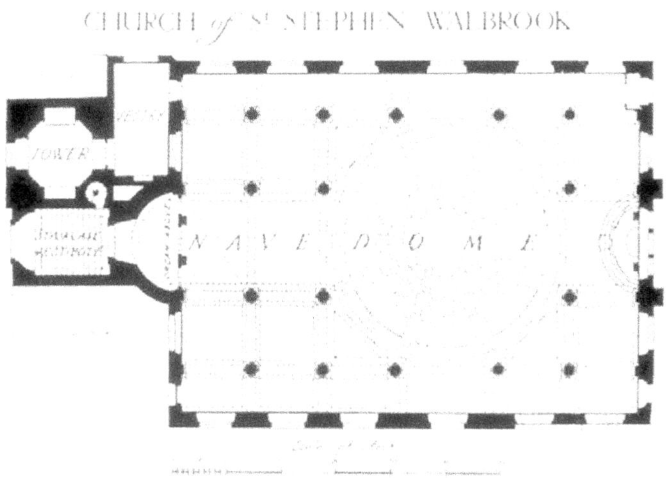

Fig. 8. St Stephen Walbrook, plan.

Envoi

Three separate features may be distinguished in Wren's use of the dome. Of these, the first is structure. Wren and Hooke had a clear command of the fact that the shape of a masonry dome should be dictated by "mechanics" (that is, geometry), and the two lowest domes of St Paul's conform to this principle. The third skyline dome is of timber and lead, and is an "architectural" expression of the design.

The second feature is indeed that of "architecture", in which the dome is appreciated as capable of surmounting much larger spaces than those made possible by the use of Romanesque or Gothic vaults, or by the use of timber trussed roofs. Thus a completely different feeling can be given to a church, as at Ely, St Stephen Walbrook and St Paul's. The domes of St Stephen Walbrook and St Mary Abchurch are of lath and plaster, and are completely non-structural. They have the nature of stage scenery used to represent something real, and there is good precedent for this in Renaissance architecture. There was no difficulty, for example, which prevented the stage designer Inigo Jones from becoming an outstanding Renaissance architect.

Finally, Wren used domes in an almost postmodern way, adding them to his designs as non-functional decorative elements.

References

[1] J. Bentham, *The history and antiquities of the conventual and cathedral church of Ely, from the foundation of the monastery, A.D. 673, to the year 1771*, Cambridge: 1771.
[2] J. Maddison, *Ely Cathedral: Design and Meaning*, Ely Cathedral: 2000
[3] A. V. Grimstone, Building Pembroke Chapel, Pembroke College, Cambridge, 2009
[4] J. Heyman, "Wren, Hooke and partners", in S. Huerta, (ed.), *Proceedings of the First International Congress on Construction History*, Madrid 2003, Madrid: Instituto Juan Ferrera, 2003, pp. 3-9.
[5] C. Wren (ed.), *Parentalia*, London: 1750.
[6] J. Heyman, "Hooke's cubico-parabolical conoid", *Notes Rec. R. Soc. Lond.*, Vol. 51, Issue 1, 1998, pp. 39-50
[7] G. Higgott, "Geometry and Structure in the Dome of St Paul's Cathedral" in A. Gerbino and S. Johnston, (eds.) *Compass and Rule,* New Haven and London: Yale University Press, 2009, pp. 155-169.
[8] J. W. P. Campbell, *Building St Paul's*, London: 2008
[9] J. Heyman, "Hooke's cubico-parabolical conoid" (Note 4)
[10] J. Heyman, "Wren, Hooke and partners" (Note 3)

Mixing concrete – remarks on the early history of modern concrete

Stefan M. Holzer
Universität der Bundeswehr München, Munich, Germany

> "At the beginning of this century, [...] the popular lecturer on the subject had a standard – one might almost say ritual – phrase with which to open his discourse. 'Concrete', he would say, 'is one of our oldest building materials, having been frequently employed by the Romans. The dome of the Pantheon, built nearly eighteen hundred years ago, is [...] reliable evidence of its durability and strength.' [...] Much more remarkable than these Roman precedents was the total neglect of concrete construction during the intervening period." [1]

This paragraph, the opening statement of Peter Collins' 1959 book on the early history of concrete, is a pretty good summary of the "popular" ideas about the history of concrete even today, more than half a century later. Collins himself did a remarkable job in highlighting the close historical links between formwork for traditional rammed earth construction and concrete. Furthermore, it is part of the general concept of the history of concrete to refer to the history of scientific investigation of hydraulic lime and artificial cements. The names typically quoted include Smeaton, Vicat, Aspdin. This is generally considered enough of early background, and subsequently the historiographer of concrete typically moves on quickly to the invention of reinforced concrete.

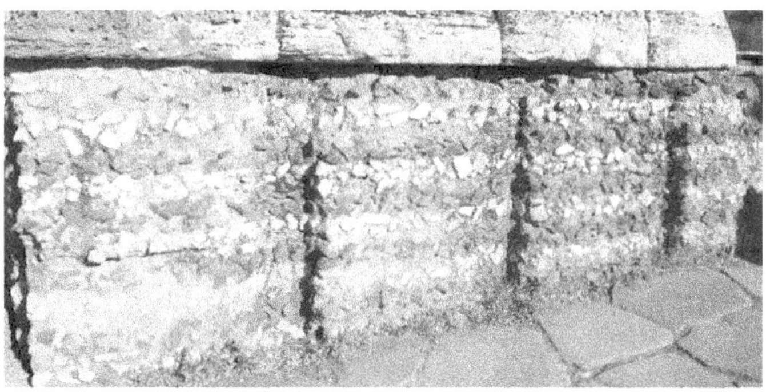

Fig. 1. Typical structure of Roman concrete: alternating layers of mortar and rubble (foundation of the arch of Titus, Rome, 81 AD)

The early history of concrete, that is to say, the development from 16th century Vitruvianism to the middle of the 19th century, has received very little attention since Collins' groundbreaking monography. This present article aims to provide some additional insight into this topic. We propose to start with the observation that modern concrete is not only characterized by the use of hydraulic lime/cement (hardening under water and achieving higher strength than ordinary lime mortar) and the fact that it can be cast in any arbitrary form, but also by an important detail of the production process. Both Roman concrete and the common "rubble infill" of medieval and early modern multi-leaf masonry consist of alternating strata of rubble stone and mortar (Fig. 1; here, layers are particularly visible because of the alternating use of bright and dark rubble); unlike with modern concrete, the two constituents were not mixed beforehand, but cast separately, covering layers of sizeable stones with fairly liquid mortar. Consequently, the distinction between "Roman concrete" and "rubble masonry" with more or less regular courses of stones is rather difficult. Many historical sources from the 16th century onwards do mention concrete-like constructions, but it is only rarely specified precisely how the mixture of mortar and rubble was achieved. Early descriptions include, e.g., Giovanantonio Rusconi's paraphrase of Vitruvius, where concrete is suggested as a material for foundations (details not found in Vitruvius' original text): "The foundations should be wider than the walls, and they should be filled with a mixture of stones, lime, and sand." [2]

As opposed to the "Roman concrete", modern gravel concrete is mixed and cast as a single homogeneous mass, and exhibits no layered structure or at least is not expected to do so. The modern term "béton" which today still denotes concrete in French and German, is typically traced back to Bernard Forest de Bélidor's *Architecture hydraulique*. [3] An attentive re-reading of Bélidor's text reveals that the "concrete" which he describes as being used occasionally for subaquatic foundations and other marine works (notably, a jetty built at Toulon in 1748 [4]), is more related to the "masonry-like" Roman concrete than to modern concrete. Indeed, Bélidor's term "béton" denotes a coarse mortar, including gravel pieces as big as an egg. [5] However, the final construction has alternating layers of even bigger rubble stones and "béton": "One creates a layer of it, ten to twelve inches thick, [...] and afterwards one distributes rubble stones or stones of a medium size (the biggest ones not exceeding one forth of a cubic foot) over its entire surface. [...] When this layer of stones has been covered by a new layer of béton, one starts the next layer of stones." [6] It took nearly half a century to develop Bélidor's revival of Roman concrete into something more comparable to modern concrete. For foundations under water, Bélidor's method was then criticized with the following argument: "This method had the great drawback that one might place too much mortar in some places, and too little in others, particularly when one was pouring the material at a great depth of water, where one could not watch the process very well." [7] These problems led to the casting of readily mixed "concrete" in the modern sense. However, casting a homogeneous mass of concrete, keeping to a fixed proportion of the constituents, and at the same time preventing the cement from being washed out of the mixture on its way through a considerable depth of water, required a set of procedures and machinery which still had to be developed and perfected. This article will show some of the main steps of that development.

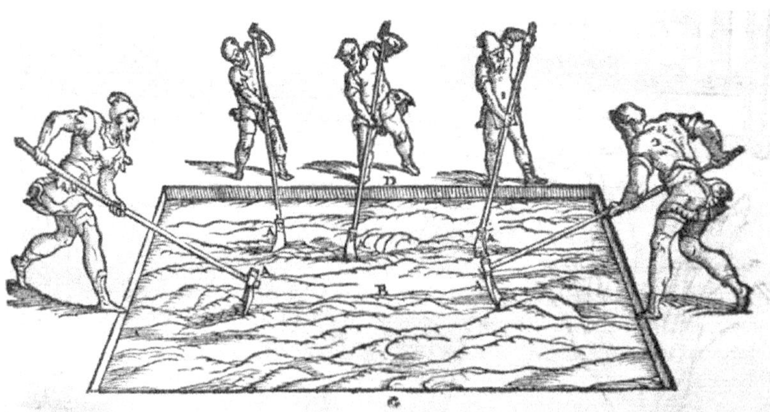

Fig. 2. Traditional technique of mixing mortar (Rusconi 1590, p. 101)

The traditional technique of preparing mortar made use of picks and rakes. Depending on the method, either quick lime or slaked lime was employed. Next, sand and puzzolana were added. The procedures described by early modern literature differ only in detail, particularly with respect to the mixing of dry or wet substances. Manual mixing of small quantities of humid mortar is depicted in many medieval and early modern illustrations, including one in Rusconi's above-mentioned book of 1590 (fig. 2). [8] This elementary technique was definitely somewhat insufficient once one required large quantities of mortar, inefficient once the "mortar" contained not only sand but also bigger pebbles or rubble stones, and unreliable once constant proportions of the constituents were required. Nevertheless, it remained common well into the middle of the 19th century, when concrete had already found widespread use as a material for the foundations of canal locks.

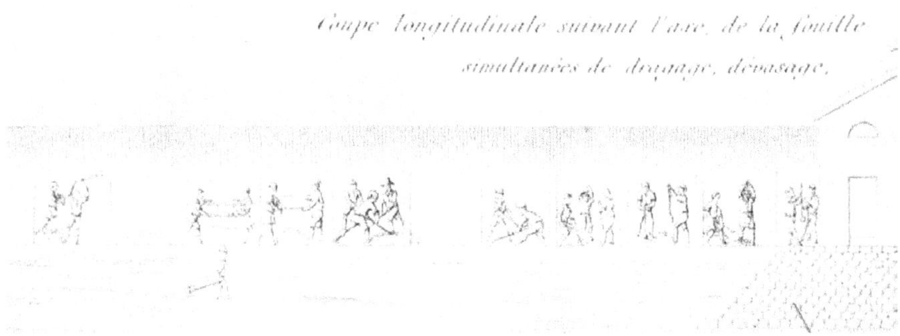

Fig. 3. Concrete production on an 1825 canal lock construction site (Beaudemoulin 1829, plate II, detail)

Concrete foundations for canal locks were probably an early 19th century French innovation. The first detailed publication on this subject is Louis-Alexandre Beaudemoulin's monograph of 1829, minutely describing the techniques employed for the construction of a canal lock at Huningue, Alsace, in 1825. [9] Beaudemoulin's book triggered a series of further publications on other applications of concrete to canal lock construction, notably the 1832 journal papers of Magdelaine [10] and Mary [11], covering locks built as early as 1812-15, and the German paper of Zimmermann [12] on a lock built in 1828. A report on similar early Italian experiments in concrete lock construction [13] went largely unnoticed outside Italy.

Beaudemoulin was probably the first to organize the mixing of the concrete in a systematic way, relying heavily on a large human workforce (fig. 3): "The concrete has been produced according to the method of M. Vicat, with the help of tamps and without the addition of water, on a roofed boarded platform, by a workforce of 180 labourers, divided in groups of four: Each group processed one third of a cubic metre at a time and delivered four such charges per day." [14] Beaudemoulin adds: "The use of rakes results, in general, in much less perfect mixing than the use of tamps of cast iron, weighing 4.5 kg. The tamps are operated vertically in such a way that no worker can dispose of the action which one demands from him." [15]. Instead of working with cast iron tamps, Beaudemoulin's German contemporary Zimmermann rather relied on labourers equipped with clogs, stomping the mortar: "This method of preparation of mortar requires sturdy footwear, because the cement mortar is chemically very aggressive. [...] The clogs common in our region are very suitable for the purpose, and in other regions, they would have to be procured." [16] Mixing the mortar with the rubble (crushed brick in Zimmermann's case) was effected in wooden troughs, once again employing labourers equipped with clogs for compacting, and picks and shovels for overturning the mass. [17]

Thirty years later, Johann von Mihálik still found a construction site arrangement similar to Beaudemoulin's convenient for the mass production of concrete for a large canal lock close to Bezdan (today, Serbia). First, mortar was mixed, employing tamps similar to Beaudemoulin's: "Three labourers, moving steadily in a circle, mix the lime and mortar to perfection by diligent tamping; one man throws the spreading mass back to a heap under the tamps, using a shovel." [18]. The gravel – wetted and spread out at the start of a boarded "mixing alley" – is added in the next step:

> "Next, two labourers, standing close to each other, [...] gradually move the heap of gravel and mortar with rakes along the mixing alley; one single other labourer follows them, throwing lost mortar particles back to the moving heap with a shovel. Once the béton has reached the end of the plank roadway, it will be sufficiently mixed. This procedure forces the labourers to mix the concrete thoroughly since they have to move it exclusively by means of rakes over a distance of 48 feet." [19]

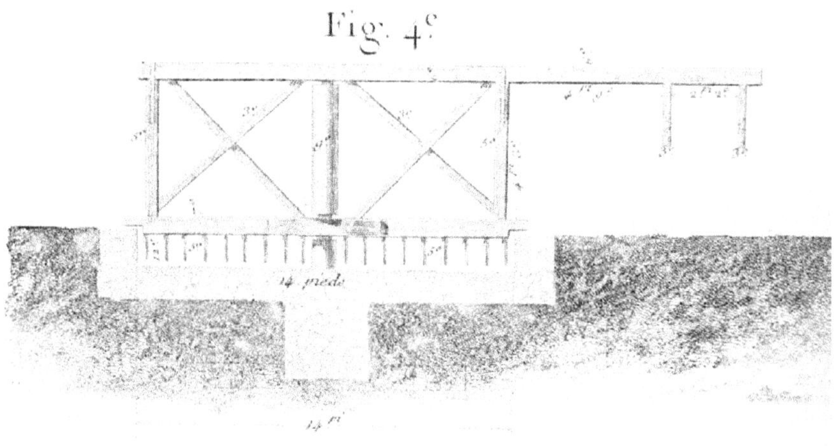

Fig. 4: Horse-capstan operated mortar mixing machine applied by Perronet in 1772 (Perronet 1788, pl. XIX, detail)

Alternatively, mixing machines could be employed. Jean-Rodolphe Perronet had already used a mortar mixer as early as 1772 when building the bridge across the Seine at Neuilly (Fig. 4): "This machine was moved by a horse which was led by a man; the duration of the horse's shift was two hours." [20] However, mixing machines became more popular only in the 1830s. F.-M. Lebrun, in 1843, discussed a horse-operated mortar mixer, much smaller than Perronet's, and remarked: "Various techniques for the fabrication of mortar are in use. On big construction sites where one typically requires large amounts of mortar, one generally employs a horse-capstan with one or several wheels which turn around in an annular masonry trough into which the material is thrown successively. [...] The mortar prepared in this way is generally well mixed." [21]. However, it seems that Lebrun himself did not use such a machine in his own early concrete constructions, but rather relied on manual mixing.

A similarly designed, but perfected mortar mixer, operated by a horse-capstan, and combining the actions of rolling wheels, as well as rakes and a "plough" dragged by the machine on a circular course, was tried by Johann Mihálik in the mid-1850s. [22] Mihálik modified the machine in such a way as to permit continuous operation, including filling and emptying.

Gotthilf Hagen gave a fairly comprehensive overview of state-of-the-art mortar mixing machines in 1841:

> "When a considerable amount of mortar is required, as is typically the case in concrete production, one sometimes employs special mortar mixing machines. They have, however, the drawback that they require more water, which leads to quality losses [...]. They consist typically of a hollow, horizontally arranged cylinder of wood or iron. There is a horizontal revolving iron axis, equipped with mixing blades, which is turned by a steam engine or some other force [...]. By contrast, the mixers which are common in France have a vertically arranged cylinder." [23]

Indeed, mixers consisting of an upright wooden barrel and a vertical, rotating axis equipped with arms were employed in the late 1830s by several French engineers. Most notably, a variety of mechanical devices for the preparation of mortar and concrete was introduced during the construction of the harbour basin at Algiers (1833-44). First, a jetty was constructed, submerging huge concrete blocks and also casting large sections of concrete in-situ. These works were conducted by Léopold-Victor Poirel who published his experiences in an impressive monograph in 1841. He remarked on his barrel mortar-mixer: "This way, one manages to prepare 10 cubic metres of mortar in three hours." [24]. The upright axis of Poirel's mixer was turned by 10 convicts by means of levers 6m long; including all the other hands, Poirel's machine required a total of 28 people. Poirel noted that his machine could also have been horse-operated, but since cheap human force was readily available, he preferred to use it. He also commented that the barrel-type machine had the advantage of being capable of being relocated, installing it close to the place where the mortar was used. Both Hagen's and Poirel's mortar mixers operated in a non-continuous mode, finishing one charge after the other.

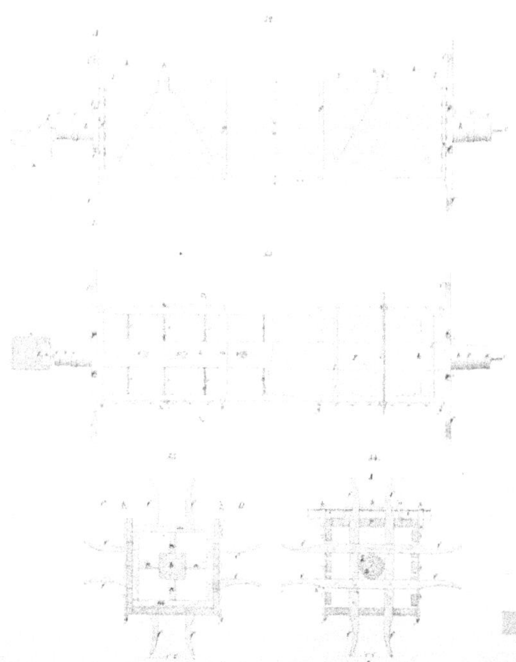

Fig. 5. Rotating drum for mixing concrete (Mihálik 1858, pl. IV, detail)

All the machines described so far mixed only the mortar, but not the gravel or rubble aggregates and the mortar. However, such machines were developed soon afterwards. Hagen, in 1841, gives a detailed description of a concrete mixer employed, as he claims, on several lock construction sites on the river Ruhr, Germany, in the late 1830s:

> "This machine is constituted by an octagonal prismatic wooden drum which is attached to a revolving axis. This axis is arranged horizontally and moved by a steam engine. When the drum starts to rotate, the concrete inside is overturned much in the same way as when thrown up with a shovel. The drum is 6 feet long and 3 feet in diameter, and it turns 9 times in a minute. [...] Mixing of one filling lasts 18 minutes." [25].

A similar "mixing box", but with a square section, was later also tested by Johann von Mihálik (Fig. 5). [26] The 1870 edition of Hagen's book discussed an improved version of the concrete mixer: "The performance of the machine was not great because of the frequent and long breaks in its operation. Therefore, machines should be preferred which permit continuous operation. This is achieved by arranging the axis of the revolving drum in an inclined position, leaving both ends open. The material is entered through the upper end, and the finished concrete leaves the drum at the lower end." [27]

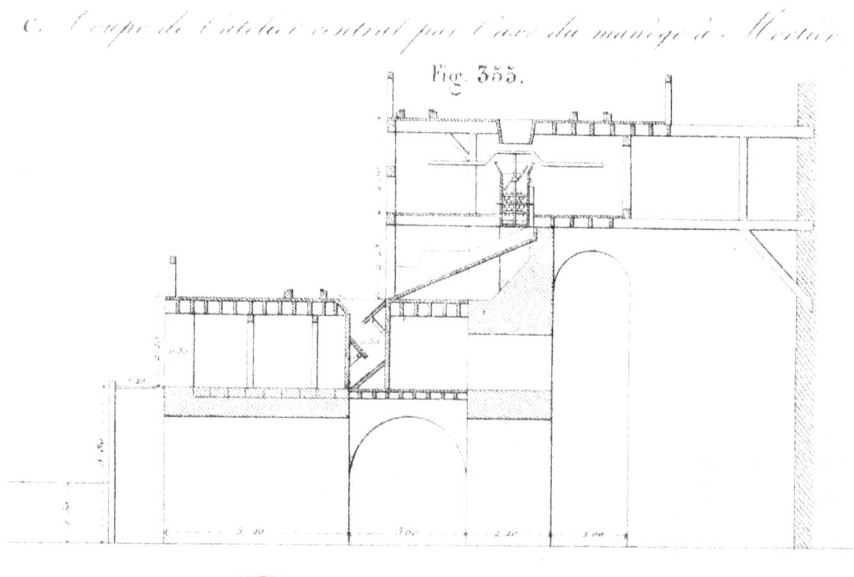

Fig. 6. Section of the Algiers port concrete field factory. At top right is the capstan-operated mortar mixing drum. From there, an inclined chute leads downward to the left where the aggregates are added, and mixed with the mortar in a vertical shaft (Debauve 1870, pl. 47, detail, copied unchanged from Annales des Ponts et Chausées, 1844).

After 1841, the Algiers port site was developed into what may well be called the world's first industrial concrete factory (Fig. 6). The factory was organized in such a way that the material (arriving at a considerable height above the construction site level) followed a natural course from top to bottom, eliminating needless lifting operations as far as possible, and that the material arrived systematically from one step of the treatment to the subsequent one. The mixing of the mortar was still achieved much in the same way as earlier, either manually or in the mixing-drum; however, the mixing of the concrete was now replaced by dropping the mortar and the aggregates into a vertical shaft equipped with a set of baffles:

> "In the interior of the wooden chute, there are three inclined planes, made from planks. These planks, installed at different levels, are inclined in opposite directions, and they force the material thrown in at the top to move from one side to the other during its descent. This ensures that the constituents turn and mix. In order to obtain suitable concrete, the shaft must always remain three-fourths full. Therefore, one keeps constantly throwing in aggregates and mortar at the top; each load descends according to the emptying at the bottom, is mixed, and exits at the bottom where it is squeezed out by the pressure of the semi-fluid mixture on top of it." [28]

As late as 1872, the Algiers factory was obviously still considered so up-to-date that both the text and plates were copied from the *Annales des Ponts et Chaussées* without any change to the handbook *Manuel de l'Ingénieur des Ponts et Chaussées*. [29]

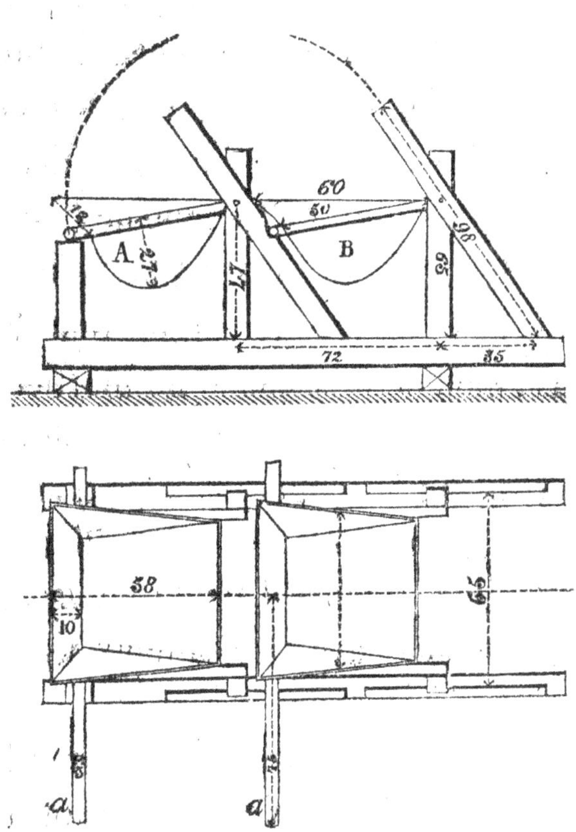

Fig. 7. "coffer machine" for mixing mortar and aggregates to produce concrete (Claudel/Laroque 1850, p. 174).

The barrel-type French mortar mixer was also applied to mix the mortar with the aggregates. The contemporary literature reports that such a machine was used in the fortification works of Paris. [30] However, a totally different, simpler and perhaps more efficient concrete mixing apparatus is described by J. Claudel and L. Laroque in 1850 (cf. Fig. 7):

> "The 'coffer machine' is the one which one ordinarily employs on important construction sites to produce concrete. It is composed of ten cast iron coffers of the form and size shown in the figure. Its operation requires six to ten labourers (depending on the desired rate of operation), half of them on each side of the machine. On one end of the machine, a boarded platform is established on which the gravel and mortar are spread out with wheelbarrows. The workers throw the mixture with shovels into the first coffer A, which, when conveniently filled, is overturned and emptied into the second one. The first coffer is then restored to its original position and charged anew. In the meantime, the second coffer is emptied into the third [...]. The material will be conveniently mixed and provide a good concrete when it has passed the ten coffers. The machine should be arranged as close as possible to the place where the concrete is needed. The economic advantage which this machine offers over manual mixing by rakes is negligible; however, the material is more efficiently and more rationally mixed by the machine." [31]

Finally, once the concrete was ready, means had to be found to bring it to the bottom of the foundation under water without any danger of washing out of the cement, and to ensure that the layers of concrete were of constant thickness throughout , and that subsequent layers would bond well with preceding ones. For this purpose, two techniques evolved quite early: the use of a funnel, reaching from the water surface to the bottom, or, alternatively, transport of the concrete in a closed box or bucket which could be opened or turned over by a tackle once it had reached its destination.

The funnel is the older idea. It can be traced to the late 16th century. Buonaiuto Lorini, in his treatise on fortification, describes the construction of underwater foundations for jetties and moles. While he does not mention concrete, but only constructions employing huge blocks of natural stone, he has nevertheless a section on how to fill the gaps between the blocks with mortar ("Ordine da far andar la calcina sotto l'acqua"):

> "When it is then necessary to fill the gaps which remain between the stone blocks, particularly in the center of the foundation, with lime mortar mixed with small stones, and of course also when it is necessary to fill breaks in the stonework, or hollow areas below old works, one can easily use a funnel for the purpose, that is to say, a chute enclosed on all sides by planks, except at both ends, as long as required; this chute is lowered with one end to the seafloor, where one wants to fill up hollows, and by the other end, which must have a somewhat wider mouth, one drops the material; it will arrive at the desired location, and the flow of the water will be prevented from carrying away the lime, and one will see that one is creating the works not only easily, but also safely." [32]

Unfortunately, Lorini's text is not illustrated by a drawing, but the description is clear enough. The oldest description – and also illustration – of the concrete bucket, on the other hand, seems to be Bélidor's. [33]

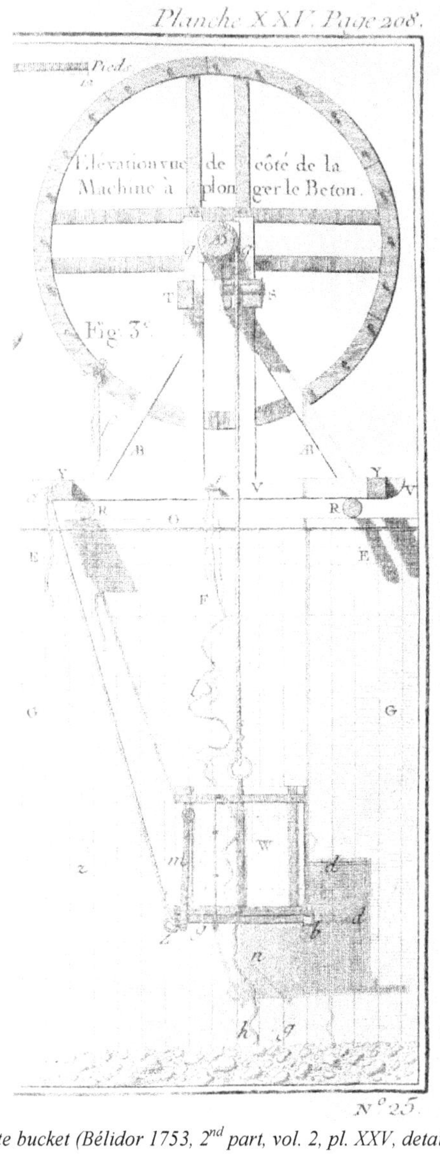

Fig. 8. Bélidor's concrete bucket (Bélidor 1753, 2nd part, vol. 2, pl. XXV, detail)

Belidor's concrete bucket (Fig. 8) is a wooden chest closed on all sides, equipped with a trap door at the bottom, which can be opened via a tackle. Bélidor's concrete bucket is big enough to hold one cubic metre of concrete. He advised not to make the box any bigger than four feet in every direction because otherwise handling would become too difficult. The use of Bélidor's bucket was also recommended in Gauthey's famous textbook on bridge building. [34]

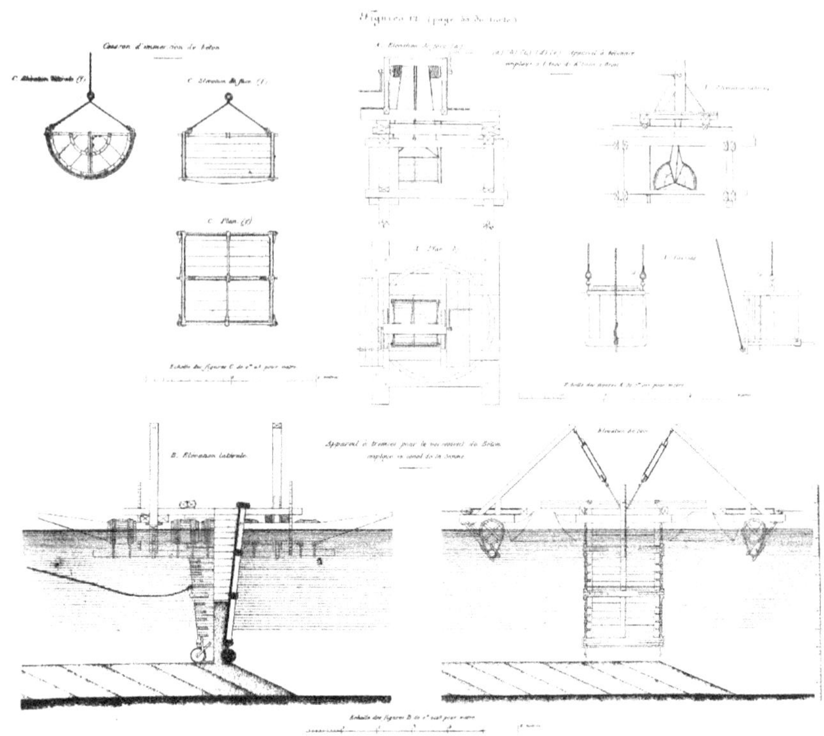

Fig. 9. Various methods of transporting concrete to the bottom of an underwater construction pit: concrete buckets vs. chutes (Sganzin 1839—41, plate 3, detail)

In the technical literature of the first third of the 19th century, many discussions on the advantages and disadvantages of the two techniques – bringing concrete to the bottom by either the chute or the bucket – can be found. We do not have enough space here to go into details, but we refer the reader to Fig. 9 for an overview of mechanisms employed for the purpose (from [35]). In the posthumous 1839–41 edition of Mathieu-Joseph Sganzin's *Cours de Construction*, a concise conclusion on the question of the best technique was formulated:

> "The casting of the concrete is achieved by means of long chutes; however, this method has the great drawback of bringing the material in contact with the water, for a trajectory of 7 to 8 metres, which provokes the separation of its constituents due to their different densities, and the elution of the lime. One has therefore subsituted the chutes by closed buckets which are only opened when the concrete has reached the depth, or by scoops which can be turned around a handle." [36]

Perfections of both techniques depicted by Sganzin include buckets consisting of two quadrant-shaped parts, which would permit one to open the bucket directly at the bottom ("the concrete flows out and falls from as small a height as possible" [37]) and chutes equipped with cast iron rolls to consolidate and flatten out the newly cast layer ("It can be stated as a fact proved by experience, that the success of the foundation slab of the lock of St. Valery can be attributed to the use of the new chutes with rolls, which can be moved several times over the same spot, and compress the concrete." [38]).

Concluding the present article, it can be summarized that concrete and its production technology are rooted deeply in early modern construction history, not only in rammed earth construction. Most fundamental ideas of concrete processing technology were well developed and available in the field of hydraulic engineering, decades before concrete started to expand to other engineering applications and architecture.

References

[1] P. Collins, *Concrete: The vision of a new architecture*. 2nd ed., Montreal & Kingston, 2004, p. 19.

[2] G. Rusconi, *Dell' architettura*, Venice, 1590 (reprinted almost unchanged: Venice: Nicolini, 1660), p.9. The text was supplied posthumously by Rusconi's editor, but the woodcut illustrating the concrete foundation slab is original and was created by Rusconi himself around 1550.

[3] B.F. de Bélidor, *Architecture hydraulique*, Paris: Didot, 1737–53, 2nd part, vol. 2, p. 168.

[4] Bélidor, *Architecture hydraulique* (note 3), 2nd part, vol. 2, pp.187.

[5] Bélidor, *Architecture hydraulique* (note 3), 2nd part, vol. 2, pp.186: "pebbles whose size must not exceed the size of a hen's egg."

[6] Bélidor, *Architecture hydraulique* (note 3), 2nd part, vol. 2, pp.186–187.

[7] Treussart, *Mémoire sur les mortiers hydrauliques et sur les mortiers ordinaires*, Paris: Carillan-Goeury, 1829, p. 161.

[8] Rusconi, *Dell'architettura* (note 2), p. 101.

[9] L.-A. Beaudemoulin, *Recherches théoriques et pratiques sur la fondation par immersion des ouvrages hydrauliques, et particulièrement des écluses,* Paris: Carilian-Goeury, 1829.

[10] Magdelaine, "Description des procédés employés de 1812 à 1815, pour la fondation, et notamment pour le coulage de l'aire de béton, du barrage éclusé de Saint-Valéry-sur-Somme", *Annales des Ponts et Chaussées,* 1832, 1er sem., pp. 52–65.

[11] Mary, "De l'emploi du béton dans la fondation des écluses", *Annales des Ponts et Chaussées,* 1832, 1er sem., pp. 65–105.

[12] Zimmermann, "Einige Bemerkungen über die Festigkeit, Mischungs-Verhältnisse und Zubereitung des Bétons, oder des Mauerwerks aus klein geschlagenen, mit Mörtel untermengten Steinen, dessen man sich zuweilen, um Fangdämme und Wasserschöpfen zu sparen, zur Fundamentirung von Bauwerken unter Wasser bedient", *Journal für die Baukunst*, no. 3, 1830, pp. 1–32.

[13] C. Parea, "Memoria sul metodo tenuto nella condotta delle opere del Canale di Pavia, e specialmente della fabbrica dell'ultimo suo sostegno presso il fiume Ticino", pp. 269–284 in G. Bruschetti (ed.), *Istoria dei progetti e delle opere per la navigazione-interna del Milanese*, Milan: Bernardoni, 1821.

[14] Beaudemoulin, *Recherches* (note 9), p. 16.

[15] Beaudemoulin, *Recherches* (note 9), p. 17.

[16] Zimmermann, "Bemerkungen" (note 12), p. 28.

[17] Zimmermann, "Bemerkungen" (note 12), p. 28–29.

[18] J. v. Mihálik, *Praktische Anleitung zum Béton-Bau für alle Zweige des Bauwesens,* Vienna: Stöckhölzer & Hirschfeld, 1858, p. 113.

[19] Mihálik, *Praktische Anleitung* (note 18), p. 114.

[20] J.-R. Perronet, *Description des projets et de la construction des ponts de Neuilli, de Mantes, d'Orléans, de Louis XVI, etc.*, Paris: Didot, 1788, p. 121.

[21] F.-M. Lebrun, *Traité pratique de l'art de bâtir en béton*, Paris. Carilian-Goeury, 1843, pp. 67–68.

[22] Mihálik, *Praktische Anleitung* (note 18), pp. 118–130.

[23] G. Hagen, *Handbuch der Wasserbaukunst*, part I: *Die Quellen*, Königsberg i. Pr.: Bornträger, 1841, pp. 768–769.

[24] L.-V. Poirel, *Mémoire sur les travaux à la mer,* Paris: Carilian-Goeury, 1841, p. 32.

[25] Hagen, *Wasserbaukunst* (note 23), part I, p. 777.

[26] Mihálik, *Praktische Anleitung* (note 18), pp. 131–137.

[27] G. Hagen, *Handbuch der Wasserbaukunst*, 3rd ed., vol. 2, *Brunnen, Wasserleitungen und Fundirungen*, Berlin: Ernst & Korn, 1870, p. 326.

[28] Krantz, "Note sur divers procédés employés pour la fabrication du mortier et du béton sur les chantiers du port d'Alger", *Annales des Ponts et Chaussées,* 1844, 2nd sem., pp. 300–331, here pp. 310–311.

[29] A. Debauve, *Manuel de l'ingénieur des ponts et chaussées,* 4ème fascicule, *Exécution des travaux*, Paris: Dunod, 1872, pp. 203–214.

[30] M. Becker, *Allgemeine Baukunde des Ingenieurs* (=*Handbuch der Ingenieur-Wissenschaft,* vol. 1), 2nd ed., Stuttgart: Mäcken, 1857, p. 49.

[31] J. Claudel and L. Laroque, *Pratique de l'art de construire*, Paris: Carilian-Goeury, 1850, p. 174.

[32] B. Lorini, *Le Fortificationi*, 2nd ed., Venice: Rampazetto, 1609, p. 193.

[33] Bélidor, *Architecture hydraulique* (note 3), 2nd part, vol. 2, pp.185.

[34] E.-M. Gauthey, *Traité de la construction des ponts*, vol. 2, Paris: Didot, 1813, p. 287.

[35] M.-J. Sganzin, *Programme ou résumé des leçons d'un cours de constructions*, 4th ed., Paris: Carilian-Goeury, 1839–1841, vol. 1, pp. 55.

[36] Sganzin, *Cours de construction* (note 35), vol. 1, p. 55.

[37] Mary, *Cours de routes et de ponts, professé à l'Ècole Centrale des Arts et Manufactures*, lithographed manuscript, 1841–42, p. 108.

[38] Magdelaine, *Annales des Ponts et Chaussées*, 1832 (note 10), p. 60.

The British cut clasp nail

Chris How
Retired structural/heritage engineer, Australia

The cut clasp nail was the most common type of cut nail in general use in Britain for more than a century, and its different forms provide clues to dating building stages between 1815 and 1920. Nails found in a building present a useful matrix of dating information arising from their individual chronologies. This paper, based on field and literature research, identifies the various forms of British clasp nails with the aim of adding to the various dating spectra that nails provide.

British cut nails owe some aspects to American ingenuity, but they quickly took a different developmental path. During the 1790s, the Americans had concentrated on developing a range of lighter cut nails suited to softwoods. These offered possible ways to resolve nailed connections in Britain where there was an increasing reliance on imported softwoods. Cut nails, however, for some time posed a problem for both the British home market and the growing demands of the colonies, since the American product lacked the necessary strength for use in hardwoods. Australia in particular rejected these American cut nails as they broke in the hard eucalypt timbers. [1] The British solution to the common problems then went through several datable stages, traced below.

Clasp nails

The first definition of a wrought clasp nail in a British context comes from Neve in 1703, listing two sizes as "Strong, suitable for oak, and other Hard woods", and, "Long, proper for any fine building with Firr, or other soft wood: The clasping of the head brings them into little compass, and admits of them sinking into the Wood, makes the Work smooth, and will admit a smoothing plane to go over them when drove." [2]

The two types of clasp nails described overlap in their weight range as the "strong" clasp nails reduce in size. The quintessential feature is the clasping action of the nails from two thin, downward facing "wings" extending out from the sides of the head on the nail's major axis. In Britain and Europe, quantities of larger nails are found with pointed raised heads, but lacking the "wings" feature. These are not true clasp nails. Both of the types described by Neve have been found in investigations into old buildings and building timbers, but finds so far suggest that the "strong" version was more widespread in the first half of the eighteenth century than the lighter "long" type. By the 1780s this latter type was used in large numbers in Georgian townhouses in London. [3] In both types the extended "wings" taper out, enabling them to bend backwards into the fibres of the timber as they are driven home and provide withdrawal resistance. The tapering of the wings results in a zone of weakness in this part of the head and nails are occasionally found without one or both wings as these have broken off.

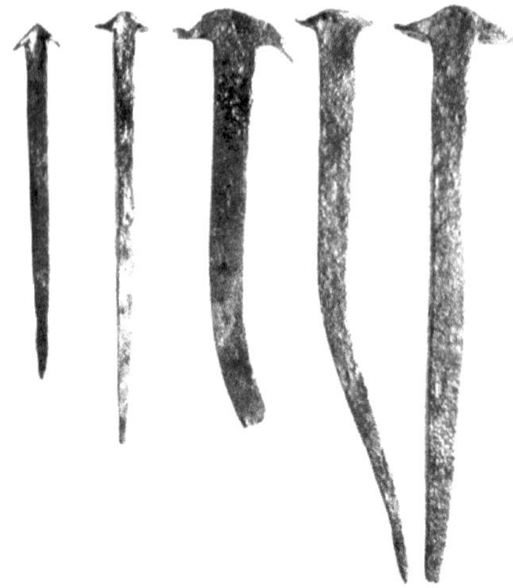

Fig. 1. Unused "long" clasp nail from Weald and Downland Museum with a used version circa 1799, Westminster. A broken "strong" version from 1732 with two used five inch (125mm) clasp nails circa 1750, all from Somerset (author).

Earlier finds have appeared in Germany and a wrought clasp nail located in pine floorboards was found in a Bauhaus, dendro-dated at 1505, at Beutelsbach near Stuttgart (Fig. 2, top). [4] Also in Southern Germany, lighter versions of the "long" clasp nail have appeared in large numbers in buildings reconstructed after the Thirty Years War, approximating to the 1650s (Fig. 2). [5] Similar slender clasp nails continued to appear in Germany in the 1750s and 1850s, more slender than the British wrought versions. [6]

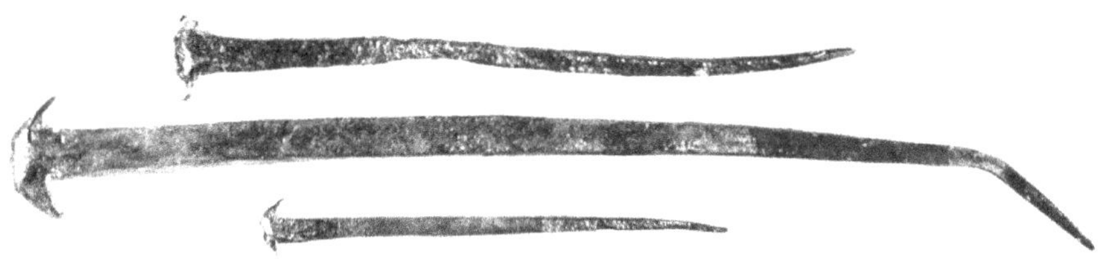

Fig. 2. Wrought floorboard clasp nail, 118mm, from Beutelsbach, Stuttgart, 1505, with a 166mm sprocket (aufschiebling) nail and a batten nail from Schloss Wolfegg, Oberschwaben, (author).

The "strong" version does not appear in the classic French works of the time [7], [8], and its equivalent is presented by Diderot as having a solid triangular head. Finds in the Ardèche and Provence verify his drawings. In place of the lighter clasp head, smaller nails are shown by Roubo as T-heads, and by Diderot as having a half-round top to the head. Both are missing the "wings" that create the clasping action. A single, double-clenched, clasp nail was taken from a 1788 New Orleans house in French Louisiana, but it may have been of imported origin. [9] The clasp nail form probably entered England during late Jacobean times, and the Continental clasp nail was reproduced in England in sturdier form from mid- to late-1600s.

The advent of British cut nails

In Europe, Britain, and North America, cut nails are likely to have started as simple headless glazing sprigs, cut cold from thin sheet as mentioned in a document of 1692: "naylls to sett the glas in the wood windos." [10]

In 1790, Thomas Clifford of Bristol registered two patents as "Machinery for the Manufacture of Nails". [11] The first relied on pressing heated nail rod into shape; the force being applied either by shaped rollers or, more significantly, by pressure dies. In the patent wording Clifford introduced the concept of using pre-profiled iron strips. His other patent involved a stamping out process, similar to buckle production, and nails were thereby pressed out of thin sheet iron into a hollow bed. The significance here is that punching shear was adopted, cutting into an under-well below. This anticipated the rising bed technique later used in producing floor brads. The stamping technique had been in use for cutting buttons and buckles from around 1728 in Birmingham. [12] By 1792, French's factory at Winebourn, Staffordshire, was making nails by both of these processes. [13] Cut lining brads were noted to have been in production in 1803, [13] and they have appeared in Wells [14] and the Welsh Marches [15] from c.1800. They were advertised in the Sydney Gazette of 1809 [16] and had featured in a tender document of 1812 for the Halifax garrison of Nova Scotia in four sizes: 1¼ inch to ½ inch (31mm to 12mm) (Fig. 3). [17]

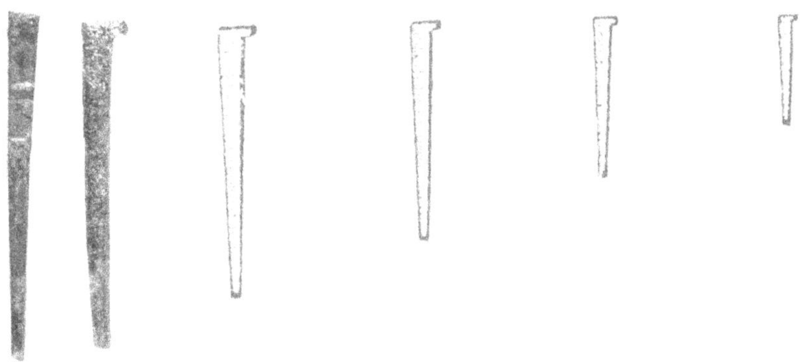

Fig. 3. Headless and spur-headed floorboard nails from an 1806 house, Dartmouth (author). Cut brads shown in 1812 tender.

Spur-head floor brads of the same shape, but 2 inches (51mm) long and twice the sheet thickness, have been recovered from an 1806 house in Dartmouth, Devon. [18] An Australian archaeologist maintains that this type of cut floor brad is found in dated contexts in Sydney from the late 1790s: "the simple cut brad became the first (cut nail) practical success". [19] An American authority dates their appearance in America to 1805–10. [20]

American cut nails

With some justification, the Americans claim that machine production of cut nails was an American invention. From a simple beginning, the lever-operated guillotine machine patented by the prolific Jacob

Perkins in 1795 was developed by various independent inventors into a series of efficient cut nail fabrication devices. [21] By 1810 water powered nail machines had been developed, although American nail making predominantly relied on manual machines.

Nails of a primitive sort could always be cut manually by a blacksmith from heated barrel hoop or other strip iron and several instances of this resort to basics are recorded in the 1788 colony at Botany Bay. [22] Perkins' later machine took the process one step further and cut tapering shanks from a heated iron ribbon. In order to maintain a taper to the proto shank, the ribbon either had to be wiggled through a shallow horizontal arc or flipped over between each cut. Perkins' machine was entirely manual, but recent research has shown that at "black heat" of around 500°C, the shear force for a typical nail shank drops to roughly 300 kilograms. Using leverage, this is easily attainable. [23] The device significantly speeded up the cutting of basic nail shanks. These simple headless brads have been found in Britain both of iron and steel, and were usually used in flooring or specialist work (Fig. 3, nail 1). In America they would be re-heated and headed separately with either conventional facetted heads or bulbous oval heads formed with a hand-held die. [24] Later, similar examples found in Britain and Australia suggest that the blank shanks were machine-headed by an oliver fitted with a die, which gave either a rounded oval or a somewhat oval facetted head. In 1792, the American inventor Nathaniel Read used two foot driven hammers to attempt a primitive head. [25]

In 1807, Thomas Odiorne and several associates set up three major nail plants. The machines were different from the Nathaniel Read versions in that the heads were no longer struck but headed under pressure. [26] This basic principle remains the same to this day, and later machines sold to the British around 1814 were linear descendants of Odiorne's machines. By then a patent for a cut nail machine had been enrolled in England four years earlier. [27]

Cutting techniques: problems and identification

Early American cut nails had an inherent problem as the rollers producing the iron ribbon were narrow, due to power availability and roller technology. The shanks, simple narrow wedges, with lengths dictated by the narrow ribbon's width, were thus cut at right angles to the longitudinal grain of impurities caused by the rolling action. Thus, cut transversely, the shank bending strength was significantly reduced and the risk of fracture increased. None of the early American cut nails were deemed capable of clenching, a necessary attribute for door nails, shutters and boatbuilding which kept in being the traditional hand-wrought nail and in particular the wrought clasp nail. [28] American researchers believe it was not until 1840 that the first longitudinally-grained shanks appeared in the United States, [29] although in Britain the method had already been filed for patent in 1810 by Dyer, an associate of Perkins. [30]

Side- and Face-clamping

Until 1830 American cut nails were side-clamped (or "side-pinched") by a lever-activated vice prior to heading. [31] The cramping of the vice, being applied to the sides of the nail shank, tended to crumple the outer layers and led to a loss of strength just below the head. This was recognized by the nail machine inventors and from 1811 a small proportion of face-clamped nails started to appear alongside side-clamped samples. [32] One American furniture expert maintains that side-clamped nails continued until 1848. [33] These display a waist below the neck which, while enhancing the nail's holding strength, weakens the head attachment. A rare sample of a side-pinched nail was found at Percival Terrace in Brighton, Sussex, built in 1845 (Fig. 5, nail 1), its "necking" suggesting it came from an early machine. This waist gradually disappears in British side-clamping as it was found that adequate grip could be obtained by setting the vice jaws essentially parallel (Fig. 5, nail 2).

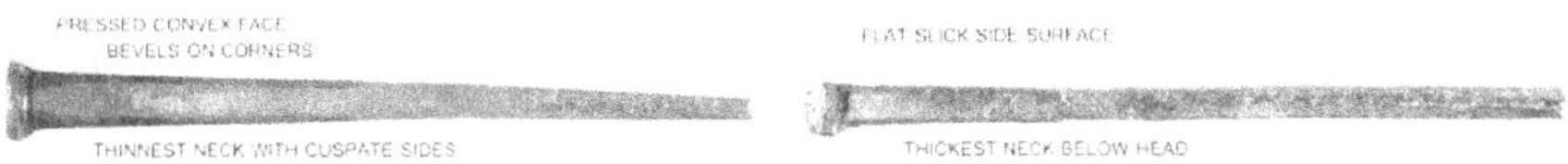

Fig. 4. An American face-clamped cut nail of 1870, left. The face-clamping and bevels extend for about 40% of the shank. British cut nail of 1860 side view, right. The side-clamping extends for about 30% of the shank (author).

In American face-clamping, the top 40 per cent of the nail was grasped by two identically indented, vice-like concave jaws that pressed bevels onto the front and back corners (Fig 4, nail 1). [34] The British persevered with side-clamping apart from on tacks, small nails and the later "T-patent" profiled cut nail by Cordes & Company (the Ewbank manufacturers). [35] From this point there is a divergence in American and British development.

Separate British cut nail development

The rapid growth of steam power in Britain by 1805 [36] allowed wider sheets and thicker sheets to be rolled which could then be cut into transverse strips for nail feed stock. The nails thus produced were stouter, with fibres longitudinally aligned, and more suitable for hardwoods. Using steam-power side-clamping of the thicker shanks was still practical and the initial British cut nails strongly resemble heavier forms of the early American versions (fig. 5, nail 1).

Fig. 5. "Necking" effect on an 1845 side-clamped cut nail from Brighton, left. The right hand nail dates from 1852 and shows more parallel sides below the head. Shear burrs can be seen on both (author).

The first step away from ovoid domed heads (Fig. 5) was to an asymmetrical head which, from above, shows only three facets, the fourth a mere blip at the rear as the face is forced into a significant swelling. This type appears c.1835 (Fig. 6) but quickly disappears and none have been found after 1845–1850. At this stage there was no apparent attempt to develop clasp "wings", and lateral projections seem to relate to attempted faceting of the domed heads.

Fig. 6. Side and front view of two 1835 British cut nails. Both have a slight twist due to separation from the guillotine blade and display the eccentric swelling of the nail neck (author).

By 1850 or so, the head eccentricity problems seem to have been fully resolved and the die shape changed to two convex facets that left two clear concave indents on either side of the nail centre. This created the high crown, as previously seen above on the wrought version of the clasp nail (Fig. 1, nail 1). It also deflected outwards two pressings of decreasing thickness to form reasonable facsimiles of the wrought clasp nail "wings" (Fig. 7).

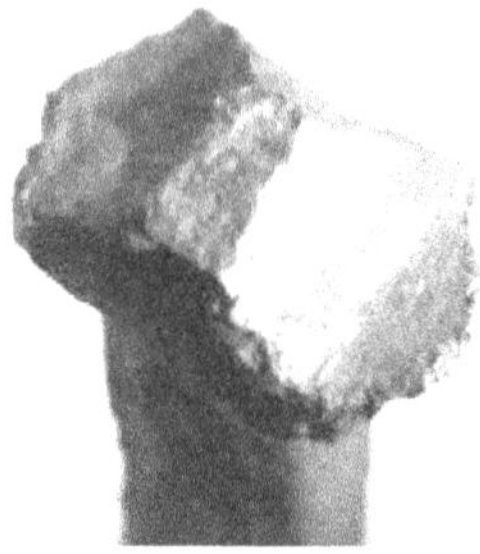

Fig. 7. Cut clasp nail of 1858 showing the development of the clasp "wings" (author).

In addition to the cut clasp nail facsimile, less frequently a rough "bull head" version occurred, lacking the clasp "wings". Around 1880 a further change was the reduction of the "wings" to nominal bumps or "lugs" on either side (Fig. 8, left). This preceded the change to steel cut nails, which became available around 1895, after the advent of steel wire nails.

Fig. 8. Cut iron clasp nail with reduced "lugs" from 1870 and a steel version, c.1905 (author).

Cut clasp nail sequences

The steps towards a facsimile clasp pattern of British cut nail went through five main stages:

1814	Machine-headed, robust cut nails introduced, following the American pattern (Fig. 5).
1825- 1844	Transition machine-made, off-centre facetted head and showing only three facets (Fig. 6).
1825- 1844	Horizontal wings with two cuspate facet pressings to the head to early prototypes gaining full true horizontal "wings" by c.1840 (Fig. 7).
1870- 1908	Less pronounced side wings, which reduce to rounded protuberances on each side and continue to appear as iron cut nails at least until 1908 (Fig.8, nail 1).
1900- 1960+	Steel cut nails gradually replace their iron counterparts. The protuberances reduce to rounded "lugs" as a norm (Fig. 8, nail 2). A version with fully developed horizontal "wings" appears much later.

Conclusion

Development of the British cut clasp nail underwent several identifiable stages of change between 1814 and 1960 which can be dated. These changes help to further define the spectrum of dating information available from other nail types and are valuable assets in validating the dates of construction sequences of buildings in which they are found.

Acknowledgements

In addition to those mentioned below, the author wishes to thank Joe Thompson and Roger Champion of the Weald and Downland Museum. Appreciative thanks are due to Professor Miles Lewis of The University of Melbourne for his long-term encouragement, advice and support. Thanks are also due to Dr Roger Luebbers for his generous advice on photographic and artefact treatment.

References

[1] G. F. Train, ed. E. A. & A. Potts. *A Yankee Merchant in Gold Rush Australia: Letters of* 1853–1855. Melbourne: William Heinemann Australia Pty Ltd, 1970, p.25.
[2] R. Neve, *The City and Countrey Purchaser, and Builder's Dictionary*. London: Rising Sun, 1703, p.211. (1st edition published under the pseudonym T.N. Philomath)
[3] Robert Ayton, Conservation Officer, Westminster City Council, supplied Regency nails and photos, 2006.
[4] Jens Bode, restorer, Stuttgart, supplied early German floor nails from Beutelsbach, 2007.
[5] Dr Wolf Münzner supplied nails and information, 2009.
[6] Jens Bode, restorer, Stuttgart, supplied later clasp nails from Heutingen and Nürtingen, 2009.
[7] D. Diderot, (ed.), *L'Encyclopédie: ou, Dictionnaire raisonné des sciences, des arts et des métiers par une société de gens de lettres* (35 vols), Paris, Briasson et al. Vol. 3, 1769. p.137, plate 1.
[8] J. Roubo, *L'Art de Menuisier* (réimpression de l'édition de 1769–1775, 4 vols), Léonce Laget, Paris: De L'Academie Royale des Sciences, 1976 (1769). Pt 2, chap. 8, fich. C.
[9] J. D. Edwards and T. Wells, *Historic Louisiana Nails*. Baton Rouge: Geoscience, 1993, p.4, fig. 2.
[10] Stewards accounts 1692–1695, Levens Hall, Westmorland. <www.nationalarchives.gov.uk/a2a/records.aspx?cat=055-ddca1/4>
[11] Patent no. 1762, 31 July 1790 & patent no. 1785, 30 December 1790. British Patent Office.
[12] R. McLean, "Introduction: Matthew Boulton, 1728–1809", in S. Mason, *Matthew Boulton: Selling What All the World Desires*, Newhaven, CT: YUP, 2009, pp. 1–6.
[13] Rees, *The Cyclopedia: or Universal Dictionary of Arts, Sciences and Literature*. London: Longman, 1819, sv "Nails".
[14] Adam Wilson, restoration specialist, recovered these nails from a Wells Diocesan house.
[15] Richard Jones, restorer & collector, supplied the cut brads.
[16] R. V. J. Varman, *Bricks and Nails: Building Materials as Criteria for Dating in Sydney and Environs from 1788* (unpublished thesis, University of Sydney, 1993).
[17] Susan Buggey, "Royal Engineers Office Halifax: List of Nails and Spikes required for the Service of the Office of Ordnance", in "Supplying Building Materials to the British Army in the Colonies", *APT Bulletin*, vol. 8, no. 3, 1976.
[18] Adam Wilson, pers. comm., 2009.
[19] Varman, *Bricks and Nails*. (Note 16) Journal of Lieut. Philip Gidley King, R.N. 17th November 1788, Norfolk Island, p.148.
[20] L. H. Nelson, "Nail Chronology as an Aid to Dating Old Buildings", *History News,* vol. 23, no. 11, 1968.
[21] Edwards and Wells, *Historic Louisiana Nails,* (Note 9) pp.15, 16, 17.
[22] Varman, *Bricks and Nails*, (Note 19) p.172.
[23] K. Ryzewski and R. Gordon, "Historic Nail-making Techniques Revealed in Metal Structure", *Historic Metallurgy*, vol. 42, part 1, 2008. p.53, fig. 6.
[24] Ryzewski and Gordon. *Historic Nail-making,* p.57.
[25] M. K. Phillips, "Mechanical Geniuses and Duckies, A Revision of New England's Cut Nail Chronology before 1820", *APT Bulletin,* vol. 25, nos. 3/4, 1993, p.7.

[26] Edwards and Wells, *Historic Louisiana Nails,* (Note 9) p.17.

[27] Patent no. 3365, Patentee J. C. Dyer, enrolled January 1811. British Patent Office.

[28] J. Loveday, *The Rise and Decline of the American Cut Nail Industry,* (note 28) Westport, Connecticut: Greenwood Press, 1983. p.13.

[29] Edwards and Wells, *Historic Louisiana Nails,* (Note 9) p.18.

[30] Phillips, *Mechanical Geniuses and Duckies*, p.7: British patent no. 3365.

[31] Loveday. *The American Cut Nail Industry,* p.18, fig. 8.

[32] Edwards and Wells, *Historic Louisiana Nails,* p.17.

[33] E. Welch, "Technical Systems for Dating Country and Primitive Furniture", *Journal of Antiques and Collectibles,* August 2002. <www.journalofantiques.com/Aug02/businessaug02.htm>

[34] C. How, "The American Cut Nail", in Miles Lewis (ed.), *The Pacific Connection Seminar,* Melbourne, Australia, 2009, pp.81, 82.

[35] C. How and M. Lewis, "The Ewbank Nail", in Karl-Urgen Kurrer et al. (eds.), *Proceedings of 3rd International Congress on Construction History,* vol. 2, Cottbus, Germany, 2009, p.836.

[36] R. Knight, *Britain against Napoleon, 1793 -1815*, Allen Lane, London, 2013, pp.352, 353.

The developer, the builder, his contractors and their tradesmen

Conor Lucey
Irish Research Council Elevate (Marie Curie Co-Fund) Postdoctoral Fellow, University of Pennsylvania & Trinity College Dublin, Ireland

Since the publication of Sir John Summerson's seminal *Georgian London* in 1945, our understanding of the 'typical' eighteenth-century terraced house and its construction has grown exponentially. In recent decades historians have addressed its myriad social, material and economic contexts including, but not limited to, emerging capitalist modes of building production and the corresponding impact on labour process; the methods and materials of individual building and decorating trades; and the financial independence and social mobility achieved through property speculation.[1] Others have endeavoured to broaden the scope of enquiry beyond aristocratic housing, or have considered the historical implications of house building for early modern urbanization.[2] However, beyond describing the *principle* of contracting labour, and the importance of site organisation and managerial competence, the particulars of the contracting and subcontracting system – the social mechanism by which houses and terraces were actually realized – have yet to be fully elucidated. With particular reference to Dublin's Mountjoy Square and tributary streets, this paper will consider the following questions: how were houses built by contract – who contracted with whom? Did commerce or mutuality define these working relationships? How did the practice of contracting and subcontracting impact on the scale of business portfolios? Who were the most successful figures in the city's house-building industry? Synthesizing archival evidence from a variety of sources, in particular the accounts of the measurer Bryan Bolger (c1758–1834), this paper will examine the correlation between building networks and building houses.[3]

Contractors and subcontractors: the organisation of labour

Far from being 'disinterested in and disconnected from matters of finance and monetary accumulation',[4] historians are now agreed on the entrepreneurial character and economic incentives of the early modern building operative/tradesman. But just as the foremost asset of the master builder was credit worthiness and capital resources, so too was credibility and reputation in business. Elizabeth McKellar and Peter Guillery have independently stressed the importance of mutuality among building tradesmen in London throughout the seventeenth and eighteenth centuries,[5] and the formation of the 'Society of Master Builders, Wrights and Masons' in Edinburgh around 1790, composed of operatives from outside the city corporations, indicates a shift towards methods of professional organisation conducive to the business of house-building.[6]

The three categories of payment for building work outlined by Sir Christopher Wren in 1681 – 'by the great, by measure, by the day' – coexisted throughout the eighteenth-century and were related to different structures of labour employment and organisation. Dublin's house building industry operated 'by measure' and 'by the great' in tandem: 'by measure' describing a piece-work system, where tasks completed by a tradesman were measured and valued; 'by the great', or 'in gross', referring to the undertaking of the whole building operation by a master builder, who in turn sub-contracted the particular tasks his firm was unable to provide.[7] Although recent (unpublished) doctoral scholarship concerning the Irish construction industry suggests that 'there are few instances of the use of measurers by contracting artisans in eighteenth century accounts',[8] this is demonstrably contradicted by reference to the historical record; in fact, regardless of the profession/trade of the Georgian developer – be they builder/tradesman, merchant or gentleman – the method of evaluating the work of individual contractors on site was almost exclusively 'by measure'. Moreover, although Linda Clarke has argued that this system of after-measurement was 'an impediment to capital formation within the building process as a whole', and that the introduction of all-trades contracts at the end of the eighteenth-century established the building sector as a 'single production process' thereby undermining the measure and value system,[9] the picture that emerges from the Bolger papers, concerning house-building in Dublin between 1787-1817, is more complex.

TABLE 1: Sample of contracted labour at various sites in Dublin, 1789–1796.
Source: National Archives of Ireland, Bryan Bolger papers, 2/476/15–16.

Developer/Builder	Contractor	Location	Date	Type	W&M	W
Stapleton, Michael	Walter Casey	Mountjoy Sq	30.12.1789	St, Br		•
Stapleton, Michael	John Prendergast	Mountjoy Sq	11.05.1790	St, Br		•
Stapleton, Michael	James Leaver	Mountjoy Sq	12.12.1793	Ca		•
Salmon, William	John Prendergast	Mountjoy Sq	19.07.1791	St, Br		•
Salmon, William	Charles Hanley	Mountjoy Sq	10.01.1793	St, Br		•
Kildahl, Nicholas	Jane Law	Eccles St	25.10.1792	St	•	
Kildahl, Nicholas	Charles Hanley	Mountjoy Sq	31.10.1792	St, Br		•
Sweeny, John	Robert Arthurs	Mountjoy Sq	04.12.1794	St, Br	•	
Sweeny, John	Gillegan & Byrne	Mountjoy Sq	20.01.1796	Pl		•
Archdall, Edward	John Prendergast	N Gt George's St	16.07.1789	St, Br		•
Archdall, Edward	William McCormick	N Gt George's St	25.07.1789	Sl	•	
Archdall, Edward	Benjamin Hallam	N Gt George's St	13.09.1790	Pa	•	

St = stonemasonry; Br = bricklaying; Ca = carpentry; Pl = plastering; Sl = slating; Pa = painting.

Table 1 presents a sample from this source, itemizing contracted labour at a variety of locations across Dublin measured by Bryan Bolger between 1789 and 1796. Here, Michael Stapleton, a plasterer and established figure in the city's house building industry, had contracted for 'work only' (W); William Salmon, a building materials supplier, also contracted for work only at his houses—unsurprising, given the nature of his business. However, Nicholas Kildahl, a builder and property developer, is shown to have contracted for labour and materials (W&M) with some tradesmen but not others; and developers not directly involved in the building trades, such as John Sweeny and Edward Archdall, operated in a similar manner.[10] Indeed, despite a burgeoning literature devoted to building quantities and costs – created so that 'every Gentleman concerned in Building, may have the Value of his money, and that every Workman may have the worth of his Labour'[11] – the professional arbitration of a measurer was clearly necessitated by the intricacy of this *modus operandi* (and the building process in general). A revealing note of 1802, written by the carpenter James Sinnott to Bryan Bolger, concerning a house in Thomas Street, states:

> 'Sir, You'll be so kind as to price this work yourself as I am not well acquainted with the price of workmanship only & perhaps what I have priced, some is to[o] high & some to[o] low.'[12]

An account of work (furnishing materials) by the plasterer and painter Andrew Callnan (d.1804), described as 'for Mr John Price at Mr Hall Kerchoffer's new Buildings in Temple Street' and measured in March 1792, illustrates the role of the measurer in quantifying such an arrangement between a developer, a master builder and a contractor.[13] Hall Kirchoffer was from a renowned family of Dublin cabinet-makers and was by the 1790s actively speculating in Mountjoy Square and its environs. John Price is listed in the *Dublin Directory* as a timber merchant between 1787–92, but as a 'Builder & Carpenter, Dealer & Chapman' in property transactions for the same period.[14] It is clear then that Kirchoffer – the developer – had engaged Price to direct the building works; Price, in turn, had contracted with Callnan for the plastering.[15] Further examples abound. A measured account of November 1792 for carpentry executed by a Mr Brown,[16] most likely Thomas Brown of Thomas Street, on behalf of the plasterer and developer Charles Thorp 'at various places',[17] records work at 'Mr Burton's Sackville Street',[18] 'Mr Dick's – Great George's Street',[19] 'Mr Cornwall's – Stephen's Green',[20] and '2 Houses in Ely Place'.[21] Other related bills record that Laurence White and William Quinn supplied 'Mountain Stone' and 'stone and bricklayers works' respectively for Thorp at a number of the same locations.[22] In this instance, Thorp was manifestly both property developer and master builder, contracting labour but typically supplying materials.

Mapping working relationships

The purpose of this paper is to examine the complexity of labour organisation on the 'typical' building site, revealing the range of actors involved in realizing Dublin's celebrated Georgian domestic architecture. At the outset, it is clear that relationships between builders and contractors were forged in response to a variety of circumstances. On property leases made by the bricklayer John Prendergast in the 1780s and 90s, for example, the carpenter James Bowden and the timber merchant John Derrick appear as witnesses to the legal contracts: Derrick was likely a supplier to Prendergast's company; while Bowden, a building contractor in his own right, may have been employed on site in some capacity.[23] Family businesses, on the other hand, often had the means, both in terms of labour, plant and materials, to readily operate 'by the great': in 1784, the timber merchants John and Charles Pentland advertised that they 'build Houses &c. by contract or otherwise';[24] while in 1789, Benjamin, Joseph and Russell Doyle of Earl Street pledged to continue their late father's 'building business'.[25] Various members of the Thorp family, listed collectively at premises in North Great George's Street in the early 1800s, provided 'building', carpentry, plastering and painting.[26]

The structure of these larger building businesses can be determined by reference to a series of related measured accounts, dated 1798–1800, itemising the various works 'done for Mrs Preston at her house in Mountjoy Square', the present number 25, and including stone and bricklaying, slating, carpentry, plasterwork and painting.[27] That all building and decorating works are described as 'by Frederick Darley', suggests that the various specialist tradesmen engaged on site were either operating directly on behalf of Darley (from a renowned family of stonecutters/master builders), or constituted the regular employees of the Darley building firm. More significantly, these accounts contradict the notion that 'the typical use of the in-gross agreement – such as in the building of speculative terraced houses – did not need the services of a measurer, or bills of measurement'.[28] While the individuals involved are not enumerated, an unrelated bill of 1792 records carpentry 'done for Fredrick Darley Esqr. at Mrs Prestons House in Rutland Square' – evidently an earlier residence – as being executed by the carpenter James Bowden, who may have been the unnamed contractor employed in this capacity at Preston's new house in Mountjoy Square.[29]

Bowden, in fact, provided carpentry for other building contractors but also managed a business in his own right. In December 1790, Bolger measured carpentry work by Bowden for the stonecutter/developer Henry Darley at 39 North Great George's Street; while at 12 North Great George's Street, measured in June 1791, Bowden's 'sundry works' for John Meares Esq. extended to bricklaying, carpentry, slating, plastering, painting and sinking foundations, suggesting the management of an extensive workforce of subcontractors.[30] The latter is particularly significant: Bowden was contracting for the entire works and not just for his own trade; in both instances he was 'furnishing materials'. (Figs 1–2)

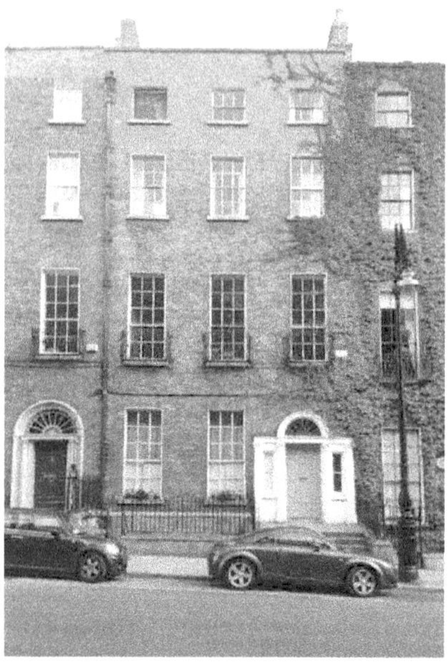

Fig. 1. James Bowden, bricklaying, 39 North Great George's Street, 1790. All photographs by the author.

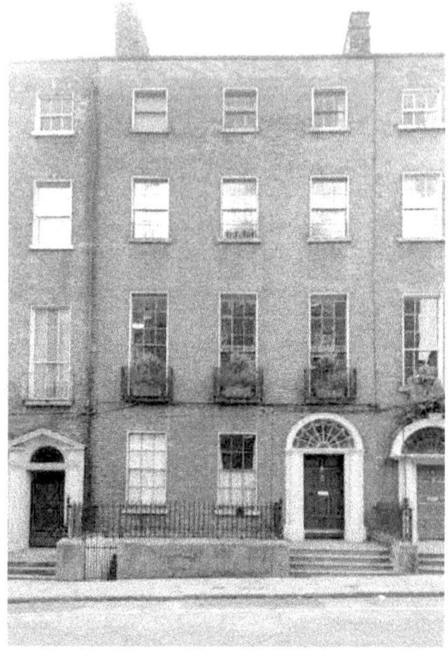

Fig. 2. James Bowden, 'sundry works', 12 North Great George's Street, 1791.

A burgeoning standardisation in house-building production, coupled with the nature of speculative property development, meant that 'there was scope for tradesmen to operate in a more autonomous manner as their skills were directly transferable from site to site.'[31] For this reason, contractors working in particular areas of the city were often employed by multiple developers/master builders. Measured accounts for the present 12 and 13 Mountjoy Square, begun in 1793/4 and subsequently leased to Counsellor John Sweeny and Alderman Samuel Reed (plumber and developer) respectively,[32] indicate that in 1794–5 both Sweeny and Reed engaged the services of the bricklayer Robert Arthurs.[33] (Fig. 3) Between 1789 and 1793, Walter Casey executed stone and bricklayers works for Michael Stapleton, plasterer, and John Howard, merchant, at their 'new Buildings', 43–44 and 48–49 Mountjoy Square.[34] (Fig. 4)

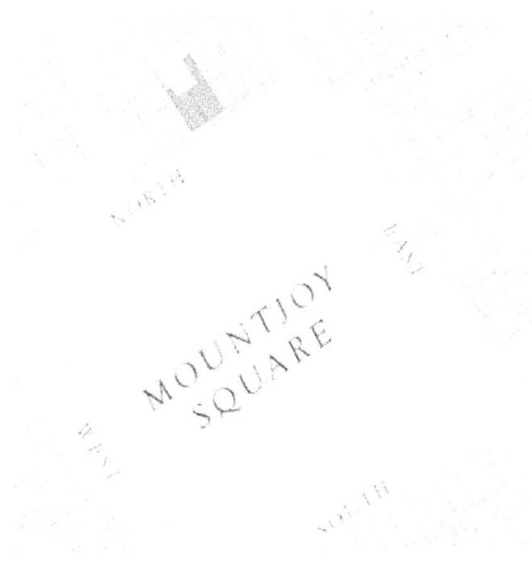

Fig. 3. Robert Arthurs, bricklaying, 12–13 Mountjoy Square, 1794–95. Illustrations by the author.

Fig. 4. Walter Casey, stone and bricklaying, 43–44 and 48–49 Mountjoy Square, 1789–93.

The bricklayer William Burke was employed by timber merchant William McDonough at 46 and 51–52 Mountjoy Square in 1791, and in the same year by Joseph Handley, stonecutter, at 50 Mountjoy Square. During 1791–92, Burke's services were also in demand by the developers Edward Archdall, John Price, Joseph Lowe, William Norton Barry and Arthur Darley at their respective buildings in nearby Temple Street, and by the carpenter/developer John Russell in Gardiner Street. (Burke was also busy on the south side of the city in the same years, at Hendy, Gibson & Donnellan's sites in Baggot Street and Merrion Square South.)[35]

Business partnerships: commerce and mutuality

A retrospective description of the practices of the principal 'carpenters and builders' in Dublin c1800, enumerating 156 employers, records that the plasterer Charles Thorp, the carpenter and builder Nicholas Kildahl, the bricklayer William Burke, and the carpenter John Russell were all in 'very extensive' business, each regularly employing thirty men. (Table 2)

TABLE 2: Sample of prominent house builders in Dublin in 1800.
Source: Royal Irish Academy, MS 4.B.31, 'Report on the Trades and Manufacturers of Dublin' (c1834).

EMPLOYER	**Number of EMPLOYEES**
Charles Thorp	30
Frederick Darley	20
William Pemberton	10
Nicholas Kildahl	30
Gustavus Hume	20
Messrs Wildridge	30
John Russell	30
William Burke	30

Others, like the builder and stonecutter Frederick Darley, and the builder and bricklayer William Pemberton, also active in Mountjoy Square, employed 20 and 10 men respectively. While others in the carpentry business apparently did not venture into the world of speculative house building – that is to say, their names are not recorded in the rent rolls of leaseholders at these prestigious locations – according to this same report, 'many Men who from the savings of their earnings as Operatives embarked in the Building line & were able in a few years to retire on independent properties'.[36]

While property ventures could prove devastating to individual builders, one way to lessen the financial risk attendant on speculative development was to enter into a formal partnership. Such enterprises were well established by the early eighteenth-century and confirm the competence of tradesmen operating within an industry 'which drew on investment from a wide range of external sources and was one of the larger and most significant sectors of the metropolitan economy'.[37] Examples abound: in 1771, the bricklayer and builder Edward Gray formed a business affiliation with the carpenter James Swinton and the mason Henry Barrell for building houses in Grafton Street, Mayfair in London.[38] In a similar manner, the present 81–87 Merrion Square were erected on ground leased by Lord Fitzwilliam in 1789 and 1791 to William Hendy, John Gibson and John Donnellan, described as 'Carpenters and Builders & Copartners in the Trade of Carpenters and Builders'.[39] A memorial of a 'deed of partition', dated 7 February 1793, describing the dissolution of their combined interests in this property, records that seven houses had been built 'by the said Partners and let to different Tenants'.[40]

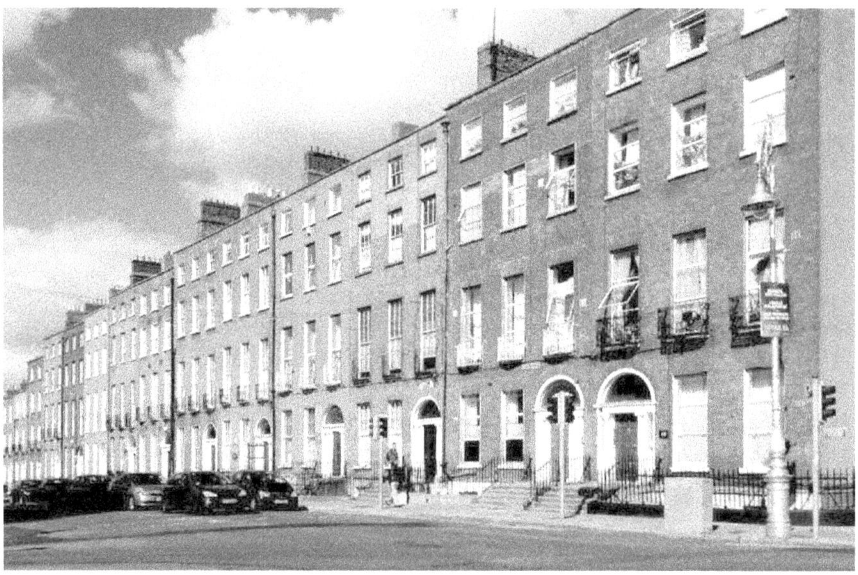

Fig. 5. Belvedere Place and Mountjoy Square East. Photograph by the author.

More specific detail is available for the building of 1–10 Belvedere Place, on ground originally leased by Lord Mountjoy to John Johnston Esq. in 1791, revealing the complexity of labour organisation on a single terrace of houses.[41] (Fig. 6) Here the principal building operatives were the carpenter John Russell[42] and the bricklayer John Prendergast.[43] (Table 3)

TABLE 3: Contracted labour at Belvedere Place, Dublin, 1796–1800. Source: National Archives of Ireland, Bryan Bolger papers, 2/476/15–16.				
House No.	**Leaseholder/Builder**	**Carpentry**	**Bricklaying**	**Lease date**
1	Henry Saunders Esq.	Russell	Prendergast	–
2	Russell	Russell	–	1796
3	Prendergast	Russell	Prendergast	1797
4	Johnston	–	–	1797
5	Johnston	Russell	Prendergast	1798
6	Russell	Russell	Prendergast	1797
7	Prendergast	Russell	Prendergast	1800
8	Johnston	Russell	Prendergast	1798
9	Edward Dalton Esq.	–	–	1804
10	Johnston	Russell	–	–

Table 3 enumerates the leaseholders of the individual lots at Belvedere Place, from which we can determine that both Russell and Prendergast were active as building speculators in their own capacities. Significantly, the property transactions for these houses, from the records of the Registry of Deeds, reveal that Johnston was also granting mortgages to Russell and Prendergast, apparently to enable their individual business interests.[44] But if we look at the minutiae of measured accounts for the individual houses, an even more intricate picture emerges. As well as building and leasing properties at Belvedere Place, Russell provided carpentry for Johnston's houses on the same street; Prendergast, who built and leased 3 and 7, was evidently working in a similar capacity for Johnston. It comes as no surprise then to find that Russell and Prendergast also contracted with one another.[45]

For merchant and gentlemen developers not directly involved in the construction industry, a different type of business partnership was required: a deed of November 1792 recording earlier 'articles of agreement', dated December 1790, made between John Manent of Crossboyne, county Mayo, 'Gentleman', Leopold Ferry, a 'perfumer' of Marlborough Street, and the bricklayer William Burke. This document first itemizes the various leases made to the three individuals, and then further recites that five houses had been built in Gardiner Place 'at the Joint Expence of said three parties'. Two of these houses, corresponding to the present 10 and 11 Gardiner Place, were built on lots originally leased to Ferry and Burke respectively, although Burke was evidently responsible for managing the construction of both properties simultaneously given their negotiation of different street levels and an arched passage to the stable lane.[46]

Conclusions

House building during the Georgian era was a multifarious exercise, requiring literacy, financial competence and the ability to manage social relations of labour production. But predicated on a growing standardisation within the building industry, and facilitated by the 'measure and value' system, building and contracting in turn offered unprecedented opportunities for realizing 'a good property': as we have seen, some of the most dynamic firms worked in their own capacity as builder developers, that is 'by the great', but also contracted with other builder developers for their particular trade specialism. In this context, a business like Frederick Darley signals a paradigm shift in the labour process: contracting by the great, and employing a more or less permanent and relatively large body of tradesmen in all the principal building crafts, it anticipates the system of wage labour that would characterize the building industry in the nineteenth century.[47]

In closing, it is worth reflecting briefly on the impact of these building networks on architectural form. While it is certainly true that uniform terrace elevations gained currency in architectural discourse during the latter part of the eighteenth-century, and the construction and fitting out of the town house relied increasingly on components prepared offsite, these builder contractors, building for themselves and for others in these exclusive parts of the city, had an equally important role in determining the celebrated qualities of visual coherence and civic urbanity that characterizes Georgian domestic architecture across the British Isles. A fitting coda to a discussion of eighteenth-century building process and how it gave shape to our cultural patrimony is found in the Dublin builder John Russell's death notice of 1823, which noted that the 'elegant outlet which bears his name owes its origin and present splendid appearance to his enterprising spirit, and few of his profession have contributed more to the improvement of this City.'[48]

References

[1] E. McKellar, *The Birth of Modern London: The Development and Design of the City 1660–1720*. Manchester and New York: Manchester University Press 1999; L. Clarke, *Building Capitalism: Historical Change and the Labour Process in the Production of the Built Environment*. London: Routledge, 1992; J. Ayres, *Building the Georgian City*. New Haven and London: Yale University Press, 1998; D. Rilling, *Making Houses, Crafting Capitalism: Builders in Philadelphia, 1790-1850*. Philadelphia: University of Pennsylvania Press, 2001.

[2] P. Guillery, *The Small House in Eighteenth-Century London: A Social and Architectural History*. New Haven and London: Yale University Press, 2004; W.C. Baer, "Is speculative building underappreciated in urban history?" *Urban History*, vol.34, no.2, 2007, pp.296-316; J.R. Ward, "Speculative Building at Bristol and Clifton, 1783–1793", *Business History*, vol.20, no.1, 1978, pp.3–18.

[3] For an overview of this archive see E. McParland, "The papers of Bryan Bolger, Measurer", *Dublin Historical Record*, vol.25, no.4, 1972, pp.120-131.

[4] McKellar, *Birth of Modern London*. pp.94–95.

[5] 'The contracting system even in the speculative world helped maintain labour relations and an organization of manpower seemingly not that different from the old guild system.' McKellar, *Birth of Modern London*, p.109.

[6] A.R. Lewis, *The Builders of Edinburgh's New Town 1767–1795* (Ph.D. thesis, University of Edinburgh, 2006), p.189.

[7] By the 1780s and 90s, master builders, that is, those who undertook buildings through individual contracts, 'virtually dominated the supply of housing'. Clarke, *Building Capitalism*, p.74.

[8] A. Gibney, *Studies in Eighteenth-Century Building History* (Ph.D. thesis, Trinity College Dublin, 1997), p.37.

[9] Clarke, *Building Capitalism*, p.82.

[10] Master builders enjoyed 'the opportunity to profit from both the materials, which they purchased wholesale, and the labour'. Ayres, *Georgian City*, p.37.

[11] P.L. Hodgson, *The Modern Measurer*. Dublin: John Parker, 1793, p.3.

[12] National Library of Ireland, Bryan Bolger papers, 2/476/15–16, bundle 'M'.

[13] Bolger papers, bundle 'P'.

[14] Registry of Deeds, Dublin (hereafter 'RD'), 466/111/295759.

[15] Callnan employed five full-time operatives c1800.
[16] Thomas Brown, carpenter, is listed in the *Dublin Directory* (hereafter '*DD*') at 153 Thomas Street from 1780.
[17] Bolger papers, bundle 'T'.
[18] A bill for carpentry, stonecutting, bricklaying, painting & plastering 'by Charles Thorp' for William Burton Esq. 'at his House in Sackville Street', measured in December 1792 (Bolger papers, bundle 'B').
[19] Samuel Dick, Director of the Bank of Ireland, resided at 34 North Great George's Street in 1798. *The Georgian Society Records of Eighteenth-Century Domestic Architecture and Decoration in Dublin* (4 vols) Dublin: Dublin University Press, 1969 (1909–12), vol.1, p.34.
[20] Robert Cornwall resided at the present 97 St Stephen's Green in 1781–92. *Georgian Society Records*, vol.2, p.97.
[21] Thorp built 19–22 Ely Place *c*1790. Casey, *Dublin*, Buildings of Ireland vol.3, New Haven and London: Yale University Press, 2005, p.518.
[22] Bolger papers, bundle 'T'. No directory listings for White or Quinn suggest they were not operating large businesses.
[23] RD 380/485/257078, 408/296/270595 and 454/421/292859. Derrick: 4 Cross Lane, 1791 (*DD*).
[24] *Dublin Evening Post*, 3 April 1784.
[25] *Dublin Chronicle*, 17 October 1789.
[26] Charles Thorp (alderman & builder), Charles Thorp Junior (plasterer & painter), Thomas Thorp (carpenter), and George Thorp (High Sheriff & builder), all at 32 North Great George's Street in 1805 (*DD*).
[27] Bolger papers, bundle 'P'.
[28]Gibney, *Building History*, p.55.
[29] Bowden: 24 Bolton Street, 1791; 84 Marlborough Street, 1792–94; 26 Dominick Street, 1795 (*DD*).
[30] Bolger papers, bundles 'D' and 'M'.
[31] McKellar, *Birth of Modern London*, 110.
[32] RD 484/173/306886 and 494/134/322243.
[33] Bolger papers, various bundles. Arthurs: 7 Mountrath Street, 1796 (*DD*).
[34] Bolger papers, bundles 'H' and 'S'.
[35] Bolger papers, various bundles. McDonough: 124 North King Street, 1792; Handly, 113 Great Britain Street, 1793 (*DD*).
[36] Royal Irish Academy, Haliday MSS 4.B.31, 'Report on the Trades and Manufacturers of Dublin', *c*1834, pp.112–137.
[37] McKellar, *Birth of Modern London*, p.68.
[38] R. Garnier, "Grafton Street, Mayfair", *Georgian Group Journal*, vol.13, 2003, p.227.
[39] Hendy & Donnellan: 'Carpenters & Builders', 10 Baggot Street, 1793; Hendy, Donnellan & Gibson, same address, 1794 (*DD*).
[40] RD 473/317/303209.
[41] John Johnston may be the architect of the same name recorded in the "Dictionary of Irish Architects 1720-1940" <www.dia.ie>.
[42] Russell: Gardiner Street, 1792; Mountjoy Square, 1793-4; Gardiner Place, 1795–1800; 11 Russell Place, 1801–24 (*DD*). In 1834, Russell was described as: 'Very extensive in 1800, he built a great part of Mountjoy Square, Russell St, Russell Place, &c.[…]This man at the time of the Union was considered very wealthy'. Haliday MSS 4 B 31.
[43] Prendergast: 101 Great Britain Street, 1788; Britain Court, 1793–1800 (*DD*).
[44] The site of 2 Belvedere Place, for example, was leased from Johnston to Russell on 1 August 1795

(RD 483/487/316505); on 18 August, Russell, 'in consideration of £250 lent to him by Johnston', mortgaged the property with an undisclosed condition of redemption (RD 499/219/316506); on 22 October 1796, Johnston reconveyed the ground to Russell on payment of the £250 with no interest (RD 502/477/326115); in November, Russell set the 'partly built' dwelling house to Thomas Williams Esq. (RD 506/1/327056).

[45] Prendergast worked for Russell at Circular Road; and was working in Mountjoy Square and Grenville Street in 1790–92.

[46] RD 462/90/294484, 434/234/280433 and 447/38/287108.

[47] Clarke, *Building Capitalism*, p.69.

[48] *Freeman's Journal*, 21 June 1823.

Dreaming of a cementless future: the story of silicalcite

Maris Mändel
Estonian Academy of Arts, Estonia

Today calcium silicate boards are mostly known as fireproof insulation material. However, in the Soviet Union, calcium silicatewas were widely used for large construction elements as an important alternative to concrete panels. After World War II, the acute shortage of building materials in the USSR, resulting from the devastation of war and the rapid growth of industrial cities, contributed to the developement of low-cost and local materials. Building materials made of lime and sand became a promising field of developement. Lime and sand were cheap and their deposits plentiful all over the USSR. Besides widespread calcium silicate bricks, large blocks, panels and other products came into use. Silicalcite, a new material invented in Estonia[1], was perhaps the most intriguing and ambitious among other sand and lime–based materials. Despite its significant role in the Soviet Union at the time, silicalcite has remained unknown among researchers of construction history outside of the Eastern Bloc. The aim of the current paper is to fill in this gap and to provide insight into the fascinating story of silicalcite, its rise and fall, and to explain the novelty of the technology, as well as to discuss its position in construction history.

A brief overview of developement of sand and lime–based building materials

The history of sand–lime products began with what are called mortarbricks, a mixture of sand and lime that is left to harden in the air. The manufacturing process took 6 to 18 months, and their strength was relatively low. These bricks were manufactured in England by the middle of the nineteenth century.[2] An important milestone in the development of sand–lime products was a patent granted to Dr. W. Michaelis in Berlin in 1881, proposing to harden mixtures of lime and sand by using high-pressure steam. However, it took almost two decades for the new technology to be applied on a large scale because an industrial autoclave had to be invented first.[3] At the turn of the century, already a number of calcium silicate brick factories were in operation in several countries, and growth of the sand–lime brick industry continued until World War II.

In the nineteenth century, improvements in calcium silicate brick technology were focused on devising better lime-slaking equipment and methods; for example, the primitive metal boxes were replaced with a mixing drum. Experiments with grinding the sand and adding various compounds to the mixture became major research topics in the twentieth century. For example, volcanic ash and ground ceramics were used in small amounts in mixtures to increase the strength of the final product; also, calcium silicate bricks became popular whereby lime was partly replaced with ash. The first theoretical work showing the positive effect of ground sand in the production of calcium silicate bricks was published in 1904 by Stöffler, followed by industrial tests from Kramer the next year. Tischler, Ernst and Leduc showed in their work that grinding sand increases the percentage of soluble SiO_2, which is essential for the formation of calcium hydrosilicate. This knowledge led to the development of various calcium silicate products that had much better properties than did the traditional calcium silicate brick. Because the properties and usage of the new material were similar to those for concrete, it was named silicate concrete. In 1925, Steinag, an Austrian plant, began to manufacture blocks, boards, stairs and pipes from silicate concrete. The sand and lime were both separately ground in a ball mill and then mixed and moistened. The mixture was kept in a silo until the lime was slakened. The products were formed using a hydraulic press. In the United States, Grande Brick Co. and Saginaw Brick Co. manufactured foundation blocks from calcium silicate. However, large-scale production of silicate concrete products was not widespread during the period between the world wars.[4]

In the USSR, large blocks made from calcium silicate were produced in a calcium silicate brick factory in Moscow in the 1930s. During the same time, experiments to develop lightweight calcium silicate blocks were carried out by Russian researchers.[5] Aerated concrete had been produced first in Sweden in 1929. The lightweight blocks, made of lime, cement and sand, became known as Ytong stones; aerated concrete blocks without lime gained popularity as Siporex. Many countries were manufacturing both Ytong stones and Siporex before World War II. Production and usage of aerated, autoclaved concrete blocks continued in the West during the postwar period; one of the most successful companies was the German Hebel.[6] In the Soviet Union, much effort was expended to develop cementless large construction elements. The mass production of large silicate concrete construction details began in the Lysychansk factory during the period from 1952 to 1953, and soon thereafter, elements from silicate concrete, both dense and lightweight, became common construction materials manufactured in numerous plants across the USSR.[7]

Silicalcite—a new construction material

A special kind of silicate concrete was invented in 1949 by an Estonian construction engineer, Johannes Hint. The new material was named silicalcite to clearly indicate the difference in technology compared with ordinary silicate concrete. The core idea of the new technology was to use a disintegrator for producing sand–lime building materials. A disintegrator is a grinding mill that refines materials by collision, unlike conventional grinding technology (e.g. ball mill or vibro mill). The central part of the disintegrator consists of two rotors that revolve in opposite directions, with six concentrically placed rows of impact elements, so-called fingers, to grind the mixture of sand and lime (Fig. 1). The disintegrator was originally invented by Thomas Carr in 1859, but until Hint's proposal, it was used for grinding only very soft and nonabrasive materials (e.g. chalk or coal) or for mixing materials.[8]

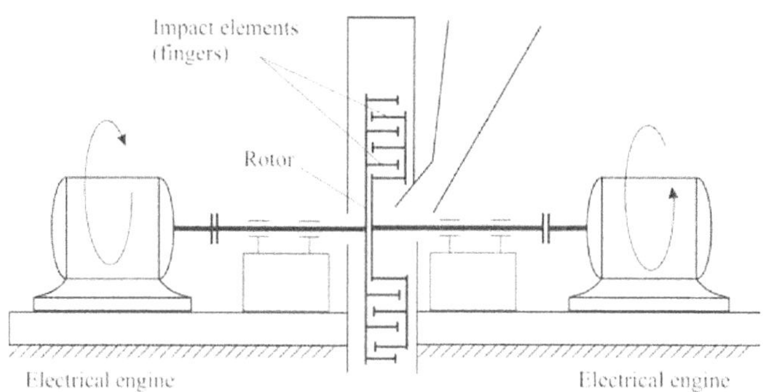

Fig. 1. A schematic drawing of a disintegrator. Source: Drawing by M. Mändel

The disintegrator was claimed to be able to "mechanically activate" materials ground in it. The process of refining sand grains in a disintegrator resembles peeling them, which creates a larger surface area, thus accelerating the chemical reaction between the sand and the lime as well as improving the strength of the final product. Compared with traditional calcium silicate products made of unground sand, the silicalcite products were two to three times stronger and required 20% to 30% shorter treatment in an autoclave.[9] The silicalcite products were claimed to have equivalent properties or to be even stronger and more durable than the construction materials made of cement binder but were approximately 25% cheaper.[10] A great advantage of silicalcite technology was in the simplification of the production process. In a disintegrator, moistened sand and lime were ground together so that no additional mixing equipment was needed; the mixture went directly from the disintegrator to the molds.

The beginning of silicalcite production and its spread in the USSR

Preparatory exploration works for producing silicalcite began in a Tallinn calcium silicate brick factory, "Quartz," in 1949, and in a couple of years, an experimental silicalcite plant was established. The factory produced a variety of construction elements, made of both dense and lightweight silicalcite, with and without reinforcement. The most common products included reinforced ceiling panels, lightweight blocks for inner and outer walls, lintels, beams and columns, but also roof tiles and paving blocks of silicalcite were produced for some years. The first building of silicalcite elements was erected in 1954.[11] A research institute for silicalcite, responsible for designing silicalcite plants across the USSR, was estabilished in 1961 based on the experimental plant in Tallinn.[12]

The large-scale production and consumption of silicalcite products were greatly helped by a new Soviet construction policy. The official architectural style of the Stalinist regime had been a pompous pseudoclassicism, and the construction methods had remained traditional. At the All-Union Conference of Construction in 1954, the new leader of the Party, Khrushchev, launched a sharp attack on extravagant architecture and outdated construction technology of the Stalinist era. In one of his speeches, the industrialized building methods were declared the correct and only possible way to proceed.[13] After that speech, the widest possible application of standardized designs and prefabricated elements was strictly monitored. Concurrently, only a few factories were capable of producing prefabricated panels or blocks in the USSR at that time.

Silicalcite seemed like an ideal material in these conditions. Its technology was already worked out and ready for use. Setting up a silicalcite factory was quick, cheap and easy. Fifteen new silicalcite plants were established in the USSR in 1957; by the end of 1958, there were 21 plants,[14] and in the beginning of the1960s, more than 40 plants were operating in the USSR.[15] There were whole towns built from silicalcite; the best-known of these is Chaykovsky in Perm Krai. Because of the simple and compact production technology, some mobile plants were also established. The first of these mobile plants had been a passenger ferry, that was converted into a floating silicalcite factory in the Oka River. The same solution was adopted on some railway lines.[16] Awarding the Lenin Prize[17] to inventors of silicalcite in 1962 was a clear indication of the significance of the new building material in the USSR.

Use of silicalcite in Estonian architecture

Silicalcite production began in the Stalinist era; therefore the first designs of silicalcite buildings were controversial, assembling honestly exposed large blocks with traditional architectural forms and even decorative elements.[18] The first buildings constructed from silicalcite elements were detached houses (Fig. 2). Architects intentionally emphasized the horizontal joint junctures on the unplastered walls to highlight the industrial construction method of the building. Compared with traditional construction, the time it took to build a silicalcite house was revolutionary; installation of silicalcite elements took approximately 32 hours.[19] Holding a strong propagandistic value, the houses were used as the main prize of a govermental lottery in the 1950s.[20]

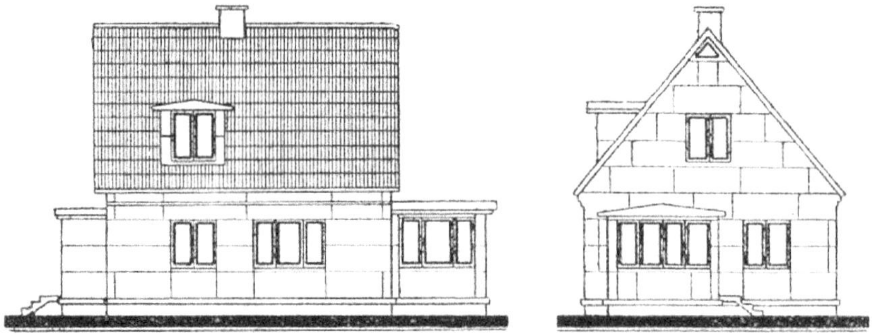

Fig. 2. Design drawings for detached house made of silicalcite. Source: Silikaltsiit – uus ehitusmaterjal, (Note 11) 1957, p. 13.

In Estonian architecture, silicalcite became truly noticeable by the end of the 1950s, when it was used for constructing *khrushchyovkas*, apartment buildings using a standardized design that were widely built in the Eastern Bloc. Because of joint and surface defects, the appearance of the unplastered *khrushchyovkas* made from silicalcite blocks was slovenly and unaesthetic, eliciting severe disapproval from local people and architects (Fig. 3). Therefore, since the mid-1960s, silicalcite blocks and panels in residential and public buildings were used only with an additional coating. Commonly, these houses were plastered, which remarkably reduced the time-saving effect of industrialized construction.

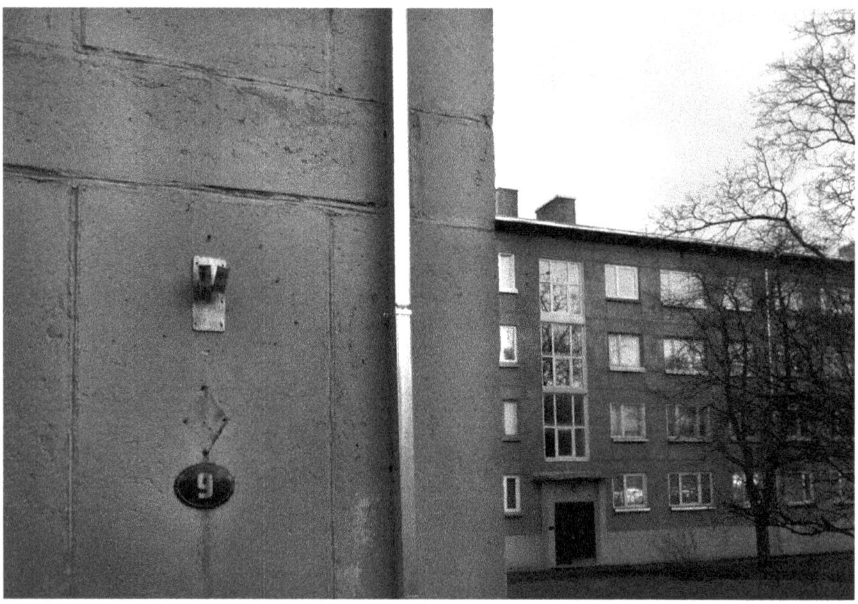

Fig. 3. Apartment buildings, so-called khrushchyovkas, made of silicalcite in Tallinn. Photo by M. Mändel, 2013.

Silicalcite became a common construction material for housing in rural areas in Estonia during the period from the 1970s to the 1980s because two silicalcite factories (in Palivere and Aravete) belonged to a construction organisation of collective farms. The thermal resistance of lightweight silicalcite satisfied the requirements of the Soviet building codes, and therefore, no extra layer of thermal insulation was needed for silicalcite walls.[21] Most buildings made from silicalcite were standardised designs, built using unified construction elements. A rare exception was an experimental structure of a bus stop shelter in Tallinn, a striking building with a fan-shaped canopy made of 20-mm-thick reinforced silicalcite elements glued together (Fig. 4).[22]

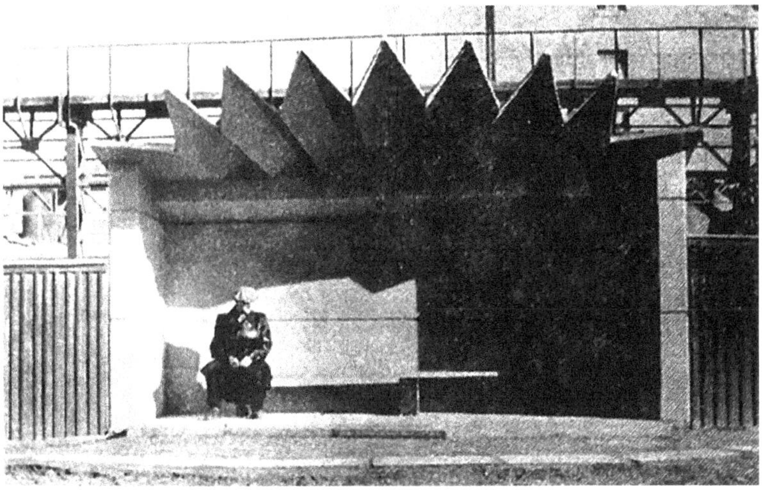

Fig. 4. A bus stop shelter in Tallinn. Source: Ehitus ja Arhitektuur, (Note 21) 1964.

Export of silicalcite technology

The spread of silicalcite was not limited to the USSR; the patented technology was successfully licensed to some capitalist countries. This was quite an exceptional case, considering that usually the USSR would import construction technologies from the West. For example, the French Camus technology for producing reinforced concrete panels, which was extensively used for mass housing across the USSR until the 1990s, was imported by the USSR concurrently with the licencing of silicalcite abroad.

In 1958, a governmental export agency, Raznoexport, received an inquiry from a group of Italian businessmen from Milan, asking for technical details about silicalcite.[23] Industrialized construction was an undeveloped field in Italy in the 1950s, especially when compared with neighbouring France.[24] The lack of local influential companies producing prefabricated elements somewhat explains the Italians' interest in Soviet technology. A contract for licencing silicalcite technology to the Italians was signed two years later, in 1960. The negotiation process included an Italian visit to the Tallinn experimental silicalcite plant as well as testing the compability of Italian raw materials with silicalcite production.The licence was sold for 925,000 USD, of which 20% was intended to be paid upfront and the remaining 80% within 10 years. In return, the Italians were supposed to receive detailed drawings for two plants, including the drawings of special equipment, such as the disintegrator and the rights to sell silicalcite products in Italy.[25] The first silicalcite plant, with 60,000m^2 annual production capacity, was estabilished near Naples in 1961. To manufacture lightweight silicalcite products, a pozzolan from Bacoli was used instead of sand.[26] The Italian company further developed the technology and patented an improved disintegrator in the late 1960s.[27] In fact, a company with the same name still exists near Naples and produces prefabricated elements although with a different technology.[28]

The second licence of silicalcite was sold to a Japanese company, Asahi Chemical Industry,[29] in 1961, and a plant was built near Tokyo. Delegations from Germany, Brazil and Great Britain, among other countries, visited the experimental silicalcite plant in Tallinn. For promotional purposes, Raznoexport created a marketing brochure in three languages that presented silicalcite as the future building material (Fig. 5).[30] Silicalcite was also marketed under the name *laprex*, referring to "king of stones" in Latin. However, the best promoter of silicalcite was Johannes Hint himself. His natural skill in public relations and willingness to send abroad booklets with technical data about silicalcite was an effective means of promotion but irritated govermental organizations in the USSR.[31]

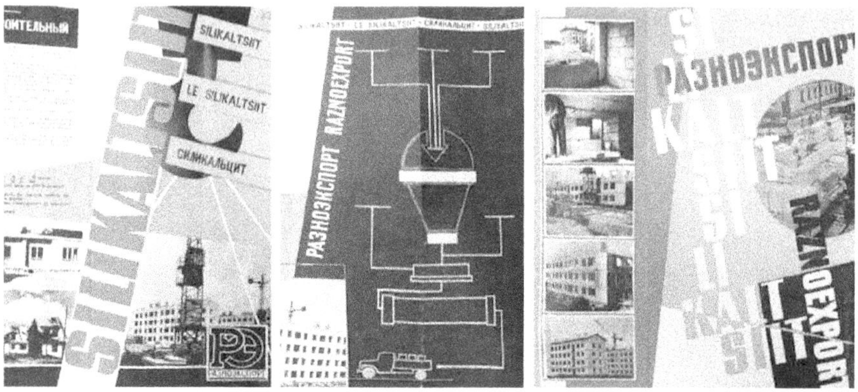

Fig. 5. Fragments from a brochure promoting silicalcite, published by Raznoexport. Source: Estonian State Archives (Note 23).

The fall of silicalcite and its causes

Johannes Hint dreamed that one day silicalcite would replace concrete. History has shown that it remained only a dream. The Research Institute of Silicalcite was reorganized and renamed the Research Institute of Silicate Concrete by 1966.[32] The institute continued making pilot batches of sand and lime from various regions and designing new silicate concrete plants across the USSR, but the disintegrator was replaced by a ball mill.[33] Hint, having a strong belief in the disintegrator and its ability to activate materials mechanically, did not give up and founded a special design office for developing disintegrator technologies. Economically, it was a major success, mostly because of the enormous sales of AU, a food additive made in the disintegrator. However, such enterpreneurial success could not be tolerated in a Communist country. In 1981, Hint was arrested and found guilty of abuse of office; nevertheless, the trial has always been considered politically motivated. Johannes Hint died in prison in 1985 and was fully exonerated posthumously in 1989.[34]

Why did the expected conquest of the world by silicalcite not happen? Hint himself blamed the opposition from cement industrialists, who, according to Hint, were afraid of the major success of silicalcite and the potential loss of their position. Hint was a popular hero in Estonia, and silicalcite is often presented in the top 10 of Estonian inventions, in the same position as are the Minox camera and the Schmidt telescope. At the same time, it has to be admitted that, despite Hint's success in setting up the silicalcite plants in the USSR and abroad, his creative ideas were never accepted among local

Estonian civil engineers. There could have been a certain amount of jealousy, but some objective disadvantages of silicalcite cannot be denied. What was usually left unsaid in advertising brochures for silicalcite was the fact that the impact elements of the disintegrator broke down quickly and therefore needed to be changed frequently. For example, at the Tallinn experimental silicalcite plant, the impact elements made of non-alloyed steel lasted only eight hours; after that number of hours, the plant had to be stopped and the rotors changed.[35] The production process, of course, took longer, but more importantly, operating costs significally increased. As such, it turns out that although silicalcite plants were relatively inexpensive to establish, the operating costs were higher than in plants that produced concrete panels. Theoretically, it would have been possible to make impact elements from more durable material, but that would have been expensive to do. In addition, frost resistance and some other properties of silicalcite were not as good as they were claimed to be. A widespread opinion held among many Estonian engineers was that the extraordinary test results of silicalcite were falsified.

Conclusion

Silicalcite, although not as brilliant as claimed, definitely filled a purpose during the time of its use. Because the USSR was experiencing an acute shortage of building materials in the 1950s, the quality of silicalcite was not an issue. The technology of producing silicalcite was simple and compact enough to allow the building of plants in a short turnaround across the USSR and to make an abundance of inexpensive building materials suitable for industrial building methods. The idea of producing silicate concrete products by using a disintegrator was discarded by the mid-1960s, but the manufacturing of large construction elements using only sand and lime continued until the collapse of the Soviet Union. Nevertheless, trends in the building materials industry have shown quite clearly that products based on cement or lime and cement have become winners. The dream of a cementless future remains only a dream.

References

[1] Estonia was incorporated into the USSR during the period from 1940 to 1991.

[2] H. O. Weller, *Sand-Lime and Other Concrete Bricks*. London, 1921, p. 3.

[3] T. R. Ernest, S.W.Parr, *A Study of Sand Lime Brick*. Urbana: University of Illinois, 1912, p. 12.

[4] J. Hint, V. A. Kuzminov, *Proizvodstvo silikalcita i ego primenenie v žilišnom stroitelstve* [Production of silicalcite and its application in residential construction]. Tallinn: Gosudarstvennyj naučno-tehničeskij komitet Soveta Ministrov Estonskoj SSR, 1958, pp. 7–12.

[5] ibid., p. 12.

[6] B. G. Hellers, B. R. Schmidt, "Autoclaved aerated concrete (AAC)—the story of a low-weight material" pp. 63–70, in *5th International Conference on Autoclaved Aerated Concrete*, Bydgoszcz, 2011.
<http://www.5icaac.utp.edu.pl/index.php?option=com_rokdownloads&view=folder&Itemid=35> [accessed on 28 December 2013]

[7] Hint, Kuzminov, *Proizvodstvo*, (Note 3) pp. 12–3.

[8] J. Hint, "Dezintegrator" [Disintegrator], *Silicalcit: Bjulleten Naučno-Tehničeskoj Informacii* [Silicalcite: Newsletter of Scientific and Technical Information]. no. 7/8, 1960, pp. 45-6.

[9] A. Veski, "Eesti NSV kohalikud ehitusmaterjalid" [Local building materials of Estonian SSR], in B. G. Skramtajev (ed.), *Ehitusmaterjalid* [Building materials]. Tallinn: Eesti Riiklik Kirjastus, 1951, pp. 547–8.

[10] E. S. Krasuskij, *Silikalcit – mestnyj strojtelnyj material* [Silicalcite – a local building material], Rostov: Rostovskoe Knižnoe Izdatelstvo, 1959, p. 11.

[11] H. Joosti, J. Hint, H. Ivand, *Silikaltsiit – uus ehitusmaterjal* [Silicalcite – new building material], Tallinn: Eesti Riiklik Kirjastus, 1954, pp. 3–4.
[12] T. Allikas, *Silbet*. Tallinn: Valgus, 1986, p. 3.
[13] N. Khrushchev, *Industriaalsete ehitusmeetodite laiaulatuslikust juurutamisest, ehituse kvaliteedi parandamisest ja maksumuse alandamisest* [About extensive introduction of industrial construction methods, improvement of construction quality and reduction of costs (Transcription of a speech)]. Tallinn: EestiRiiklik Kirjastus, 1955, pp. 6–28.
[14] Hint, Kuzminov, *Proizvodstvo*, (Note 3) pp. 4, 13.
[15] J. Hint, *Mysli o silikalcite*. Tallinn: Naučno-issledovatelskij i proektnyj institut silikalcita UPSM SNH Estonskoj SSR, 1963, p. 11.
[16] Krasuskij, *Silikalcit*, (Note 10) p. 9.
[17] The Lenin Prize was one of the most prestigious awards in the USSR, presented to individuals for accomplishments relating to science, literature, arts, architecture and technology.
[18] Estonian State Archives, coll. T-14, inv. 4–6, no. 42002, Design drawings for silicalcite building, 1953.
[19] Joosti, Hint, Ivand, *Silikaltsiit*, (Note 10) pp. 11–23.
[20] M. Kalm, Eesti 20. Sajandi arhitektuur [Estonian 20th Century Architecture], Tallinn: Sild, 2001, p. 328.
[21] T. Kolk, "Pilguheit Antsude ja Maiede kerkimise aega" [A glance at time of construction of "Antses" and "Maies"], in *Äripäev: Ehitus*, 1. June 2005, pp. 24–5.
[22] P. Varep, "Silikaltsiidist ootepaviljon" [Bus shelter from silicalcite], in *Ehitus jaArhitektuur*, no. 1, 1964, pp. 32–5.
[23] Estonian State Archives, coll. R-2320, inv.1, no. 3, Correspondence about selling patent for production of silicalcite, 1958–1961.
[24] T. Iori, "Prefabrication and industralization made in Italy", in F. Graf, Y. Delemontey (eds.), *Architecture industrialisée et préfabriquée: connaissance et sauvegarde/ Understanding and Conserving Industrialised and Prefabricated Architecture*. Lausanne: PPUR, 2012, pp. 86–95.
[25] Estonian State Archives, coll. R-2320,inv. 1, no. 3, Correspondence (Note 23).
[26] Estonian State Archives, coll. R-2320,inv. 1,no. 22, Correspondence about licencing production of silicalcite, 1961–1962.
[27] US Patent No 3411724. Cage type disintegrator with blade shaped impacting members, particularly suited forprocessing hard materials <http://www.freepatentsonline.com/3411724.pdf> [accessed on 28 December 2013]
[28] Sviluppo Silicalcite website <http://www.svilupposilicalcite.it/> [accessed on 28 December 2013]
[29] The same company begun to produce aerated autoclaved concrete of Hebel's brand in Japan in 1967.
[30] Estonian State Archives, coll. R-2320, inv. 1, no. 3, Correspondence.
[31]H. Kaljulaid, "Johannes Hint: Ühe XX sajandi ausa mehe tragöödia", in *Kultuur ja Elu* [Culture and Life], no. 3, 2006, pp. 38–43.
[32] Allikas, *Silbet*, (Note 11) p. 3.
[33] According to Aino Sevastjanova, employee in Institute of Silicate Conrete in 1964–1991.
[34] Kaljulaid, *Kultuur*, (Note 31) pp. 38–43.
[35] I. Kleis, P. Kulu, *Solid Particle Erosion: Occurrence, Prediction and Control*. Springer, 2007, p. 177.

Transferring construction technology and conflicts of interest - a century ago

Robert C. McWilliam
Independent, Chartered Engineer, Edinburgh, UK

The paper arises from the author's role in the Editorial Board of the Biographical Dictionary of Civil Engineers, which is currently examining the period from 1890 until 1920. Many ambiguities seemed to have been tolerated through this period. Individual biographies bring out interesting examples. The example selected is Thomas Aitken (1856-1918). He was able to have four concurrent careers all associated with building and maintaining highways at a time when catering for motor vehicles began to be considered. His achievements in the "silent majority" of British civil engineers, who had little to do with the elites in Westminster with their influential imperial connections, were remarkable. His four careers as County Surveyor, technical author, consultant and contractor resulted in certain 'conflicts of interest' – but it was a very effective way of transferring technology in a hurry. Was he simply a man of his time or is there a case for more contemporary changes in construction technology being accelerated by concentrating on outcomes while ignoring bureaucratic niceties?

Thomas Aitken was born in Edinburgh, on 28 February 1856, and educated at the city's High School and Heriot-Watt College. While at college in 1872 he became an engineering pupil of the North British Railway Company, then Scotland's second largest railway. At the end of his pupilage he became an assistant and then a resident engineer in the northern district of the Company's system, until he was 26 when he was appointed road surveyor of Fife. He was still in that post when he died at the age of 62 on 27 May 1918.

At first sight Aitken's was a simple career, but its longest strand as a paid official was only one of four strands of a diverse engineering career. [1,2] The second strand was as a technical author, which he began at the age of 30 and resulted in texts still in use several decades after his death. The third strand as a consultant began a year later. At the age of 52 Aitken was one of two consulting engineers "of international renown" hired by a leading American industrialist to advise on a prototype motorway to be built privately then handed over to the state when completed. The fourth, and nowadays perhaps most surprising, strand began in 1906 when at the age of 50 he became the Managing Director of a road surfacing contractor, which numbered his employers in Fife amongst its customers.

The Fife peninsula lies on Scotland's eastern seaboard. The rapid expansion of coal-mining and its supporting industries doubled Fife's population between 1850 and 1900 to over 300,000. [3] By the late 1850s the peninsula's ports and main centres of population had been linked by a network of local railway schemes that had survived from the many formulated during the 1845 railway mania. [4] Few places were more than a dozen kilometres from a railway of sorts. All heavier industry was served by branch railways. From the age of 19 until he was 26 Aitken surveyed, set-out and superintended the construction of short branch lines and tramways to collieries and quarries and latterly superintended additional tracks to the main railways. By 1882 both the second Tay Bridge and the even more famous Forth Bridge were in progress and a main east-coast trunk railway line was being established between the two major bridges. [5]

In 1882 Aitken opted for a seemingly quiet life – as County Road Surveyor. In 1882 the decline of turnpike roads was well advanced as longer distance traffic transferred to the parallel railways. [6] County councils had begun to improve roads for the agricultural and local road traffic which now ran to the railway stations by former cross-roads. In the words of earlier Parliamentarian, as railways took over the long distance traffic county roads required to be renewed such that "the veins of traffic (became) arteries, and the arteries veins". [7] For the first five years of his appointment Aitken concentrated on the tasks before him as a paid official – the first strand of his career. During this time he undertook his most scenic road project. [8] It followed the best tradition of Thomas Telford's practice by running obliquely down the Garlie Bank, now the A916 south of Cupar, Fife, to "link two valleys to replace a route with gradients as steep as 1 in 10 while saving 1½ miles with a maximum gradient of 1 in 30 suitable for mixed traffic" (horse and traction engine). [9] At the age of 30 he began the second strand of his career - writing what he called "pamphlets" on topics such masonry bridges and rock-drilling. The third strand of his career as a consulting engineer can be traced to 1887 when he commenced his "concurrent private practice" with a small sewerage scheme for the Burgh of Ladybank. It was probably with this third strand in mind that he applied for Associate Membership of the Institution of Civil Engineers (ICE) in 1893 at the age of 37. His application made much more of his first ten years with the North British Railway than his subsequent eleven years as a Road Surveyor. [10] Two years later he was to share his more recent experience in his paper on the maintenance of macadamised roads in the Minutes of Proceedings of the ICE. [11]

Five years later the first edition of his seminal work was published, "Road Making and Maintenance: a practical treatise for engineers, surveyors, and others with an historical sketch of ancient and modern practice", by the London publisher, Charles Grifffin and Co. Ltd. [12] Griffin was a major publisher. Its extensive catalogue of Standard Engineering Publications included the 20th edition of "A Manual of Civil Engineering" initiated by Professor William MacQuorn Rankine – which had been the best established British text to include road-works to date. Aitken's 456-page text dealt with the issues that concerned county councils, whose responsibilities for highways had recently increased in scope. In this first edition Aitken described "mixed traffic" as horse and steam traction and dealt with road alignment, bridgeworks, stone-breaking and the metalling of roads. At that time the county councils involvement in roadworks included local quarrying, stone-breaking, rolling and keeping the surface clean. The latter was important then as bituminous materials were not in routine use.

Aitken's position in the first strand of his career was enhanced in 1900, when he was elected President of the Road Surveyors' Association of Scotland. In 1902 he also attempted to transfer to senior membership grade of the Institution of Civil Engineers. He failed after re-submitting his 1893 application with brief additions referring to his work in Fife, his being consultancy work for "several County Councils in Scotland and Ireland" and his 456-page book. His list of sponsors had been extended. These now included consulting engineers in Scotland and Ireland and the mechanical engineering expert Professor Hele-Shaw, who confirmed the worth of Aitken's book. [13] In the following year Aitken re-applied successfully with additional support and also provided a much more detailed list of both his publications and other works. [14] These included the reconstruction of the main road between Belfast and Lisburn and his detailed planning for one of the last independent railways proposed to be built in Scotland, the un-built Falkland Light Railway. Light railway projects were abandoned as motor vehicles gained popularity.

The automobile also enabled Aitken to launch the fourth strand of his career. Motor vehicles brought the "dust problem" to the unsealed roads of Europe during the dry summer of 1905. [15] This problem was addressed by Aitken, who developed a machine to force under pressure tar or other viscous liquid into the surface of the roads to both "overcome the dust nuisance and at the same time add materially to the life of the road". Aitken's pneumatic tar sprayer was entered for the U.K. Road Improvement Association's prize at an international competition held near Windsor in England, in May 1907. [16] Three other competitors had brought completed machines to the trial, all of these other machines were failures for one reason or other. Aitken's was the only machine that met the stipulated requirements. [17] Aitken gained both a gold medal and a prize of 100 guineas.

Aitken transferred his patent rights to the Taroads Syndicate Ltd. From 1906 until his death Thomas Aitken was listed as manager and representative in Scotland for the Taroads Syndicate Ltd., sometimes known as the Aitken Taroads Syndicate. It was among the first tenants of a new office block at 29 St.Vincent Place, Glasgow, which continues to be one of the prime commercial buildings in Glasgow. [18] Then the building housed the offices of several influential Scottish engineers including John Strain, during the second phase of his career, centred on the Lanarkshire Steel Company, who helped initiate the British Standards movement [19] and Albert Richard Brown, whose role as intermediary and guide for his connections with Japan were to influence the industrialisation of the Far East. [20]

Aitken's machine was patented not only in Great Britain, but also in United States as well as France, Germany, the Netherlands, Belgium, Australia, South Africa and other countries. Several of these foreign patents were sold within a few years, while the Taroads Syndicate continued to exploit the technology with plant on his principles sold all over the world. Most machines were on steam wagon whose steam coils could be used to keep tar melted. [21]

Also in 1907 the second edition of "Road Making and Maintenance: a practical treatise for engineers, surveyors, and others with an historical sketch of ancient and modern practice" was published with 540 pages, 84 pages more than the first edition. [22] The extra pages dealt with the attributes of motor cars, citing extensively from, inter alia, Professor Hele-Shaw's work as well as new chapters on dust prevention and the advances in the analysis of rock for use in road-making. Shortly before his death Aitken had received the proofs of a large part of a third edition from the same publisher. Aitken's book has become known to a whole new audience following the publication in 2006 of Roger Hutchinson's "Calum's Road", now also a play and in development as a movie - about the crofter, who single-handedly built 2 miles of road through hostile terrain with pick, shovel, wheelbarrow and frequent reference to a sixty-year old copy of Aitken's seminal book. [23, 24] Aitken's texts were the primary reference source for professional engineers in Britain until Edward Arnold & Company established their Roadmaker's Library series in the mid-1930s, notably "Principles of Road Engineering" by Collins and Hart. [25] It cited extensively from American research on roads undertaken in the 1920s. In the early years of the twentieth century it was America that sought road-making expertise from Europe. In February 1908, Thomas Aitken made a special request to Fife County Council for an Assistant for six months so that he would be able to be absent from the County and that he was willing to contribute to the remuneration of the assistant. [26] The council agreed to pay half of the total proposed payment of £55 for the assistant.

Coleman duPont was a member of the American state of Delaware's leading industrial family, an automobile enthusiast, a leader in the U.S. Good Roads Movement, and serious student of roadways in Europe and the United States. He offered to construct a visionary superhighway the length of the state of Delaware. [27] He planned a highway design unlike anything that had ever been built. It would have had central lanes for high-speed automobiles, and flanking lanes for trolleys, heavy motor freight, horses and horse-drawn vehicles. In 1908 duPont brought to Wilmington "two consulting engineers of international renown, Thomas Aitken from Scotland and Ernest Storms from Belgium".

The Boulevard Corporation Act, as passed by the Delaware General Assembly authorized a corporation, known as Coleman duPont Road, Inc., to construct the highway the length of the state. As each section of ten miles was completed, it was to be conveyed to the state free of charge. DuPont himself was chief engineer. Litigation interrupted construction but the first 20 miles of the road was completed and presented to the state in 1917. It was built to a smaller scale than originally proposed by duPont, but was an example of the most modern highways in the U.S. at that time. The highway bypassed towns and was connected to them through spur roads. Widening of what is now U.S. Route 113, began only three years later.

On his return to Fife in September 1908 the Surveyor's official response to damage to property arising from tar-spraying in a village was "that the work of tarring the road was done by contract with the Taroads Syndicate Ltd. And suggested that the letter be forwarded to that syndicate (which was done), with the request that it be dealt with, and nothing further has been heard on the subject". [26] Aitken's personal involvement in the Syndicate was no secret, but was not acknowledged in any official documents in Fife Surface-spraying of roads with tar was credited in the Council's records in May 1909 "of forming in-situ tarmacadam (which) has proved very successful and has been greatly appreciated by those residing in houses adjoining the highway, further work of this nature will be carried out over this summer more particularly in villages through which much wheel traffic passes". Throughout the period from 1906 until 1910 Aitken continued to publish a series of articles on his system, which was intended to meet the requirements of highway authorities in Britain to both preserve roads and prevent dust as motor-traffic increased. [28-34]

Aitken's annual report to the County Council in 1912 noted that "vehicular traffic is still on the increase necessitating the application of larger quantities of metalling and the employment of additional labour". [35] He continued "the surface spraying of certain roads with prepared tar has had good effects, and has been greatly appreciated by the occupiers of houses situated near the highways and by travellers generally. It is very noticeable where roads are so treated that the number of pot-holes, caused principally by motor-cars, are practically non-existent compared with untreated portions of roads adjoining. The grant provided by the Imperial Road Board will be principally allocated to making tar-macadam through the villages situated on the "selected routes" and also for surface spraying at different places". At that time Aitken's annual salary as the Road Surveyor was £325. However the conflicts of interest apparent between Aitken's 1st and 4th strands were becoming more obvious. Lessons about such conflicts, which had been learned in other public works over the previous century, were being learned in Fife. [36] In July 1915 the County's Treasurer reported that there were "considerable difficulties in passing accounts incurred to the Taroads Syndicate, certified by the Road Surveyor". [35] The contracts that the Syndicate was awarded had been offered in competition but other contractors seldom submitted bids and if they did they were always undercut.

On 27 May 1918 Thomas Aitken died. The local newspaper reported "in the death of Mr. Thomas Aitken, there passed away an official of long standing in the county of Fife and of wide reputation in his profession as a road surveyor. To the efficiency with which he carried out his work as road surveyor frequent testimony has been borne. A stranger on a cycling tour told the present writer that in all the counties he had visited he had not come across roads equal to those in Fife". [37] The article continued to outline of his career including a reference to the formation of the "Taroads Syndicate, Glasgow, with Mr. Aitken as managing director."

On 1 April 1919 William Boyd's contract as Road Surveyor was confirmed at a salary of £250 per year rising at £20 per year to a maximum of £400, but he was not allowed to undertake work for anyone else. [35] Boyd remained a full-time paid official in Fife until he retired in 1944. One of his sons, also called

William, became the Resident Engineer for Fife's first motorway, the M90, which was completed using much mature technology in 1970. [38]

References

[1] Obituary – Thomas Aitken, *Engineering*, 1918, vol. 105, p. 646.
[2] Obituary –Thomas Aitken, *The Engineer*, 1918, vol.125, p.495.
[3] W.R. Stephen, *The industrial archaeology of Fife* (PhD thesis, University of Strathclyde, 1975).
[4] A.W. Brotchie and H. Jack. *Early Railways of West Fife*. Catrine: Stenlake Publishing, 2008.
[5] W.S. Bruce, *The Railways of Fife*. Perth: Melven Press, 1980.
[6] Silver, *The Roads of Fife*. Edinburgh: John Donald, 1987.
[7] "Bill to alter Turnpike Trust Laws, etc.", *The Times*, 12 June 1839.
[8] R.A. Paxton and J.S. Shipway, *Civil Engineering Heritage: Scotland – Lowlands and Borders*. London: Thomas Telford, 2007.
[9] R.C. McWilliam, "The Garlie Bank: a case study, Assessing the historical significance of routes", ICE PHEW conference, London, 2004, pp.3-7.
[10] Candidates for election to Associate Member, ICE Archives, London, 1893.
[11] T. Aitken, "The maintenance of Macadamised Roads", *Minutes of Proceedings of the Institution of Civil Engineers*, 1895, vol.122, pp.215-223.
[12] T. Aitken, *Road Making and Maintenance: a practical treatise for engineers, surveyors, and others with an historical sketch of ancient and modern practice*. London: Griffin, 1900.
[13] Candidates for transfer to Member, ICE Archives, London, 1902.
[14] Candidates for transfer to Member, ICE Archives, London, 1903.
[15] W.P. Judson, *Road preservation and dust prevention*. New York: Engineering News Publishing, 1908.
[16] R. Jeffreys, *The King's Highway: an historical and autobiographical record of the developments of the past sixty years*. London: Batchworth Press, 1949.
[17] R.A. Whitehead, *A Century of Steam-rolling*. Shepperton, Surrey, Ian Allan, 1975.
[18] E. Williamson, A. Riches, and M. Higgs, *The Buildings of Scotland – Glasgow*. London: Penguin, 1990.
[19] R.C. McWilliam, *BSI: the first hundred years 1901-2001*. London: The Institution of Civil Engineers, 2001.
[20] L. Bush, *The Life & Times of the Illustrious Captain Brown: a chronicle of the sea and Japan's emergence as a world power*. Tokyo: Tuttle, 1969.
[21] D. Rayner, *Steam Wagons*, Princes Risborough: Shire, 2003.
[22] T. Aitken, *Road Making and Maintenance: a practical treatise for engineers, surveyors, and others with an historical sketch of ancient and modern practice,* second edition, London: Griffin, 1907.
[23] R.A. Paxton and J.S. Shipway, *Civil Engineering Heritage: Scotland – Highlands and Islands*. London: Thomas Telford, 2007.
[24] R. Hutchinson, *Calum's Road*. Edinburgh: Birlinn, 2006.
[25] H.J. Collins and C.A. Hart, *Principles of Road Engineering*. London: Arnold, 1936.
[26] Fife County Council minutes FCC/7/10/1/3. Fife Archives, Markinch.
[27] J.B. Rae, "Coleman duPont and his road", *Delaware History*, 1975, vol.16, no.3, pp. 171-183 (Spring-Summer).
[28] T. Aitken, "Adaptation of highways to modern traffic", Road Surveyors' Association of Scotland 22nd. AGM, County Council Rooms, Peebles, pp.2-12 (19 July 1906).
[29] T. Aitken, "Construction and maintenance of roads", in *Proceedings of the Incorporated Association of Municipal and County Engineers,* 1907, vol. 33, pp.246-257.

[30] Forbes, "Results of tar-spraying experiment", Road Surveyors' Association of Scotland 23rd. AGM, County Council Rooms, Edinburgh, pp.27-28 (28 June 1907).

[31] T. Aitken, "The application of tar to macadamized roads", *Engineering News (US),* 1907, vol. 58, no.8, pp.206-207 (22 August).

[32] T. Aitken, "The construction of macadamised roads suitable for modern traffic", *The Surveyor and Municipal and County Engineer*, 1907, vol.31, pp.584-587 (6 December).

[33] Anon., "The 'Aitken' system of road making", *The Surveyor and Municipal and County Engineer,* 1910, vol.37, p.147 (28 January).

[34] T. Aitken, "A few remarks in regard to surface-spraying of roads with tar and making in situ tar-macadam", *The Surveyor and Municipal and County Engineer*, 1910, vol.37, pp.698-699 (20 May).

[35] Fife County Council minutes FCC/7/10/1/4. Fife Archives, Markinch.

[36] Jarvis, *The Liverpool Dock Engineers.* Stroud: Sutton, 1996.

[37] The late Mr. Thomas Aitken: an eminent road-maker, *The Fife Herald & Journal (Cupar),* 1918, no.2787. p.2 (29 May).

[38] R.C. McWilliam, "Scotland", in W.J. McCoubrey (ed.), *The Motorway Achievement, vol.3, Building the Network,* London: Thomas Telford, 2009, pp.469-564.

Spatial unification of Romanesque churches in Zamora and its influence on the buttress system

Francisco Javier Rodríguez Méndez
High Polytechnic School of Zamora (University of Salamanca), Zamora, Spain

Summary

In addition to the Cathedral in Zamora there are twenty-two Romanesque churches registered, erected between the eleventh and the thirteenth centuries. The most common typology – visible in twelve of these churches – is a single nave, with the body wider than its chevet and without a visible transept. Of the other ten churches, originally with three naves, only one has remained intact. The other nine at some point underwent what we refer to as "spatial unification" by removing the interior columns in the interests of providing more light and clarity.

The churches of San Juan and San Ildefonso showcase this perfectly, because of their size and the diversity of solutions. The purpose of this paper is to study the consequences brought about by "spatial unification" of the buttress system, the tectonic structure and the configuration of Zamora's churches. Ultimately, this research includes archiving and planimetric activity, as well as focusing on structural stability.

Introduction

At present, besides the cathedral, there are a total of twenty Romanesque churches surveyed in the city of Zamora, mostly built between the eleventh and thirteenth centuries. The most abundant church typology – recognizable in twelve of the twenty churches – is the single nave church with the body wider than the chevet and without a visible transept. Of the remaining ten churches, originally built with three naves, only one has remained intact, the other nine being subjected at some point to what we call "spatial unification" which could be defined as the merging of the naves of a church by the elimination of internal columns for the sake of clarity. It is not, of course, a phenomenon specific to Zamora. The church of Santiago in La Coruña is a clear example of a church with three naves which, after a fire and subsequent remodelling, gave way to a fully Gothic pointed diaphragm arches structure. Why this solution was used repeatedly in Zamora – and not in other neighbouring capitals such as Salamanca – is not an easy question to answer. Perhaps the explanation lies in the fact that, unlike those other capitals, Zamora's development was very localized in the eleventh and twelfth centuries. Churches built during this period might have been sufficient to meet the needs of future generations in both capacity and layout within the city. The few examples of Gothic architecture, today completely ruined or disappeared, consisted of monasteries located in neighbourhoods outside the walled city. This new architecture, more spacious and bright, influenced the transformation of churches inside the walled enclosure. As it was impossible to replace them with new ones – either for lack of funds or lack of adjoining lots in an already overfilled town – alteration was chosen instead.

Besides the Cathedral of Zamora, only the church of Santiago del Burgo preserves its original three-naves layout. We will therefore make use of it in order to show the original state and characteristics of the remaining nine churches that are still standing and which formerly had three naves. Its floor plan is rectangular with the high chapel barely standing out from the nave. Its inner dimensions are of just 11.9 metres in width and 26.0 metres in length. Its chevet includes three rectangular chapels in the "Ancient Spanish"[1] style (term coined by Gómez Moreno) – the prevailing style in Zamora. The cross-section shows the slender proportions of the naves. In the central nave we find a three-to-one proportion (Fig. 1). Pillars are square and have semicircular adjoining pilasters. Their excessive size is, according to Gómez Moreno, the reason behind the fact that "they were destroyed in other churches to keep just one nave"[2].

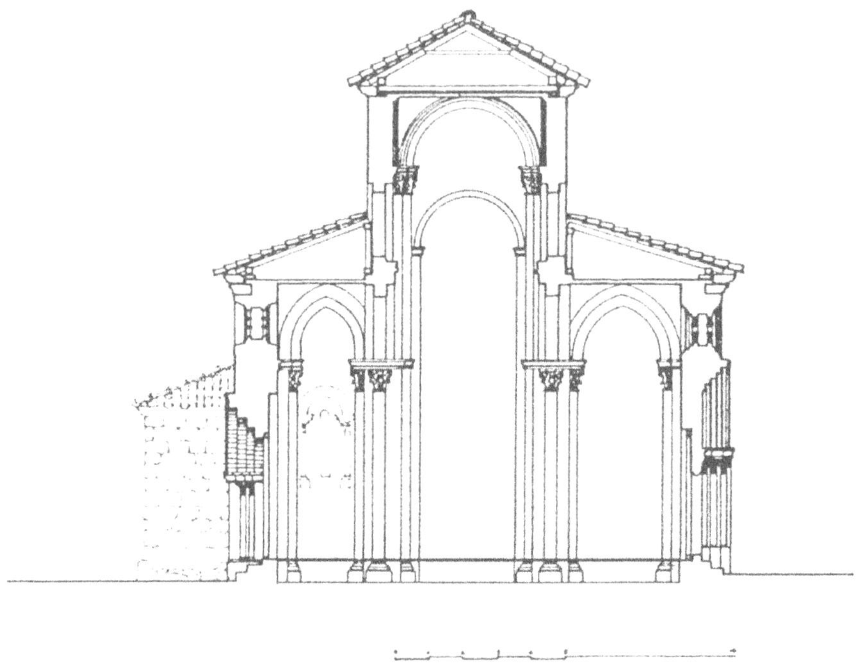

Fig. 1. Church of Santiago del Burgo, Zamora. Cross-section. Source: Enciclopedia del románico en Castilla y León. Zamora, p. 443.

Most of the nine Romanesque churches (seven to be precise) where the three initial naves were merged into one can be included in the "general case". To describe the general case I will use as an example one of the seven churches, the church of Santo Tomé. During the fifteenth century, a refurbishment was carried out and the four internal columns were replaced by two semicircular diaphragm arches. For Arturo Zaragozá, this system was used for economical reasons because it was the cheapest in terms of the quantity of wood needed for its construction[3]. Diaphragm arches and wooden roofs as a construction system had been used with great frequency for a long time in Catalan and Galician churches and they then disseminated throughout the whole Spanish Peninsula[4].

The "spatial unification". The case of San Ildefonso

This church, 18.8 metres wide and 33.7 metres long, is the largest of the Romanesque churches still standing in the city of Zamora (excluding the Cathedral). Originally, its interior layout was of three naves in four sections completed with semi-circular apses, very similar to the current layout of the church of Santiago. It remained as such until it was profoundly altered in the fifteenth century. The barrel vaults, supported by side and transverse arches (which laid over columns) were replaced by ribbed vaults covering the whole width of the church. Elements found in the church of San Ildefonso show a style which is midway between the late Gothic and the early Renaissance: elongated ribbed vaults which together compose a semicircular barrel vault of 18.4 m in diameter counterweighted by lunettes.

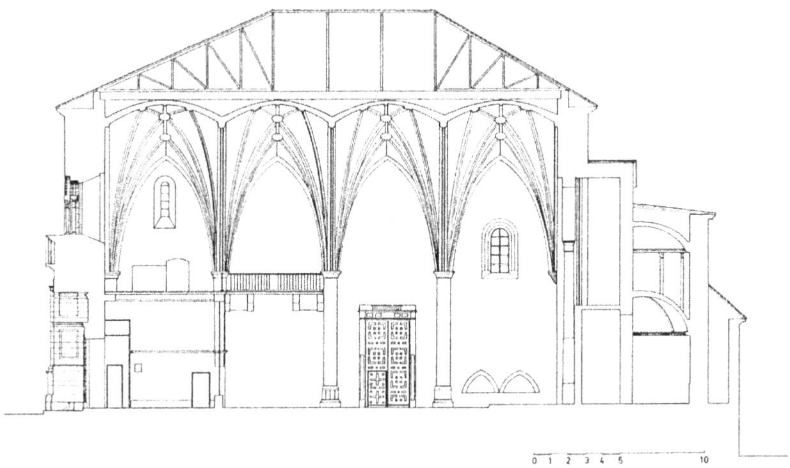

Fig. 2. Church of San Ildefonso, in Zamora. Longitudinal section. Source: Enciclopedia del románico en Castilla y León. Zamora, p. 430.

It is reasonable to believe that the architect in charge of San Ildefonso's works was aware of the process followed in Gerona[5] and that the architectural solution therein employed (i.e. single nave covered with ribbed quadripartite vault) influenced his design. However, the longitudinal section of the church in Zamora (Fig.2) shows that each barrel section is, in essence, a toroidal vault with a longitudinal horizontal axis and an arch with a diameter of 10 metres. Figure 3 shows a modelling of the toroidal vaults and transverse lunettes. Unlike in the case of a barrel vault, the double curvature of a toroidal vault splits the pushing forces into both the longitudinal transverse directions of the church. The transverse pushing forces are biased toward the buttresses by the action of the lunettes, and the longitudinal pushing forces compensate one another. This is the natural evolution of ribbed vaults which were at that time (the end of fifteenth century) according to Arturo Zaragozá "an ancient and experimented solution" being gradually replaced by "new groined vaults in which the edges and the element made just one body"[6]. This is the case in the vaults of San Ildefonso, very similar in geometry to the sail vaults found in the church of Santo Domingo in Medina de Rioseco, designed by Juan de Nates nearly a century later[7].

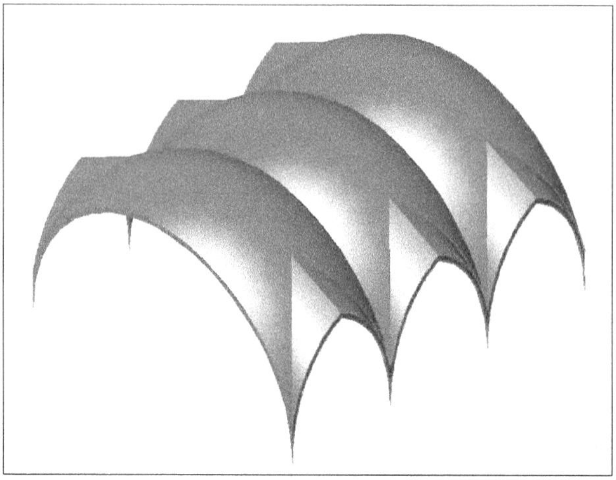

Fig. 3. Church of San Ildefonso, in Zamora. Modelling of the vaults.

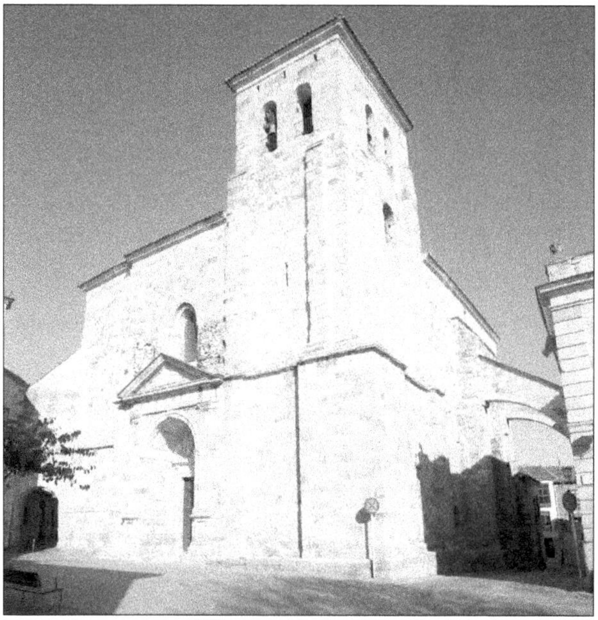

Fig. 4. Church of San Ildefonso, in Zamora. Exterior. View from the west. Source: Proyecto cultural Zamora románica.

The roof covering consisted of a filling on the vaults' extrados until it was level with the slopes of the peaked roof. All of this entailed greater loads than in the case of a roof covering made using diaphragm arches with the aggravating factor that in this case the distance the arches must cover was some 18 metres. The counterweight system needed to ensure stability affected nearby buildings. Figure 4 shows how, in order to counterweight the thrusts of the new roof covering, it was necessary to prop the church up over two adjacent palaces. The shoring was achieved without affecting the surrounding streets by using flying buttress arches in the south (right side of the figure) and through a solid corner buttress in the northwest. Later on, the new roof covering of wood trusses forced the eave height to be raised above the toroidal vaults' keystones to allow the tie rods to pass over them.

The "spatial unification". The case of San Juan

The church of San Juan, called "de Puerta Nueva" [of the New Entrance], was near the wall of the first defensive enclosure and of the new entrance opened therein in order to connect with the eastern extension of the city. By the beginning of the sixteenth century an adjacent stretch of wall had already been demolished for some time, and the Main Square was placed in the esplanade that opened in front of the Church[8]. The church thereafter lost its free-standing nature and was gradually surrounded by buildings by the side of the square. Figure 5 shows the state of the church by 1890 with its tower – the only visible element – dominating the Main Square. The upper body or bell tower was dismounted in 1898 because it was in danger of collapsing. Currently the church stands alone again (Fig. 6) as a result of a controversial demolition of one side of the square's buildings which leaned against the church's walls.

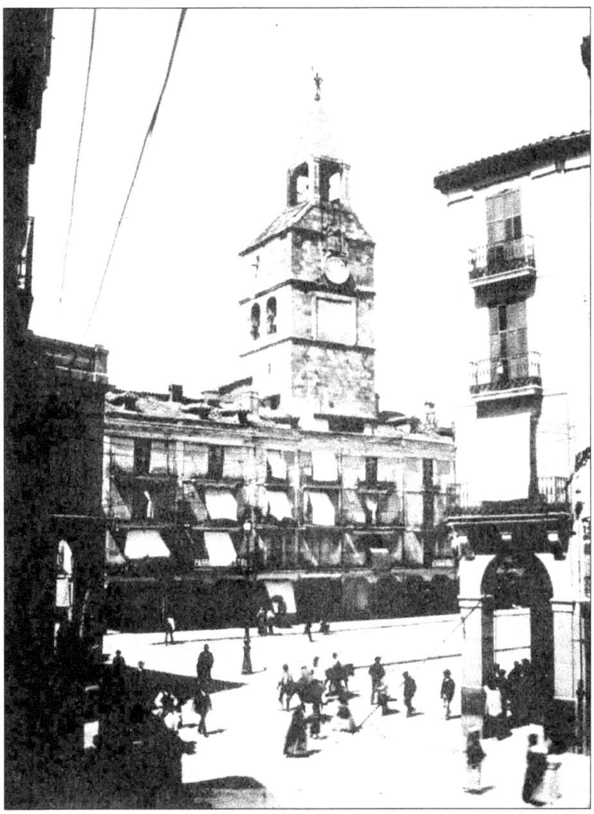

Fig. 5. Church of San Juan, in Zamora. Eastern View, c. 1890.

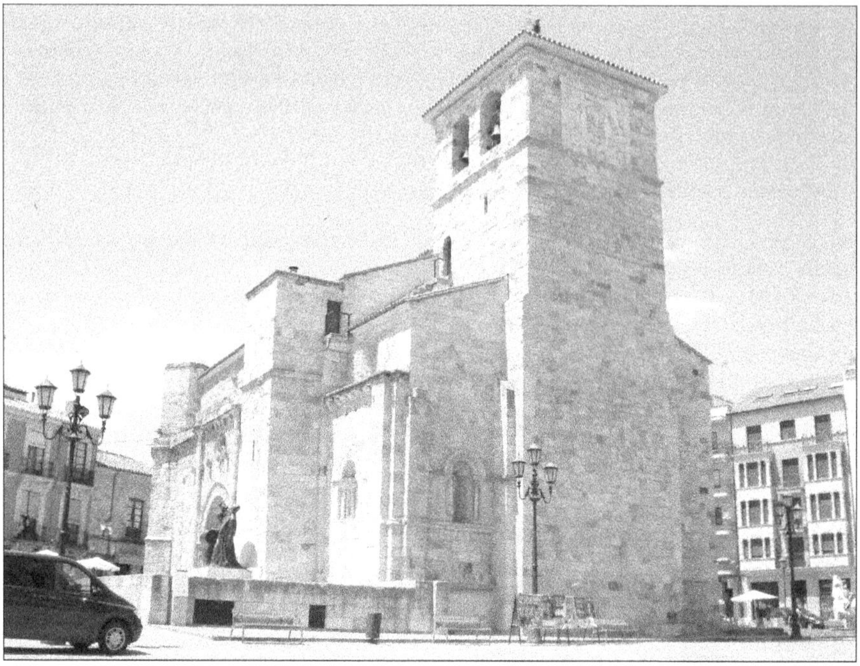

Fig. 6. Church of San Juan, en Zamora. Eastern View, current state.

The transformation experienced by the church of San Juan cannot be seen only as the result of the urban evolution of its surroundings. The changes in its internal structure, also influenced by external circumstances, were the most decisive factor in the final configuration of the church. As for the twelfth century Romanesque church with three naves, all that remains are the chevet structure and the north and south façades which define a floor plan of 14.4 metres in width and 25.3 metres in length. In the words of Gómez Moreno "inside the church nothing of the source elements can now be recognized (…), the ancient naves have transformed into a single one with two huge arches running along it, supported by Gothic pillars"[9]. The pair of big side arches with a span of 14 metres is the most eye-catching element inside the church of San Juan (Figs 7 & 8). Also the tower, built upon the chancel, corresponds to an uncommon type of tower for the Romanesque period in Zamora. It is certain that the current tower is not the original one, but doubts arise as to whether its location is the original. Historians disagree on this issue. Whereas for Jaime Nuño the tower's location is the original one ("it can be assumed that the tower had had from the beginning the same current location and characteristics, moreover given the proximity of the wall on that side"[10]), Guadalupe Ramos suggests to the contrary when she states that "we are surprised by its location because there is no other case in the Romanesque in Zamora where we can find a tower in the chevet"[11]. I will examine these two subjects in depth in the following paragraphs.

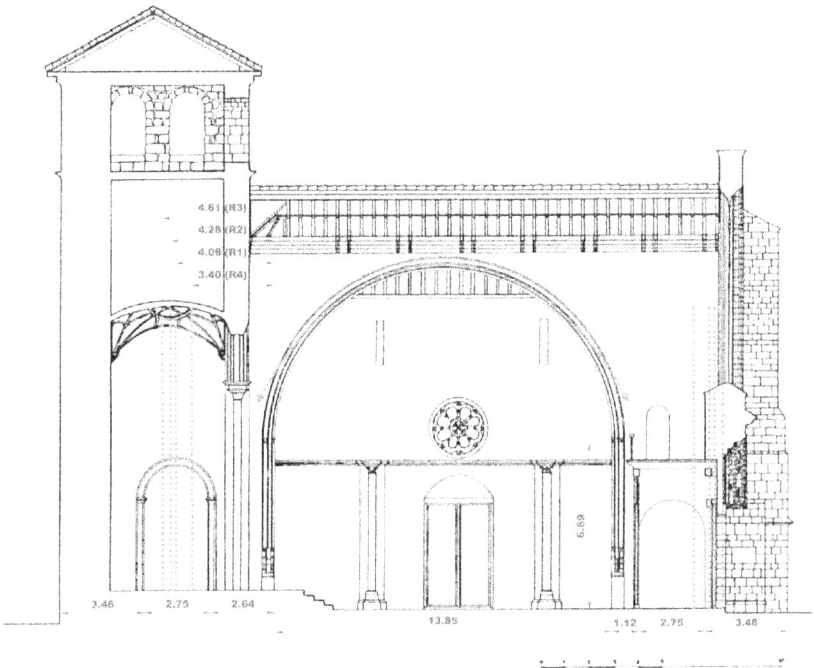

Fig. 7. Church of San Juan, in Zamora. Longitudinal Section. Source: Enciclopedia del románico en Castilla y León. Zamora, p. 462.

When contemplating the elegant solution used inside the church, the question that arises is this: why have diaphragm arches laid out transversally to the longitudinal axis of the nave not been used in the church of San Juan as they have in most other "unified" churches in Zamora? To explain the reasons behind such a choice it is necessary to take into consideration the urban evolution of this particular area. With the construction in the thirteenth century of the second walled enclosure, the stretch of the first wall running in front of the church lost its functionality and was finally demolished. Houses were gradually built around the perimeter of the church until it was completely enclosed. This situation remained unchanged until recently.

The enclosed layout of the church prevented the thrusts' counterweight to come from the sides as happens with other renovated churches. This is the most likely explanation for the longitudinal layout of the arches which replaced the rows of columns. The longitudinal layout made it possible to counterweight the thrusts on both sides of the arch by means of large blocks of the same thickness, drilled only into the essential wall voids (Fig. 8). Whereas at the foot of the church the huge buttress walls delimit the inner hallway and support the upper gallery, they frame the main chapel and support the tower in the chevet. The advantages of this layout are twofold. First, both the choir gallery and the tower in the chevet contribute with their loads to provide stability to the church as a whole. Moreover, this solution did not affect adjacent buildings (nor did it at the time affect the wall) and it could be applied wholly from within the church. Thus, in our opinion, it is reasonable to think that the tower layout over the chancel is not the original one and that the tower must have been built at the same time as the side arches as a solution for improving counterweight, avoiding at the same time works on the outer side of the building. The uncommon buttress located at the south-west corner, (Fig. 8) hides within it a spiral staircase which may well be the remains of the ancient tower which from that position could also have served as a watchtower.

The successive renovation works undergone by the church of San Juan have been carefully studied by Luis Vasallo Toranzo in his article dedicated to the intervention of Rodrigo Gil de Hontañón[12]. Vasallo recognizes that "the work done by Rodrigo Gil is difficult to define" because of the difficulty in interpreting "obscure and fragmented circumstances" and due to the later ruination of the building. This being so, the precious documentation recovered from the Zamora Provincial Historical Archive provides an understanding of a highly confusing process.

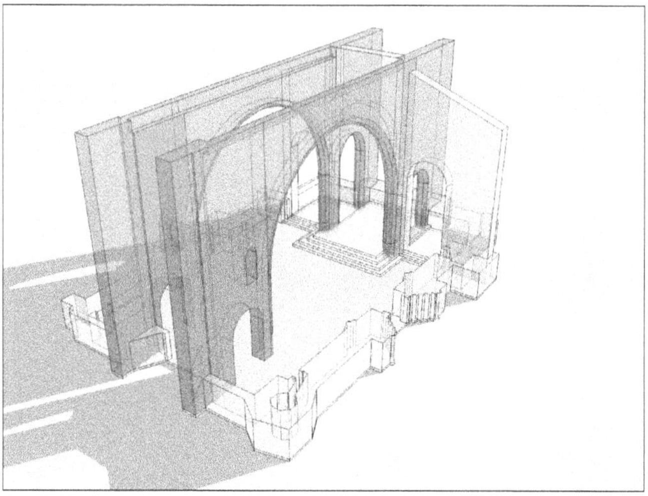

Fig. 8. Church of San Juan, in Zamora. Modelling of side arches.

The starting point is an undocumented intervention at an unknown date (we will suppose it dates from the end of the fifteenth century) in which the four central pillars were removed and replaced by two large side arches. From the early sixteenth century the church, already renovated, and its tower, needed to be repaired frequently. Vasallo explains that "in 1531 the tower and one of the side arches (the one located at the epistle side) were in danger of crumbling" and that because Rodrigo Gil was at that moment in the city of Zamora working on a different project, he was required by the City Council to take care of its reparation. Gil de Hontañón, who was barely thirty and was starting his career as an independent professional, was already a celebrated and esteemed architect. Once the architect's fees – considered by all as excessive – were accepted, the agreement was executed in September 1553.

In June 1532 Rodrigo Gil completed the works. Prior to delivery and subsequent payment, the council demanded an inspection of the works by independent architects[13]. The report's conclusion – which was repeated a year later with the same results – was favorable for Rodrigo Gil, except with regard to the execution of a particular detail. According to the authors of the report, the toral pillar had not been sufficiently dismantled, while some cracked elements were visible and not sufficiently bonded. It would therefore be necessary (as specified in the conditions) to fasten again the newly-built side arch to the arch situated between the two chapels, at a height of eight courses. In the opinion of the previously mentioned author, Luis Vasallo, "Rodrigo Gil's intervention in the church of San Juan was not very successful." Although the architect went as far as to pawn "his life and personal assets" to ensure the future integrity of the church, in 1559 the tower collapsed. It should be clarified that, in Gil de Hontañón's defence, this time the affected side was the gospel side, whose side arch had not been touched in the repair works of 1531. In particular, this side arch had been re-built by different stone-carvers "in the image and likeness of the arch of the epistle side" (which had been re-built by Gil in 1531) and both arches are still in place today. The reconstruction of the tower over the chancel (regardless of this location being the original one or, as suggested by Ramos de Castro and supported by us, the new one) ended in 1579.

It would be inexcusable not to apply the geometric rules attributed to Gil de Hontañón in the case of the side arch used by him in the church of San Juan. For the implementation of the four geometric rules attributed to the great Spanish architect we have followed the interpretation of Santiago Huerta[14]. The rules are based on (sometimes somewhat convoluted) geometric constructions which provide the abutment's thickness according to the span of the semicircular arch they support. The thickness obtained by applying each of the four rules are, respectively, 4.06 metres (1st rule), 4.28 metres (2nd rule), 4.61 metres (3rd rule) and 3.40 metres (4th rule). The outer edges of the abutment which result from the application of the four rules are represented in figure 7. The thrust lines have also been included, using graphic statics and joining the intersection points of the resulting lines from the thrust of each voussoir to the contact planes between the voussoirs.

Several conclusions can be drawn by observing the drawing and the lines obtained. The first is that the arch molding is extremely thin: all graphic constructions have an arch molding of the same size as the abutment but here it is only of 2 Castilian feet, i.e., not even one-sixth of the smallest abutment (3.4 metres resulting from rule nº4). But from an aesthetic point of view and also from a construction viewpoint, an arch with a 3.4 metre molding could not be adopted. It is the reduced load the arch must support that makes a 0.56 metres moulding viable as only a couple of courses cross over the side arch's keystone. On the other hand, the Mudejar wooden roof, whose load the arch must also support, uses the lightest roof solution possible.

The thickness of the abutments situated on both sides of the arch are below 3.4 metres which is the measure required by the fourth rule. Although it is questionable that in this unusual case the rules were applied, the thickness measurement calculated by means of the fourth rule is confirmed by the first rule of Derand (published by Derand in 1643 but frequently used in the early Renaissance) which, if applied, has a result of 3.46 metres, almost the same result as the previous one reached using Gil's rule[15]. Indeed, 3.46 metres is the exact measure of the church porch's blocks located on both sides of the arch. This leads us to believe that the architect made sure that the holes drilled on the abutments left room at their edges for blocks of the magnitude required by Derand's rule. The line of thrust, because the arch molding is so thin, exceeds its limits and reaches up to the haunches even though the arch supported very reduced loads. Stability is ensured on the left side by the vertical loads coming from the tower and on the right side thanks to the weight of the gallery and the western end.

Conclusions

There is no doubt that in Zamora, by the end of the fifteenth century, the idea of carrying out spatial unification of the old Romanesque churches gained shape and momentum. Of the ten churches which today remain standing and which originally had three naves, only one has kept its original layout: the church of Santiago del Burgo. The other nine churches have suffered alterations which eliminated the inner pillars in order to create a clear space dedicated to religious worship. The general procedure was used in seven cases and consisted of replacing rows of pillars with transverse diaphragm arches. In the church of San Ildefonso the spatial unification was achieved through the construction of stellar vaults with toroidal geometry. This solution generated great horizontal thrusts and it was necessary to prop up the church over the adjacent palaces using buttresses and flying buttresses. Finally in the church of San Juan de Puerta Nueva, a lighter solution was chosen, consisting of replacing the rows of pillars with large, longitudinal side arches. The counterweight elements could not stand above the adjacent façades' perimeter because of the enclosed layout of the church and according to Guadalupe Ramos de Castro it was necessary to change the tower to its current location over the chancel in order to contribute its own weight to the building's stability.

References

[1] M. Gómez Moreno, *Catálogo Monumental de España. Provincia de Zamora.* Madrid: Ministerio de Instrucción Pública y Bellas Artes, 1927. p. 150.

[2] Gómez Moreno, (Note 1) p. 150.

[3] A. Zaragozá Catalán, *Naves de arcos diafragma y techumbre de madera en la arquitectura civil valenciana*, pp. 551-555 in S. Huerta (ed.), *Actas del Primer Congreso Nacional de Historia de la Construcción.* Madrid: I. Juan de Herrera, 1996.

[4] L. Torres Balbás, *Naves cubiertas con armadura de madera sobre arcos perpiaños a partir del siglo XIII*, *Archivo Español de Arte*, Vol. 34, Madrid: C.S.I.C., 1960. pp. 13-43.

[5] S. Huerta, *Mecánica de las bóvedas de la Catedral de Gerona*, pp. 179-204, in M. Llorens (ed.), *Seminari sobre l'estudi i la restauració estructural de les catedral g'otiques de la corona catalana-aragonesa, Girona 2001.* Girona: Universitat de Girona, 2004.

[6] A. Zaragozá Catalán, *A propósito de las bóvedas de crucería y otras bóvedas medievales*, *Anales de Historia del Arte*, Vol. Extr., 2008, pp. 99-126.

[7] S. Huerta Fernández, *Mechanics of masonry vaults: The equilibrium approach*, pp. 47-69 in P.B. Lourenço (ed.), *Historical Constructions*, Guimaraes, 2001.

[8] F.J. Rodríguez Méndez, *Plan Director de las murallas de Zamora. Algunas aportaciones al conocimiento del Monumento*, pp. 557-570, in *Actas del IV Congreso Internacional "Restaurar la Memoria", Valladolid 2004*, Valladolid: Junta de Castilla y León, 2006.

[9] Gómez Moreno, (Note 1) p. 159.

[10] M. A. García Guinea y J. M. Pérez González (dirs.), *Enciclopedia del románico en Castilla y León. Zamora.* Aguilar de Campoó: Fundación Santa María la Real, 2002. p. 465.

[11] G. Ramos de Castro, *El Arte Románico en la provincia de Zamora.* Zamora: Diputación Provincial, 1977. p. 176.

[12] L. Vasallo Toranzo, *Una nueva obra de Rodrigo Gil de Hontañón en Zamora: la reforma de la iglesia de San Juan de Puerta Nueva*, *Boletín del Seminario de Estudios de Arte y Arqueología*, Vol. 59, Valladolid: Junta de Castilla y León , 1993, pp. 313-320.

[13] Vasallo, (Nota 12) p. 317-318.

[14] S. Huerta Fernández, *The medieval 'scientia' of structures: the rules of Rodrigo Gil de Hontañón*, pp. 567-585 in A. Becchi (ed.), *Towards a History of Construction.* Berlin: Birkhäuser Verlag, 2002.

[15] Huerta, (Note 14) pp. 571-572.

Surviving late payments: strategies of Christopher Wren's masons from Burford

Melody Mobus
PhD, BA, FRICS, architectural historian, UK

Introduction

Late payments have long been the bane of the construction industry. This thorny problem became a serious issue in the rebuilding of the City of London after the Great Fire of 1666, and remained endemic in major building projects such as the Queen Anne churches, Greenwich Hospital and Blenheim Palace. Of these the latter two are now UNESCO World Heritage Sites.

It was not until the mid-1930s that the general problem of dilatory payments masons experienced in London in the seventeenth and early eighteenth centuries was recognised by Douglas Knoop and G. P. Jones who concluded, 'heavy indebtedness to mason-contractors appears to have been the rule rather than the exception'.[1] Howard Colvin took up the discussion some forty years later in respect of the Crown as employer, noting 'only craftsmen with extensive capital had the financial stamina to wait for payment'.[2] More recently, James W.P. Campbell explored profit margins for mason-contractors at St Paul's Cathedral, where payments were seriously delayed.[3]

This paper looks at the problem through the experience of three dynasties of stonemasons from Burford, Oxfordshire: the Strongs, Kempsters, and Beauchamps. After the Great Fire, these stonemasons went to London and took on many late paying contracts. The Strongs, working two quarries near Burford, were established masons and stone suppliers of regional importance. By contrast, Christopher Kempster, also in possession of a Burford quarry, was the first of his dynasty. Ephraim Beauchamp, born in 1661, was the latecomer to this group. The outstanding masons were Thomas Strong and his brother Edward, whose son Edward succeeded him; Christopher Kempster and his son William; and Ephraim Beauchamp. They rapidly rose to become Wren's foremost masons. Of the fifty-one churches rebuilt in the City of London after the Great Fire, these three dynasties were involved in almost half of them. All were mason-contractors at St. Paul's Cathedral, the Strongs for over forty years having the distinction of laying the first stone and the last.[4]

Structure of the paper

This paper focuses on the period immediately following the Great Fire up to the late 1720s, by which time the rebuilding was finished, the Queen Anne churches were almost completed, and the major Burford Masons ceased to be active. Selecting some of their most tardy paying, and most prestigious projects, this paper looks at funding, why payments fell into arrears, and most particularly, the Burford Masons' strategies to stay financially solvent. It draws on primary sources such as Christopher Kempster's recently resurfaced day book, William Kempster's London day books and accounts, the Edward Strongs' memorandums and letter-books, and legal documents and bank accounts, almost all hitherto untapped. It will be shown that these leading mason-contractors helped expedite the completion of hugely ambitious projects not only through their craft, but also through financing them by default.

Outstanding projects, outstanding arrears

The City churches, St. Paul's, Greenwich Hospital and the Queen Anne churches were all to be funded from an increase in the tax on coal landed in the Port of London. However, this revenue almost immediately turned out to be both insufficient and unreliable. Shortly after its introduction in 1670, reimbursement for earlier church rebuilding resulted in payments for more currently contracted work to fall seriously behind.[5] Revenue plummeted by around a third when the Dutch, and later the French, blockaded coal deliveries to London,[6] a state of affairs which continued up to the 1690s.

Funding for Greenwich Hospital was augmented by various contributions, including £2,000 annually from the Crown, and from a sixpence levy on sailor's pay. Both sources, however, rapidly dried up. Only four years into their contracts at Greenwich, the Strongs were owed £1680. Their men went on strike, forcing the employing commission to release £500 to pay them. By 1720, the Strongs were still owed some £8,500 for work carried out five years earlier.[7] For all of these publicly funded projects, small advances were reluctantly issued in the form of imprests, or tallies to be cashed in later. Discrepancies in the records have prevented an assessment as to how far these concessions actually offset arrears. What is clear, however, is that contracting craftsmen were left to bridge the remaining gap themselves.

Among the Burford Masons' late paying contracts were two palaces: Winchester in Hampshire, paid for by Charles II, and Blenheim in Oxfordshire, initially financed by a grateful nation for the Duke of Marlborough's success against the French. At Winchester, when work ceased upon the king's death in 1685, the craftsmen were left with some £4,250 arrears.[8] At Blenheim, the situation was far more serious. Following Queen Anne's rift with her long-term companion Sarah, Duchess of Marlborough, payments came to a halt. Discontent among the workmen rose to the extent that in 1710 they smashed the newly crafted stone capitals in the colonnade. When Treasury official Samuel Travers visited the site later that year, he was so moved by the 'many poor starveing people' laid off work that he gave them £300 on his own credit.[9] The Strongs continued on the main pile until 1712, by which time their arrears, with interest, had reached some £12,500, around a million pounds in today's money.[10] Efforts to recover their dues culminated in them taking the Duke to the highest court of appeal, the House of Lords. Although successful, the Strongs nonetheless had to shoulder the arrears for more than fifteen years.[11]

Rates of Pay

It might be expected that late paying clients compensated their contractors with higher rates than the building trade at large. Yet the Burford Masons took on several late paying contracts at reduced rates, probably to secure outlets for their stone and keep them in the market for other commissions. At St. Stephen Walbrook, for instance, Thomas Strong and Christopher Kempster were prepared to accept lower rates than those paid to Thomas Cartwright at St. Mary-le-Bow.[12] For their south-east section of St. Paul's, Christopher Kempster and Beauchamp were prepared to work at the rates their predecessor Edward Pierce had refused.[13] At Greenwich, Clerk of Works Nicholas Hawksmoor confirmed that his rates were much lower than those of other public works in London,[14] and significantly, by 1716, the Strongs were the only mason-contractors tendering.[15] John Vanbrugh justified his rates at Blenheim as being 'considerably less than what was paid, either in the Queens Works, or others of any Consideration, about Town'.[16] The extent to which payment, when it came, represented profit has proved difficult to assess owing to incompatible accounts: comparison of William Kempster's meticulous records of expenditure on his workforce with the building accounts for St. Paul's has thrown up widely differing and inconclusive results.[17]

The Burford Masons' Strategies for Survival

The Burford Masons went up to London as soon as legislation in February 1667 permitted provincial craftsmen to work on the rebuilding. Back home, they appear to have enjoyed prompt payments, smaller clients sometimes settling in local tokens or even bartering their trade. Christopher Kempster, for instance, as payment for his building work for the Burford tailor William Kinge, received 'a pair of drawers a vest and a pair of breeches a coat and awestcot' for his wife, and gowns, petticoats, undercoats and waistcoats for his five daughters.[18] The credit-based economy they came up against in London demanded more challenging survival skills.

Contractors were left to find their own means of dealing with late payments. The immense task of rebuilding the City fell in a time when banks were in their infancy, credit being provided mainly by goldsmiths and private individuals. Records of six emergent banks (including the Bank of England, where Edward Strong junior ran an account for twenty-five years) have not revealed any of the Burford Masons raising credit from them, using them instead as clearing banks only, over a period of forty years.[19] Casting the net beyond banks, sources have brought to light six strategies for survival which helped the Burford Masons overcome the problem of interim finance.

(i) Dynastic Organisation.

The Burford Masons' large families enabled them to deploy family labour to optimum use. Some, like Valentine Strong, remained in the Burford area, administering the extraction and shipping of quarry stone, and undertaking local masonry work. Others oversaw major contracts, like John Kempster substituting for his father Christopher at Winchester Palace, freeing him to run his contracts on City churches.[20] Later family masons such as Oswell Strong, Edward Kempster, and Edward Beauchamp simply remained employees of the Strongs and Kempsters in London, doubtless more tolerant of late payments than their colleagues.

Family also provided credit. William Kempster's day books reveal a striking example of this practice showing his father's indispensable financial support to keep him afloat when he took over the contract for the south-west tower of St. Paul's in 1700. In the absence of any payment from the Commission for the first fourteen months, Christopher forwarded money weekly to cover outgoings on William's team of approximately twenty-five men.[21]

(ii) Ownership of Quarries.

Unlike London masons, the Burford Masons possessed quarries, three yielding quality freestone. The infrastructure for transporting Burford stone along the River Thames had been well-established over the centuries. This smoothed the way for the Strongs and Kempsters to start shipping their stone to London within only a few months of the 1667 rebuilding legislation. An entry in Christopher's day book, which I have dated to summer 1667,[22] records 'the first parsell of stone delivered at London', noting payments from leading London masons Edward Marshall, Thomas Cartwright and Thomas Knight.[23] Thomas Strong at this time was extracting 'stone at the Quarrys which he had command of, & sent the same to London, & sold great Quantities to other Masons'.[24] As payment usually included the cost of transport, the rest, apart from the outlay for quarrying, was profit. From the earliest years of the rebuilding the Strongs and Kempsters were capitalising their firms from this commodity, which went a long way to support them in leaner times. As much as 40 percent of the Strongs' earnings at St. Paul's and 45 percent at Blenheim Palace represented payment for stone.[25] Although Ephraim Beauchamp did not own a quarry, he was drawn through his sister's marriage into the Strong family and selling their stone. He branched out as a merchant of more recherché stone, like Swede marble.[26]

(iii) Partnerships.

In order to spread financial risk, the Burford Masons increasingly entered into partnerships. This strategy had the further advantage of freeing heads of firms to take on contracts elsewhere. The Burford Masons formed partnerships for at least twenty-five contracts. For their first one in London, in 1672, Christopher Kempster and Thomas Strong joined forces to build St. Stephen Walbrook; this happened to be the first contract let to a partnership for any City church. After Thomas's death in 1681, his brother Edward took over, declaring to the Commission that he and Kempster were 'Partners in receipt of the rest of the money due'.[27] Kempster went on to form another partnership with Edward Strong for the foundations of Winchester Palace, for which they even ran a joint account.[28] The longest partnership, between Kempster and Beauchamp at St. Paul's, lasted seventeen years. At Greenwich Hospital, Beauchamp was a joint contractor with the Strongs.

The Burford Masons also formed partnerships outside the three families. Of particular importance was that of Edward Strong junior who partnered bricklayer Edward Tufnell for building six Queen Anne churches. After Tufnell's death in 1719, Strong partnered fellow mason Christopher Cass for three more church contracts.[29] Lack of documentation suggests that many partnerships were informal arrangements. By the 1720s, however, partnerships appear to have become more commonplace among building contractors and were often placed on a legal footing. Edward Strong junior's arrangement with London masons Christopher Cass and Andrews Jelfe, for instance, was legally dissolved in 1728, all three having been 'copartners in a joint Stock'.[30]

(iv) Diversification of Client.

Some of the Burford Masons' most elite clients, the Crown and the Commissions, turned out to be the tardiest payers. By broadening their client base, the Burford Masons increased the likelihood of receiving prompt payments from some to offset arrears from others.

Clients proving reliable in their payments included institutions, such as Oxford colleges (Trinity and Christ Church), the corporation of Abingdon Borough, Berkshire, for whom Christopher Kempster built County Hall, and London's St. Bartholomew's Hospital. Of their aristocratic clients such as the Earl of Chandos at Cannons, Middlesex and Lord Willoughby in Queen's Square, it is likely that at least some were prompt payers. William Kempster, whilst engaged at St. Paul's and on several City churches, also operated a successful business as a City mason. His day books show payments for work on coffee houses (such as All Saints' and St. James's), public houses (The Haunch of Venison in Bow Lane and The Crowne in Milk Street) and for businessmen (such as Mr Austin the winecooper, Mr Ikey the tobacconist, and Mr Palmer the surgeon on Garlick Hill).[31] Running concurrent contracts also afforded the Burford Masons greater flexibility in deploying their large workforces to financial advantage. Comparison of the Strong teams at St. Paul's and Greenwich, for instance, throws up a dozen masons in both teams to have been used on both sites.[32]

(v) Diversification of Trade.

Both the Kempster and Beauchamp families retained their by-occupation as yeoman farmers in Burford, a fallback position in times of slack. Another country-based sideline was the lucrative sale of malt to a thirsty London which possessed no malt works at the time.[33] In 1699, one shipment alone yielded the Kempsters over £8,500 profit in today's money. [34]

The Burford Masons soon sharpened their competitive edge by widening their craft base into bricklaying, securing the Strongs at least seven of their City church contracts and Beauchamp his commission for the College of Arms. Although bricklaying was essential for speculative house building, the only indication of their involvement is Edward Strong's ownership of two dwellings in Wren's Bridgewater Square development, where other St. Paul's contractors also held property.[35]

(vi) Generating Income from Capital.

In the later seventeenth century, granting interest-bearing credit became a significant form of income. By the late 1680s the Burford Masons had built up sufficient capital to enter the newly-emerging mortgage business. Edward Strong senior and Ephraim Beauchamp, for instance, provided mortgages to Ephraim's brother Benjamin, a carpenter, enabling him to develop eighteen tenements in Gardiner's Lane near Whitehall and Earls Court, Soho.[36] Edward Strong junior also became a provider of mortgages, granting several in Threadneedle Street (including the Crown Tavern opposite the Royal Exchange),[37] in prestigious Old Burlington Street,[38] and for developing three houses in Hanover Square.[39] By the late 1720s, whilst still engaged on Queen Anne churches, he was granting bonds to Kings College, Cambridge, and loans for the Rye Harbour project in Sussex.[40]

When work almost came to a halt at St. Paul's in the 1690s due to mounting arrears, Wren's idea of paying interest on the sums owed effectively converted the Burford Masons' £5,450 arrears into bridging loans. The Burford Masons maximised their interest by buying in other contractors' arrears.[41] The Strong family placed an additional £3,000 on interest-paying loan for the purpose of securing their future employment at St. Paul's, effectively becoming co-financiers of the Cathedral.[42] Edward Strong senior generated income not only from loans, but also by becoming an active investor in property, accumulating by the 1720s ownership of at least twenty houses in London and several more in St. Albans and the Burford area.[43]

Conclusion

The Burford Masons drew heavily on their country roots to counter the endemic problem of late payments. Stone capitalised their firms, and their large and skilled dynastic base enabled them to mobilise both labour and financial support. In London, they spread the risk by forming partnerships and by continuously broadening their client base and trade, commonplace methods today, but at that time, these were significant steps in changing their provincial medieval practices into recognisably modern firms.

In principle, late paying clients could have secured sufficient credit to settle their dues promptly, yet the problem was transferred to the contractor. The Burford Masons effectively became financiers, subsidising their clients in order to continue working and selling their stone. In meeting these financial challenges the Burford Masons adapted to the new credit-based economy, emerging as entrepreneurs. In so doing, they helped expedite the completion of many seminal buildings, thereby securing their integrity of design. How would St Paul's Cathedral look today if it had taken a hundred years to complete instead of forty?

References

[1] Douglas Knoop and G.P. Jones, *The London Mason in the Seventeenth Century*. Manchester: Manchester University Press and the Quatuor Coronati Lodge, 1935, p.50.

[2] H.M. Colvin (ed.), *The History of the King's Works* (6 vols). London: HMSO, 1963-82, Vol.5, p.44.

[3] James W.P. Campbell, "Building a Fortune: The Finances of the Stonemasons Working on the Rebuilding of St Paul's Cathedral 1675-1720", pp.303-304, in *Proceedings of the Third International Congress on Construction History, Cottbus 2009,* Berlin, 2009.

[4] Sir John Soane's Museum [SM], Letter Book of Edward Strong junior, 1730-1740.

[5] Paul Jeffery, *The City Churches of Sir Christopher Wren.* London: Hambledon, 1996, p.48.

[6] Stephen Porter, *The Great Fire of London.* Stroud: Sutton, 1996, p.130.

[7] Knoop and Jones, *London Mason,* p. 51.

[8] A.Bolton, D.Hendry et al., *The Wren Society.* Oxford: Wren Society, 1923-43 (20 vols), vol XVIII, p.74.

[9] David Green, *Blenheim Palace,* London: Country Life, 1951, pp.119, 121.

[10] Green, *Blenheim Palace,* p.150.

[11] Oxford, Bodleian Library [Bodl.], MS. Top Herts c. 2, Strong family papers.

[12] London Metropolitan Archives [LMA], CLC/313/J/002/ MSS25539/1 fol. 36, & 25539/2 fol. 82, building accounts.

[13] LMA, CLC/313/1/A/002/MS 25622/002, minute book.

[14] The National Archives [TNA], PRO ADM 67/2, minute book.

[15] TNA, PRO ADM 67/4.

[16] Geoffrey Webb (ed.), *The Works of Sir John Vanbrugh*. London: Nonesuch, 1928, p.187.

[17] Kempster v Wrigglesworth, TNA, PRO C 106/145; LMA, CLC/313/1/B/003/MS 25473.

[18] Burford Tolsey Museum [BTM], Kempster day book.

[19] Bank of England Archives, C 98/2549-2653.

[20] RIBA, WRE/1/1, f.236, Wren family papers.

[21] TNA, PRO C 106/145.

[22] Melody Mobus, 'The Burford Masons and the Changing World of Building Practice in England 1630-1730' PhD Thesis, Open University, 2012, pp.86-87.

[23] BTM, Kempster day book.

[24] SM, Strong letter book.

[25] Ibid.

[26] LMA, CLC/313/1/B/021/MS25493, craftsmen's payments.

[27] LMA, CLC/313/J/012/MS25544/7, imprest books.

[28] TNA, PRO C 106/145.

[29] Lambeth Palace Library, MSS 2701 & 2714, book of works & miscellaneous papers.

[30] British Library, Stowe MS 412, no. 77, f.187, release upon settling accounts.

[31] TNA, PRO C 106/145.

[32] Knoop & Jones, *London Mason,* p. 62.

[33] Peter Borsay, *The Eighteenth Century Town*, London: Longman, 1990, p.150.

[34] TNA, PRO C 106/145.

[35] LMA, CLC/521/ MS02461, indenture.

[36] LMA, MSS 0/382/1 & 3, lease and counterpart.

[37] TNA, PRO C 111/194, Ashpoole v. Denn.

[38] F.H.W. Sheppard (ed.), *Survey of London*. London: Athlone, 1963. vol. 32, p.508.

[39] TNA, PRO C 11/2014/16.

[40] SM, Strong letter-book.

[41] LMA, CLC/313/1/B/003/MS 25473/30-34.
[42] Bodl., MS. Top Herts c.2.
[43] TNA, PROB 11/596, Edward Strong' senior's will; Bodl., MS. Top Herts c.2.

Concrete Platforms in the North Sea – a golden era for development of advanced concrete structures

Jan Moksnes
Independent Consulting Engineer, Norway

Synopsis

A total of 25 concrete Gravity Base Structures of different designs and sizes, supporting large drilling, production and storage facilities, have been delivered to international oil and gas industry in the North Sea from 1973 to 1995.

From the early stages of being prototype structures with no history related to offshore oil and gas production, the concrete GBS is now a proven concept with known characteristics and impressive service records. National and international Codes and Recommendations for design and construction have been developed over these years, and much progress has been made with respect to design, materials and construction methods to ensure satisfactory performance and durability in very severe marine environments. The early North Sea platforms have been in operation for 40 years and the concrete hulls will perform with little maintenance for many years to come.

The history of the concrete GBS illustrates the ability of the construction industry to recognize opportunities and meet new and challenging demands, to adapt to a new set of clients to work together to develop Codes to ensure safe and predictable performance.

Introduction

Large quantities of oil and gas were discovered in the North Sea in early 1970's and the operators started to examine their options for offshore installations to extract the new found wealth. For the construction industry this turned out to be a tough challenge, having to deal with:

- ✓ Very demanding and hostile climatic conditions
- ✓ Water depths up to 300 m
- ✓ Wave heights up to 30 m
- ✓ Fields located up to 200 km from shore
- ✓ Deck weight at tow out up to 50 000 tons
- ✓ Displacement during tow up to 1.5 mill tons
- ✓ Storage capacity for up to 2 mill barrels of oil
- ✓ Tow - out in a short selected summer weather window
- ✓ Accurate installation at the field in a few days with little or no site preparation
- ✓ Fast track deliveries
- ✓ Multidiscipline activities with different cultures and traditions

Prior to the early 1970's all marine oil and gas installations in deep water worldwide were made of steel. Concrete structures were absent. At the time there was a severe shortage of equipment for marine operations and heavy lifting and this favoured robust, floating structures that could be constructed inshore and installed directly on the sea floor. In France, Britain, Holland and Norway this sparked considerably activity and resulted in the development of several concepts for so-called concrete Gravity Base Structures (GBS).

The GBS concept

The essential elements of the concrete GBS concepts include:

- Heavily reinforced and post tensioned high strength concrete hull
- Construction in a dry dock and in a floating position moored at a sheltered inshore location. Detail design and construction performed in parallel, often under a single EPC (Engineering, Procurement and Construction) contract, saving time and ensuring that constructability is given proper attention.
- The topside produced in parallel with the GBS. The two items are mated by float over at a deep water inshore location and the completed unit towed to the field in a selected weather window. The GBS is fitted with steel or concrete skirts under the base, designed to penetrate the sea bed to capable soils, eliminating the need for offshore piling.
- The large GBS caisson offers floating stability and large load carrying capacity at float out, stability during operation and the capacity for significantly increased topside weight once installed. The caisson offers storage capacity for oil enabling production to continue in bad weather and the conductors and riser pipes are sheltered inside the concrete shafts.

The structures are predominantly of the wide caisson base and slender shafts configuration, the shafts being tapered to a minimum cross section at the water line. Some have a large diameter caisson fitted with a Jarlan perforated wall to dissipate the wave energy (Fig.1).

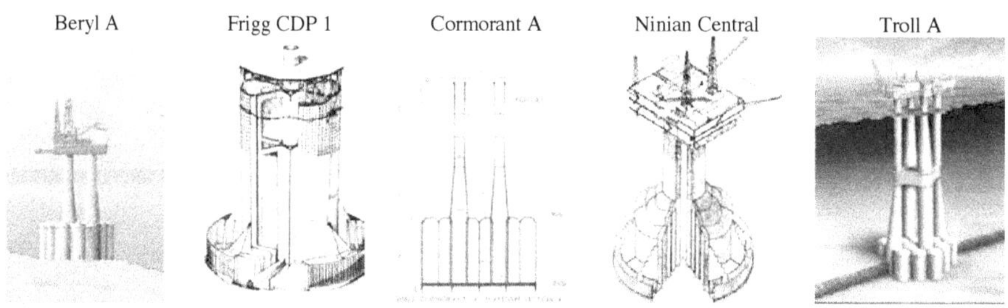

Fig.1. A selection of GBS structures installed in the North Sea

The history

The breakthrough for concrete came when Phillips ordered a large storage unit for the Ekofisk field in 1971. This pioneering structure was designed by Doris of France. A dry dock facility was quickly established in Stavanger by Norwegian Contractors (NC) and the Ekofisk caisson was delivered on time in the summer of 1973. The Stavanger yard put NC in pole position for new contracts and contributed to the subsequent development and growth of the city.

The Ekofisk project gave rise to a sudden demand for new concepts in a new setting driven by the multinational oil companies with clear business objectives and open minds to innovative solutions. Invitations to tender were issued for a number of concrete platforms in the British and Norwegian sector of the North Sea. The GBS pioneer were the Beryl (Mobil) and Brent (Shell) projects, followed in quick succession by the Frigg, Dunlin and Statfjord fields, in water depths of 100 – 150 m [1]. All these oil fields are located east of Shetland and close to the dividing line between the UK and Norwegian continental shelf.

The forecast in the early 1970's was a possible demand for some 6 to 8 concrete platforms a year, sparking considerable (and overly enthusiastic) activity for new concepts and a search for suitable dock sites. The successful contenders for the early contracts were Norwegian Contractors, McAlpine / Sea Tank Co, Andoc and Howard Doris. Yards were developed in Stavanger and Åndalsnes in Norway, at Ardyne Point and Loch Kishorn in Scotland, in Rotterdam and in Stromstad in Sweden. Later on dock facilities were also developed at Hunterston, Graythorp, Portavadi and Nigg Bay in the UK.

Norwegian Contractors (now Kvaerner CS) was a Joint Venture of the 3 largest civil contractors in Norway joining forces to share resources, risks and rewards. The JV gained the support of a national engineering and scientific network and the cooperation of the authorities resolving the formalities involved in establishing the necessary dry dock facilities.

NC became a major player in this field and delivered a total of 16 GBS platforms, all but two from their Stavanger yard, taking advantage of an efficient dry dock facility, deep fjords and towing drafts up to 230 m. Slender shell structures and advances in high strength concrete and slipform technology made it possible to achieve the desired weight and floating stability to cope with increased water depths and heavy topsides.

Tab. 1. The North Sea GBS platforms delivered during the period 1973 – 1995.

Nr	Year Install	Field		Client	Contractor	Water Depth	Concrete vol. m^3
1	1973	Ekofisk	N	Phillips	Doris/NC	71 m	80 000
2	1975	Beryl A	UK	Mobil	NC	118 m	52 000
3	1975	Brent B	UK	Shell	NC	140 m	64 000
4	1975	Frigg CDP1	N	Elf	Howard Doris/ NC	104 m	60 000
5	1976	Brent D	UK	Shell	NC	140 m	68 000
6	1976	Frigg TP1	N	Elf	Howard Doris/Skanska	104 m	49 000
7	1976	Frigg MCP-01	N	Elf	McAlpine/Sea Tank Co	94 m	60 000
8	1977	Dunlin A	UK	Shell	Andoc	153 m	90 000
9	1977	Frigg TCP2	N	Elf	NC	104 m	50 000
10	1977	Statfjord A	N	Mobil	NC	145 m	87 000
11	1978	Cormorant A	UK	Shell	McAlpine/Sea Tank Co	149 m	120 000
12	1978	Ninian Central	UK	Chevron	Howard Doris	136 m	140 000
13	1978	Brent C	UK	Shell	McAlpine/Sea Tank Co	141 m	105 000
14	1981	Statfjord B	N	Mobil	NC	145 m	140 000
15	1984	Statfjord C	N	Mobil	NC	145 m	130 000
16	1986	Gullfaks A	N	Statoil	NC	135 m	125 000
17	1987	Gullfaks B	N	Statoil	NC	141 m	101 000
18	1988	Oseberg A	N	Norsk Hydro	NC	109 m	116 000
19	1989	Gullfaks C	N	Statoil	NC	216 m	244 000
20	1989	N Ravenspurn	UK	Hamilton B	John Laing	42 m	9 800
21	1993	Sleipner A	N	Statoil	NC	82 m	77 000
22	1993	Draugen	N	Shell	NC	251 m	85 000
23	1996	Harding	UK	BP	Costain/Taylor Woodrow	109	37 000
24	1995	Troll A	N	Shell	NC	303 m	245 000
25	1999	South Arne	DK	Amerada Hess	Brown & Root	60 m	35 000

Some key data for an early and a late North Sea GBC project are presented in Table 2, illustrating the developments in size, complexity and schedule over the 20 year period.

Tab. 2. Key data Beryl A (1975) and Troll A (1995).

Beryl A: ***The pioneer***. Field discovered September 1972, Contract award July 73, GBS installed June 75. Dry dock 9 m. Water depth 120 m. Towing draft 85 m. 19 no 20 m dia. cells caisson, 3 no 96 m tall shafts, 3.5 m steel skirts, 64 000 m^3 concrete (50 MPa), rebar density 180 kg/m^3
Troll A: ***The giant***. Field discovered 1979, Contract award March 1991, GBS installed May 95. Dry dock 14 m. Water depth 303 m. Towing draft 227 m. 19 no 32 m dia. cells caisson, 4 no 300 m tall shafts braced at mid height, 40 m skirt piles, 245 000 m^3 concrete (85 MPa), rebar density 400 kg/m^3

Construction

The construction of a GBS is performed partly in a dry dock and partly afloat at a sheltered location where the draft is controlled by ballasting. Some main aspects of construction are illustrated in Fig. 2, 3 and 4.

Fig. 2. Dry dock construction; *Fig. 3. Shaft slipforming;* *Fig. 4. Deck mating.*

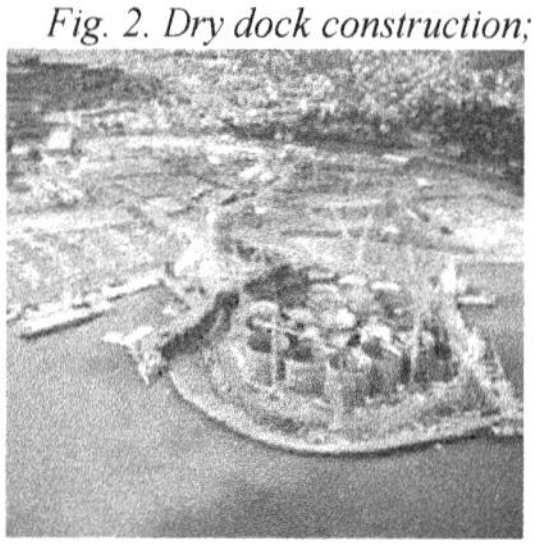

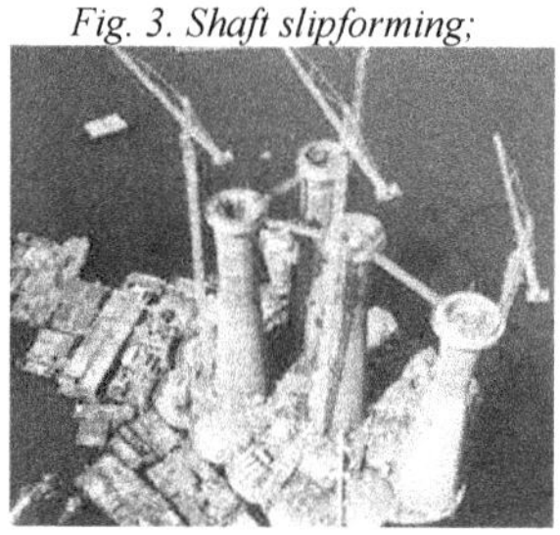

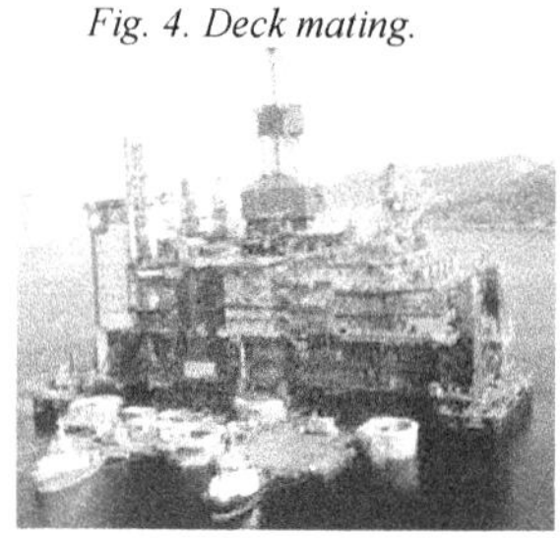

Fig. 2 shows the Gullfaks C caisson in the 14 m deep dry dock near Stavanger (1986).The base consisted of 24 no 28 m dia. cylindrical concrete cells (skirt piles) with a footprint of 16 000 m2. The 42 m tall bottom section was floated out of the dry dock by the innovative use of a large cushion of compressed air. Once out of the dock the air cushion was released and the structure was towed to the inshore construction site at its natural draft of 34 m.

Fig. 3 shows shaft slipforming at the inshore construction site. All 4 shafts slipformed together at a rate of about 2 m / day. Note floating rig and bridges connecting the 4 shafts.

Fig. 4 illustrates a critical moment: Deck mating by float over. The GBS is submerged to within 6 m from the top of the shafts and then de-ballasted to pick up the deck.

A large volume of mechanical equipment (conductors, J tubes, pumps, valves, cables instrumentation etc) is installed in the GBS hull during construction. Typical total weights steel would be in the region of 3 000 – 5 000 tons. Close coordination between the civil and mechanical disciplines and keen prioritizing of activities were the keys to timely progress.

Marine engineering is an important aspect of the projects; floating stability and weight management need to be considered during all stages of construction. Water ballasting is an ongoing operation to maintain the desired draft and freeboard during construction (Fig. 4) and under tow. The Gullfaks C platform during tow to field in 1989 at a draft of 210 m, was carrying a topside weight of 50 000 tons and had a displacement of 1.5 mill tons.

The concrete GBS's are examples of large multidiscipline projects embracing the fields of civil, mechanical and marine engineering and imposing strict demands on modern project management and execution [2]. Detailed work planning and quality assurance programs were implemented and taken to a new level through the GBS era. The offshore projects took a leading role in developing modern project management systems as the focus shifted from the product to the process and then to the Total Quality Management approach of today.

Innovation

The prototype projects in the early 1970's were examples of major innovations, developing concepts to meet the demands of the oil- and gas fields and mobilizing the courage and belief to enter into significant contractual commitments. The demands for GBS structures in ever deeper water and more difficult soil conditions led to further developments in the design and construction procedures and in the associated Codes. A number of major R&D programs were executed in the GBS era, being supported by the oil companies and by the construction industry with a desire to perform and deliver. Some of the developments are described below:

Concrete technology

The quality requirements for strong and durable concretes are well understood and accepted. The essential requirements relate to sound constituent materials, maximum water / binder ratio, proven freeze / thaw resistance, satisfactory workability, and, for slipforming, predictable setting time commensurate with the anticipated lifting rate. The site procedures include requirements to placing and compaction, absence of cold joints, adequate curing and weather protection and strict attention to a minimum cover to the reinforcement of 50 mm.

The concrete quality improved considerably through the North Sea GBS era. The onus was on constructability and durability as well as design strength. In the early 1970's the 45 MPa/ medium workability concrete specified for the Ekofisk caisson was a tough challenge, whereas in the 1990's a 80 MPa pumpable concrete mix could be delivered with confidence. The main factors in this evolution were the development of high strength cements and efficient modern chemical admixtures, fully automatic batching plants and comprehensive quality assurance programs.

The North Sea concrete platforms have been in service since 1973 and field tests and observation such as chloride tests on drilled out core samples have revealed no significant signs of deterioration or steel corrosion [4].

Geotechnical engineering

The sub soil conditions at the different oil fields vary from relatively dense sand to soft clays.

The early platforms were founded directly on the sea bed and stabilized by shallow (3 – 4 m) steel and concrete skirts. Cement grout was injected between the concrete base and the sea floor to ensure a moderate and acceptable bearing pressure. The penetration of the skirts, amounting to a few meters only, was achieved by sluicing ballast water into the storage caisson. The foundation design, with large environmental forces and cyclic loads and settlements, entailed new challenges in terms of soils testing and theoretical analysis modelling. For some of the later platforms soft clay was encountered to much greater depths and the concept of skirt piling was developed to overcome an otherwise almost prohibitive problem.

For Gullfaks C (1989) the sea bed consisted of normally consolidated soft clay and layers of loose silty sands which would normally entail piling or soil strengthening. The problem was solved by means of tall cylindrical concrete skirts installed beneath the lower domes during construction in the dry dock (Fig 2) and made to float out as an integral part of the structure and penetrate 22 m into the seabed through additional ballasting of the caisson. The concept was called *skirt piling* and was enhanced by an accelerated soil consolidation progress achieved by soil drain filters attached to the skirts and extracting ground water. The weight of the platform was transferred to the soil as skirt tip resistance and skirt wall friction and the GBS settlements were brought within acceptable limits.

The concept was subsequently applied to Troll A (1995) where 36 m deep skirt piles were required to achieve sufficient bearing capacity and stiffness. A major issue in this case was controlling the evacuation of trapped water inside the skirt compartments during penetration. The entire penetration procedure was completed within 104 hours from touchdown.

Skirt piling has been successfully applied to a number of platforms and sea bed anchor installations in the North Sea and elsewhere. Skirt piles is an interesting example of how scientific knowledge coupled with model testing, competent analysis and monitored performance can lead to milestone developments in challenging projects. The tall skirt piles led to the innovative out-of-dock application of large air cushions illustrated in Fig.2.

Slipforming

Slipforming is a highly efficient method of constructing tall concrete structures and became an essential feature of all the GBS construction programmes (Fig.3). Rates of construction of several meters per day of varying geometrical shapes and cross sections and containing multiple inserts and openings were achieved within strict geometrical tolerances.

The essential elements of successful slipforming are a well designed concrete mix with uniform and predictable properties, a distribution system ensuring uniform and timely delivery of all components, and a strong and durable slipform assembly lifted at a predetermined speed commensurate with the setting time of the concrete mix. Mock-up trials were commonly performed on site to verify the choice of methods and materials and included the spectacular 50 m tall leaning tower in Stavanger verifying inclined slipforming.

Slipform construction technology made great progress during the GBS era. The examples below illustrate the versatility of the slipform method:

Gullfaks C (1987). The caisson slipform operation; the 24 no. 28 m diameter cylindrical cell caisson made up a gross area (footprint) of 16 000 m^2. This was lifted through a height of 56 m by a total of 1800 hydraulic jacks. The operation lasted 42 days, consuming 114 000 m^3 of 65 MPa concrete, 24 000 tons of reinforcing steel and more than 7000 steel embedments.

Draugen (1992). The single cylindrical shaft was slipformed through a height of 242 m. The diameter varied from 41m at the base to 15 meter at the water line and flared out to a rectangular shape 22 x 22 m at the top.

Reinforcement and post-tensioning

The concrete GBS's were designed to resist extreme loading conditions and were extensively reinforced and post tensioned. The constructability aspects were given the necessary attention through close coordination between design and construction and mock up tests to identify problem areas and train the operators.

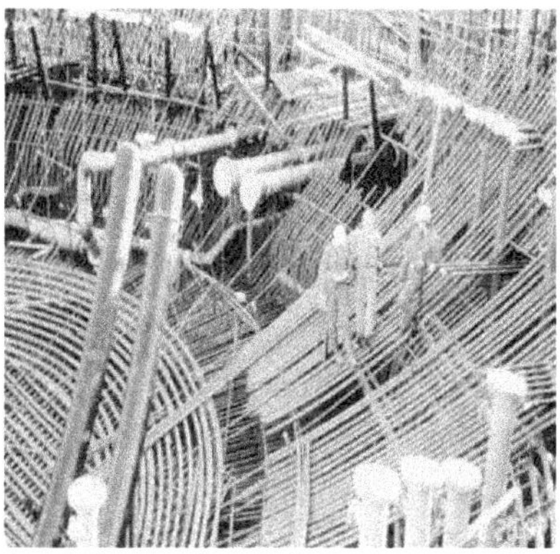

Fig. 5. Dense rebars and pipe systems

For Troll A GBS the average rebar density was 400 kg/m3. The post tensioning consisted of 6400 large cables tensioned by total force of 2.3 million tons, demonstrating the power of post tensioning!

Innovative T-headed bars (ordinary bars with a square plate welded on to the ends) were found to be invaluable on several projects to replace large numbers of ordinary small diameter stirrups in densely reinforced areas. Screw type rebar couplings were developed and used to eliminate conventional lap splices and ease congestion and adopted at construction joints and in temporary openings to eliminate long protruding bars.

Health, Safety and Environment (HSE)

Perhaps the most significant innovation in the GBS era came in the field of health and safety. The offshore industry did not focus on HSE in the early days of North Sea developments and the numbers of accidents and fatalities were totally unacceptable. The attitudes started to change in the 1980's and by the 1990's a paradigm shift occurred. HSE is now the number 1 priority in all major international oil companies and contractors. Documented HSE attention and performance is an essential parameter for all operators and the emphasis has moved from rules and regulations to positive attitudes and HSE being totally integrated in all activities.

Analysis Tools

In addition to the foundation design developments mentioned above the concrete platform projects brought about evolution and improvements in analysis tools such as finite element applications, wave load calculations, dynamic analysis with soil/structure/hydrodynamic interaction and laboratory testing methods. The engineering toolbox of today has greatly benefited from the challenges that were met and resolved during the GBS projects.

Codes and Standards

The offshore concrete structures led to major developments in design and construction procedures and in the associated Codes and Standards. The FIP report "Recommendations for the design and construction of concrete sea structures" [3] was first published in 1973. This was essentially a state of art review of current knowledge and aided the initial dialogue between Clients, Contractors and Authorities. The Norwegian Code NS 3473 "Design of concrete structures" was established as a credible basis for the design. Today ISO 19903 "Fixed offshore concrete structures" [6], evolved from the early documents, constitutes the basis for most project specifications.

The international concrete federations (FIP, CBE, ACI) played an important role in the development of the new breed of concrete sea structures. FIP, located at the Institution of Structural Engineers in London, set up working groups and published a number of reports related to offshore concrete structures [3,4,5].

National and international Codes and Recommendations for design and construction have been developed over these years, and much progress has been made with respect to materials and construction methods to ensure satisfactory performance and durability in very severe marine environments [5]. Whereas the early North Sea platforms were designed for a service life of 25 – 30 years they are still in good condition and capable of prolonged service. In principle a concrete GBS can be de-ballasted and refloated upon depletion of the field.

Concluding remarks

The GBS era in the North Sea culminated with the giant Troll A in 1995. The sudden decline may be contributed to several factors including the changing needs of the oil industry due to advances in sub-sea technology, the presence of a network of sea bed pipelines eliminating the need for intermediate storage and to keen competition from floating installations.

The GBS era brought about major advances in analysis and design, in materials and construction methods, in project planning and the management of large multidiscipline projects and in the approach to health and safety.

In retrospect, the North Sea GBS era was a window of opportunity which was seized by the civil construction industry and generated business for some and new knowledge for many. For the city of Stavanger and for Norway the concrete GBS became a catalyst for the new offshore industry and the associated growth and prosperity.

Since 1995 concrete GBS's have been delivered in Australia, Philippines, Canada and Russia and the concrete GBS is a strong contender for further developments in the arctic.

References

[1] New Civil Engineer Special Review *"North Sea Oil"* London, May 1974.

[2] Moksnes et al: *"Concrete platforms: history, technical breakthrough and future"* OTC Paper 7630, Houston 1994

[3] FIP Recommendations. *"Design and construction of concrete sea structures"*. 1973/1985

[4] FIP State of the Art Report *"Durability of concrete sea structures in the North Sea"* 1996

[5] Fib Bulletin 50. State of the Art Report *"Concrete structures for oil and gas fields in hostile marine environments"*. 2009

[6] EN-ISO 19903 *"Petroleum and Natural Gas Industries - Fixed Offshore Concrete Structures"* 2006

Timber hyperbolic lattice structures: Shukhov revisited

Ekaterina Nozhova
Institute of Historic Building Research and Conservation, IDB ETH Zurich, Switzerland

Summary

In 1899 the polymath engineer Vladimir Shukhov (1853–1939) was awarded a patent in Russia covering the principles he had devised for hyperbolic lattice towers. In their construction, these hyperbolic towers are an assembly of straight elements connected together by complex three-dimensional joints, whose full geometry is difficult to discern in the extant (orthographic) drawings. Until now, only structures made of metal were thought to have been built. A research group at the IDB ETH Zurich has found presumably the only surviving wooden structure built according to Shukhov's structural principles. In the course of investigations into the design of timber structures in Soviet Russia, another (unrealised) project for a timber hyperbolic water tower from 1932 was unearthed. This paper sets out to analyse these unique structures.

Introduction

The geometry of hyperboloids

The inclusion of a chapter on hyperboloids was conventional for the manuals on descriptive geometry by the end of the nineteenth century; either circular or elliptical in plan, they were examined along with the other solids of revolution and curved surfaces. The form was analysed not just for purely theoretical reasons, as hyperboloids were widely applied for of the design and manufacture of gears. There are five basic parameters (Fig. 1a), which describe the geometry of a hyperboloid: the lower radius, R; the upper radius, r; the height, H; the rotation angle, φ; and n, the number of struts that form the body of revolution.

Challenges in constructing metal hyperbolic towers

By the late nineteenth century, attempts had been made to realise lattice hyperboloids structures, however, both known projects remained on paper. These were the design proposals for the towers made of Mannesmann tubes (between 1890 and 1895) [1] and the design for the Great Tower for London by Thos V. Trew from 1890. [2] These were rather abstract sketch designs, based on clear geometrical principles. Ultimately, hyperbolic lattice structures were realised for the first time by the Russian engineer Vladimir Shukhov in 1896. [3] In the period 1896–1930, around forty towers were built. Of particular note were the 75-metre-high Adjigolsky Lighthouse (1912), the 150-metre-high Schabolovskaya radio mast (1919–22) [4] and 128-metre-high NiGRES tower (1927–29).

According to the patent, awarded to Shukhov in 1899, the straight elements formed "[...] a steady structure, strongly resistant to external forces at a reduced material consumption." Indeed, the inherent steadiness of a double-curvature is supplemented by the principle of a network structure comprising many equal members. [5]

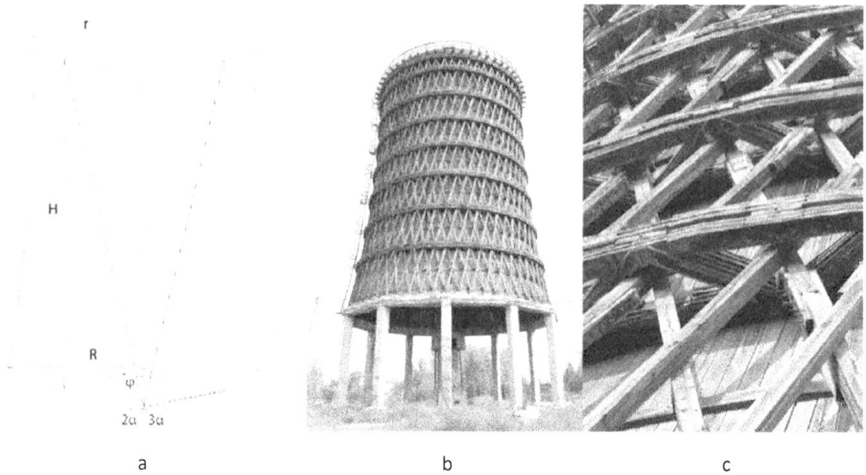

Fig. 1a: The defining parameters of a hyperboloid (source: author); 1b,c: The cooling tower at Severouralsk (source: author)

The shift to the use of timber in industrial structures in the 1930s

Timber was widely used in Russia for industrial structures, both in the nineteenth century and at the beginning of the twentieth. Renewable, easily worked, and with an availability guaranteed by state control [6], it was also supported by the newly established forestry education system. As timber requires little energy in manufacture (mostly the energy needed to saw it into logs and planks, which at the time could equally be done by manual labour) it was placed amongst the principal construction materials for industrial production and warehouses at the beginning of the era of the five-year plans. [7] In construction, a mass substitution of timber for steel and concrete occurred in the interwar period, and not only in Russia. With metal supplies prioritised for military purposes, together with the urgent necessity to develop and expand industrial production, there was a demand for advanced timber industrial structures, which could be reproduced throughout the country. In 1927, the State Research Institute of Structures was founded; it had a department responsible for timber structures, which edited the first Technical Norms and Conditions in 1931. It is important to underline that in Germany, where the problem of material shortage and its optimal use was also of vital importance, the Preliminary Specifications on Timber Constructions of the Reichsbahn were issued in 1926: it was a precursor of DIN 1052 from 1933 [8]. The issue of regulations initiated a well-founded, appropriate approach to constructing with timber, but in reality the transition was time-consuming and needed serious reorganisation in the construction industry.

Wooden cooling towers

In 1931, together with *Kotloturbina*, an organisation that united the heat and power plants of the country, the State Research Institute of Structures produced a 28 m high cooling tower, constructed as a timber shell. [9] This shell consisted of two diagonal layers of planks, reinforced by a system of vertical posts and horizontal rings. Besides the risk of decay at the junction between the two solid layers of planking, the structure was generally not steady enough and required an additional rigid diaphragm at the top, made of steel rods, in the model of a bicycle wheel. Tests on a scale model had shown that without the diaphragm, the structure deformed when loaded to only half the rated maximum loading, whereas with the diaphragm the, structure could sustain loading two and a half times greater than the rated one. This indicates that the construction of the upper rigid diaphragm was essential for the stability of the structure.

In 1938, a hyperbolic cooling tower based on Shukhov's principles was built in Kharkov. Neither photos nor drawings of this tower remain. We only know about this structure from an article published in 1938 by one of its constructors. [10] In the project, the steadiness of a hyperbolic lattice structure was advantageously applied for a 36-metre-high tower and eliminated the need to use additional structural elements such as cables or metal cover plates.

In 2012, a cooling tower was discovered in Severouralsk (Fig. 1b,c), which once belonged to a plant that provided heat and energy to local bauxite mines and the surrounding industrial city. It was built presumably in 1948, and according to the same construction principles as the one in Kharkov. Archival research and attempts to find out the name of the constructor of the cooling tower were unsuccessful. It is highly probable that the cooling tower was built by one of Shukhov's former colleagues who served a term of imprisonment in one of the camps in the Nizhny Tagil district; which would not have been unusual, for it is known that in the late 1930s many highly qualified specialists were imprisoned for political reasons.

In June 2012, a group of researchers from IDB ETH Zurich (Dr. B. Irmler, S. Liebscher and E. Nozhova) took measurements of the cooling tower and documented its basic elements [11]. The radius of its base ring is 11.52 m, the radius of its upper ring is 9.55 m, the height of the wooden structure is 27.75 m, and $\varphi = 45°$. The body of the tower is formed by 60 pairs of overlapping struts, 150 x 150 mm in section. The position of rings coincides with every second overlap of the struts.

Analysis of the struts

The struts are sequentially rotated in order to fit the hyperbolic surface (Fig. 2a). In the cooling tower, the angles are made at the joints between the elements that form the struts, and are realised with the help of specially cut cover planks (Fig. 2b). Ideally, at each point of overlap the strut should rotate 3.5 degrees, but as we see in the pictures, each second overlap is taken into account, which coincides with the position of rings. This partially explains the considerable deviation from this deduced ideal angle (Fig. 2a). Also, the actual lengths of the strut segments between the points of overlap differ considerably between the lower and the upper parts of the tower, so the rotation along the axis is regular and uniform only in horizontal projection.

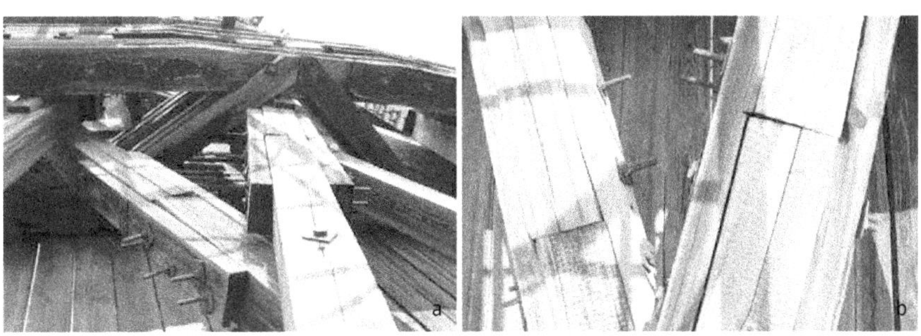

Fig. 2a, 2b: The lattice of wooden struts; consecutive rotation and extension at the junctions (source: author)

Analysis of the rings

The rings of the tower are layered structures; each ring consists of eleven layers (Fig. 3a,b): two belts of five layers of planking (approximately 250 x 50 mm in section) and a middle belt of 100 mm deep. The struts are joined to the ring through the elements of this middle belt. As there are sixty pairs of struts, each ring has a form of a sixty-sided polygon with inner angles of 174°, and the planks of the ring are cut to fit the geometry of the belt. The front and the back rings are linked by diagonals, in a manner similar to the construction of the rings between the sections in the Schabolovskaya and NiGRES towers. The vertex of the triangle formed by the diagonal links always coincides with the position of the radial axis, which might have served to control the exactness of construction; the inaccuracy of the secondary elements forces one to consider that elements were processed on-site.

a

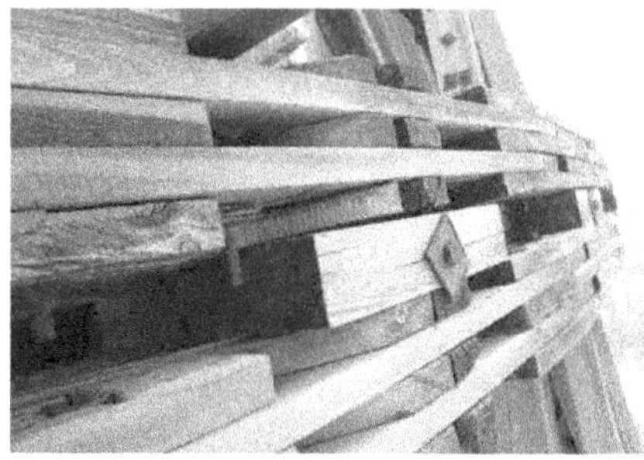

b

Fig. 3a: Strut junctions and the ring structure (survey drawings made by the IDB ETH team); 3b: The ring structure (source: author)

Analysis of the strut overlaps

The detail of the strut overlap is exceptionally important: it partially explains the proportions of the ring. The struts are inclined to the horizontal surface; the projection of the struts on the upper surface of the ring, which is approximately 60 mm high, differs from the projection on the lower surface of the ring (Fig. 3a). It was essential to take into account these margins to avoid the overlap of the strut with the diagonal elements of the ring; this consideration has led to the certain depth of the ring section. Another bottleneck of the volumetric detail was the necessity to link the vertical surfaces of the ring to the inclined surfaces of the strut. This was done with the help of trapezoidal wedge (Fig. 4a,b), where one surface of the wedge is vertical and another parallel to the inclined strut. The mirrored element was squeezed behind the strut overlap to link it to the back belt of the ring.

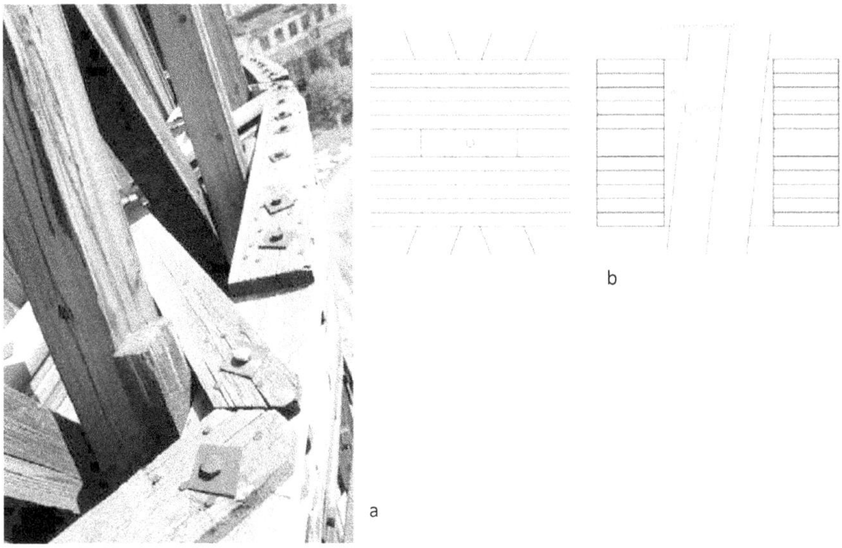

Fig. 4a: The strut overlaps and wedges (source: author); Fig. 4b: The scheme of the strut overlap (source: author)

The precision of the structure

When we closely analyse the details of the tower, we see that the only characteristic that needed to be controlled was the position of the radial axis of the strut overlaps. The geometry of the separate details does not need to be precisely controlled: for it does not define, but rather is a result of the general geometry. That is the reason why the inaccuracy of the separate elements is relatively significant.

As an experiment, we have tried to obtain the geometry of the curve by two different methods. Around five hundred points were measured with the help of tachymeter LEICA TPS 1200. On the basis of these measurements, we could restore the dimensions of each ring; this was the first method used to define the curve of the hyperboloid.

We had the tachymeter measurements of the first and of the eighth rings; the same rings were measured manually so we knew the distance from the ring outer surface to the strut overlap and thus could obtain upper and lower radii. As we also knew the height and the rotation angle, we could calculate the radii of the tower at the points of the strut overlaps, and thus obtain a second curve. We also needed to calculate margins, or the dimensions of the wedges that were mentioned above. When we summarised the calculated radius and wedge margins, we obtained numbers to build the second version of the hyperboloid. Here, we skip detailed calculations [12] and adduce the results: the deviation between the two curves was between 16 and 44 mm. Even taking into account all possible factors, from the measurement and calculation tolerances to wood shrinkage, the precision of the structure was extremely high.

Kovalchuk's water tower

The second timber hyperbolic lattice tower design considered here was developed in the period 1929–32 at the State Research Institute of Structures by the engineer Kovalchuk. [13] The water tower (Fig. 5a,b) carries the reservoir of 40 m^3; its height is 13.0 m, has 16 struts (8 x 2), its weight is 5,200 kg. The reservoir`s total mass together with the water is 43,000 kg. A brief description of the project mentions material economy, assembly of replaceable prefabricated parts and safety among the advantages of the tower. For the safety, the argument is supported by an experiment: a model of the scale 1:10 was tested at the institute's laboratory.

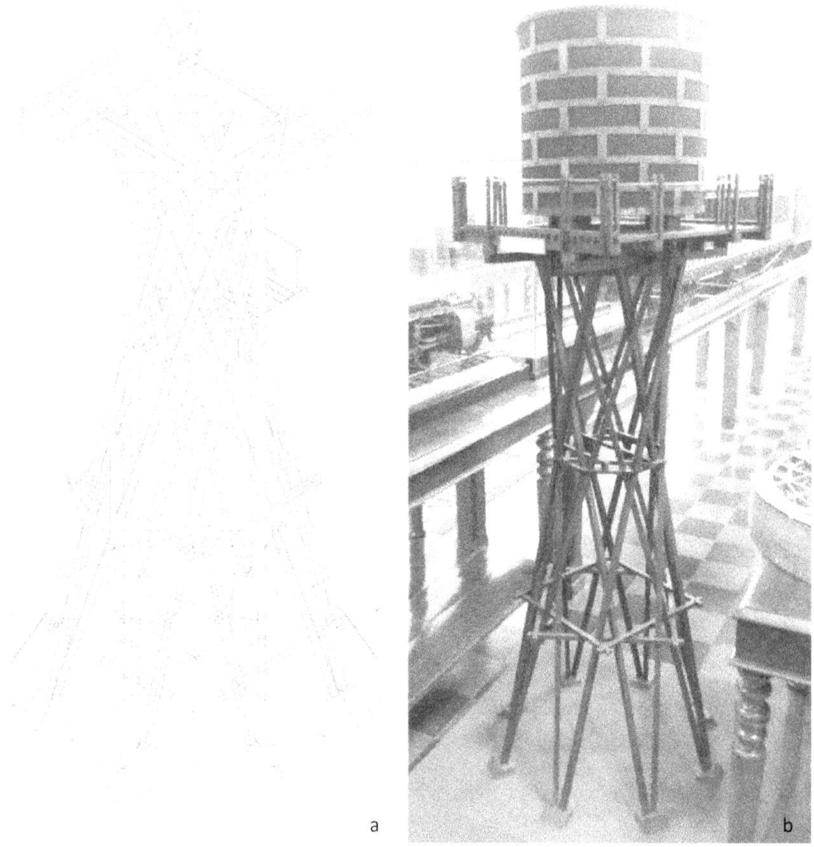

Fig. 5a: Axonometric view of the water tower based on working documentation (source: author); Fig. 5b: The model of the tower at the State Museum of Rail Transport in St. Petersburg (source: author)

Analysis of the struts

The tower struts are square in section, 150 x 150 mm. Each strut consists of three sections, linked by bolted halved-and-lapped joints. In the given structure, the necessity of rotating is compensated by the tower geometry: the tower is octahedral in plan, which can be represented as a system of two crosses, rotated 45° to one another (Fig. 6a). The structure of the first and the upper ring repeats the same logic of two superimposed rectangular systems, rotated at 45°: so, the struts and the elements of ring they are linked to are perpendicular in plan.

The tower design has a trick: the hyperboloid opens upwards (Fig. 6d). It was probably designed like this to get a wider supporting platform for the reservoir. To achieve this and to stay within the above described system of two superimposed rectangular grids, the upper octahedron has to be rotated at 22.5° and its vertices in plan need to coincide with the points of the first strut overlap (Fig. 6b). By a different arrangement (Fig. 6 c,e), the hyperboloid would not open upwards and, more importantly, each strut would only have four, not five, overlaps with the struts of the opposite direction, which might affect the stability of the structure. This shows the proportions of the tower are of critical importance.

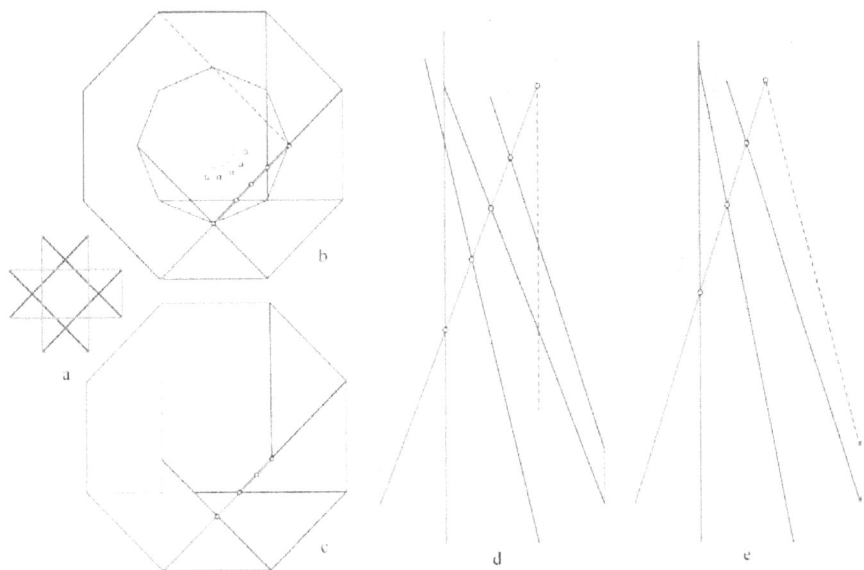

Fig. 6a: Octahedron as a system of two superimposed crosses; 6b: Upper ring is rotated at 22.5°; 6c: Upper ring is not rotated; 6d: Axonometric scheme of the tower with a rotated upper ring; 6e: Axonometric scheme of the tower with an upper ring parallel to the lower ring (source: author)

Analysis of the rings

The tower has three rings, and these rings are each designed differently. The lowest ring consists of two squares rotated at 45°. The second ring is in the narrowest part of the tower, just after the second strut overlap. It consists of eight specially tailored spacers, which direct and tie together four struts in their different orientations (Fig. 7a). It seems that spacers played an important role during the assembly: the second strut section ends up exactly after the second ring, so, in all probability, the strut pieces assembled of the first two sections were fixed in their correct positions with the help of ring elements. It means the elements could not be tailored by dimensions taken at the construction site, but rather that they were precisely calculated and prefabricated, and this precision was critical for the geometry of the structure. The resulting octahedron is strengthened by the structure that supports the ladder (Fig. 5a). The upper ring consists of four layers; each layer is a square, rotated at 45°. The struts are linked to the elements of the upper ring with a cross-lap joint (Fig. 7b): this weakens both elements less than a halved-and-lapped joint, and also transfers the stress better. However, the prefabrication of the incisions had to be done very precisely and would have been time-consuming both for the draughtsman and carpenter.

The geometric complexity of the tower elements leads one to assume that at least some parts of the structure were tested in the model. It is known that the models of all the timber structures published in the *Atlas of Timber Engineering Structures* were built at scales between 1:5 and 1:200 and exhibited at the Permanent all-USSR Building. This collection was later distributed among different museums and institutions. By chance, the water tower model was recently discovered at the State Museum of Rail Transport in St. Petersburg (Fig. 5b) – it has a mark of the workshop that made the models for the Exhibition, so the dating to 1931-33 is reliable.

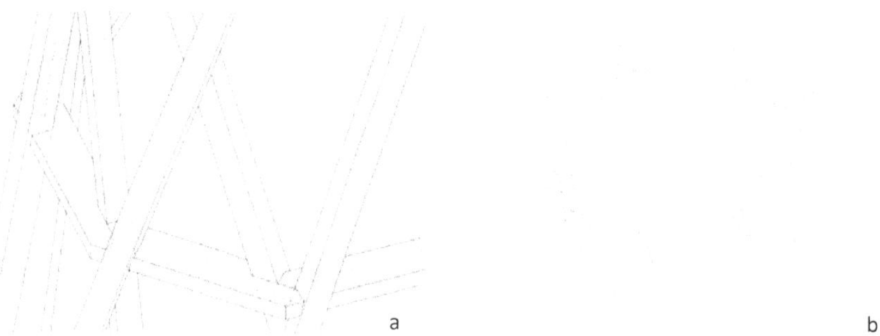

a b

Fig. 7a: The structure of the second ring (source: author); 7b: The upper part of the tower struts: inclined cross-lap joints (source: author)

Analysis of the strut overlaps

Each strut has five overlaps with the struts of the opposite direction (Fig. 6d); the upper, fifth overlap, is in reality rather implicit and lies within the upper ring. In plan, the angle between the struts at the first overlap is 90°, at the second 135°, at the third 180°, and at the forth 225°; thus the solution for the second and fourth overlap should be similar.

The problem of the strut overlap has influenced the structure of the plan (Fig. 8 a-d). If we superimpose the struts of the actual thickness (150 mm) on the symmetrical scheme (Fig. 8a), the third overlap, where the struts go parallel in plan, turns out to be problematic: one strut would need to be shifted 150 mm (Fig. 8b). If we were to do this, we would create a considerable distance between the struts, which might complicate the first and the second overlaps. The solution in this case would be to shift the radial axes: each axis is moved 83.5 mm from the centre (Fig. 8c). If the struts of the opposite direction are symmetrical along the axis, the geometrical intersection of the two struts is still too "deep" (Fig. 8e): it involves a considerable part of the strut section, and so if realised, might weaken the strut. The challenge is to reduce this intersection and at the same time to reserve enough overlap surface. This is done by the slight vertical and horizontal shift of the strut in plan, correspondingly 50 and 25 mm (Fig. 8 d,f). Pure geometric intersection of the struts gives a complicated section (Fig. 8 f,h). To avoid weakening the struts, the joint was designed in the form of a stretched birds mouth, which takes away only a small part of the section (Fig. 8 g,j).

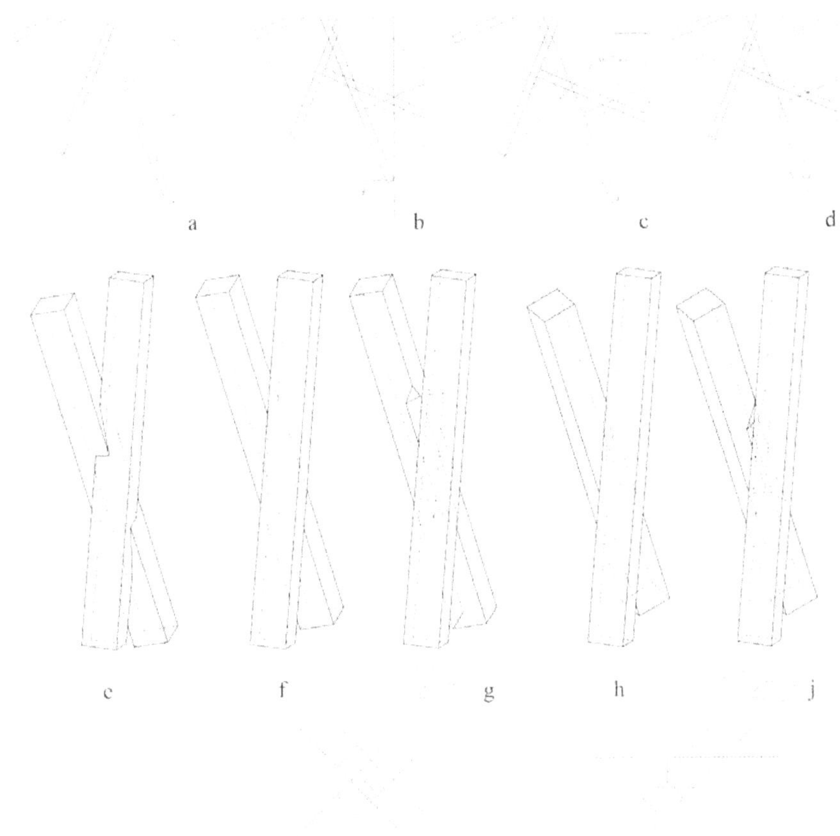

Fig. 8a–d: The consecutive steps of the plan development; 8 e,f: First strut overlap. Pure geometric intersection of the tower struts by the different plan versions; 8g: Its solution in wood: bird mouth joint; 8h: Second strut overlap. Pure geometric intersection; 8j: Its solution in wood (source: author)

A prototype for mass production?

The design was apparently intended for mass production and even recommended to "reduce the price of water supplies" [14]. However, several features cast doubt on this intention. As shown in the analysis of its details, the precision and variety of the joints used was very high and demanding, and so only prefabrication of the whole set made the construction feasible. In production terms, this would mean transportation of the whole structure and assembly on-site on the basis of detailed plans. Functionally, the proportions of water towers are dictated by reservoir volume, terrain and many other factors: this explains the variety of proportions in Shukhov's metal towers. Following this logic, the design should be adoptable and flexible; qualities that this structure has little of: for with every change in the proportions each overlap and each link of the strut to the ring would have to be redrawn. As we saw, the necessity of bending is compensated in this tower by its geometry: it means that with only 16 struts (8 x 2) the problem is eliminated. All these facts bring us to important conclusion: this unique design was not suited for mass production; it was rather a sophisticated attempt and the challenge of translating the structural principle to timber construction. But even in the form of an engineering exercise, it is very complete and elaborated, and despite the lack of flexibility, it represents an advanced engineering approach to design.

Conclusions

These two towers, built according to the same principles as their metal siblings, demonstrate the limits and advantages of the general method being applied to another construction material. As the two towers analysed are of different scales and purposes, they illustrate a range of different constructive solutions for the complex three-dimensional joints.

An important methodological conclusion, which refers directly to timber as a construction material, is revealed by the comparison of the two structures. The water tower was regarded as a prefabricated product. It should have been serially produced as a set of marked details, which would have required a high degree of precision in detailing and (with a high probability) mechanised production. The analysis of how the strut overlap joints influenced the structure in plan reveals that the engineers that designed the tower were thoroughly familiar with carpentry.

As the comparison between the measurements and calculations show, the cooling tower was built very precisely, which on first sight might seem to be a contradiction, as the inaccuracy of its details and the degree of freedom by elaboration of repetitive elements is relatively significant. We do know nothing of the tower constructor and assume that the project was realised by one of the qualified imprisoned engineers. He might have had a clear idea about the limited availability of qualified workers – and so the details of the tower were designed in such a way that general parameters were controlled and maintained and secondary details could withstand greater tolerances. In this case, the engineer foresaw the problem of the construction site and compensated for the lack of skilled workers with a sophisticated design approach and an elaborate design.

Shukhov's structures are being often characterised as "light" or "economic", meaning the reduced material use. However, their main characteristics are inherent steadiness, resulting from their double-curved form, reinforced by the network structure, and the uniformity of structural elements. As metal towers were constructed purely of standard profiles, so timber towers were assembled of conventional sections and blocks. And similarly, timber elements demanded precise processing, and the guarantee of their availability was higher than for special elements such as castings, cables or cover plates. At a time when Russian industry had to compensate for a lack of capacity and technology, these characteristics for structures were highly desirable. In this context, one can find the *raison d'être* for these sophisticated and demanding structures that were developed by Shukhov and those he inspired.

Acknowledgements

This research, financed by SNF (Swiss National Foundation), has been conducted under the aegis of the general D-A-CH Project: *Konstruktionswissen der frühen Moderne – Schuchovs Strategien des sparsamen Eisenbaus* [Construction knowledge of early modernism – Shukhov's strategies for economical iron structures], which unites the research groups of IDB ETH, Innsbruck University and Technical University Munich.

References

[1] M. Beckh, *Hyperbolische Stabwerke. Schuchovs Gittertürme als Wegweiser in den modernen Leichtbau.* Munich: Detail, 2012, pp. 20–21.

[2] F. Lynde. *Descriptive illustrated catalogue of the sixty-eight competitive designs for the Great Tower for London.* London: Industries, 1890, pp. 34–35.

[3] An experimental prototype was built at the Steam-boiler Factory of Alexander Bary before 1896, but neither photos, nor drawings of this tower are available. Thus, the water tower, built for the All-Russian Exhibition in Nizhny Novgorod in 1896 is reputed to be the first structure of this type. See also: Graefe R. et al., *Vladimir G. Šuchov 1853–1939: die Kunst der sparsamen Konstruktion.* Stuttgart: Deutsche Verlags-Anstalt, 1990.

[4] U. Hassler, E. Nozhova, "Paradoxien der Geschichtsschreibung zur Baukonstruktion: Eine kurze Untersuchung am Schabolovskaya-Sendeturm in Moskau", *Stahlbau* vol.82, No.7, 2013, pp. 483–493.

[5] The situation when in 2005 the NiGRES Tower in Nizhny-Novgorod luckily sustained the loss of sixteen struts, brilliantly illustrates the steadiness of hyperbolic structures.

[6] In 1800, forests that belonged to the works and were seen as a fuel resource, were assigned to Berg-Collegium, which was under control of the Finance Ministry.

[7] The first five-year nationwide centralised plan was accepted for the period 1929–1933.

[8] M. Seraphin, "On the origin of modern timber engineering" pp. 1845–1854, in Santiago Huerta (ed.), *Proceedings of the First Congress on Construction History. Madrid 2003*, Madrid: Instituto Juan Ferrera, 2003.

[9] K. Dubiaga,V. Vachurkin, *Spravochnik injenera-proektirovchika promichlennih soorijenij [A handbook on industrial structures for the designing engineer].* Moscow/Leningrad: ONTI, 1937, pp. 637–656.

[10] I. Mishustin, "Postrojka gradirni s bashnej novoj konstrukzii" [The Construction of the Cooling Tower with a Stack of Novel Design], *Stroitel` [The Builder]*, no. 1, 1938, pp. 52–57.

[11] E. Nozhova, "The wooden cooling tower". In: Cruz, Paulo J. Da Sousa (Ed.): *Structures and Architecture: Concepts, Applications and Challenges.* Leiden: Balkema, 2013, pp. 155–156, CD-ROM: pp. 519–526.

[12] Full calculations to be published in the PhD thesis of Ekaterina Nozhova.

[13] G. Karlsen, G. Zwingman, *Atlas inzhenernih derevjannih konstrukzij [Atlas of Timber Engineering Structures]*, Moscow/Leningrad: Gosstrojizdat, 1933, pp. 155–159.

[14] K. Dubiaga,V. Vachurkin, *Spravochnik injenera-proektirovchika promichlennih soorijenij [A handbook on industrial structures for the designing engineer].* Moscow/Leningrad: ONTI, 1937, pp. 657–668.

Old Ennisnag Bridge

Dermot O'Dwyer and Ronald Cox
Department of Civil, Structural and Environmental Engineering, Trinity College Dublin, Ireland

Summary

This paper begins with a brief synopsis of previous work on Irish medieval bridges, particularly the work of O'Keeffe and Simington on stone bridges, before detailing some recent discoveries.

The evidence for the forms of Irish medieval bridges and the techniques used in their construction comes from a variety of sources, including surviving structures, archaeological investigations and, in some limited cases, early texts. The paper presents a case study of the medieval bridge at Ennisnag. Ennisnag is approximately six miles south of Kilkenny on the old road from Kilkenny to Waterford. The bridge crosses the King's river close to where it flows into the river Nore.

The paper considers the situation of the bridge and relates the characteristics of the bridge to features found in other early medieval bridges. The bridge is one of very few bridges that retain evidence of the wattle centring used to form the barrel vault. This technique was also used to construct vaults in ecclesiastical buildings, castles and keeps. The link with ecclesiastical buildings is of particular importance given that many surviving medieval bridges are located close to monastic sites.

Introduction

Irish Stone Bridges: History and Heritage by Peter O'Keeffe and Tom Simington is the seminal book on Irish stone bridges [1]. The book presents an overview of masonry bridge construction in Ireland before dealing with a large number of important bridges in some detail. The authors discuss most of the medieval bridges in Ireland that had been identified when their book was written.

The main method of crossing rivers and streams was to do so at a ford. The Gaelic word for ford is *ath* and this word is incorporated in many Irish place names [1]. In many cases *ath* is qualified to indicate the type of ford, for example in Dublin in Gaelic is Baile Átha Cliatha, the town of the hurdle, or wicker, ford.

The Irish Annals contain references to early bridge builders but it is assumed that these were timber bridges [1]. Crannogs were a very common form of settlement in Ireland in ancient times and therefore the construction techniques necessary to build timber bridges across rivers would have been well understood.

The earliest archaeological evidence of an Irish bridge was found in 1993 beside Clonmacnoise, a major monastic site in Co. Offaly. The remains of this bridge, which crossed the river Shannon, comprised two parallel rows of oak posts, each over 50ft in length. Dendrochronology has been used to date these posts to circa 804 AD [2].

One of the best examples of an early timber bridge is contained in O'Kelly's paper on the remains of a timber bridge over the river Cashen in Co. Kerry [3]. Unfortunately, when this bridge was found, in 1961, radio carbon dating and dendrochonology was not as reliable as it is today and so the date of this bridge remains unknown.

There are some early clapper bridges in Ireland but they are relatively rare, and the most well known example near Louisburg in Co. Mayo is now believed to be a recent English import [4]. The presence of stone-arched bridges is usually assumed to post-date the arrival of the Normans (1169), or Cistercians (1148) [1]. (Cresy made a similar comment on bridges in Britain [5]). This is logical in many ways, although it must be remembered that there was a long tradition of building in stone that existed hundreds of years before the Normans. Ireland's many round towers and the early monastic settlements on Skellig Michael off the coast of Co. Kerry are among the best examples of pre-Norman stone construction.

The earliest Irish stone medieval bridges are thought to date from the late twelfth- or early thirteenth-century. Many other bridges can be dated back to the fourteenth and fifteenth centuries. The more significant bridges were built on well-established routes and close to traditional fording points. Thus, the bridge at Bennettsbridge, which was constructed in the fourteenth century, was constructed within about a half mile of the historic ford at Ballyredin. Medieval bridges over major rivers have tended to have been destroyed by flood or replaced by newer bridges so that few original structures remain.

One method of locating many of Ireland's medieval bridges, or medieval bridge sites, is via early maps. Ireland's earliest complete survey was the Down Survey that was carried out in the early 1650's [6]. This survey mapped the country at County, Barony and Parish level. The survey did not always record bridges where they were present, but in most cases the key crossings were recorded. The number of bridges recorded in the Down Survey is, in fact, relatively few, the majority being located in major towns and cities and on key routes.

The majority of Ireland's oldest surviving bridges are of moderate span, tend to have semicircular arch barrels, have a large pier to span ratio and have pointed cutwaters on the upstream face. There are, however, notable exceptions, and some of the very earliest bridges, such as the bridge at Trim in Co. Meath, have pointed arch profiles similar to some medieval English bridges. Irish medieval bridges generally have arch barrels that are not comprised of individual parallel ribs.

Although O'Keeffe and Simington's book captured the majority of medieval Irish bridges, there have been some discoveries since it was published in 1991. These more recent discoveries have a number of sources. During the recent spate of motorway building in Ireland many sites underwent detailed archaeological investigation. Other archaeological investigations were the result of flood relief schemes. It is now normally mandatory to employ an archaeologist on any major infrastructure scheme.

In addition to archaeological work, the increased interest in history and heritage has in some areas provided support for systematic surveys of the riverbanks that have identified bridges that have previously escaped attention. For example, a survey of the river Nore upstream of Kilkenny recently located a simple beam bridge on a tributary of the Nore (Fig. 1). This structure may not be old, but the bridge form is.

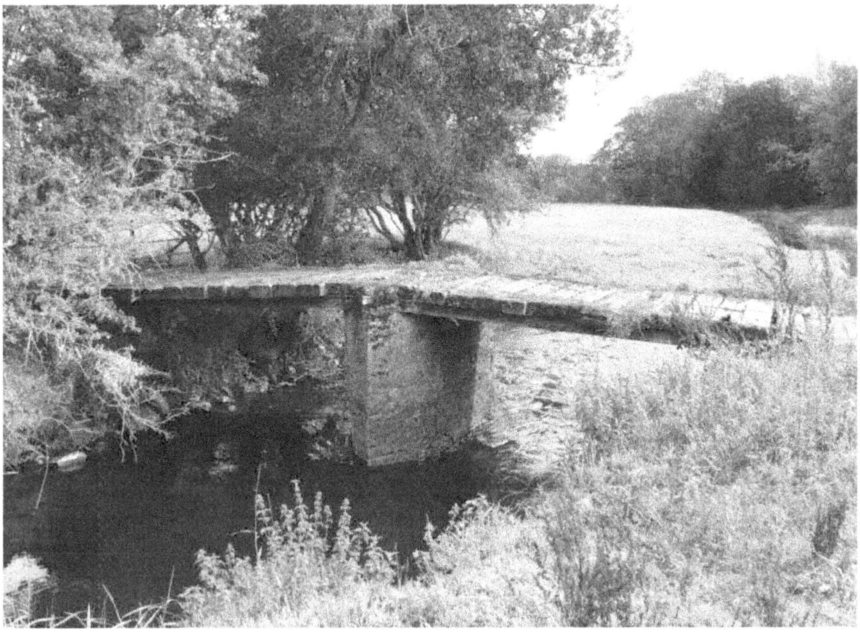

Fig. 1. A plank bridge - This ancient bridge form is often signalled in Irish place names by the Gaelic word clár, which means 'a plank', (e.g. Clarinbridge near Galway City [7]).

Bridges in the Kilkenny region

Kilkenny is a medieval city that predates the twelfth-century Norman invasion [8]. However, with the coming of the Normans its importance grew. Kilkenny is situated in the south east of Ireland on the river Nore and is approximately thirty miles north of the port of Waterford.

Ted Ruddock included a section on the bridges of this region of Ireland in his key work on arch bridges [9]. However, he concentrated on the group of architecturally refined bridges that were built on the Nore following the serious flood of 1763, which washed away or damaged all the significant medieval bridges on the river.. There are gripping accounts of bystanders venturing on to John's Bridge in Kilkenny to watch the floods that washed away Green's bridge, which is half a mile upstream. Unfortunately as the crowd stood on the bridge it too was washed away with the loss of fourteen lives [10].

In 2003 flood relief work was carried out by the Office of Public Works on the section of the river Nore running through Kilkenny. In conjunction with this work a very detailed archaeological investigation of the riverbed was carried out [10]. This investigation located and excavated the pre-1763 late-medieval bridges close to the sites of the present Green's and John's bridges. The excavations provided a detailed picture of the techniques used to construct the foundation of these bridges and unearthed a large number of metal shoes used to tip the timber piles [11].

The structures in these medieval foundations are very similar to the foundation of Essex Bridge, constructed in Dublin in 1676, and described by George Semple in his seminal book, *'On Building in Water'*, in which he describes the construction of its replacement between 1753 and 1755 [12].

The Kilkenny section of the river Nore has also undergone a series of detailed surveys carried out by the County Council's Heritage Officer. These surveys have identified a number of little-known bridges, such as the simple crossing shown in Figure 1. Previous authors have speculated that many early bridges were of this form, i.e. simple piers spanned by simple beams.

However, one group of bridges that has received relatively little attention to date are the bridges on the King's river, which is a tributary that flows into the river Nore approximately six miles south of Kilkenny. The King's river runs West-East and the Down Survey (1654-1656) maps show bridges over the river at Callan, Newtown, Kells and Ennisnag (Fig. 2a). Callan was a major town and Kells (Co. Kilkenny) was the site of a large Augustinian Priory, which was founded in the first decade of the thirteenth century [13]. Ennisnag is the crossing closest to the Nore and is situated beside the townland of Annamult (ford of the weathers in Gaelic [13]), which traditionally was the fording point of the ancient road heading south from Kilkenny towards Waterford [14]. The bridge in Callan was replaced in 1818, but the remaining three bridges were found to be worthy of study, not least because they were not discussed in any detail by O'Keeffe and Simington. This paper concentrates on the bridge at Ennisnag.

Ennisnag Bridge

The old bridge at Ennisnag over the King's river, which lies approximately six miles south of Kilkenny and lies on the original road between Waterford and Kilkenny, is located very close to the historical route from Dublin toward Cork and the South. This route from Dublin crossed the river Liffey at Kilcullen, continued to Leighlinbridge, where it crossed the river Barrow, proceeded towards Gowran, about six miles to the east of Kilkenny before crossing the river Nore at Bennettsbridge. Each of these crossings had a bridge in early medieval times and these are described in O'Keeffe and Simington.

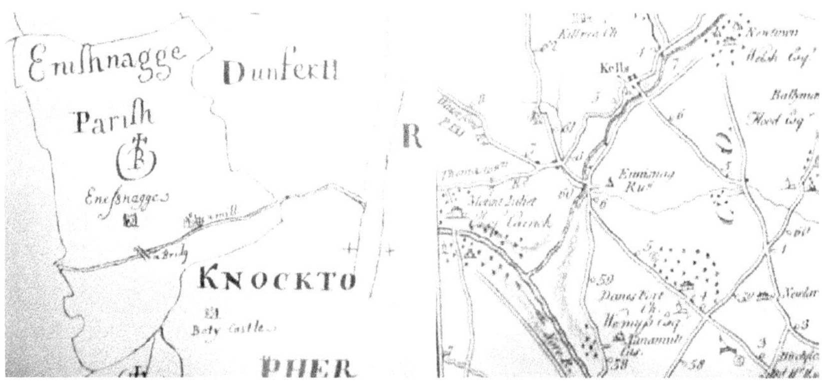

Fig. 2. (a) Down survey of 1656 showing a bridge in the parish of Ennisnag; (b) Taylor and Skinner's map showing the Kilkenny to Waterford (North down) road crossing the King's river at Ennisnag.

Ennisnag Bridge is clearly indicated, and named, on Moll's '*New Map of Ireland*' (1714), the earliest map of Ireland's road network. Taylor and Skinner's *Maps of the Roads of Ireland* (1778) shows the exact location of the bridge (Fig. 2b). Ennisnag is also a little less than four and a half miles from Jerpoint Abbey and lies on the road between Jerpoint and Kilkenny. Jerpoint Abbey was founded by the Cistercians in approximately 1158 [13]. The road from Ennisnag to Bennettsbridge continues alongside the river Nore to Kilkenny. Carrigan states that this old road was called "Boher-na-monnach", the Friar's road, and it was reputedly part of the highway that formerly ran from Jerpoint Abbey, by Annamult and

Grange-Kilree, to Kilkenny, the Bishop's see. Grange-Kilree had been gifted to the Abbey of Jerpoint by Felix Bishop of Ossory and this grant was officially confirmed in 1185 [13]. Thus there is clear evidence that Ennisnag Bridge lay on the route that the monks from Jerpoint took to their own lands in Grange-Kilree and on to Kilkenny.

Ennisnag has recently been completely bypassed by the M9 motorway to Waterford. In fact the medieval Ennisnag bridge had already been bypassed in 1827 when a three-span ashlar masonry arch bridge was constructed downstream of the original bridge.

The downstream face of the medieval bridge (Fig 3) shows a picturesque bridge that appears to date from the eighteenth century.

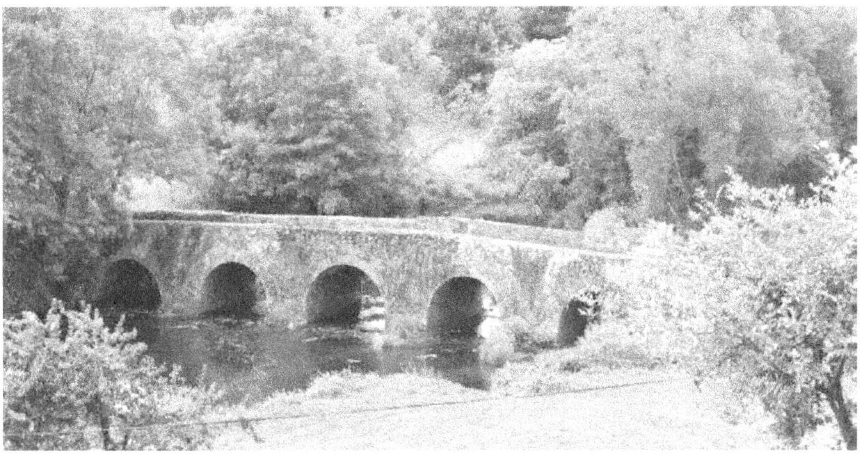

Fig. 3. Downstream view showing the eighteenth-century extension.

However, the upstream elevation indicates that this bridge is likely to be much older (Fig. 4). The bridge's upstream and downstream faces are markedly different because the bridge was later widened by 6ft 4in. The original bridge had a width of only 12ft, the view shown in Figure 3 being part of the widening. O'Keeffe and Simington's research indicates that bridges of this width are typically fifteenth century or earlier. The dimensions of the six spans, from left to right as they appear in Figure 4, with the pier widths in brackets are, 8ft (5ft 2in) 12ft (9ft 10in) 12ft 4in (7ft 3in) 14ft 6in (7ft 2in) 12ft 9in (7ft 6in) and 13ft 4in. Thus the ratio of the span widths to pier widths is approximately 2:1. The parapet walls are approximately 19in wide and the arch barrels are all roughly semicircular. There is a relieving arch of 7ft span in the causeway on the northern bank and a second relieving arch in the causeway on the southern bank. The short spans of the arches and the low span to pier ratios match O'Keeffe and Simington's observations of other Irish bridges that have been dated to the eleventh and twelfth centuries.

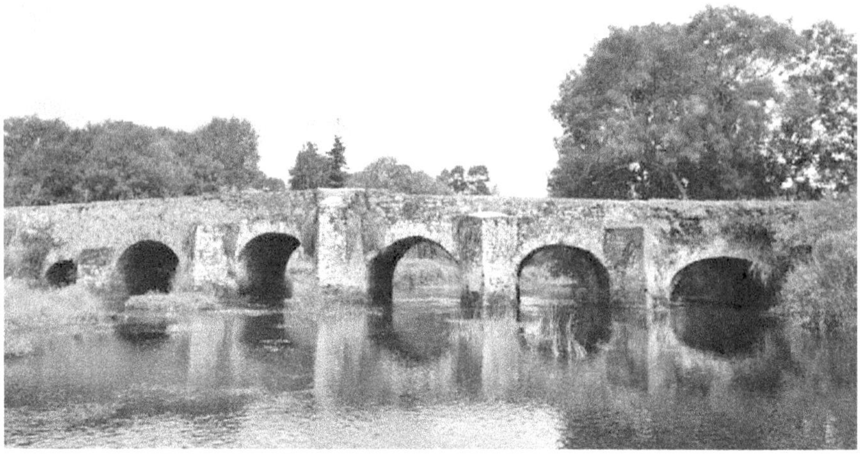

Fig. 4. Upstream view of the Old Ennisnag Bridge showing the six main spans of the original bridge.

Figure 5 shows the underside of one of the spans, in which the break in the arch barrel between the original structure and the later widening is clear. The photograph also shows the wattle marks in the mortar.

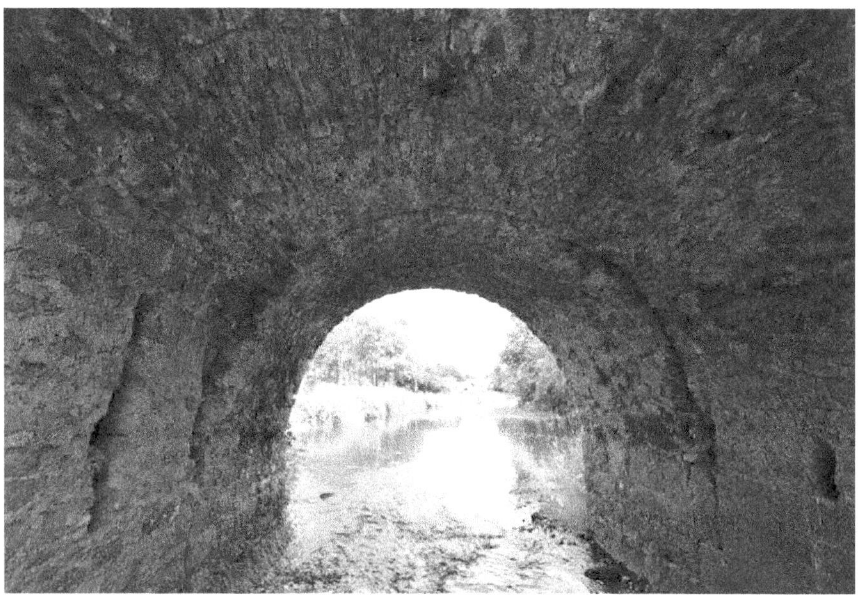

Fig. 5. Underside of an arch showing the discontinuity between the original bridge and the eighteenth-century extension and the wattle marking in the mortar on the intrados.

These marks arise from the use of wattles to cover the centring upon which the original arch was formed. Such wattle marks are often associated with vaulting in Irish castles constructed in the fifteenth and sixteenth centuries [15]. There is evidence that this construction technique dates back to the twelfth century or earlier [16]. However, similar wattle markings were found on pieces of a collapsed arch that came from the pre-1763 John's Bridge in Kilkenny [17]. Dendrochronology carried out on the piers of that late medieval bridge gave dates from the sixteenth and seventeenth century. Wattle markings are also to be found in the vaulting in Jerpoint Abbey [16].

There is a vaulted floor in Kilkenny Castle that was constructed using wattle centring (Fig 6). The castle was a major Norman stronghold and was constructed between 1192 and 1213. Although wattle centring is commonly found in historic Irish castles and churches it is very rare to find bridges that have retained this feature.

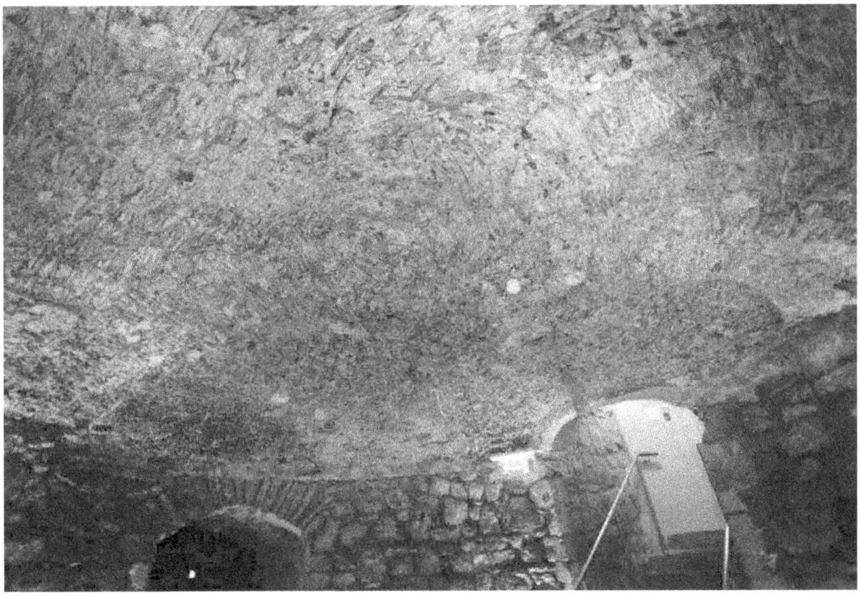

Fig. 6. Soffit of a floor vault in one of the original round corner towers of Kilkenny Castle.

Establishing the age of many early bridges is problematic, and Ennisnag is no exception. In the absence of written records, bridges may be dated by examining their construction details. The bridge at Ennisnag has a number of features that indicate that it may be dated to as early as the twelfth century. The narrowness of the bridge, the semicircular shape of the arches, the ratio of the widths of the piers to the spans, the triangular upstream cutwaters (one of which is extended to form a pedestrian refuge), the wattle marks and the use of field stone to construct the arch barrels, when taken together indicate that the original portions of the structure could date from as early as the twelfth century. A strong clue that the bridge is very early comes from the nature of the arch rings. Figure 7 shows that the arch rings of three of the main central spans have not been 'turned', that is the stones have not been laid in a radial pattern. O'Keeffe and Simington (1) identified this feature in bridges that date to the thirteenth century or earlier. This pattern of arch construction is also to be seen in pre-Norman Irish churches.

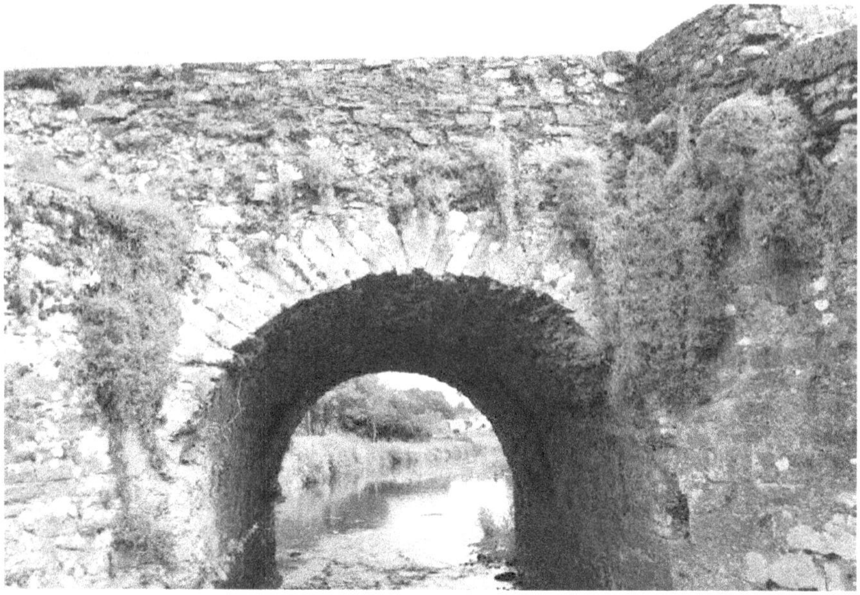

Fig. 7. Arch ring showing non-radial voussoirs.

Finding written evidence of the existence of a bridge prior to the Down Survey is a matter of luck. In the case of Ennisnag, such evidence has recently been identified. The Norman invasion of Ireland was precipitated when Diarmaid Macmurrough, the Irish King of Leinster, was deposed by a rival. Diarmaid crossed the Irish Sea seeking support. He ultimately came to an agreement with Richard de Clare, 2nd Earl of Pembroke, known as Strongbow. Strongbow supported Diarmaid's cause and eventually married Diarmaid's daughter Aoife and inherited the kingdom of Leinster. Strongbow and Aoife had a daughter, Isabel de Clare, who was married to William Marshal, 1st Earl of Pembroke, who at the end of his life acted as protector for the boy-king Henry III and regent for his kingdom.

William Marshal was responsible for building Kilkenny Castle and establishing the Norman section of Kilkenny City, known as English Town. Marshal also had extensive dealings with the Church and some of the documents describing his dealings with the clergy survive, or at least their contents have been recorded.

Bishop Hugh De Rous, who was the Bishop of Ossory from 1202-1218, exchanged lands with Earl William Marshal early during his term. The details of these land transfers are recorded in the Bishop's Charter and the Earl's acknowledgement. Transcripts of both documents, which were originally recorded in the *White book of the Diocese of Ossory*, were translated and published by the Reverend Graves in the *Journal of the Kilkenny Archaeological Society* (1859) [18]. Both documents locate a parcel of land with reference to the bridge of Ennisnag.

When translated the relevant section reads as follows:

> *......with the assent and consent of the Countess Isabella, my wife, I have now given and granted to the said Bishop and his successors eight carucates of land in places useful and suitable to him, viz., Bally-sly as three carucates, and Growin as four carucates, with the church benefice of the said land, and one carucate of land on one side of the bridge of Insnak towards the south (... pontis de Insnake versus meridiem...), to be possessed for ever....*

Of course it is possible that the bridge mentioned in the document was replaced later by the current structure. However, the very early features make it possible that sections of the Ennisnag bridge date from this period. This would place Ennisnag amongst the oldest bridges in Ireland.

As with most medieval bridges there are clear signs that the bridge had been repaired at various times. The spandrel wall on the upstream face of the most southern arch, which can be seen on the left-hand side of Figure 4, shows an arch ring with an elliptical profile. However, the shape of the arch expressed on the spandrel wall does not match the shape of the arch barrel behind.

The potential confusion that can be caused by repairs is perhaps best illustrated by the medieval bridge at Kells two miles upstream from Ennisnag (Fig 8). This bridge also appears on the Down Survey, its ratio of span to pier thickness, semi-circular arch barrels, pointed upstream cutwaters and narrow width also suggest that it is early medieval. However, the inset in the figure shows the bridge's down-stream aspect, which is clearly an extension, as at Ennisnag. In this case the extension has elliptical spans and each arch has approximately twice the span of the adjacent original structure.

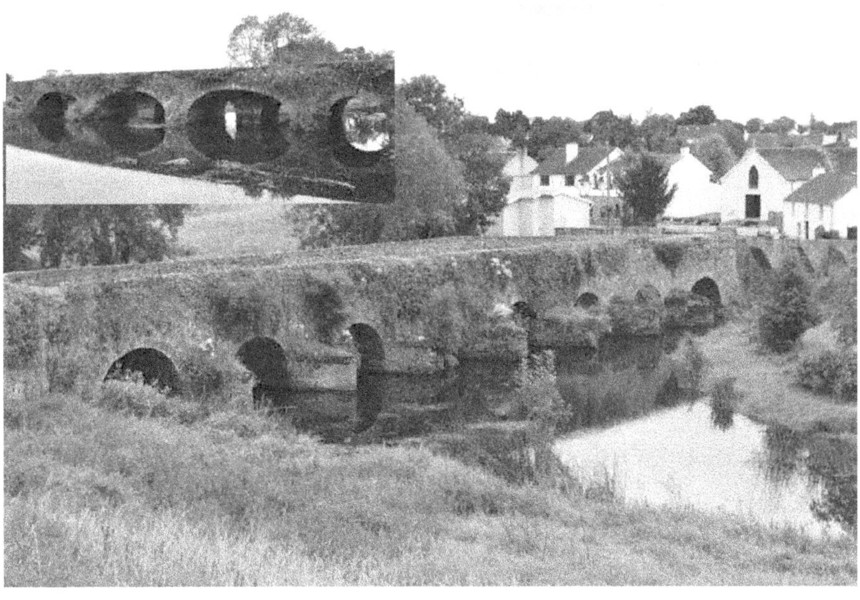

Fig. 8. Medieval bridge at Kells Co. Kilkenny. The insert shows the downstream aspect indicating that the bridge has been widened with an extension in a style typical of the late-eighteenth or early nineteenth century.

Later repairs are not the only source of confusion. The central pointed cutwater extends to the top of the parapet to form a pedestrian recess. This may have been essential before the bridge was widened. However, this feature is frequently misinterpreted as a toll point and the bridge is sometimes listed as a toll bridge, despite the fact that the road it carried was never a turnpike. This confusion may have arisen because sections of the Kilkenny to Waterford road south of Ennisnag were part of a turnpike road.

One interesting, and as yet unexplained, feature of Ennisnag Bridge is the evenly spaced recesses in the arch barrels (Figs 5 and 7). These recesses are at approximately the spacing and locations that one would expect to find corbelled stones for supporting centring. These recesses all start at the approximate springing point of the bridge and are likely to have been used to support centring. It is hoped that other examples of this feature may be found in Ireland, Britain or the Netherlands, as there is a possibility of a Flemish connection. Shortly after the Normans a group of Flemish natives settled in Kilkenny. They were described as tradesmen and mechanics and have traditionally been associated with the construction of the original medieval Great Bridge, now called Green's bridge, in Kilkenny [9].

Conclusion

Three old bridges on the King's river that were shown in the Down Survey appear to have survived from medieval times. This paper concentrated on the bridge at Ennisnag because recent written evidence shows that there has been a bridge on this site since the very early years of the thirteenth century. The bridge has many features that suggest that parts of the bridge could date from the twelfth or thirteenth century, though this is speculation. Most other medieval bridges in Co. Kilkenny were destroyed in historic floods on the river Nore. However, the King's river carries a significantly lesser flow and it is possible that bridges on this river may have survived. Ennisnag Bridge escaped renovation because it was bypassed in 1827 rather than replaced.

The fact that the bridge lies along the old route from Jerpoint Abbey to the Abbey's lands at Grange-Kilree and on to Kilkenny may be significant. Jerpoint Abbey was founded about 1158 and probably largely complete by 1185. The mention of pontage in the 1207 Charter of William Marshal granting privileges to the first generation of Anglo-Norman Kilkenny burgesses could be taken to indicate that bridges had been constructed in Kilkenny by that time [19].

The recesses in the arch barrels were probably related to the centring used in the construction of the bridge, but no other examples of such recesses have been found. The extensive wattle markings on the underside of the bridge are very rare and it is to be hoped that this feature can be preserved.

References

[1] Peter O'Keeffe and Tom Simington, *Irish Stone Bridges: History and Heritage*. Dublin: Irish Academic Press, Blackrock, Co., 1991.

[2] Ronald Cox and Michael Gould, *Ireland's Bridges*. Dublin: Wolfhound Press, 2003.

[3] Michael J. O'Kelly, "A Wooden Bridge on the Cashen River, Co. Kerry", *The Journal of the Royal Society of Antiquaries of Ireland*, vol.91, no.2, 1961, pp. 135-152.

[4] Ronald Cox and Philip Donald, *Ireland's Civil Engineering Heritage*. Cork: The Collins Press, 2013.

[5] Edward Cresy, *An Encyclopaedia of Civil Engineering: Historical, Theoretical and Practical*. London: Longman, Brown, Green and Longman, 1847 (Facsimile edition, London: Thomas Telford Limited, 2010).

[6] Micheál Ó Siochrú, "Down survey of Ireland Project", *History Ireland*, Issue 2 (March/April), vol.21, 2013.
[7] P.W. Joyce, *The Origin and History of Irish Names of Places*, Second Series, Dublin: McGlashan & Gill, 50 Upper Sackville Street, 1875.
[8] John Bradley, "Kilkenny Past and Present", *Irish Arts Review (2002-)*, vol.25, no.2, Summer, 2008, pp. 108-113.
[9] Ted Ruddock, *Arch bridges and their builders, 1735-1835*. Cambridge: Cambridge University Press, 1979.
[10] John Hogan, *Kilkenny; The Ancient City of Ossory, The seat of its Kings, the see of its Bishops and the site of its Cathedral*. Kilkenny: P.M. Eagan, 1884.
[11] Ian W. Doyle, "The Lost Bridge of Kilkenny City: John's Bridge, 1765-1910", *Archaeology Ireland*, vol.17, no.1, Spring, 2003, pp. 8-12.
[12] George Semple, *Treatise on Building in Water*. Dublin: J. A. Husband, 1776.
[13] William Carrigan C.C., *The History and Antiquities of the Dioceses of Ossory*. Dublin: Sealy Bryers & Walker, 1905.
[14] John Hogan, "Topographical and Historical Illustrations of the Suburbs of Kilkenny", *The Journal of the Kilkenny and South-East of Ireland Archaeological Society*, vol.3, no.2, 1861, pp. 355-387.
[15] Colin Rynne, *Industrial Ireland 1750-1930: An Archaeology*. Cork: Collins Press, 2006.
[16] Roger Stalley, *The Cistercian Monasteries of Ireland: An Account of the History, Art and Architecture of the White Minks in Ireland from 1142 to 1540*. London & Newhaven: Yale University Press, 1987.
[17] Ian Doyle et al., "River Nore (Kilkenny City) Drainage Scheme: Archaeological Excavations at John's Bridge, Kilkenny 2001-2003", on behalf of the Office of Public Works, Unpublished Report, Margaret Gowen & Co. Ltd.
[18] James Graves and John G.A. Prim, "The History, Architecture, and Antiquities of the City of Kilkenny", *The Journal of the Kilkenny and South-East of Ireland Archaeological Society*, vol.2, no.2, 1859, pp. 322-331.
[19] John Bradley and Tom Brett (Photography), *Treasures of Kilkenny, Charters and Civic Records of Kilkenny City*. Kilkenny: Kilkenny People Printing Ltd., 2003.

John Rennie's Dublin cast-iron trusses

Dermot O'Dwyer, Jay Cummins and Ronald Cox
Department of Civil, Structural and Environmental Engineering, Trinity College Dublin, Ireland

Summary

This paper examines the genesis of the early cast-iron trusses designed by John Rennie (1761-1821) for the Stack A Warehouse, in Dublin's Dockland. The Stack A building (now known as the CHQ building) is located beside George's Dock near Dublin's International Financial Services Centre. The building was erected in 1820 and the trusses used to support the roof are an early example of the use of iron for roof trusses. The building was designed as a fireproofed warehouse and was used as a bonded tobacco store for many years.

The Stack A trusses are of particular interest for two reasons. Firstly, they illustrate the move from timber to cast iron, shown by the adoption of joints that are derived from traditional timber construction. Rennie had designed and constructed similar warehouse buildings with both timber and iron roof trusses previous to his Dublin designs and a comparison with these earlier structures gives an insight into the development of the later cast-iron trusses. Secondly, the Dublin trusses were constructed before the widespread use of truss analysis techniques, such as those developed and publicised by Bow and others. However, static equilibrium was well understood by 1820. John Rennie had received lectures from John Robison (1739-1805) when he attended the University of Edinburgh and remained a friend and correspondent until Robison's death. Robison published on methods of calculating the internal forces in wooden frames and on the static analysis of masonry arches and there is evidence that Rennie collaborated with Robison on some of these works.

This paper investigates the extent to which Rennie is likely to have applied calculation in the design of the Dublin trusses. The analysis and design of Rennie's trusses is considered against the development of similar structures and the general advent of engineering calculation that was underway at that time.

Introduction

The roof trusses in the CHQ building (Fig. 1), formerly called the Stack A Warehouse, were designed by John Rennie (1761-1821) and constructed in 1820 [1]. The building is located beside George's Dock and Dublin's International Financial Services Centre. The building was one of a number of bonded warehouses constructed as part of the development of the enclosed Custom House Dock area in the early part of the nineteenth century.

The CHQ building was originally designed to store tobacco, wine and tea. It was designed as a fireproof building in which cast iron and wrought iron were used to replace all combustible elements including the roof and column members, windows and doors [2].

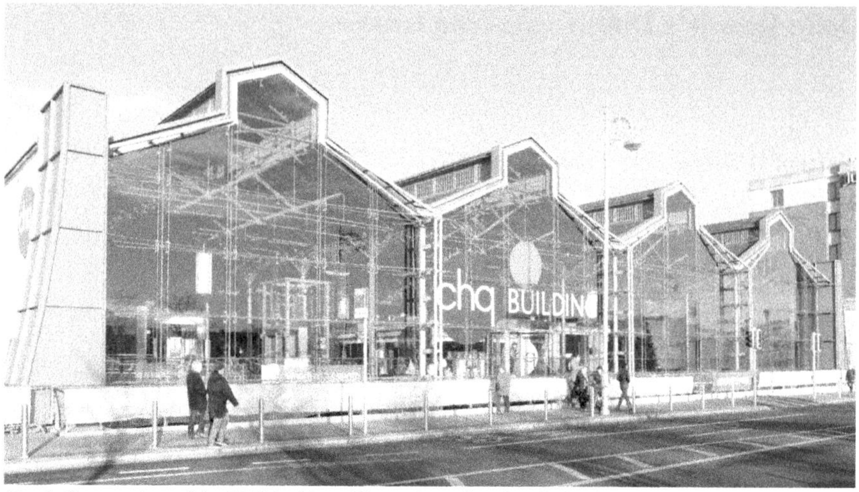

Fig. 1. Current view of the CHQ building. The modern glass facade displays the cast-iron trusses of the original roof system.

When originally constructed the building was approximately 500ft long and 160ft wide: a floor area of 80,000sqft. The structure comprised four bays supported internally by three rows of twenty-six cast-iron columns. The cast-iron columns, some of which serve as down pipes, rest on granite pillars that continue through the floor to the vaults below.

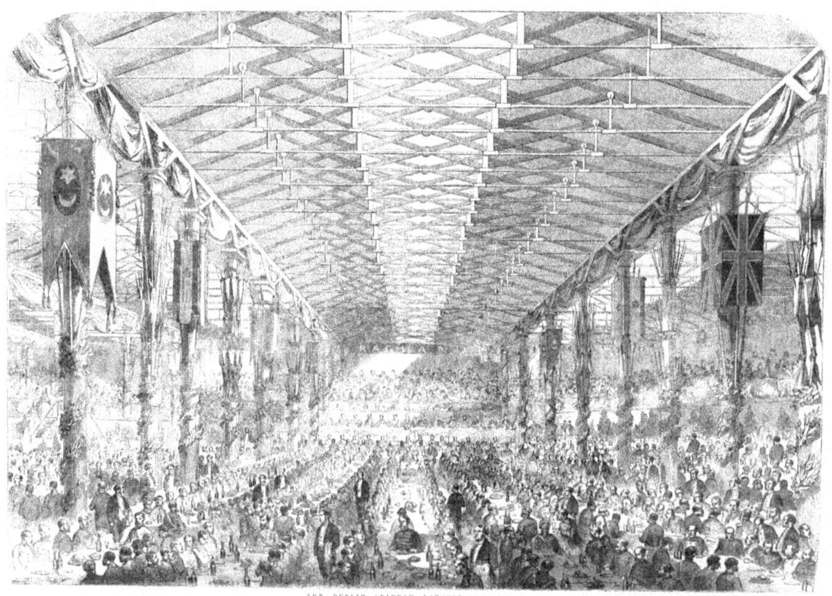

Fig. 2. The Crimean Banquet, the columns and beams being painted red, white and blue for the occasion [3].

These large dimensions facilitated its use in October 1856 as a venue for a celebratory banquet for Irish veterans of the Crimean war (Fig 2). The banquet hosted over 3,000 veterans and was a sumptuous affair with each guest receiving a quart of porter and a pint of port or sherry [2, 3].

The building underwent a number of alterations during its lifetime. In 1871 a mezzanine floor was installed and in 1884 one bay was removed to accommodate the expansion of Custom House Quay.

By the 1980s the building had fallen into a dilapidated state, but it was successfully revitalised between 2002 and 2007 by the architectural firm Michael Collins Associates acting on behalf of the Dublin Docklands Authority. Michael Collins Associates' work on the Stack A building was hailed as an exemplary conservation project and they were awarded the Royal Institute of Architects of Ireland (RIAI) best conservation/restoration project award in 2008 and the best conservation project at the Irish Planning Institute National Planning Awards in 2008. The refurbishment of the Stack A building cost approximately €40 million.

Description of roof truss system

The roof trusses and the structural system that supports them are largely unchanged and are the building's key visual feature. Figure 3 gives some indication of the length of the building and shows the cast-iron columns and the intermediate cast-iron arch-shaped beams that connect the column heads. The large horizontal glazed roof panels are not original.

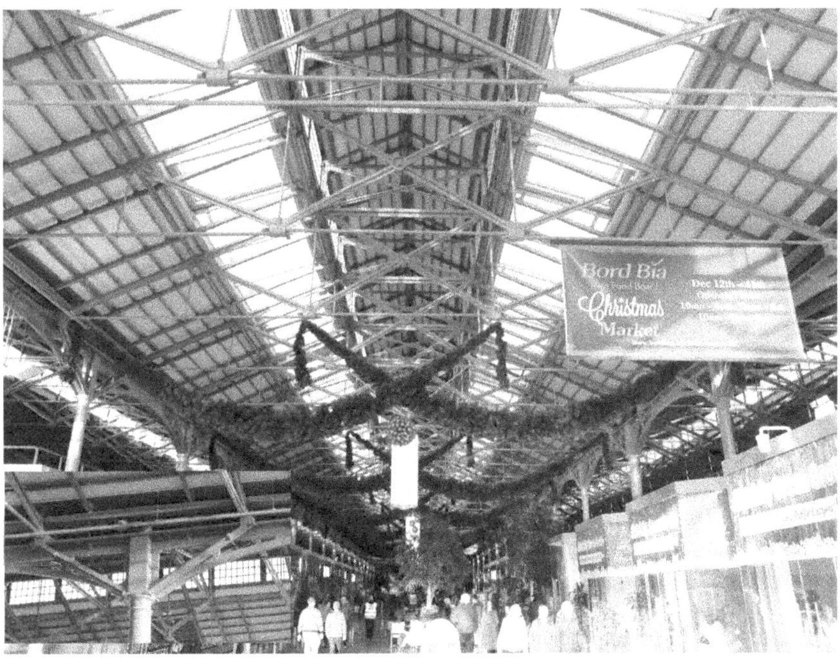

Fig. 3. Internal view of one of the four aisles in the Stack A building. The roof trusses are supported by both cast-iron beams and a strut-and-tie system, which are supported by cast-iron columns.

The insert in the bottom left hand corner of Figure 3 shows the column head detail. The detail shows that the trusses are supported by the cast-iron beams and by a strut-and-tie system. Casey [4] states that the strut-and-tie system was added later to deal with cracking in the cast- iron arches. If this is correct then the detailing of these ties suggests that they were added very shortly after the roof was completed. However, in the absence of firm evidence, the fact that the line of intersection of the rafter and the horizontal tie intersect almost directly above the strut does not support the hypothesis that the struts were a later addition.

Figure 4 shows a clear view of one half of the truss arrangement that is used. The trusses are fabricated from two member types, namely; round solid bars and cast-iron members that have cruciform sections. The round bars are intended to resist tensile loads.

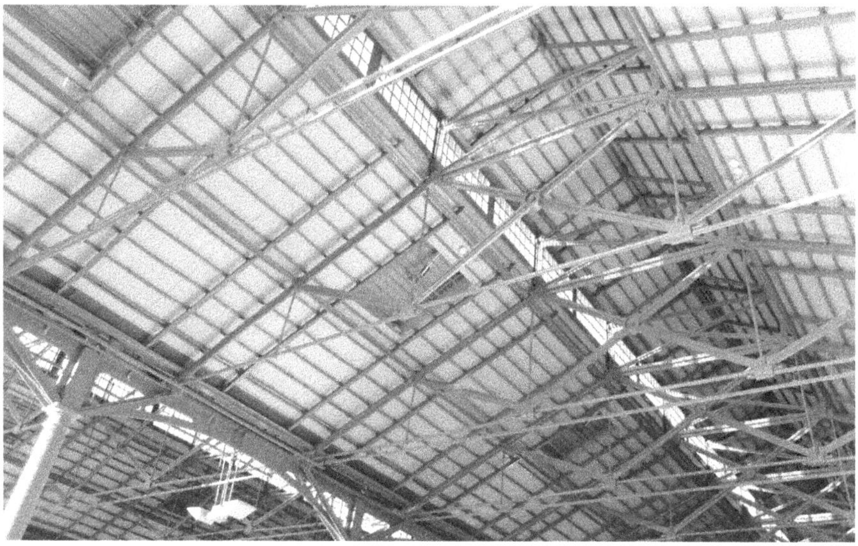

Fig. 4. Showing a clear view of one half of the standard truss. The truss is comprised of members with a solid round section and cast-iron members with a cruciform section.

This can be seen from the connection detail evident in Figure 5. This figure shows that the ends of the solid round bottom tension chord are joined using a tusk tendon joint. The use of such a typical timber connection highlights that this is an early iron roof. The vertical round members are also designed to resist tension only. This is clear from the detail used to secure the top of these members, which can be seen in the background of Figure 5.

Some of the joints between the cast-iron members are complex. Most of these joints have been developed so that the members meet at a pin. Although there are plenty of cast-iron roof members that are carrying loads in bending, the roof members have been detailed in such a manner that all the loads applied to the roof trusses are applied at the nodes of the truss. Reference will be made to these features later.

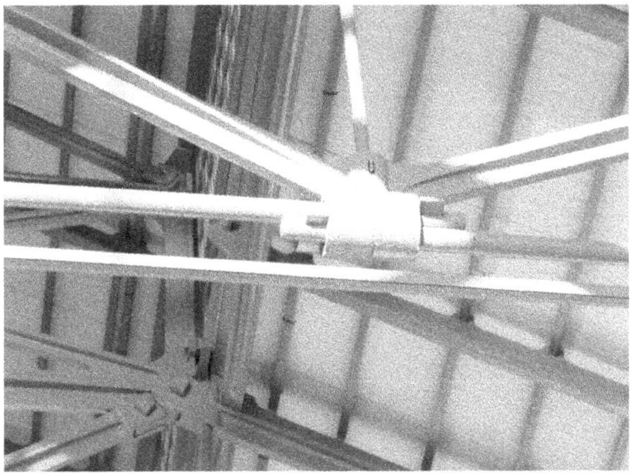

Fig. 5. The foreground shows the tusk tendon tension joints that prevent the solid circular joints from being pulled clear of the joint. In the background the level of attention that was paid to fabricating articulated joints may be seen.

Structural analysis of pin-jointed trusses

Many engineering historians take the 1840s as the period when engineers began to use static analysis to design statically determinate trusses [5]. Although all the essential theoretical understanding had been in place before and must have been well understood by analysts, such as Navier, it was Whipple and other engineers who began to publicise their methods. Culman's 1866 work *Die Graphische Statik* is often taken as the founding text of graphic statics and Maxwell's contemporaneous work of 1864 covered similar ground [5]. Finally in 1873 Henry Bow developed the methodology that was taught to engineering students until quite recently [7].

The 1840s are a logical starting point for static analysis because this was the decade in which railway construction boomed and which saw the construction of significant engineered large-span iron structures. However, this view is not strictly accurate. John Robison lectured natural philosophy in the University of Edinburgh from 1773 to 1805, the year in which he died. He was highly regarded as a physicist and wrote a number of works on mechanics and other topics, which comprised the substance of the course of lectures he delivered in the University [8]. He also wrote a number of articles on mechanics for the 3rd edition of the Encyclopaedia Britannica. This material was gathered, edited and published in 1822 as "*A System of Mechanical Philosophy, by John Robison LL.D*" by Sir David Brewster, who had been a student of Robison's. This work gives a very good overview of Robison's knowledge of structural analysis. The section on carpentry dealing with the analysis of pin-jointed frames demonstrates the use of graphic statics to calculate the forces in the members of a truss. Volume I of this work also contains chapters on strength of materials, roof construction and the design of arch bridges and centring [9].

Anyone who has taken an elementary course in statics and who reads Robinson will recognise his approach. His text is written for a student and he introduces concepts gradually using the same analogies that lecturers use when teaching the subject today. For example, he encourages the reader to identify whether a member is in tension by considering whether it could be replaced by a rope. He is clear that these techniques can be applied to identify the magnitude and sense, tension or compression, of the forces in the members and to calculate the support reactions. Robison is very clear that for analysis the joints should be considered free to rotate,

> *The joints are supposed perfectly flexible, or to be linked compass joints; the pin of which only keeps the pieces together when one or more pieces draws or pulls.*

Figure 6(a) below, from Robison [9], shows an elementary pin-jointed structure. The procedure proposed to calculate the forces in members A-C and A-D is as follows. Draw the line AE scaled to represent the magnitude of the applied load B. Then draw lines EG and EF parallel to the members A-C and A-D respectively. The lengths of the sides of the parallelogram AGEF give the magnitudes of the compressive forces in the two members: FA, AE and GA form a force polygon. Robison followed this simple abstract example with a less abstract example that might represent the queen post in a timber roof truss, Figure 6(b).

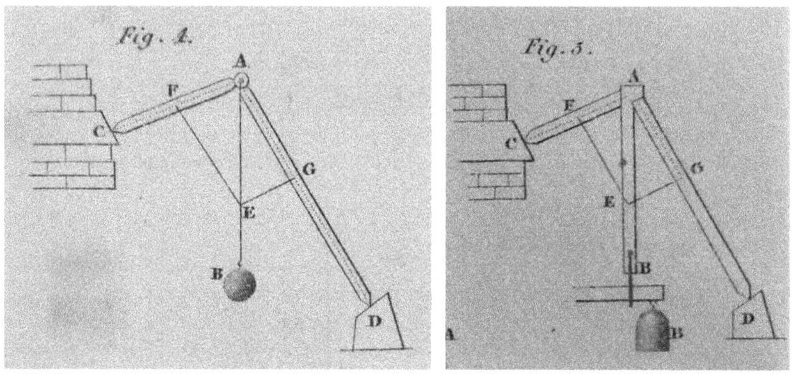

(a) (b)

Fig. 6. (a) Shows the use of a graphical method to calculate the forces in members AC and AD. (b) repeats this analysis but relates the theoretical analysis in (a) to a real analysis of the forces in the members at the top of a queen post [9].

His treatment of pin-jointed trusses developed to a general method that could be applied to frames. The construction is less complex than it appears but is less elegant, or obvious, than the graphic statics procedures developed by Bow.

Robison was fully aware of the key previous work on the analysis of structures ranging from beams to arches. His treatise also includes detailed analyses of beam bending and his writings on strength of materials include tables of the tensile and compressive strength of common materials. Robison is one of many engineers and mathematicians, such as Peter Barlow and Thomas Young, who were starting to investigate the tensile and crushing strength of materials. It is difficult at a distance to identify whether the availability of tables of material strengths gave impetus to the analysis of structures or whether the analysis of structures required accurate material strengths. It may be that both evolved simultaneously. However, it is clear that John Robison had knowledge of both and it is clear that this information was shared with John Rennie.

John Rennie and Robison

The work of Robison might be little more than a footnote in this paper: a contemporary but little known body of work on structural mechanics. The majority of English engineers of Rennie's period had very little higher education and many had a distrust of mathematicians and theoreticians. This was not the case with John Rennie.

John Rennie was born in Phantassie in East Linton in Scotland in 1761. He was the youngest of nine children and his father was a moderately prosperous tenant farmer. At the age of twelve he left the local school and became apprentice to Andrew Meikle a renowned millwright who was a tenant on the Phantassie estate. Rennie worked with Meikle for two years before it was decided that he should continue his studies at the public school in Dunbar. He was an outstanding student and by coincidence his academic prowess was recorded in David Loch's *Essays on the trade and Commerce, Manufactories and Fisheries of Scotland* [10]. Rennie then returned to work for Meikle until, when he was eighteen, he set up in business on his own. He later supported himself during his three years, 1780-1783, at Edinburgh University where he studied under Robison. He and Robison became friends and Rennie visited and corresponded with Robison for the rest of his life. There is substantial evidence that Rennie collaborated with Robison on the entries that Robison wrote for the Encyclopaedia Britannica on arches, carpentry, roofs, water-works, on the resistance of fluids and the flow of water in channels [10, 11]. This would have made sense in that, by this time, Rennie was one of the leading engineers in Britain [12] and his input as an experienced practitioner, who also had a firm knowledge of theory, would have been of great assistance to Robison.

Therefore, given Rennie's involvement with Robison, down to his collaboration in the very article on truss analysis, it is certain that Rennie was capable of analysing simple statically determinate roof trusses. A further argument that he did analyse some of his structures can be inferred from the fact that he constructed a testing machine. His son, George Rennie, carried out an extensive series of material tests [13]. George's remark in the opening paragraph,

> *The knowledge of the properties of bodies, which come more immediately under our observation, is so instrumental to the progress of science, that any approximation to it deserves our serious attention. The passage over a deep and rapid river, the construction of a great and noble edifice, or the combination of a more complicated piece of mechanism, are arts so peculiarly subservient to the application of these principles, that we cannot be said to proceed with safety and certainty, until we have assigned their just limits.*

makes the purpose of such experiments clear.

Fireproof construction

Rennie's use of cast iron in the Stack A building must also be seen in the context of fireproof construction. Rennie's career as a consulting engineer had been launched by his work on the installation of one of Bolton and Watt's steam engines in the Albion Mills. This enterprise was an outstanding success that established Rennie. However, a few years later, in 1791, the Albion Mill was destroyed in a fire. This was one of many similar fires that occurred in this period. Fires in multi-storey mills frequently resulted in the total loss of the building and the increased frequency of mill fires led to increased insurance costs.

William Strutt was a progressive young man who was acquainted with members of the Lunar Society, including Boulton and Watt. More importantly he was a member of a mill-owning family and had an engineering bent. In 1781 the Strutt family lost its Nottingham mill to fire and this was followed by the loss of its Darley Abbey mill in 1788. In response William Strutt used fireproof construction when he built their new six-storey mill at Derby and their four- storey warehouse at Milford. Both buildings were completed in 1793. In 1795 he built a six- storey mill at Belpar in which he used lightweight jack-arch floors. Strutt's buildings used cast-iron columns and timber beams, which were fire-proofed [5].

Strutt was followed by Charles Bage who was the first to use cast-iron beams at Castle Foregate Mill in Shrewsbury, completed in 1797. Bage was also the first to use calculation to size the cast-iron columns and beams he used. Boulton and Watt's seven-storey fireproof mill in Salford, completed in 1801, was the first to use circular hollow columns [5].

Rennie's earlier warehouses

Stack A was Rennie's last warehouse and it represents the logical culmination of the warehouse buildings he had constructed in the West India Docks. In 1807 Rennie had written to the Royal Navy promoting the use of fireproof cast-iron warehouses [2].

> *"...upon the plan of the cotton and flax mills lately erected at Derby, Leeds, Manchester and Glasgow...."*

He used cast-iron columns in the "skin floor", these columns being formed from cruciform sections and branched towards their top to provide multiple support points to the beams they carried [14]. This pattern may have prompted Rennie's use of the strut-and-tie arrangement in the Stack A columns.

In 1813 Rennie used a cast-iron king post truss system supported by hollow cylindrical cast- iron columns in the fireproof Rum Quay Shed. The Rum Field Sheds, which were being constructed at the same time, used wrought-iron roof trusses to a patent design by Thomas Pearsall. However, this roof collapsed and was rebuilt in timber. As a result of the problems encountered with the Rum Field Sheds, Rennie's plans to construct the new Rum Warehouses in 1817 with an iron roof were refused and he had to redesign the roof as a timber queen post truss. The Stack A building is clearly a development of the Rum Warehouses with the timber queen post truss replaced by a similar cast-iron truss with braced diagonals in the centre panel.

Stack A re-examined

Given Rennie's close relationship with Robison it would be surprising if Rennie had not applied Robison's techniques to analyse the roof trusses in the Dublin docks. This is particularly the case given the cost of cast iron at the time. The cost of cast iron would have provided an incentive to ensure that the material was used as efficiently as possible.

When one re-examines the structure in the knowledge of Rennie's background in mechanics some details appear more significant. For example, consider the effort that has been extended to make the joints in the truss pinned (Fig. 4). The shapes chosen for all the beam elements in the roof that resist bending moments are also significant. For example, consider the shape of the cast-iron beams that support the stringers. Figure 7 shows that the flanges of these members have a fish-bellied shape. In a similar manner, although difficult to see from a photograph, the stringers vary in depth.

Figure 7 shows the stringers, which are supported by beams that span parallel to the roof trusses. These beams in turn bear on beams that span between the nodes of adjacent trusses. All these members have sections that vary along their span. The variation in the depth of the stringers can be inferred from the variation in the depth of the shadows they cast.

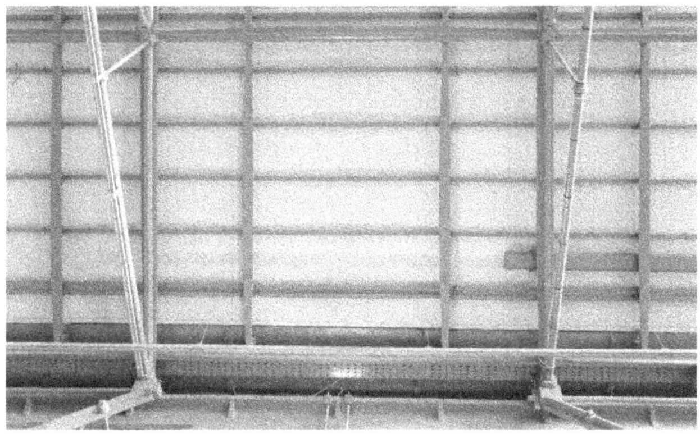

Fig. 7. Stringers and cross beams.

Figure 8 shows a dimensioned drawing of one of the trusses in the Stack A building. The upper section comprising the apex of the roof and the roof windows, members 21 – 23, can be considered as separate from the main structure. Most of the members of the truss, 1 – 6 and 15 – 20, are statically determinate and could have been analysed using Robison's techniques. However, the inclusion of the centre X brace, members 8 – 12, makes the central section of the truss impossible to analyse using statics.

It might seem surprising that the central section of the truss is statically indeterminate. However, the pattern that Rennie adopted is logical. If the central diagonal members, 9 – 13, are omitted then the truss has the same form as the queen post trusses Rennie used in the Rum Quay Warehouses. If the roof truss had been constructed in timber, with a single solid horizontal tie and with rigid joints connecting the queen posts, 7 and 14, to members 8 and 11, then the truss would have been stable. However, because cast iron was used and much effort was taken to make the joints hinged it was logical to X-brace the centre panel. These extra members were formed to resist compressive loads because Rennie could not have been certain what loads they would carry.

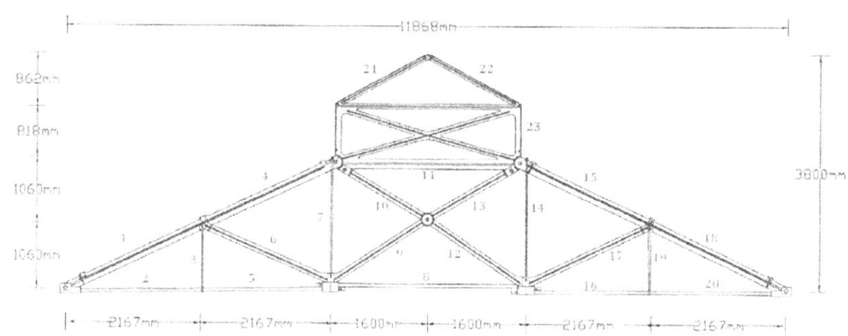

Fig. 8. Stack A truss [15].

Recent analyses of the trusses suggest that the member sizes of the statically determinate members are broadly proportionate to the loads they carry.

Conclusions

The roof of the CHQ building, formerly known as Stack A, is an early example of a cast-iron roof truss. The roof contains some structural details that highlight how its form developed from earlier timber constructions.

Although the use of graphic statics for the analysis of statically determinate trusses is generally taken to date from the 1840s it is clear that John Robison had previously developed and published a methodology for the analysis of statically determinate structures. There is clear evidence that John Rennie, who had been a student of Robison's, was consulted by Robison when he wrote some of his articles for the Encyclopaedia Britannica. Therefore it is certain that Rennie must have been at the very least familiar with Robison's methods.

An examination of the details of the Stack A roof truss supports the hypothesis that Rennie analysed the truss. It is clear that Rennie ensured that the loads applied to the main roof trusses were applied at the nodes of the truss. The joints in the truss were fabricated as hinges, as per the analysis method of Robison. The sectional properties of the beam elements vary along the members in a manner that shows that Rennie understood how the internal forces in a beam varied along the span.

References

[1] Ronald Cox and Philip Donald, *Ireland's Civil Engineering Heritage*. Cork: The Collins Press, 2013.

[2] D.J. Griffin, *Stack A, Custom House Docks; A Brief Architectural History*. Dublin: The Irish Architectural Archive, 1995.

[3] "The Dublin Crimean Banquet". *The Illustrated London News*, Nov. 8th, 1856, pp. 483-484.

[4] Christine Casey, *Dublin: The City within the Grand and Royal Canals and the Circular Road with the Phoenix Park, Buildings of Ireland*. New Haven & London: Yale University Press, 2005.

[5] B. Addis, *Building: 3000 Years of Design Engineering and Construction*. London: Phaidon Press, 2007.

[6] K. Kurrer, *The History of the Theory of Structures: From Arch Analysis to Computational Mathematics*. Berlin: Ernst & Sohn, 2008.

[7] S. Timoshenko, *History of Strength of Materials*. Dover Publication, 1983.

[8] J. Robison, *Elements of Mechanical Philosophy: being the substance of a course of lectures on that science by John Robison*, Edinburgh: Printed for Archibald Constable & Co., 1804.

[9] J. Robison, *A system of Mechanical Philosophy, by John Robison LL. D., with notes by David Brewster LL.D. in four volumes and a volume of plates*. London: Printed for John Murray, 1822.

[10] C.T.G. Boucher, *John Rennie 1761-1821: The life and work of a great engineer*. Manchester: Manchester University Press, 1963.

[11] S. Smiles, *Lives of the Engineers (Volume II)*, London: John Murray, 1904.

[12] A. Skempton, *A Biographical Dictionary of Civil Engineers in Great Britain and Ireland, Volume 1*, London: Thomas Telford, 2002.

[13] G. Rennie, "Account of Experiments Made on the Strength of Materials", *Philosophical Transactions of the Royal Society of London*, vol.108, 1818, pp 118-136.

[14] R.J. Mainstone, *Developments in Structural Form*, 3rd ed., Oxford: Architectural Press, 2001.

[15] Jay Cummins, *A Structural and Historical Analysis of the Stack A Warehouse Roof Truss*, Final year Dissertation, Department of Civil, Structural & Environmental Engineering, Trinity College Dublin, 2012.

Timber supply in colonial China, 1840-1940

Yiting Pan and James W.P. Campbell
Department of Architecture, University of Cambridge, UK

Introduction

After the Opium Wars in the mid-19th century, the British brought many of the ideas of the Industrial Revolution to Imperial China. The Treaty Ports, as the intersection of Chinese and Western cultures played a crucial role in introducing Western technology. This article shows that, perhaps surprisingly, the supply of timber provided the impetus for the wider adoption of western forms of carpentry construction such as purlin roofs with king-post trusses. It also illustrates the story of how the competition between timber suppliers ultimately led to Chinese nationalist back lash. This article brings together four sources for the first time: (a) the Chinese literature and visual records for Chinese timber trades; (b) the accounts on forestry published in the early 20th century by R. Rosenbluth (1912), N. Shaw (1914) and Y. Tang (1935); (c) early articles from *the China Year Book* (1912-1939), etc.; (d) and Chinese building construction manuals published in the late 19th and early 20th centuries.

The last few decades since 1985 have witnessed an increasing interest in the architecture of the Treaty Ports. Although general works on the appearance of the colonial buildings in China are abundant, how these buildings were constructed — and the relationship between Chinese and Western techniques— remains largely unexplored. The combined construction methods, which marked the one hundred years' Chinese building history under Western influence, had its origins in the two parallel systems: the Western and Chinese building construction traditions. This article concentrates on one aspect of this broader question: the relationship between the supply of a particularly important building material, timber, and its use in building construction.

Chinese forest trees and the native timber trades

Botanical, zoological and archaeological studies suggest that much of both Northern China (until 500AD) and Southern China (until 1300AD) was originally covered with great forests. [1] The fact that China in the 19th century had been reduced to a condition of deforestation (Fig.1) unequalled by that of any other great nation can probably be attributed to three factors: (1) The structural significance of timber: the importance of timber in construction can be traced back to the Qin Dynasty (221-206BC), when the small seal script for "structure" or "構" started to include a left-hand character " 木 ", standing for wood, and replaced the previous oracle bone script and bronze scrip ([illegible]). The great passion for timber can be seen from the extraordinary pursuit of structural timberwork in the Tang Dynasty (618-907) and the Song Dynasty (960-1279), as well as of furniture making in the Ming Dynasty (1368-1644); and it is also revealed in the accounts of the gathering of "Royal Lumber" for the Imperial Palaces in the Ming and Qing Dynasties. (2) Attitude and practice: the traditional collective belief in *Fengshui*, the balance of man and nature, and the awareness of the ecological benefits, usually yielded to more utilitarian attitudes in China. The methods of felling and rafting were wasteful (accounting for a loss of 30-35%), while replanting was rare and largely ineffective.[2] (3) Political reasons: Mr. P. C. King, former President of Tsing Hua

University, attributed the present scarcity of timber in China to 'the decay of the Chinese feudal system in the third century before the Christian era, the laissez-faire policy of the Government, and the numerous internal disturbances from which China has periodically suffered', and his view was considered at the time 'most deserving of attention among a mass of facts and theories'[3].

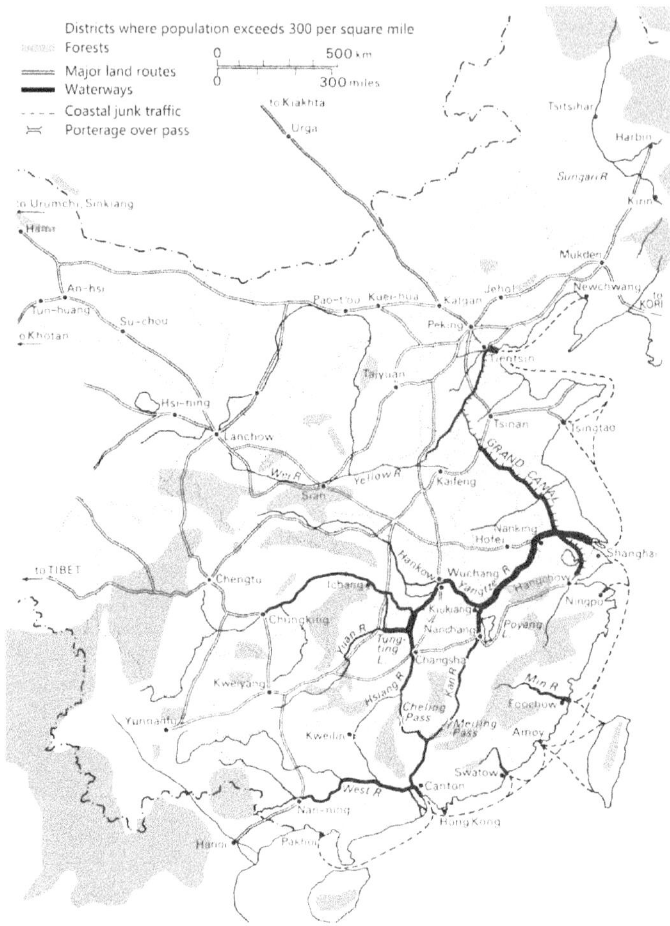

Fig. 1. Map of current forests and local communications in 19th century[4].

Whatever the causes, the direct impact on the building history is that by the 19th century good construction timber had become a scarce commodity, and the size of available trees was commonly small. Felling usually took place in the autumn and winter. Choosing an auspicious date for felling was an important issue in geomancy or *Fengshui*[5]. Trees were usually cut down with axes or saws, sometimes assisted by more people on both sides pulling on ropes tied to the saw when sawing a larger tree. In some areas, the trees were only cut half through, and left for the first gale do the rest of the work[6]. Once felled, branches were chopped off with axes, saws, knives, or adzes. A wedge was driven into the log near the centre of gravity to act as a handle, allowing a man to carry it on his shoulders and walk with it for several hours; alternatively, a whole tree might be transported by a team using ropes and carrying poles[7]. In mountain regions, a system of temporary transport structure (*jiaxiang* [架厢]), was usually constructed, through which the logs were dragged to the water head[8] (Fig.2). After the logs were gathered near the river, the names of the workshop and the mountain [店号山号] from which they came were marked on the log ends[9].The rafts were constructed in a customary manner. The augur-hands made a hole in the butt of each tree, and tied the trees to a perpendicular hard wood stick with assembling ropes made of *maozhu* (mao bamboo/*phyllostachys edulis*)[10] (the towing cables were also made of the same material). Then the rafts were floated down a convenient town at a point where the depth of the river increased. They were then rebuilt into bigger rafts to be sent down river to the principal timber markets. Contemporary illustrations show that the *Shanmu* (Chinese fir/cunninghamia lanceolata) logs that made up these huge rafts rafts were so buoyant that they could support houses and even layers of earth, allowing vegetables to be grown and pigs to be kept on board[11]. Over the months of travel, the rafts became floating farmsteads. Bamboo was also an important material. Its principal use was in scaffolding, although in some areas it was used as a substitute for wood in timber construction.

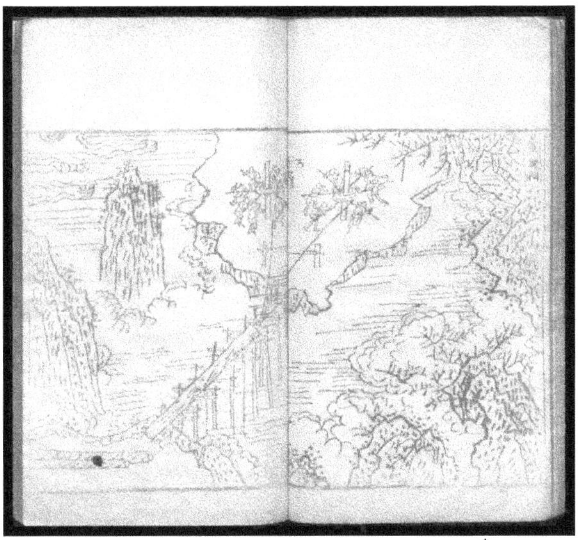

Fig. 2. "Tianche yuejian", or "sky vehicle over creek", in Xicha huicao, 16th century[12].

Because of the great difficulties of obtaining and transporting timber, an extraordinary trade network developed throughout China in the late 19th century. The timber guilds were particularly significant in the early 20th century, uniting individual merchants to compete in the new markets in open ports against the tide of imported woods. The actual practice of the timber trades varied in different regions of China. The common practice was that the raft merchants bought timber from the local villagers who paid some form of a Government tax each year[13]. Raft merchants only transported the raw logs down to the nearest markets, where they were sold on to the members of the wood merchants' guilds who carried out the selling down the line[14].

Timber was measured in *chi'h* [尺], or the Chinese foot, and a number of species were used. *Shanmu* dominated the native timber market. It was commonly supplied in thick logs or long poles with bark removed, and ranged from 0.4 to 1.5 *chi'h* in diameter. *Songmu* (Chinese pine/ *pinus massoniana*) commonly ranged from 11 to 30 *chi'h* in length, with smaller ones supplied in long logs and larger ones in sawn planks. Both *Shanmu* and *Songmu* were used in building and shipbuilding. *Bomu* (Chinese weeping cypress/*ping juniper*) was normally around 1 *chi'h* in diameter and 50 to 60 *chi'h* in length. *Bomu* was supplied mainly for the construction of temples, furniture and coffin making.[15] The price was set through a complicated calculation system, involving the measurement of the circumference at different points for different kind of woods and the guaranteed length of the log[16] (Table.1). The measure of length varied at different open ports and again with different trades[17]. *Hankow* was the largest timber market in colonial China[18]. Timber merchant's *chi'h* equalled approximately 13.3 English inches (338mm) in *Hankow*. The timber merchant's ruler was called a *tanchi* [滩尺]. There were two common types of *tanchi*: one (made of wood) was for measuring the length of the log; the other (made of bamboo) for measuring the circumference of the log. The latter could be subdivided into ten types for different discounted values of *chi'h*, named "由中人工大王主井羊非" respectively. Therefore, the price of timber also depended on the choice of *tanchi*, which had to be clarified in every timber contract.

Tab. 1. Methods of measuring different types of native timber in Hankow, 1921-25[19] (Drawing: Y. Pan)

Type	Method	Classification by sizes	Classes of timber
Shanmu (pole)	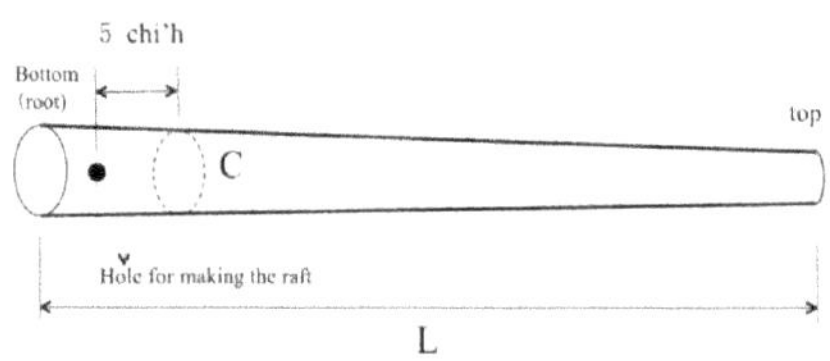	C ≤ 1*chi'h*	Budeng [不登]
		L< 30 *chi'h*, C: 1 ~ 1.5 *chi'h*	Xiaofenmamu [小分码木]
		30 *chi'h*≤L < 36 *chi'h*, C: 1.55 ~ 1.8 *chi'h*	Xiaoqianmamu [小钱码木]
		36 *chi'h*≤L <40 *chi'h*, C: 1.85 ~ 2.2 *chi'h*	Zhongqianmamu [中钱码木]
		40 *chi'h*≤L <48 *chi'h*, C: 2.25 ~ 2.5 *chi'h*	Daqianmamu [大钱码木]
		48 *chi'h*≤L <52 *chi'h*, C: 2.6 ~ 3 *chi'h*	Daqianliangmamu [大钱两码木]
		48 *chi'h*≤L <52 *chi'h*, C: 3.05 ~ 3.5 *chi'h*	Liangmaguomu [两码过木]
		L = 60 *chi'h*, C: 3.5 ~ 4 *chi'h*	Zhengmuma [正木码]
		C: 4.05 ~ 4.5 *chi'h*	N/A
		C: 4.5 ~ 5 *chi'h*	N/A
		C: 5.05 ~ 5.5 *chi'h*	N/A
		C: 5.55 ~ 6 *chi'h*	N/A
		C: 6.05 ~ 6.5 *chi'h*	N/A
		C: 6.55 ~ 7 *chi'h*	N/A
Songmu (log)	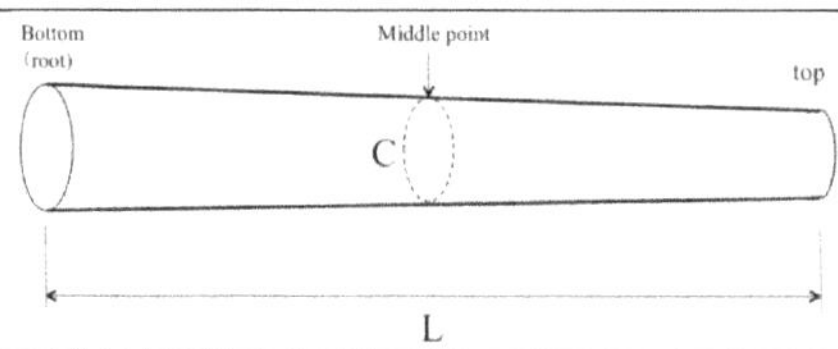	$\frac{(C - 1.5\,chi'h) \times L}{100}$ (unit: fang [方])	
Plank	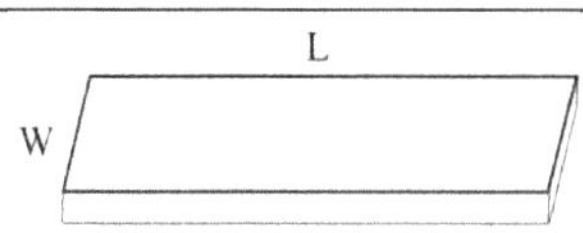	$\frac{L \times W}{100}$ (unit: fang)	

Timber supply for native and early foreign-style buildings

Chinese carpenters used many different woods in building construction. *Nanmu (nanmu/phoebe zhennan)*[20], *shantao** (hickory/*carya cathayensis*), *muhe* (Chinese gugertree/*schima superba*), *Bomu*, *jümu** (Chinese beech/*zelkova schneideriana*), *zhangmu* (camphor/*cinnamomum camphora*), and *limu** (Chinese chestnut/*castanea mollissima* or *castanea henryi*) were regarded better than *Shanmu*, *Songmu*, *yuanbo* (Chinese juniper/*Juniperus chinensis*), *wuyü** (tallow tree/*sapium sebiferum*) and *zimu** (catalpa/*catalpa ovata*)[21].

In general, softwoods were widely preferred for carpentry as they are usually large-sized, straight-textured and easy to process , more importantly, they could provide the pleasing smooth surface finish preferred in Chinese buildings. By contrast, hardwoods are of beautiful texture and wearable character, but easily deformed or cracked, and are more difficult to process, so they were normally reserved for joinery. Climate and insect attack were two critical aspects that also influenced the decision of the choice of timber. *Nanmu* is a valuable wild tree characterized by its straightness and, more importantly, the durability against insects and the extraordinary climatic range in China. *Shanmu* from Fujian Province was famous for its resistance to being attacked by white ants[22]. When the imported timber market was being developed in China, those species of trees that could not adapt to native environment were eliminated. Rosenbluth gives some goods examples: 'Oregon pine rots rather badly in China south of the Yangtze basin, it will probably be less used after five years than it has been in past. [...] Japanese wood is practically worthless on account of the rapidity with which it decays; and in the southern parts also because of the partiality which the white ant show for a diet of Japanese wood.' [23]

A remarkable feature of carpentry in China was the use of roundwood for purlins, rafters, columns and sometimes beams, constructed using the traditional building methods (Fig.3). In Southern China carpenters selected timber by the circumference. According to *Yinzao fayuan*, a building construction manual of native buildings in the late 19th and early 20th century, the carpenters in Suzhou measured the circumference of the log with a flexible bamboo strip, called *weimie* [围篾]. The width of the house, or *jinshen* [进深], was commonly used as the reference size (module) for the whole building. When the *jinshen* was decided, the circumference of each beam was known through its design ratio to *jinshen*, which indirectly ensured the depth of the beam for the span. It is worth noting that the design ratios were adjusted depending on the species and grade of wood. A reduction was applied to the ratio when superior woods were used. Use of inferior woods required only a modest upward amendment to the ratio, but demanded a careful examination of knots and other timber defects especially for beams to be used over a large span.

The trade in imported timber in colonial China probably started in the middle of the 19th century and was initially associated with shipbuilding. A comparison between the four editions of *The Chinese Commercial Guide 1834-1863* [24], as the precursor of *The China Year Book*, gives clues of the early development of this foreign timber trade in China. In the 4th edition (1856) in the list of tariffs on "Articles of Imports", no articles that could be counted as "building material" are listed. Certain "woods, viz. Ebony (Sandal-wood, Sapan-wood)" were included, but according the attached description of these articles the imported woods were used mostly for other purposes including for instance the making incense sticks. Construction timbers appear for the first time in the 5th edition (1863) of *The Chinese Commercial Guide* in the list of imported timber in the table of "Tariff on Articles of Import", which listed 6 types of masts and spars, 1 type of beams, and 3 types of planks[25].

Effect on Design

Western design of foreign-style buildings presumed the use of western construction techniques, but of course the buildings could only be realized with the material available to the Chinese builders and relied on their skills. The lack of suitable timber and the Chinese tradition of working with round timber created problems when dealing with western layouts as demonstrated in the following account of the building of an English House in Hong Kong in 1844-45: 'Some old masts [were used] as girders [...] For the roof we had provided king post trusses; and when these were set up, the carpenters began to bed the hip pieces; four clumsy round pine logs—they were tied to nothing, had no square bearing anywhere—being just bedded in the brick walls, and the purlin spars rested on them [...] Every morning some blunder stared me in the face.' [26] From Ashworth's memoir, we can see that colonial architects found working with Chinese materials and craftsman a challenge. What had started out as Western King post trusses or Queen Post trusses with squared timber and carefully worked joints became re-worked as best as could be done in the circumstances. This type of problems can be seen in a considerable number of examples including many early industrial buildings.

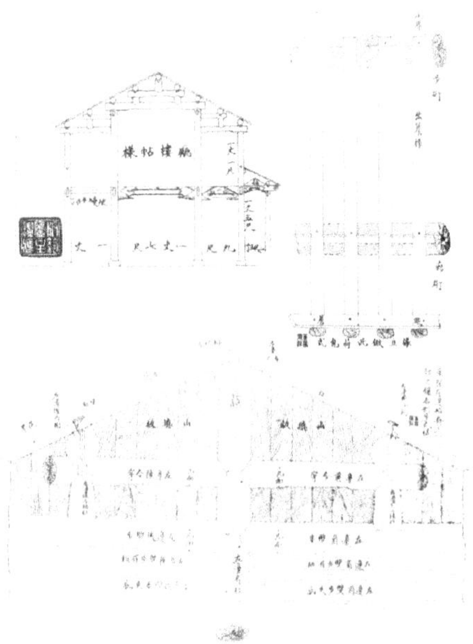
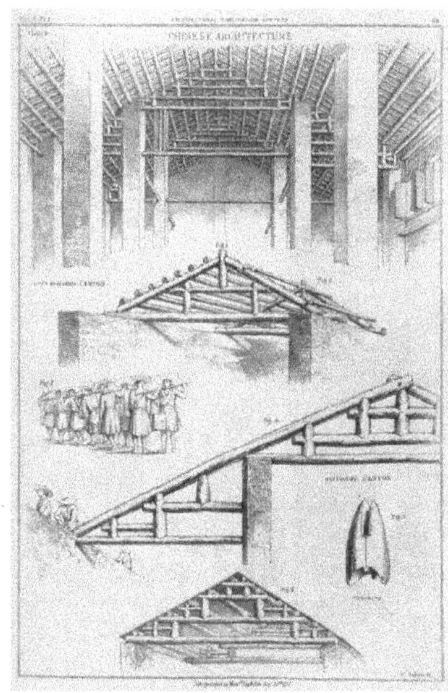

Fig. 3. (left) Original construction drawings of Yingzao fayuan: timber frame for an ordinary native house (upper left); rafters (upper right); details of the timber frame (bottom) (Drawing: C. Yao). Fig. 4. (right) Chinese timber roof construction with regional features in Canton, 1851 (Drawing: E. Ashworth).

Of course, not all such exercises were successful. Chinese carpentry, which was associated with the size and nature of native woods, had developed to cover limited spans, and its limitations became even more apparent when these techniques were applied to inappropriate forms such as those shown in Ashworth's drawings (Fig.4). The roof structure employed in the first Trinity Church (built in 1847-8) in Shanghai fell down in a storm two years later. The failure was put down to it having been too heavy for such a large span[27]. This accident probably partly explains the later preference for western roof construction and the subsequent greater attention to detail in colonial buildings in Treaty Ports.

Cost was one of the top considerations and undoubtedly influenced the architects' choice of timber. It was not only the cheaper cost of native timber over its foreign rivals that determined compared the overall cost; labour was also a significant factor. At the end of 19th century, the cost of carpenters' and joiners' work in China about half that of British craftsmen [28]. Chinese carpenters were inevitably used, and they customarily used traditional setting out methods (including the traditional *Luban* ruler) and followed the methods of they had learnt from their masters in working native woods. Their setting out and structural calculations had relied on rules of thumb handed down for generations. Imported timber was unfamiliar. Chinese patrons were wary of using it, and the market for imported timber remained relatively small despite the rising costs of Chinese timber until some time after the First World War (Fig.5) [29].

The rise and fall of imported timber

Despite its slow initial uptake there were many opportunities for foreign timber in China. One important factor was the increasing use multi-storey buildings instead of the traditional single storey courtyard designs. Multi-storey buildings required a very considerable quantity of timber for flooring, window-sashes, etc. – at least this is true for all the principal ports, which undoubtedly will spread rapidly elsewhere. One of the points continually made by western writers was the difficulty in finding out anything about the Chinese timber trade. This sparked considerable debate about whether China was able to remain self-sufficient in terms of timber supply, when the Chinese open ports were undergoing an industrial revolution. Rosenbluth apparently supported the idea that there would be an increasing demand for foreign wood[30]. But Mr. Sun Hwah Ting, the largest wood dealer at *Hankow*, believed the home supply would always be sufficient. Rosenbluth considered Sun's view biased [31]. But two decades later, in 1936, Emms could still write that: 'China at the present time is importing the bulk of her timber. This is totally unnecessary. She has more than sufficient timber for her needs standing idle within her boundaries.'[32] (Fig.5)

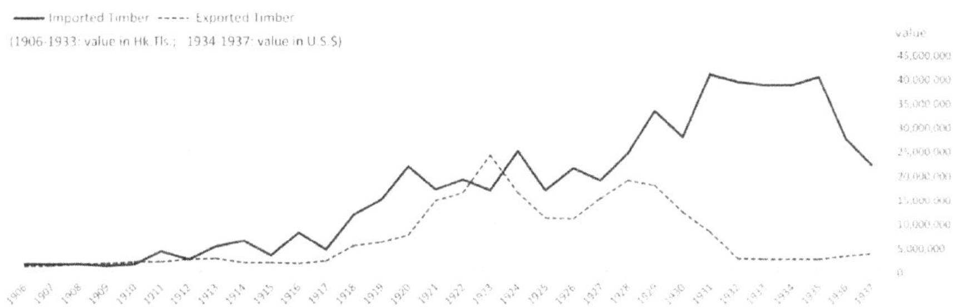

Fig. 5. Imported timber and exported timber in China, 1906-1937 (Drawing: Y. Pan).[33]

What is clear is that in the twentieth century promotion of foreign timber by foreign firms became increasingly aggressive. With the continuous promotion of the imported timber markets, types of timber became increasing varied. Foreign architects and engineers, used to working with and sizing foreign timbers, undoubtedly had an important part to play in prioritizing imported timber over native products, in the Treaty Ports and major inland cities by 1921. This was partly simply a matter of convenience: if familiar woods were specified then the construction details could be directly copied from foreign manuals-

. There were also sound commercial concerns: the strength of imported timber was known hence the sizes could be minimized, an important factor in larger projects. As a result, prior to the Sino-Japanese War, the imported timber came to dominate the Chinese building industry, particularly along the coastal and Yangtze River areas. There were however other factors in play. The foreign timber merchants worked closely with Chinese middle-men to developed local networks and facilitate demand. A good example is the case of the British owned China Import & Export Lumber Company, Ltd.[祥泰木行], which was started in 1884. A modern study has shown that they managed to establish a connection between the company and the municipal council in Ningbo which led to the government passing a law requiring all construction drawings to use the sizes and standards of Oregon pine and that all such drawings had to be approved by the council[34].

It is interesting to note that even with the increasing use of imported timber and foreign detailing, Chinese construction still managed to retain some characteristics of the older traditions. Thus, for instance, a construction textbook in 1934 shows a European truss with rounded Chinese purlins (Fig 6). This type of illustration could too easily be taken to mean that the building technology transmitted from Europe to China was not well understood by the Chinese builders. However, it probably indicated a delicate balance between the imported and native timber traders. According to a modern study, during the time when local timber markets suffered a considerable depression in face of foreign competition , some native timber suppliers offered sales that bundled Oregon pine and Fujian poles toegther[35]. Perhaps it is this sort of arrangement that is being illustrated here.

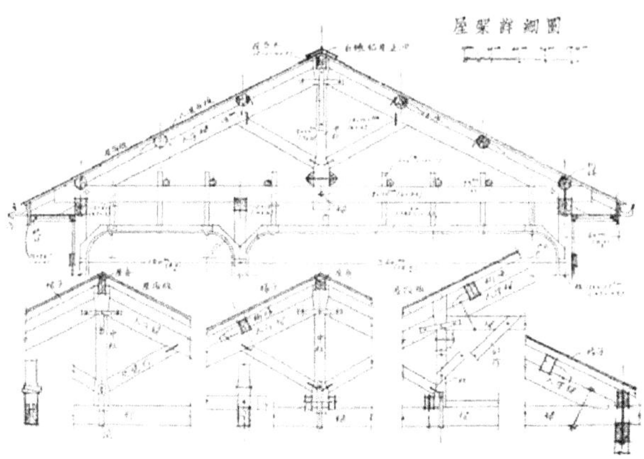

Fig. 6. A plate in a Chinese construction book published in 1934[36].

The domination of foreign woods was challenged in the second half of the 1930s. By this time it was Chinese architects and engineers (rather than foreign ones) who controlled 80%-90% of the building projects. More importantly, the Building Acts now required local products to be used as the principal building materials [37]. In fact, over two decades before the rethinking of timber supply in the building industry, the introduction of western scientific methods of forestry had awakened the central National Government the need to regulate the industry and to manage the forests. From the 1920s threat of foreign imports led the Chinese to actively research how Chinese forest trees could be exploited properly to replace foreign imports [38]. The National Government actively encouraged the consumption of native woods, requiring that 'in future, preference must be given in all contracts to firms using native timber.' [39]

Summary

This paper has sought to introduce some of the factors involved in the complex relationship between timber supply and building construction in late 19th and early 20th century colonial China. These factors included: the attitude towards forest trees; the structural importance of timber; the size and nature of timber; the climate; the cost; consumer preference; the knowledge of the standards and calculation; the change of building types; the scale and speed of the construction development in Treaty Ports; the role of timber merchants, middle men, architects, engineers, and the officials in public construction sectors; the policy of the National Government; Building Acts and regulations; and construction textbooks. As has been shown, timber had a crucial role to play in building in this period and just as the architecture was a fusion of east and west so it was made by a complex interaction between foreign and local interests.

References

[1] A. Herrmann, *Historical and Commercial Atlas of China.* Cambridge, Mass.: Harvard University Press, 1935, pp.6-7.

[2] Based on various sources: *Miaoman tuceye* [苗蛮图册页], 1786; N. Shaw, *Chinese Forest Trees and Timber Supply.* London: T. Fisher Unwin, 1914, p.92, p.96, p.120; R. Rosenbluth, "Forests and timber trade of the Chinese empire", *Forestry Quarterly*, vol.10, Ithaca, N.Y., 1912, pp.654-655; W. Lowdermilk, "Forest Destruction and Slope Denudation in the Province of Shansi", *The China Journal of Science & Art,* vol.4, 1926, pp.127-135; H. Woodhead (ed.), *The China Year Book.* London: George Routledge & Sons (1912-1919), Tientsin: Tientsin Press (1921/22-1929/30), Shanghai: North China Daily News & Herald (1931-39), 1912, pp.20-21; 1921/2, pp.69-79.

[3] *The China Year Book*, 1921/2, (Note 2) pp.69-79.

[4] This map is made by overlapping Shaw's forest map and "Trade Route" map in "Economic Trends, 1870-1911", *Cambridge History of China*, vol.11, Cambridge: Cambridge University Press, 1980, p.42.

[5] Z. Li [李浈], *Zhongguo chuantong jianzhu xingzhi yu gongyi* [中国传统建筑形制与工艺]. Shanghai: Tongji daxue chubanshe, 2006, pp.66-67.

[6] N. Shaw, (Note 2) pp.58-59.

[7] R. Hommel, *China at work.* Cambridge, Mass.: M.I.T. Press, 1937, p.219.

[8] Z. Li, (Note 5) pp.69-70.

[9]Anon., "Jilin mucai yu qi jianglai [吉林木材与其将来]", *Shiye zazhi* [实业杂志] 3, 1912, p.24; R. Hommel, 1937, p.221.

[10] The tree species mentioned for the first time are listed as: Chinese name (English name/Latin Name). In addition, hardwood is marked with *.

[11] Z. Li, (Note 5) p.70.

[12] H. Gong [龚辉], *Xicha huicao* [西槎彙草]. [Around 1522CE-1566CE]. <http://www.wdl.org/en/item/4696/#q=China&qla=en> [accessed 20 December 2013]

[13] A. Emms, "The Practice of Joinery and Carpentery amonst the Chinese of the Yanagtsze Valley", *Engineering Society of China*, 1936, p.73.

[14] R. Rosenbluth, (Note 2) p.663; Anon., "Hankou zhi mucai maoyi [汉口之木材贸易]", *Zhongwai jingji zhoukan* [中外经济周刊] 109, 1925, pp.2-3.

[15] *Zhongwai jingji zhoukan* 109, (Note 14) p.3.

[16] R. Rosenbluth, 1912, pp.663-664; T. Hua [华通斋], *Jianzhu cailiao cuoyao* [建筑材料撮要]. [n.p.]: [n.pub.], 1919, pp.39-40; *Zhongwai jingji zhoukan* 109, (Note 14) pp.6-7.

[17] T. R. Jernigan, *China's business methods and policy.* London: T. F. Unwin, 1904, pp.180-181.

[18] *Zhongwai jingji zhoukan* 109, (Note 14) p.1.

[19] B. Tang [汤丙南], "Hankou cemucai zhi fa [汉口测木材之法]", *Hubeisheng nonghui nongbao* vol.2 [湖北省农会农报] 9, 1921, pp.91-93; *Zhongwai jingji zhoukan*109, (Note 14) pp.6-7.

[20] The existing structure of Imperial Palaces built in Ming Dynasty (1368-1644) was mainly made of nanmu from Sichuan, Guangxi, Fujian and Zhejiang. According to S. Shan [单士元], *Wo zai gugong qishinian* [我在故宫七十年]. Beijing: Beijing shifan daxue chubanshe, 1997, pp.416-417.

[21] J. Zhu [祝纪楠], *Yingzao fayuan quanshi* [营造法原诠释]. Beijing: Zhongguo jianzhu gongye chubanshe, 2012, p.25.

[22]Anon., "Fujian zhi mucai [福建之木材]", *Zhongwai jingji zhoukan* [中外经济周刊] 35, 1923, p.5.

[23] R. Rosenbluth, (Note 2) pp.659-660.

[24] 1834 (1st ed.), 1848 (3rd ed.), 1856 (4th ed.), 1863 (5th ed.)

[25] S. W. Williams, *The Chinese commercial guide*. Hongkong: A. Shortrede & Co., 1863, pp.65-66.

[26] E. Ashworth, "How Chinese workmen built an English House", *The Builder* 4 (456), 1851, pp.686-688.

[27] H. Zheng [郑红彬], "Trinity Church: the Design History of an English Residents' Church in Shanghai (1847-1893) [上海三一教堂: 一座英国侨民教堂的设计史 (1847-1893)]", *Jianzhu shi* vol.33, 2014. (in Chinese, to be published)

[28]Anon., "Chinese Labour", *The Builder*, 17 January, 1880, pp.74.

[29] L. Jin[金礼], "Yangsong qingxiao yu yingshang xiangtai muhang jiaoshe'an [洋松倾销与英商祥泰木行交涉案]", *Ningbo wenshi ziliao dijiuji* [宁波文史资料第九辑], 1991.
<http://202.107.212.146:838/nbwshj/index.php?option=com_content&view=article&id=231:2011-09-09-01-42-45&catid=42:2011-09-07-00-48-41&Itemid=61 > [accessed 20 December 2013]

[30] Based on Rosenbluth's data, the foreign timber had occupied 15% of the total supply in China in 1909, sixty years after the earliest opening of Treaty Ports since 1850s. Rosenbluth predicted that, "without any extraordinary development" and "conservatively", by 1915 China would require twice as much as the present imported timber per year, and by 1925 the demand would be for five to ten times present worth of forest products imported each year. (R. Rosenbluth, 1912, pp.656-657, p.663)

[31] R. Rosenbluth, (Note 2) pp.662-663.

[32] A. Emms, (Note 13) pp.74-75.

[33] The diagram is generated with the data in the Import-Export reports from *the China Year Book*, 1912-1939 by the author.

[34] L. Jin, (Note 29)

[35] L. Jin, (Note 29)

[36] Zhonghua zhiye xuexiao [中华职业学校], *Jianzhu zhitu* [建筑制图]. Shanghai: Zhonghua zhiye xuexiao, 1934.

[37] Addressed in Building Act, no.40.

[38] Y.Tang [唐耀], *Mucai xue* [木材学]. Shanghai: Shangwu yinshuguan, 1936.

[39] A. Emms, (Note 13) pp.74-75.

Stretched templates in Gothic tas-de-charge construction

Carmen Pérez-de-los-Ríos and Enrique Rabasa-Díaz
Departamento de Construcción y Tecnología Arquitectónicas, Universidad Politécnica de Madrid, Spain

A *tas-de-charge* is the springing of Gothic vaults, where ribs are not yet separate, and their mouldings may intersect or interpenetrate. *Tas-de-charge* form part of the wall or the support, and consist of several pieces with horizontal courses, except the last one which ends with a sloped surface in order to connect with the rest of the arch. In theory, the horizontal and sloped cross-sections are different; however, this does not seem to have been a problem for Gothic builders, who either used or avoided stretched templates for the horizontal courses depending on each specific case.

The presence of this topic in treatises is not common; the only historical drawings of a Gothic *tas-de-charge* with stretched templates are contained in the manuscript of Joseph Gelabert [1], a Spanish treatise from the 17th century. Archaeologist Robert Willis made a fundamental contribution in this field: he was the first to explain the construction of this part of the vault. His analysis was based on the observation of the shape of the pieces in the vaults and their location. When dealing with disassembled details, he could also measure and study traces that still remained in the pieces. [2]

The example measured and explained by Willis will be critically analysed in this paper; as will other cases where the use of stretched templates has been attested or would have been necessary. Variables making the use of stretched templates either essential or unnecessary will also be pointed out. Some are the result of practical experiences in *tas-de-charge* building, and others of full station surveys of actual, historical buildings.

Stretching templates in 16th, 17th and 18th century treatises

Although the pinnacle of Gothic construction was already over in the sixteenth century, many stonecutting treatises continued to deal with Gothic vault construction. In the seventeenth century, the manuscript of Majorcan stonemason Joseph Gelabert pointed out for the first time the necessity of stretching templates when building a *tas-de-charge*: he showed how to do a *retirade* (retreat) in each course. However, he admitted that, due to the small size of the drawing of the templates in the manuscript, their deformation is almost imperceptible there. [3]

Gelabert suggested carving the pieces first '*en cuadrado*' (squared-up), with the container prism adapted to the intrados line (Fig. 1). Subsequently, the stretched moldings were traced on the stone. He noted that the stretching of the templates had to be done from the intrados line to the extrados line, and not the other way around. He also dedicated other parts of his manuscript to show how to deform moldings in order to stretch and skew them. The usual procedure he employed consists of adjusting a form to a reticulated frame. Next, the frame was deformed and the form redrawn in the new frame. [4] He did not consider it necessary to give more detail for the case of the *tas-de-charge*, as the profiles he used in his example were very simple. Nevertheless, the probable procedure for this case could have been similar.

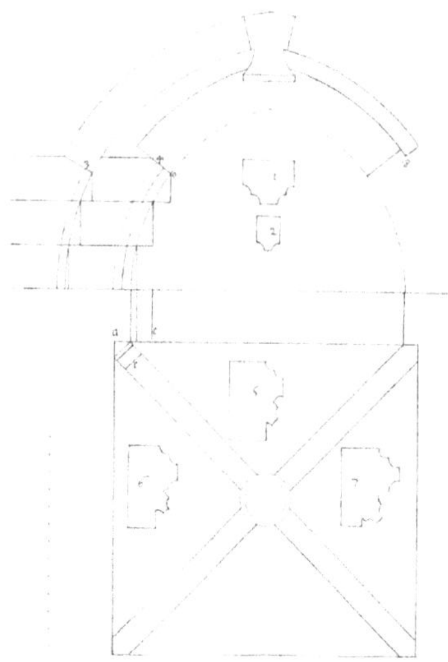

Fig. 1. Gelabert, Verdaderes, (Note 1), layout number 118 in fol. 133v.

It has to be taken into consideration that older, similar procedures to stretch or deform a flat form are known, many of them leading to oval forms. Philibert de L'Orme called the different procedures to stretch a curve *cherche ralongée*, and included one where the x- and y- coordinates of each point were found and the deformation was carried out by the change of scale along the x-axis. [5] This expression appeared in Derand's and other authors' treatises too, but in a more general sense. [6]

Dürer employed more clearly the deformation of one of the dimensions of the reticulated frame to transform a circumference into what today is called ellipse. He said that this procedure was useful for stonemasons who built vaults (he probably referred to groin vaults). Dürer also applied the reticulated frame to forms from nature. [7] In a similar way, D'Arcy Thompson spread this idea in *On Growth and Form*, 1917. [8]

Sebastiano Serlio in his first book gave the name *transporte* to changes in scale, on one or two axes, using Thales' Theorem. [9] Master mason Hernán Ruiz from Seville also used similar procedures, but he called them *transferente*. [10] Besides, as it is commonly known, Serlio showed how to draw an ellipse through an affine transformation of two circumferences. Nevertheless, it was Alonso de Vandelvira [11] that showed for first time what is now known as the transformation of a circumference into an ellipse defined by conjugate diameters, using an oblique reticulated frame; as explained by Rosa Senent. [12] Dealing with Gothic vaults, Vandelvira gave the name *revirado* to an affine transformation of a rib template. The axis of this rib remained vertical and fitted to the curvature of the vault in conditions where tilting the axis of the rib could have appeared more intuitive. For the same concept, Rodrigo Gil de Hontañón used the word *bulco*. [13] Neither of them explained how to develop the transformation. Consequently, it has been assumed that skilful use of these methods for stretching templates in *tas-de-charge* whenever required was common knowledge among Gothic master masons.

François Derand and Amédée-François Frézier, both quoted by Willis, explained how to trace Gothic vaults and deal with *tas-de-charge* construction. Derand maintained that the meeting of mouldings of several ribs could entail complex intersections, so he proposed the use of different colours, or different ways for tracing each template. This led to the assumption that all of the different templates would be placed on each bed. Derand used as an example a tierceron square vault because he claimed that other kind of vaults were contrived and confusing. When he explained how to trace the *tas-de-charge*, he did not use any drawing; he also did not make any reference to the need of stretching templates. [14]

Frézier discussed the superposition of templates in the courses, where some traces are hidden by others. He showed a drawing similar to what Willis found in a piece in Canterbury Cathedral. Although he did find some advantages in Gothic forms, he did not find them aesthetically appealing. Moreover, Frézier forced the intersection of two ribs in the *tas-de-charge* to take place in an exactly vertical plan, which would be a typical way of thinking for an eighteenth century author in its preference for 'regular' configurations. [15]

He used the term *cerche ralongée* ('l'alongement, c'est-à-dire, la cerche ralongée de leur contour') to explain how to stretch templates; but he did not indicate why it was necessary. In fact, he showed an example in which the stretching of the template depends on the angle between the lierne and the diagonal rib without any apparent relationship with the height of the courses in the *tas-de-charge* (Fig. 2). He probably suspected that stretched templates are necessary, but when he went into detail about it he started to have some doubts.

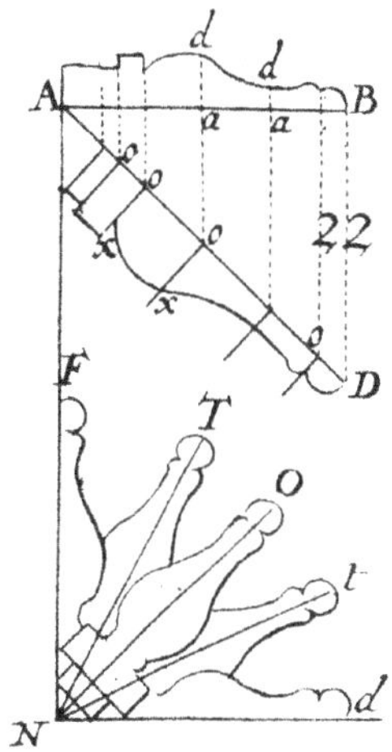

Fig. 2. Frézier, Coupe des pierres (Note 15), pl. 71.

In the nineteenth century, these flat form deformations were studied from a geometrical point of view as affine transformations, leading to interesting considerations. It should be pointed out that they already had been carried out for a long time without any problem and without the support of any theory.

Archaeology of architecture in nineteenth and twentieth centuries: contributions of Robert Willis and Luc Mojon

Robert Willis was the first to explain the way to design and carve keystones and *tas-de-charge* in Gothic vaults. He pointed out that for carving the external surface the masons would only have needed the contour of the ensemble of templates in the upper and the lower bed of each block. On these beds, the lines that represent the meeting of the ribs on the ground plan would have to be reproduced in order to place each template in the correct direction. [16]

Years later, Viollet-le-Duc and Auguste Choisy also wrote about *tas-de-charge* construction, agreeing with Willis' explanations. Viollet probably knew the French translation of Willis by Cesar Daly (1843). Consequently, in his article *Construction* in *Dictionnaire* (1859) he took for granted that the templates had to be stretched in the *tas-de-charge*. [17] Choisy, who was very young when he read Willis' book in his father's library, drew Willis' sketches again in his *Histoire de l'Architecture*. [18]

Willis said that the distance between the template and the meeting point increases in the upper beds of the pieces of the *tas-de-charge* because the rib is curved, so the stonecutter has to measure, on the elevation of each rib, where the intrados of the rib is represented, the horizontal displacement of the template in each bed. Willis realized that the length of the horizontal cuts of the ribs in these pieces should be not the same as the length in the radial cut of the last course of the *tas-de-charge*, or the radial cuts which define each of the arch voussoirs. Theoretically, a horizontal cut requires a stretching template, while a radial template does not.

Willis found an opportunity to study in detail the pieces of a *tas-de-charge* that became ruinous of the so-called Lanfranc's Tower in Canterbury Cathedral. He drew one of the beds with all the axis lines and the templates used to define the contour. With this information and the plan of the Tower [19], we have proposed a plan for the vault with radii close to six metres for the main ribs (Fig. 3). Willis said that in this case, and in others he examined, the same template was used for the horizontal and the oblique beds; and 'the mouldings were drawn in and contracted very disagreeably'. The use of the same template should neither be understood as a general strategy, nor a result of a stonemason's mistake. Willis maintained that is a proof of stonemasons' tolerance, taking into account that the *tas-de-charge* is watched from such a large distance that this adjustment is negligible. Medieval masons were more practical than Neo-Gothic builders of the nineteenth century, who insisted on an abstract approach to Gothic vaults.

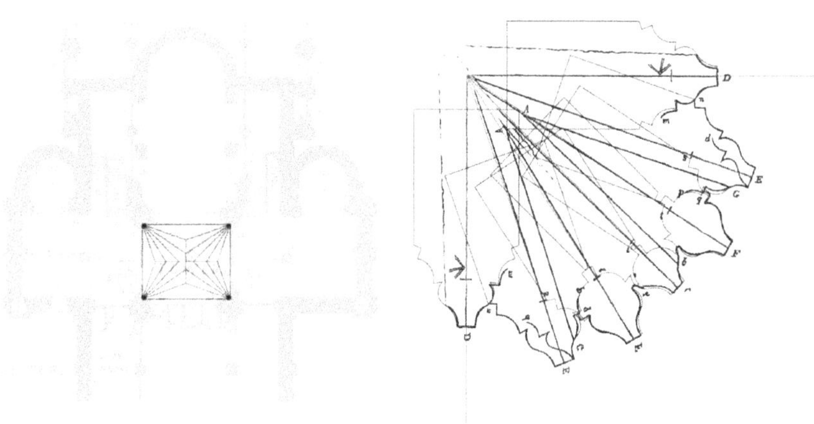

Fig. 3. On the left, hypothesis of Lanfranc's Tower plan vault on Willis plan; Willis, Canterbury Cathedral, (Note 19). On the right, complete tas-de-charge from Willis drawings; Willis, On the Construction, (Note 2), p. 8.

Willis probably thought that this use of a unique template works in vaults with a large span, that is, with large arch curvature radii. He wanted to quantify it in Canterbury, so he took measures of the courses of the *tas-de-charge* (about ten inches), the dimension of the moulding in the sloped section (five inches), the stretched dimension in the horizontal section (five and one-half inches); and the angle between this horizontal level and the point of the intrados of the rib where the sloped section is located (110 degrees). He represented this idea in a sketch showed in our figure 4, on the left. [20]

However, when dealing with the measures given by Willis, and taking into account that vault arch radii are around six metres, we realized that this geometrical construction is not possible. There are two possible solutions (Fig. 4). First, if we respect the angle and the height proposed by Willis, the length of the horizontal moulding is five and one-quarter inches, rather than five and one-half inches. Secondly, taking a correct moulding length for granted, Willis' figures could be possible if we measure the *tas-de-charge* four courses higher than the one that Willis showed. In this case, the angle should reach 120 degrees, instead of 110 degrees. We are of the opinion that Willis could probably measure properly the pieces of the *tas-de-charge*, thus we can assume that the length of the sloped and the horizontal section can easily be precise, but not the angle; where achieving precision is more difficult.

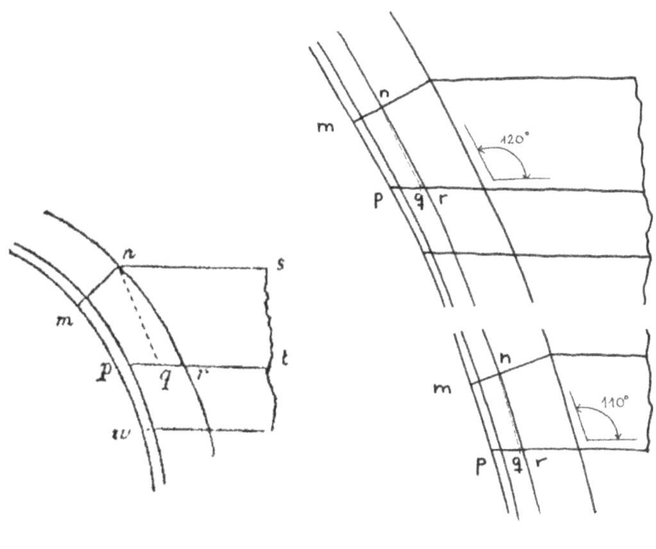

Fig. 4. On the left, Willis, On the Construction, (Note 2), p. 9. On the right, two different solutions to Willis' drawing.

Due to the difficulties of studying *tas-de-charge* beds in existing buildings, the experience of Luc Mojon (1986) is also highly relevant. [21] He measured and drew the traces on a *tas-de-charge* from an abbey church in Switzerland dated from 1365-90 that was disassembled in the 1960s. In the upper and the lower beds of the pieces, he found the traces for the arch axes and the templates for each rib. He realized that the contour in the lower bed of the first piece, formed by combining the first three rib templates, was reproduced in the beds of the upper pieces in order to have exact control of the direction of the ribs (Fig. 5).

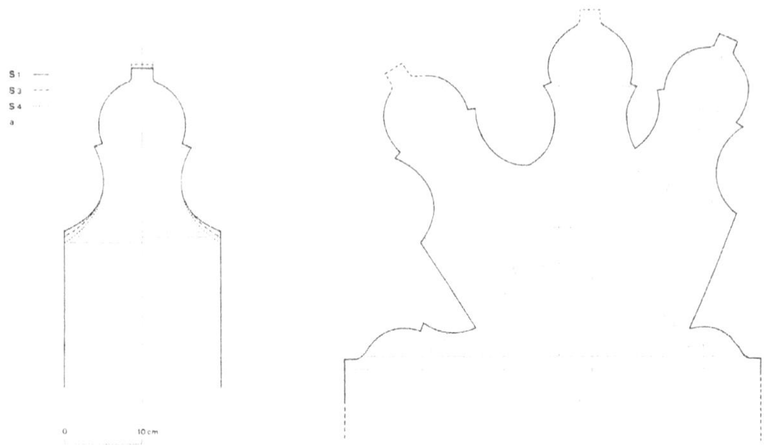

Fig. 5. Mojon, "...Tas-de-charge", (Note 21), p. 95 and p. 101.

In addition, Mojon pointed out the use of unstretched templates in the curved mouldings of the first pieces of the *tas-de-charge*, until the level at which changes of direction in the profile result in noticeable lines coming out. He claimed that stretched templates were used in the construction of this *tas-de-charge* considering his measurements of the pieces, as he showed in a drawing with all the templates superimposed.

Criteria for template stretching in Gothic *tas-de-charge*

The difficulty of finding a disassembled *tas-de-charge* has hindered the study of the use of stretched templates in their construction. Willis and Mojon have drawn disassembled pieces containing traces; additionally, we have found rib axis traces in the ruins of the Valdecristo Carthusian monastery near Valencia.

In order to counter such a lack of information, our research on the stretching of templates has made use of full station surveys in *tas-de-charge* courses of existing buildings and the experience of building examples in the Stonecutting Workshop at the Madrid School of Architecture. We have thus established some variables, which determine the necessity of using stretched templates.

The first one is vault size, which is often related to the number of horizontal courses of the *tas-de-charge*, the place where it ends and the curvature of the ribs. On one hand, in large-span vaults, with great rib curvature, the *tas-de-charge* often shows the sloped cut in the first half of the vault's height. As a consequence, the angle between the last horizontal course and the first sloped course is very small, and the same template can be used for both sections. In main nave vaults of cathedrals this procedure seems to be common. Our study is not limited to the use of the same profile in the last piece of the *tas-de-charge*: in vaults that we have measured, such as the Valldigna refectory near Valencia or the Sala dei Baroni in Naples, both with large vault spans, the same template is not used in the last piece, but it might have been used in the horizontal courses of the first half of the *tas-de-charge*.

On the other hand, in small-span vaults with small rib curvature, the use of the same template is almost inconceivable. Ribs separate at a higher point and, as a result, the *tas-de-charge* can sometimes reach up to half the height of the vault to ensure the separation of mouldings. Thus, the angle between the last horizontal cut and the sloped cut is greater. In the examples we have built in Centro de los Oficios in León and in the Stonecutting Workshop at Madrid School of Architecture, with small vaults, stretching the templates was completely necessary (Fig. 6). [22] We have also confirmed the use of stretched templates in measured examples: Narbonne cloister vaults, Valdecristo Charterhouse, La Trinidad Convent, Perpignan Chapter Hall, Lonja de Palma, Lonja de Valencia, etc. In the first *tas-de-charge* we built, similar to one in the Narbonne cloister, we did not use templates to control deformation of the mouldings. The profiles were very simple, which allowed us to take the measures of template deformation from the full size drawing of the *tas-de-charge* on the wall of the workshop. However, when tracing the second *tas-de-charge*, a scale model of the *tas-de-charge* springing from the twisted shaft in Lonja de Palma with more complicated mouldings, we found it indispensable to use templates (Fig. 7). Without them, it would have been impossible to guarantee that the deformation of the moulding in the upper bed of one piece would fit to the lower bed of the following piece.

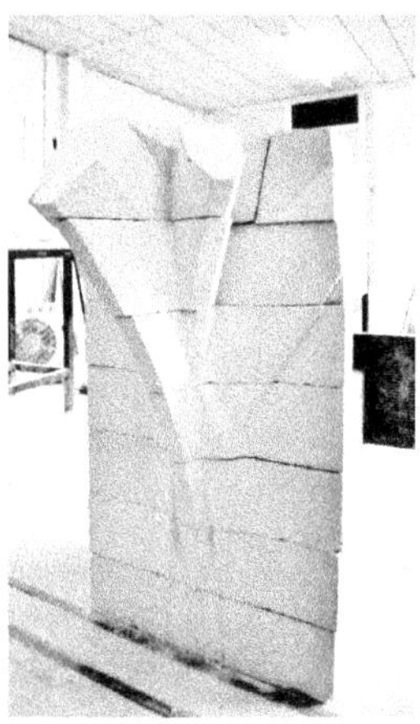

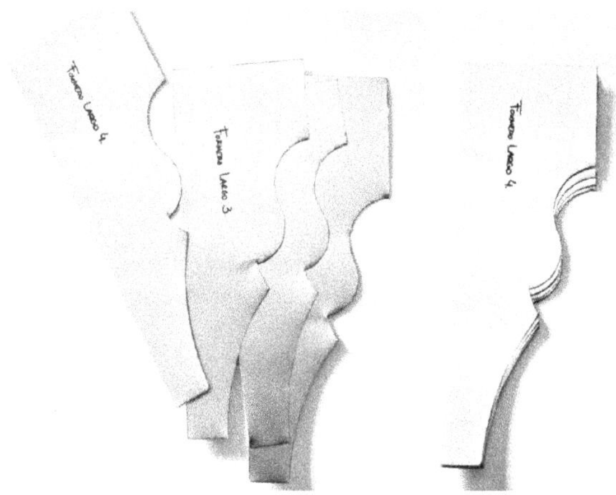

Fig. 6. (left) Tas-de-charge built in Madrid School of Architecture Stonecutting Workshop. Fig. 7. (right) Stretched templates of Lonja de Palma tas-de-charge in Madrid Stonecutting Workshop.

We found an interesting case in a *tas-de-charge* in the *Salle de Théologie* at the *Palais des Papes* in Avignon. The vault span is small and the *tas-de-charge* is quite tall, so the use of stretched templates in the second half of the *tas-de-charge* can be seen by the naked eye, and was confirmed by a topographic survey. We focused on a portion of the moulding of the transverse rib that penetrates through the diagonal rib and appears in the lower part of the *tas-de-charge*, in contact with the wall (Fig. 8). Using survey data, we realized that the stretched template did not fit in this course, and the unstretched template was used. The difference between the position of both templates is almost six centimetres, but it is impossible to appreciate through observation because it is only a small part of the rib after the intersection [23]. This is the only case in which we can confirm the use of unstretched templates in the lower courses through survey.

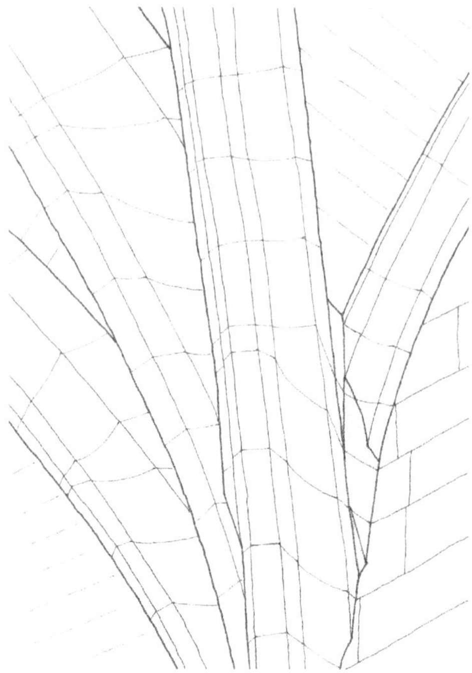

Fig. 8. Moulding of the transverse rib that penetrates through the diagonal rib in the tas-de-charge in Avignon Salle de Théologie.

Flat Gothic vaults represent a special case, due to the presence of two different arch radii in the ribs. In the *tas-de-charge*, the use of arches with small curvatures evidently enforces the use of stretched templates.

The size of the ribs in relationship to the vault also has to been taken into account. Furthermore, lesser complexity of the profile, as well as a greater number of ribs springing for the *tas-de-charge*, allow for the use of unstretched templates in the first horizontal courses. Closer intersections between the ribs mean that a smaller part of the profile appears in each course, avoiding changes in the direction of the profile. This has been confirmed in Valdigna refectory [24], and had already been pointed out by Mojon.

Finally, it should not be forgotten that a *tas-de-charge* is meant to be seen from a certain distance. As Willis said, if it is very far away from the floor, small corrections or mistakes are inappreciable; but in small vaults located in cloisters or chapels, this is often not the case. Besides, the quality of vault execution also plays a role: a better final result requires more precision, and consequently a more careful use of templates.

Conclusions

The analysis of examples and the practical experiences have led us to the conclusion that it is impossible to set a rule to decide when it is necessary to stretch templates. Each vault depends on so many variables that the resulting solution for each one is necessarily unique. Because of that, we can find unstretched templates in lower *tas-de-charge* courses, where the difference between stretched or unstretched templates would be impossible to appreciate; as well as in vaults with large span such as the main naves of cathedrals. However, from our experience, the use of stretching templates is clear in many of the buildings we have studied, and was essential in the examples we have built. Small span vaults, well-executed vaults with complex rib mouldings, etc. require an exhaustive use of stretching templates. Use of unstretched templates in certain situations would have led to an unpleasant appearance of the rib and the *tas-de-charge*, but we have not found any example of this extreme solution.

The effectiveness of the work with templates and the essential role they played in Gothic construction has also been registered in historical documentation. For example, it is known that Guillem Sagrera, master mason of Palma de Mallorca Cathedral, was in charge of the tracing of the templates for the arcs of the new vaults, which was considered a very important task in that time. [25]

Gothic construction is systematic and methodical, but the common sense of Gothic builders enabled them to follow the precise procedures to obtain a successful result in each situation, with just as much effort as necessary.

Acknowledgements

This paper is part of research project '*Construcción en piedra de cantería en los ámbitos mediterráneo y atlántico. Análisis de ejemplos construidos*', funded by the Spanish Ministry of Science and Innovation (BIA2009-14350-C02-01); and of the doctoral thesis currently being carried out by Carmen Pérez de los Ríos: '*Aspectos formales y constructivos en la obra de Guillem Sagrera: el uso de las plantillas*'. Credit should be given to Alberto Calderón González for his invaluable help in reviewing the paper.

References

[1] J. Gelabert, *Verdaderes traçes de l'Art de Picapedrer*, MS 1653, (facsimile in Palma de Mallorca: Consell Insular de Mallorca, 1977). Analyced in E. Rabasa Díaz, *El manuscrito de cantería de Joseph Gelabert*, Palma de Mallorca/Madrid: Collegi Oficial d'Arquitectes de les Illes Balears /Fundación Juanelo Turriano, 2011.

[2] R. Willis, "On the construction of the vaults of the Middle Ages", *Transactions of the Institute of British Architects*, London: RIBA, [1842] 1910, vol.1, pp. 1-69.

[3] Gelabert, *Verdaderes*, (Note 1), layout number 118 in fol.133v.

[4] Gelabert, *Verdaderes*, (Note 1), does it in layouts 81 (fol.94v), 82 (fol.96v) and 113 (fol. 127r). In layout 154 he shrinks a cornice by means of what we could name affine transformation.

[5] Ph. de L'Orme, *Le premier tome de l'Architecture*, Paris: Morel, 1567 (facsimile in Paris: Léonce Laget, 1988, and 1648 edition in Bruxelles: Pierre Mardaga, 1981) fol.55r-55v, and 56 v.

[6] F. Derand, *L'Architecture des voûtes*, Paris : Sébastien Cramoisy, 1643.

[7] A. Dürer,., *Unterweisung der Messung*, Nürnberg, 1525 (facsimile edition, Nördligen: A. Uhl, 1983) Fig. 33.

[8] D. W. Thompson, *On Growth and Form*, New York: Dover, 1992 (1st. ed. 1917).

[9] S. Serlio, *Il Primo libro d'Architettura*, Paris, 1545.

[10] H. Ruiz, el Joven. *Arquitectura*, ca. 1545. Biblioteca de la Escuela Técnica Superior de Arquitectura de Madrid. MS R.16.

[11] A. de Vandelvira, *Libro de traças de cortes de piedras*, 1575-1580, Biblioteca Nacional de Madrid MS 12.719 and Biblioteca de la Escuela T.S. de Arquitectura de Madrid MS R.10 (facsimile in G. Barbe-Coquelin De Lisle, *Tratado de Arquitectura de Alonso de Vandelvira*, Albacete: Caja de Ahorros, 1977), fol. 126v.

[12] R. Senent Domínguez, "Las bóvedas irregulares del tratado de Vandelvira: Estrategias góticas en cantería renacentista", in S. Huerta et al. (eds.): *Actas del Séptimo Congreso Nacional de Historia de la Construcción*, Madrid: Instituto Juan de Herrera, 2011, pp. 1329-1337.

[13] S. García, *Compendio de arquitectura y simetría de los templos...*, Valladolid: Colegio Oficial de Arquitectos de Valladolid, 1990.

[14] Derand, *L'Architecture*, (Note 6), pp. 177-179.

[15] A.-F. Frézier, *La théorie et la pratique de la coupe des pierres et de bois ... ou traité de stéréotomie ...*, Strasbourg/Paris: Jombert, 3 vol., 1737-39 (facsimile in Nogent-le-Roy: L.A.M.E., (1980). pp. 24-31.

[16] Willis, *On the Construction*, (Note 2), pp. 6-9.

[17] E. Viollet-le-Duc, *Dictionnaire raisonné de l'architecture française du XIe au XVIe siècle*, Paris: B. Bance, A. Morel, 1868.

[18] A. Choisy, *Histoire de l'Architecture*, Paris: Gauthier-Villars, 1899-1901, t. II, pp. 273

[19] R. Willis, *The architectural history of Canterbury Cathedral*, London: Longman, Pickering and Bell, 1845, pp. 37-38.

[20] Willis, *On the Construction*, (Note 2), p. 9.

[21] L. Mojon, "Analyse eines spätgotischen Tas-de-charge", in *St. Johannsen Saint-Jean de Cerlier. Beiträge zum Bauwesen des Mittelalters aus den Bauforschungen in der ehemaligen Benediktinerabtei, 1961-1984*, Bern: Staatl. Lehrmittelverl., 1986.

[22] C. Pérez de los Ríos and E. Rabasa Díaz, "Innovations constructives gothiques dans le Palais des Papes d'Avignon: les premiers tas-de charge aux nerfs croisés. Un premier pas vers le style gothique tardif ", In *Deuxième congrès francophone d'histoire de la construction, Lyon*, 2014, in press.

[23] E. Rabasa Díaz and C. Pérez de los Ríos, "Late Gothic as procedure expression", in D. Wendland (ed.), *Traces of Making*, Dresden, in press.

[24] C. Pérez de los Ríos and A. Zaragozá Catalán, "Bóvedas de crucería con enjarjes de nervios convergentes que emergen del muro en el área valenciana, ss. XIV-XV", in S. Huerta and F. López Ulloa (eds.), *Actas del Octavo Congreso Nacional de Historia de la Construcción*, Madrid: Instituto Juan de Herrera, 2013.

[25] J. Domenge i Mesquida, "Guillem Sagrera", in *Gli ultimi indipendenti. Architetti del gotico nel Mediterraneo tra XV e XVI secolo*, Palermo: Edizioni Caracol, p. 76.

Robert Willis' contribution to understanding the Gothic Flamboyant style

Elena Pliego de Andrés[1] and Alberto Sanjurjo Álvarez[2]
1.Universidad Politécnica de Madrid, Spain; 2. Universidad CEU San Pablo, Madrid, Spain

Nineteenth century French architectural historians, such as Viollet-le-Duc and Auguste Choisy, focused their studies of Gothic architecture on its central period, the Rayonnant Gothic, and scarcely took into account the late Gothic period.

From then on, research focused mainly on the origin of the Gothic style; and a serious analysis of the development of the late Gothic in France did not took place until the 1960s with the work of Paul Frankl and also the first monograph, *L'Architecture flamboyante en France,* by the historian Roland Sanfaçon.

Recently there have been partial studies and even dissertations about the Flamboyant; this work has been well received among a good number of researchers.[1]

Our paper analyses the contribution of the British scholar Robert Willis to the knowledge of the Flamboyant style. His early interest in this architectural style has scarcely been considered until today. In 1832 Robert Willis travelled across a great part of France, Germany and Italy. He published the results of this trip in 1835 as *Remarks on the architecture of the Middle Ages, especially of Italy.* Trying to determine the origin of the Gothic style, Willis offered observations that provide an interesting comparative review of the characteristics of the Gothic style in Europe, mainly in France, Germany, England and Italy. Probably this travel and publication led him to develop this subject further in his 1840 essay *On the Characteristic Interpenetrations of the Flamboyant Style,* that has remained largely unnoticed in the historiography of the Flamboyant style and even in the research on Robert Willis' writings until now.

The interest in French late Gothic architecture in the first decades of the nineteenth century in England

Arcisse de Caumont (1801-1873), considered to be the father of French archeology, said in 1830 that the English had studied mediaeval architecture "with more enthusiasm and success" than the French.[2] He undoubtedly referred to a group of British researchers who played an important role in encouraging archeology to be considered as a modern science.

In this context, the first half of the nineteenth century produced a group of researchers, of different disciplines, who focused part of their studies on the origin of the pointed arch as the key principle of Gothic architecture, on classification of English Gothic architecture and by extension of other European countries, principally France, Germany and Italy.

We refer to figures such as Augustus Welby Northmore Pugin, Thomas Rickman, William Whewell, John Ruskin and Robert Willis.

Among their studies, they also dealt with the analysis of the fundamental features of the French late Gothic architecture in comparison with the English Perpendicular, achieving a knowledge in some cases deeper than that reached by the French authors, usually focused or more interested in the first phases of the development of the style and in its central period, the Complete Gothic.

Furthermore, these scholars were concerned with comparing their theoretical studies with the constructed architecture on site, examining the buildings in order to obtain general rules about the development of the Gothic style, that led them to make extensive journeys through the principal European countries. Therefore, William Whewell (1794-1866) stated that he had had the opportunity of travelling through Germany to check some theories about the introduction of the pointed arch in relation to some common architectural practices in Germany.[3] There was also the general belief that a common endeavour in taking architectural notes of the examined buildings would provide important information to other scholars. As a consequence there were not only an important number of journeys but also written guidance on about the way to take these notes. Whewell included a chapter on how to take architectural written and drawn notes[4], a subject which Willis also considered, and the Cambridge Camden Society's most popular publication *A Few Hints on the Practical Study of Ecclesiastical Antiquities* (1843) concluded with a "Church Scheme" according to which every conceivable church arrangement could be noted and standardized.[5]

This passion for visiting and recording churches led William Whewell to travel for the first time to Normandy and Picardy in 1823, making a second journey from 1829 to 1831.

His third trip in 1832, accompanied by Thomas Rickman, involved visiting a great number of buildings in both French regions (Fig. 1) and was much more fruitful. The result of the study of these buildings was published as a new chapter in the second edition of the *Architectural Notes on German Churches*.[6] Here, he made reference to the late Gothic style developed in France with the term, named by Arcisse de Caumont, Flamboyant.[7] This term was used later by Rickman, Ruskin and Willis, among others. Whewell dedicated a section to the Flamboyant style, establishing a parallel between the Perpendicular or Tudor style and the French late Gothic. He describes the construction elements that are common to both styles such as "pinnacles crocketted, finialled, grouped, and formed into niche-canopies, surfaces covered with moulded panels; pierced parapets and Italianized members". But as the first difference between both styles he mentions the tracing of the arches, being the arch of four centres characteristic of the Tudor style, while the Flamboyant style more frequently employs a three-centred arch with the horizontal line arched at the ends. Another differentiating feature is to be found in the design of the tracery and the "peculiar flame-like forms" that gave name to the Flamboyant style.

Whewell also points out the richness of the Flamboyant doorways, making an accurate description of both portals of the transept of Beauvais Cathedral, the work of Martin Chambiges, that is, for him, a model of the Flamboyant construction, and in which description he is constantly searching for the differences and similarities of both styles.

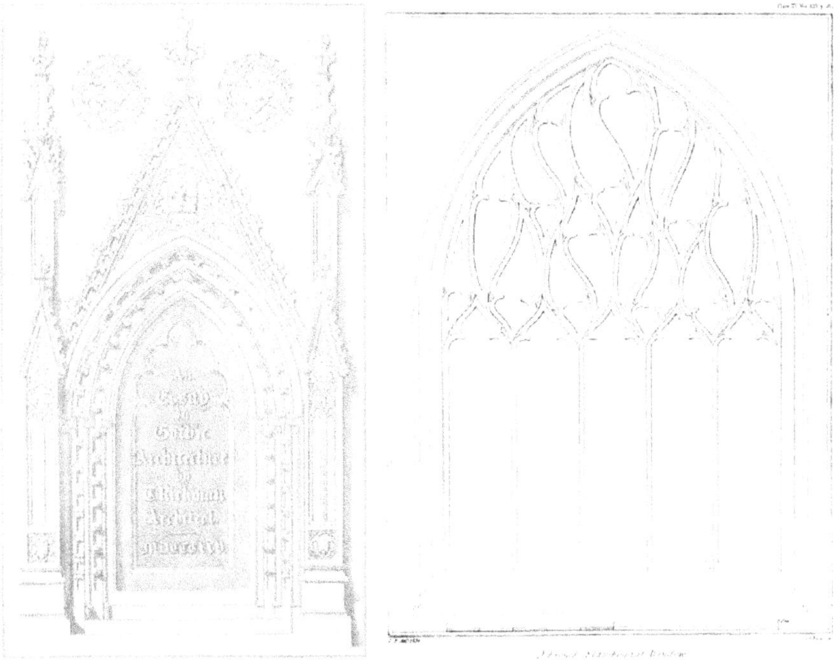

Fig. 1. A French Flamboyant Window by T. Rickman in Four letters on the Ecclesiastical Architecture of France (1834)

In 1834 Thomas Rickman (1886-1841) published *Four letters on the Ecclesiastical Architecture of France*[8], also a product of the journey made with Whewell. Rickman, 18 years older than Whewell, had published in 1817 *An Attempt to Discriminate the Styles of English Architecture, from the Conquest to the Reformation*[9], where he established for the first time a classification of the English Gothic style in Early Gothic, Decorated and Perpendicular thatremains in use today. In the last of his *Four Letters*, he analyses the development of French architecture, from the Romans to the late Gothic, comparing, as Whewell does, the French with the English architectural features. He points out that the French Gothic style from 1400 is very different from the English Perpendicular and this is the reason that he gives for using Arcisse de Caumont's term, 'Flamboyant'.

Rickman enumerates a brief number of common points between the English Perpendicular and the French Flamboyant style: the absence of capitals, running the mouldings of the piers into the arches, the interpenetration of mouldings and the piers with bases (so common in English Perpendicular), and the flattened mouldings with large hollows, but then he mentions other curiously contrasting points such as the plainness of the Perpendicular in comparison with the Flamboyant.

The reception of the Flamboyant style in the French context

In the second half of the nineteenth century in France, when Viollet-le-Duc wrote the *Dictionnaire Raisonné de l'architecture*[10], French late Gothic architecture had not been deeply studied and did not enjoy great prestige among the scholars who dedicated their writings and research to Gothic architecture in France. Viollet took part in this trend. For him, the architecture developed during the 13th century constituted the key point in the history of construction and the Gothic style from the following centuries was the result of a process of gradual decadence.[11]

In his article *Voûte* from his *Dictionnaire* Viollet provides a complete description of the English Gothic vault, comparing it to the French version, making use of the paper of Robert Willis *On the construction of the vaults of the Middle Ages*[12], written in 1842, translated into French and published in the review from Cesar Dally a year later, in 1843.[13] According to Viollet, the English Gothic follows a logical process in its development that is solid, methodic and the consequence of a long experience faithful to an established principle, while the late French Gothic - Viollet does not use the term Flamboyant -, seems to follow fancy impulses without the intervention of need or reason.[14]

Viollet professed a great admiration for the person and work of Robert Willis[15] although he probably did not know his complete writings, including about the Flamboyant style. Viollet demonstrated a deep and precise knowledge about English architecture, particularly in relation to the design and development of the English Gothic vaults, not only through Willis' work but also through direct examination of the structures themselves. This led him to profess an admiration for England and its people in addition to its architecture.[16]

As a result, and taking into account the importance of his work, in 1855 he was nominated to become a member of the *Royal Institute of British Architects* and in 1864 he was honoured with the Royal Gold Medal from this institution. In spite of his somewhat negative appraisal of the French Gothic from the fifteenth and sixteenth centuries, he took on the task of studying it thoroughly. When he analysed the portals, for example, he mentioned the most important ones of the Flamboyant such as those of the Cathedrals of Tours, Beauvais, Troyes, Sens and Senlis.[17] He also developed a detailed study of the façade of the *Hotel de Jacques Coeur* in Bourges, the entrance doors of the *Hotel de Sens* and *Hotel de Cluny* and of the doors to the stairs (Fig. 2).[18]

Auguste Choisy knew Willis' writings from a young age[19] and there is no doubt that they constituted an important foundation for his explanation of Gothic architecture in England in his *Histoire de l'Architecture*[20]. In general terms, he shared the same spirit as Viollet-le-Duc, focusing his study on the Gothic style from the twelfth and thirteenth centuries. This he considered of maximum splendour, although he also identified the main features of the style from the fifteenth century: the proliferation of ribs in cross-ribbed vaults and the absence of capitals allowing a direct connection between the ribs and the shaft. This feature contributed to developing the impost by means of the interpenetration of the ribs. Auguste Choisy also mentioned with quite meticulous attention the peculiarities of the design of members such as buttresses, tracings and rose windows in constructions from the fifteenth century. He only used the term Flamboyant to refer to the traceries or the rose windows. Indeed, Choisy's analysis of the late Gothic is limited to a few paragraphs at the end of each section of the chapter dedicated to Gothic architecture.

At the beginning of the twentieth century a debate on the origin of Flamboyant architecture took place in France, driven principally by Camille Enlart, who was in favour of an English origin[21] and A. de Saint-Paul, who supported the claim of a French origin[22]. From then until the first studies of Paul Frankl in the 1960s, the Flamboyant was mired in a long period of lethargy.

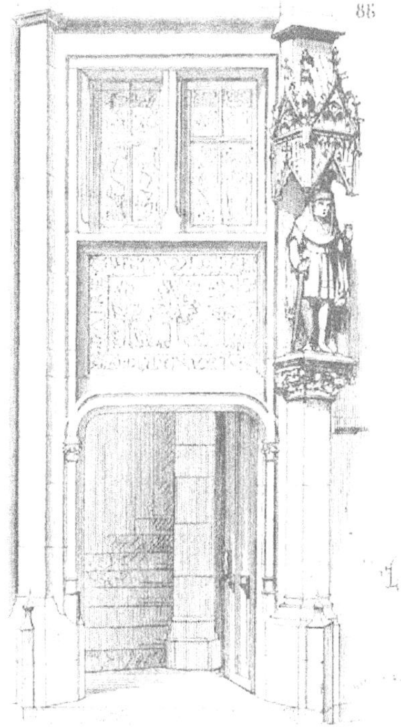

Fig. 2. Hotel de Jacques Coeur, Bourges, Viollet-le-duc, Dictionnaire raisonné de l'architecture française du XIe au XVIe siècle, 1854-1868, vol. VII, Porte, p.464.

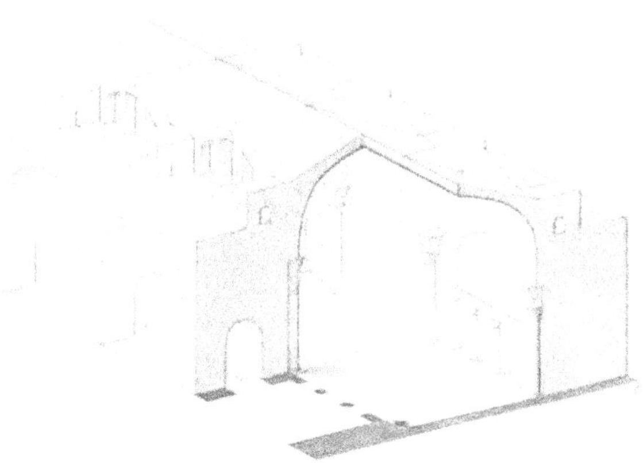

Fig. 3. Axonometrical drawing of the Baths at Rome. R. Willis, Remarks on the architecture of the Middle Ages, especially of Italy. Cambridge: 1835, plate 1, fig. 1.

Robert Willis' contribution to understanding the Flamboyant style

Robert Willis focused his study of the Flamboyant style on the interpenetrations.[23] For him this was a principal matter that led him to write an article dedicated exclusively to this subject. Until then the most important argument used by other authors to define the Flamboyant was the very special design of its tracery, that had given its name. Authors like Rickman or Whewell had already noted the interpenetrations as a feature of the style, but Willis did it in a new sense. The Flamboyant architects introduced new architecturally redundant elements to enrich their formal compositions. Willis realized that, in most cases, these elements were introduced with the single purpose of producing interpenetrations[24]. For Willis the key is the coexistence of these elements in the same place. This is, for him, the difference between the German Aftergothic and the English Perpendicular. It can be understood as an attempt to extend further the constructive Gothic principles, corresponding to each element, from a pure formal point of view, the extrusion of the template of its outline. Intersections and interpenetrations are the combination of these templates merging together, sometimes resulting in unattractive compositions, from the point of view of a mind that appreciates the pureness of the surfaces and their ability to adapt to an ideal pattern based on surfaces such as the sphere, the cylinder, or the cone, but perfectly consistent with the Gothic viewpoint. We could consider the Flamboyant as a decadent and complicate extension of the most rational style from the point of view of the construction. But a modern analysis would lead us to think that it is the consequence of this rational logic, carried to the extreme, occasionally with brilliant results that are sometimes difficult to understand.

The British authors of the first half of the nineteenth century commented on the importance of this architectural style. They analysed it and were concerned to understand its essence, but they were also imbued in the spirit that reigned in France at that time which considered these late versions decadent and degenerate[25]. Willis, however, got closer to the idea that we have now of the meaning of these constructions of the late Gothic.

The tools employed for the architectural analysis

Willis made use of drawing, both as an analysis tool and to justify or demonstrate his arguments[26]. In his *Remarks on the Architecture of the Middle Ages especially in* Italy[27] he made very synthetic axonometrical section drawings that could have influenced Auguste Choisy in the special style of drawing that he developed (Fig. 3).[28] In Willis' paper about the Flamboyant there is an extraordinary analytical drawing that he uses to explain the coexistence of architectonical members in a doorway in Nevers Cathedral. A perspective drawing shows each member that shapes the pilaster, carefully separated from the others, describing each of them more clearly (Fig. 4).[29]

Willis' graphical approach to architectonic and constructive analysis is also new. However, Whewell provides another point of view as he offers an accurate literary description of his studies rather than skillful drawings, which are not to be found in his writings.[30] On the contrary, Ruskin deals with this subject in a very different way. For him, the production of section drawings to explain this kind of intersection graphically cannot be avoided (Fig. 5).[31]

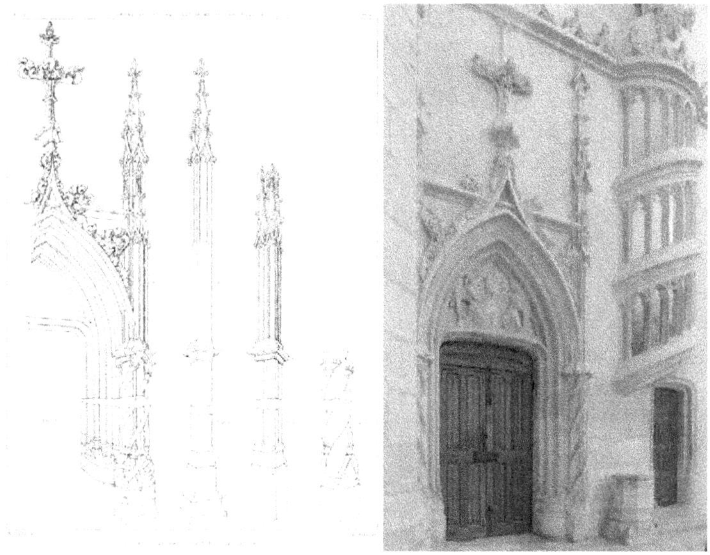

Fig. 4. Doorway of the cathedral of Nevers. R. Willis, On the characteristic interpenetrations of the flamboyant style, 1842.

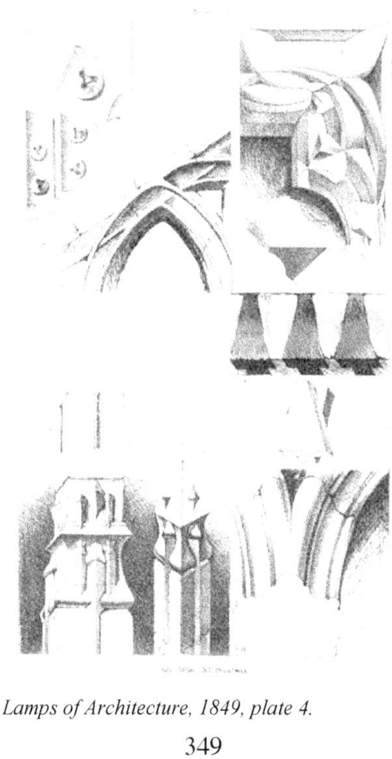

Fig. 5. J. Ruskin, The Seven Lamps of Architecture, 1849, plate 4.

These three authors each show their concern to define accurately the different architectonical members, especially the mouldings and intersections. Whewell fundamentally does it through language and Willis by means of drawing. In the explanation of the beautiful and complicated drawing of Nevers Cathedral, Willis indicates that drawing of these members have been taken with all care and precision.[32] It is possible that he took notes and full size measurements of the mouldings with a cymagraph, an instrument he invented for copying the profiles of the mouldings, [33] also mentioned by Whewell.[34]

Fig. 6. R. Willis, Description of the Cymagraph for Copying Mouldings 1842.

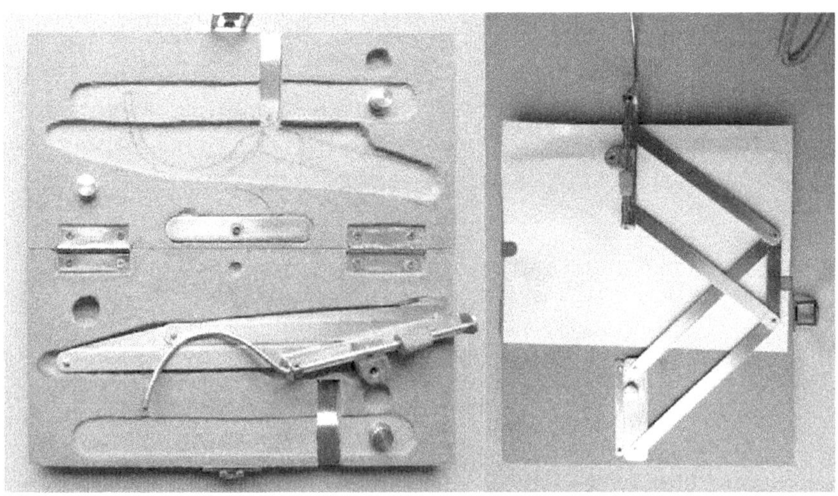

Fig. 7. Robert Willis Cymagraph. Model constructed by Domingo Pliego, 2013 (in process)

Conclusions

Flamboyant architecture was studied thoroughly by the nineteenth century British scholars. Willis' contribution to understanding this style can now be shown to be significant not only for his approach to this subject but also for the way he articulated it through graphical illustration. His analytical drawings demonstrate a surprisingly contemporary approach to research and knowledge in the field of construction history.

References

[1] L. E. Neagley, "The Flamboyant Architecture of St.-Maclou, Rouen, and the Development of a Style" in *Journal of the Society of Architectural Historians* XLVII: 374-396, December 1988; Denis Cailleaux, *La cathédrale en chantier: la construction du transept de Saint-Étienne de Sens d'après les comptes de la fabrique, 1490-1517*. Paris: Comité des travaux historiques et scientifiques, 1999; Florian Meunier, "Les cathédrales flamboyantes de Martin et Pierre Chambiges" in *Revue de l'art*, n° 138, 2002; Agnnès Bos, *Les églises flamboyantes de Paris. XVe-XVIe siècles*. Paris: Picard 2003; Étienne Hamon, *Une capitale flamboyante. La création monumentale à Paris autor de 1500*. Paris: Picard 2011; and the works of Stephen Murray, *The work of Martin Chambiges*, Courtauld Institute-University of London, Ph.D. 1973; "The Choir of Saint Etienne at Beauvais, in Journal of Society of Architectural historians, Vol. 36, No. 2, pp. 111-121, 1977; *Building Troyes cathedral: the late gothic campaign*, Bloomington, 1987; *Beauvais cathedral: architecture of transcendence*. Princeton: Princeton University Press, 1989

[2] N. Pevsner, "William Whewell and his Architectural Notes on German Churches" in *German Life and Letters*, pp. 39-48, 1968; Arcisse de Caumont, "Cours d'Antiquités monumentales", vol. IV, Paris: 1831, p. 8.

[3] W. Whewell, *Architectural notes on German Churches; with notes written during an architectural tour in Picardy and Normandy*, third edition, Cambridge: Cambridge University Press, 1842, Original preface p. 18: 'It so happened, that the churches which came under my notice in the neighbourhood of the Rhine, illustrated very remarkably an opinion which had long appeared to me almost certain, with regard to the introduction of the pointed arch. Some of the modes of building assumed in this theory, which had been only hypothetical suppositions when it was applied to English churches, were found existing as common architectural practices in Germany.'

[4] Whewell, (note 3), chapter 3 "Suggestion on the manner of making architectural notes", 'the architectural observer will become aware how completely words alone may avail to preserve and transmit distinct and adequate conceptions of an edifice. And when he has thus begun to feel the import and value of a technical language…'

[5] A. Buchanan, 'Robert Willis and the Rise of Architectural History' (PhD. thesis, London University, 1994).

[6] Whewell, (note 3), second edition 1835, "Notes on Picardy and Normandy".

[7] T. Rickman, "Four Letters on the Ecclesiastical Architecture of France" in *Archaeologia* vol. XXV, pp. 159-186. London: Society of Antiquaries, 1834, p. 179, 'Mons. De Caumont's name is *Flamboyant*, alluding to the waving of a flame; and the tracery of the windows of this style (which are the great, but not the only distinguishing feature) gives very forcibly the idea of this waving in its dividing lines."

[8] T. Rickman (note 7).

[9] T. Rickman, *An Attempt to Discriminate the Styles of English Architecture, from the Conquest to the Reformation*, London: 1817.

[10] Viollet-le-duc, *Dictionnaire raisonné de l'architecture française du XIe au XVIe siècle.* 10 vols. París: B. Bance, A. Morel, 1854-1868

[11] Viollet-le-duc (note 10), *Voute* p. 548: 'Au point de vue de la structure, l'art de XVIe siècle était, pour les voûtes comme pour le reste, à l'état d'infériorité sur les arts antérieurs.

[12] R. Willis, "On the Construction of the Vaults of the Middle Ages", in *Transactions of the Royal Institute of British Architects*, London, 1842.

[13] C. Daly, *Revue d'architecture*, vol. IV, "Memoire de M. Willis sur la construction des voutes au Moyen-Age" in *Revue Generale d'Architecture*. Paris, 1843 (translation by Cesar Daly).

[14] Viollet-le-duc (note 10), *Voute* p. 549

[15] Viollet-le-duc (note 10), *Voute*, p 525: 'Dejà en 1842, un des hommes les plus distingués en Angleterre parmi les archéologues s'occupant de l'architecture, avec le sens pratique que dans ce pays on apporte à toute chose, M. Le professeur Willis, avait publié sur la construction des voûtes anglaises du moyen âge un travail très-étendu et savamment déduit.'; Robert Mark, "Robert Willis, Viollet-le-Duc and the Structural Approach to Gothic Architecture" in *Architectura* 7 : pp. 52-64, 1977.

[16] Viollet-le-duc (note 10), *Voute*, p 531: ' Il y a, dans cette forme observée part tous ceux qui ont visité l'Angleterre, non pas un caprice, une question de goût, mais l'application rigoureuse d'un système suivi, comme nous venons de le démontrer, avec un esprit méthodique rigoureux dans ses déductions.'

[17] Viollet-le-duc (note 10), *Porte*, p. 436. Most of these portals were designed and constructed by the team of Martin Chambiges, probably the most important architect at the end of the 15th century and the beginning of the 16th in France.

[18] Viollet-le-duc (note 10), *Porte*, p. 460-465. Figs. 87, 88; *Maison*, fig. 39. Viollet agrees with Willis in analyzing the doorways and elements asociated to the spiral staircases that usually have a particular design.

[19] H. Bryon, "Construing Construction with Drawings: Robert Willis' and Auguste Choisy's Axonometric Representations of Vaulted Structures" pp. 673-681 in Nuts and Bolts of Construction History, Vol. 1, Paris: Picard, 2012.

[20] A. Choisy, *Histoire de l'Architecture*, Paris: Gauthier-Villars, 1899.

[21] C. Enlart, "Origine anglaise du style gothique flamboyant" in *Bulletin Monumental* 70 (1906), pp.38-81 and 74 (1910), pp.125-67.

[22] A. de Saint-Paul, "Les origins du gothique flamboyante en France" in *Bulletin Monumental* LXX, 1906, pp. 483-510.

[23] R. Willis, "On the characteristic interpenetrations of the flamboyant style" in *Transactions of the Royal Institute of British Architects of London*, Vol. I, Part II, London: Longman, Brown, Green and Longmans, 1842.

[24] R.Willis (note 23), 'New members are also introduced for the mere purpose of showing interpenetrations'.

[25] N. Pevsner (note 2), p. 45, 'Willis agreed with Whewell that this, i.e. the Late Gothic, was *the dissolution* of the style'; W. Whewell, "Remarks on the complete gothic and after-gothic styles in Germany" in *The Archaeological Journal*, september 1850, p. 230, 'we see the decline, perversion, and disappearance of the principles which had prevailed in the time of the complete Gothic'.

[26] R. Willis (note 23), pp. 81-82, 'Thus, as we shall see, two different bases may be given to the same shaft, or even two or more different turrets with pinnacles may be placed in an identical position on the plan, and made to interfere and interpenetrate throughout their entire height from the base upwards in a manner that defies description, and can only be illustrated by drawings.'

[27] R. Willis, *Remarks on the architecture of the Middle Ages, especially of Italy*. Cambridge, 1835.

[28] H. Bryon (note 19).

[29] R. Willis (note 23), p. 85, 'I have drawn these separately [...], on the same scale and from the same point of view as in the group; and by comparing them singly [...] the mystery of its complication will easily unravelled.'

[30] N. Pevsner (note 2), p. 40, 'he writes that skill in drawing is not absolutely necessary for records on churches, but that *profiles of mouldings...should be copied*'.

[31] J. Ruskin, *The Seven Lamps of Architecture*, New York: 1849, p. 53: 'It is impossible to render a very elaborate example of this kind intelligible, without some fifty measured sections'.

[32] R. Willis (note 23), p. 85: 'but the interpenetrating mouldings are drawn with the sufficient care and precision'.

[33] R. Willis, "Description of the Cymagraph for Copying Mouldings", in *Civil Engineer and Architects' Journal*, vol. V, July 1842, pp. 219-221. 'The purpose for which this instrument is constructed is to obtain exact drawings of the profiles of existing moulding.' According to Willis, other methods were already in use to make measurements of the mouldings (by points, measuring ordinates and distances, by means of lead or clay, or tracing the outline introducing the paper in a transverse saw cut, being this latter plan excellent but only employable in relation with ruins). The instrument consists basically in a stylus with a tracing point that is capable of following the changes of direction of the mouldings. ... 'The base of the machine is a drawing board of mahogany (3/4 in. thick), 10 1/2 in by 11 1/2 in, and made to fold in the middle like a book, for the convenience of carriage. When in use it is kept open by a button R on the lower surface.' Willis' description of the instrument is so detailed that constitute a kind of instructions that enables the construction of the cymagraph, what the authors had made in order to test it (Fig. 7).

[34] W. Whewell, (note 3), Original Preface, p. 43: 'Some of the most desiable drawings may be made sufficiently well without the draftsmans' eye and hand; for example, profiles of mouldings, which are very important elements, and should be copied.'

Dante Bini's air structures (1964-1979)

From early Italian prototypes to the Australian experience

Alberto Pugnale[1] and Alberto Bologna[2]
1.Faculty of Architecture, Building and Planning, University of Melbourne, Australia; 2. Ecole Polytechnique Federale de Lausanne, Switzerland

The construction of concrete shells has always been a difficult and expensive procedure. Preparation of complex formworks, as well as placement of curved reinforcing rods, requires experience and increases the overall cost. Building concrete shells is even more of a concern within the Australian context, where the use of simple and rapid technologies has always been a priority. In the Sixties, Italian architect Dante Bini developed and patented a form-finding and construction technique that aimed to solve this issue, the so-called "Binishells".

This paper investigates Dante Bini's system, from its conception within the Italian environment until its international success in the USA and Australia[1].

1. From Normand W. Mohr to Dante Bini: the pneumatic formwork from 'The City of Saint Francis Idealized' to the 'Architettura autoformante', 1927-2009.

In February 1927, architect Normand W. Mohr proposed to realise the Golden Gate Bridge in San Francisco as an underwater tunnel of reinforced concrete. It should have been built through a pneumatic formwork and that probably appeared (extraordinary?) to the readers of the American magazine *The Architect and the Engineer* [2]. Mohr was experienced in the manufacture of aerostatic balloons: in April 1916 he filed a patent titled *Combined aeroplane and dirigible balloon,* in which characteristics, potentialities and construction procedures of inflatable structures were clearly mastered [3]. Despite his intuition to apply such a system for the erection of circular buildings too, his ideas fell into oblivion for about six years [4]. In August 1933, *The Architect and the Engineer* was still giving space to Mohr's invention by publishing a short article in which, through a rough sketch, he further detailed his construction idea [5]. The steel reinforcement as conceived by Mohr was of traditional type and assembled above the pneumatic formwork just after inflation. Only in 1964, was Italian architect Dante Bini able to introduce a substantial innovation. He thought of realising his concrete shells with an extensible steel reinforcement, which was mounted above a deflated pneumatic formwork, and then deformed to take shape following the membrane inflation.

The assumptions made by Mohr were successfully developed again by Californian architect Wallace Neff, whose reinforced-concrete shell houses, built with inflatable and reusable pneumatic formworks, were a great success [6]. In 1934, Neff made the first drawings of a pneumatic house and in 1939 he wrote a description of his pneumatic form[7].

Neff's patent, which was filed in April 1941, represents the result of the socio-political environment of that period, which also led other architects to work on the same topic and develop low-cost houses or refuges. Such a spirit aimed to build thin-shell concrete structures of barrel and dome type? for dwellings, hangars and barracks for defence housing purposes[8]. Neff planned to insert a series of hooks into a concrete footing slab, to which he would anchor an inflatable membrane that could be shaped in order to assume a desired size and curvature. Neff's form was fabricated from an airtight and substantially non-stretchable material of sufficient strength to withstand hard usage?, as a rubber-impregnated canvas. Once inflated, the formwork was covered with a wire reinforcing mesh of approved type?; thus, with the so-called 'gunite' method, a continuous layer of about three inches in thickness (a mixture of cement, fine aggregate and water) was literally shot against the inflated membrane through a cement gun, in such a manner that the wet concrete kept its position. The concrete passed across the iron mesh and flowed directly against the rubber surface of the formwork. Prior to inurement, the exterior face of the shell was levelled with rods to obtain the final curvature and a smoother surface. After the structure's hardening, the membrane was deflated and then removed through one of the openings that had been created. The subsequent patents filed by Neff between November 1941 and June 1952 aimed to improve concrete casting, manufacture of an economic and reusable pneumatic formwork (Neff suggested nylon, cotton muslins or treated paper of the type developed at that time for parachutes) and positioning of steel reinforcement[9]. Through the employment of such a construction system, by 1941 Neff built twelve experimental houses at the so-called *mushroom village* in Falls Church (Virginia)[10]. In 1945, once he understood the potential of the invention, Neff founded a construction company, the *The Airform International Construction Company,* through which he built hundreds of reinforced concrete shell structures all over the world[11].

During the Fifties there were no further innovations. The shell structures which were born from the collaboration between Neff and architect Eliot Noyes were erected using a well-established construction technique that, in the United States, allowed for a fast production of low-cost houses[12]. In 1954, engineer and mathematician Mario Salvadori worked with Neff and Noyes to improve casting of in-situ concrete through the use of a 'concrete gun'[13].

In parallel to Neff and Noyes' realizations in the U.S., in Europe in summer 1955, Swiss engineer Heinz Isler exploited the proprieties of pneumatic structures. Scale models were used to derive the exact geometry of minimum surfaces in order to build shells with a minimum of material[14]. However, the design process innovation by Isler was not affecting construction, and therefore it was not able to eliminate timber formworks for concrete's jet.(?)

Although Bini admitted being aware of the building systems that led Neff and Noyes to build their 'bubble houses'[15], he argued that his own method was the result of casualness (?), inspired by the observation of a tennis (?) pneumatic structure and (its?) structural response under snow load[16]. In the early Sixties, Bini had the intuition to build a reinforced concrete dome using pneumatic preformed formwork, an extensible steel reinforcement and performing concrete pouring prior to inflation.

Fitting into a typically Italian post-war tradition, Bini was at the same time designer and manufacturer of his own shells, the so-called *Binishells*. In his first prototype, which was built in Crespellano, Italy on July 4, 1964, after constructing a traditional reinforced concrete footing ring, Bini arranged on the ground a framework composed of concentric steel bars connected to arcs of metal chains, above a deflated pneumatic membrane. Such bars were an empirical transfer of the ribs' geometries of Pier Luigi Nervi's *Palazzetto dello Sport* of Rome[17]. Bini's framework was then able to deform from the ground to a tri-dimensional shape of which the planar configuration was just a projection; the anchoring system of the inflatable membrane was comparable to that used for fixing traditional pneumatic envelopes[18]. A reinforced concrete dome (or, as Bini states, a shell structure in 'ferrocemento'[19]) was obtained after sixty minutes of membrane inflation. During the formwork erection, concrete was restrained by a net fence. For that reason, at the final stage it appeared grainy and not compacted. An acceptable smooth surface was obtained through an extrados'(?) hand levelling with a trowel, without applying any mechanical vibration.

The introduction of iron springs as reinforcement in the framework (the rods were inserted inside the springs, anchored to the ground only at one end and therefore free to take the shape given by the dome) took place in the fourth experimental dome, which was built by Bini in Pegola, Italy[20].

Between 1964 and 1965 Bini conducted new experiments and with his company *Binishells Spa* started exporting his patents outside Italy (Fig.1): a thirty metre diameter dome was built in Castelfranco Emilia, with the use of a lightweight reinforced concrete obtained with "Leca-type" aggregates and a new type of framework characterized by a spiral shaped mesh, interlaced with structural steel rods and bars[21].

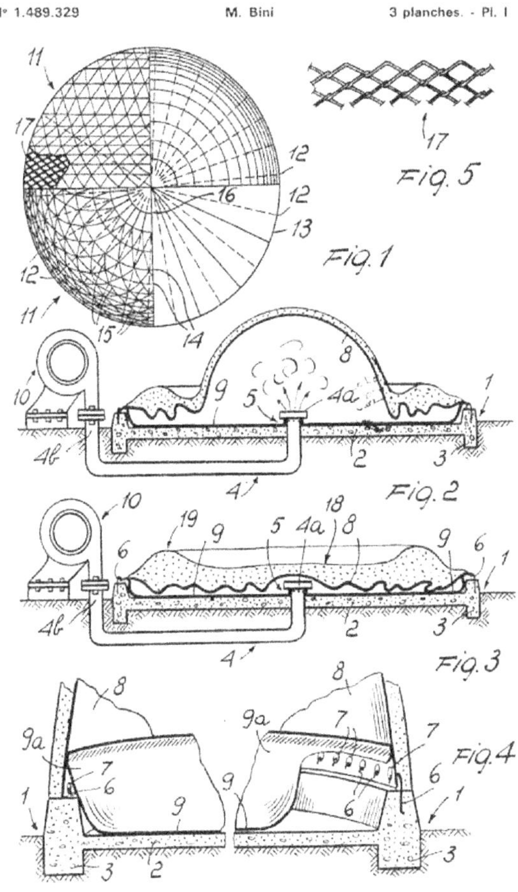

Fig. 1. 1964-1966, construction scheme of the first domes built by Dante Bini (Source: Patent FR 1489329, August 9, 1966).

The description of this construction method crossed the Ocean and at the end of 1965 the authoritative Salvadori was already aware of Bini's first concrete shell dome, which was built in Crespellano: their meeting was destined to mark a turning point in Bini's career. Following the filing of Bini's patent in the United States, a demonstrative dome was erected between May 16th and 17th 1967 at Columbia University, on the campus where Salvadori used to teach (Fig. 2)[22].

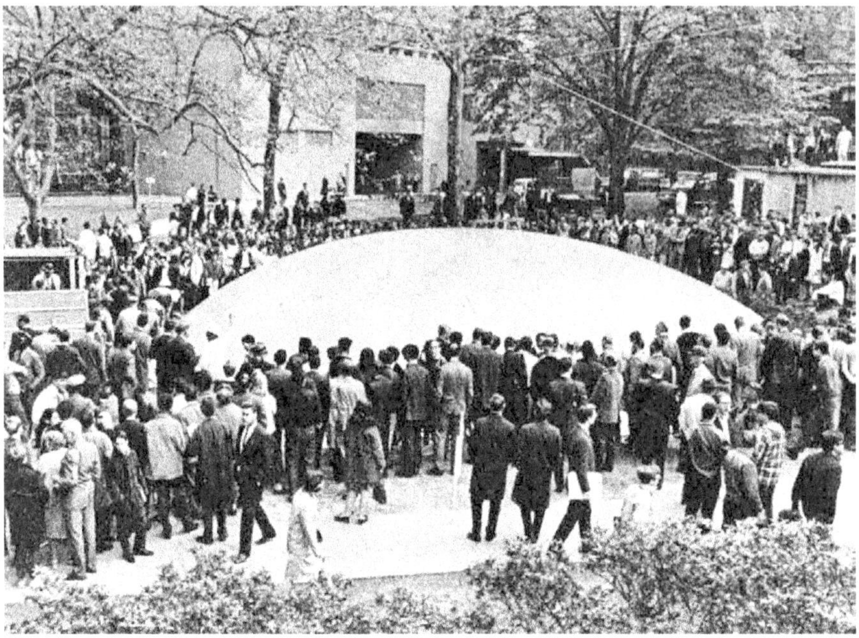

Fig. 2. May 16-17, 1967: demonstrative dome built in the campus at Columbia University (Source: The Architectural Forum, June 1967, p. 85).

On that occasion, Bini employed (perhaps according to Salvadori's advice) a second external PVC membrane, which was laid above the liquid mass of concrete poured prior to inflation. This new element facilitated the levelling process – only mechanical vibrators were rolled over the surface of the PVC membrane in order to obtain smooth internal and external surfaces (Fig. 3).

During the *1st International Colloquium on Pneumatic Structures*, which was organised at the University of Stuttgart, Germany on May 11-12, 1967 by the International Association for Shell Structures (IASS), and took place just a few days before construction of the New York dome, Isler stated that 'it would be logical to use pneumatic membranes directly as formwork. But, at present there still are great difficulties to do this'[23]. Dante Bini challenged Isler's statements in the same colloquium and presented his Italian shell structures. The publication of the proceedings included his contribution with photographs of the latest achievements on the use pneumatic formwork, completed with the second outer PVC membrane [24].

Following the demonstration in New York, Bini acquired international fame and continued to export his patents abroad and overseas. In December 1973, he finally moved to Australia in order to develop his professional activities[25].

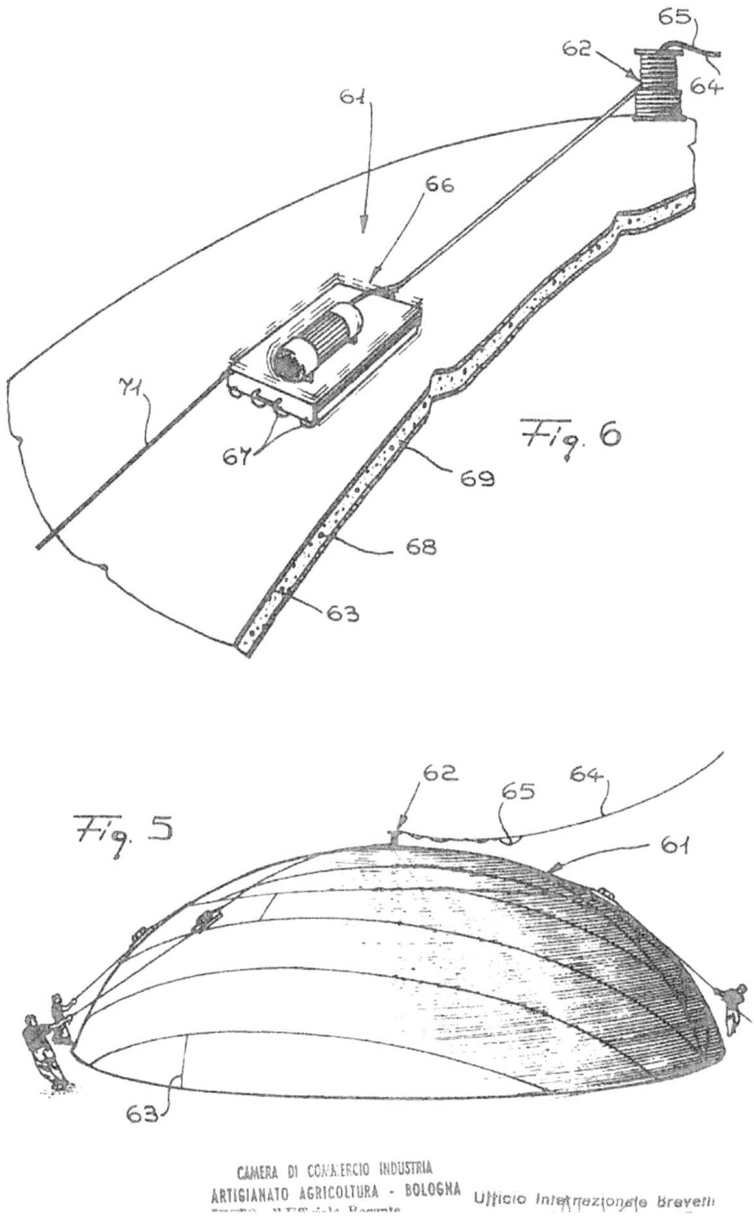

Fig 3. Construction scheme of Bini's domes: concrete is poured between two membranes and is mechanically vibrated (Source: Patent IT 853736, November 21, 1968).

2. From Italy to Australia

The international recognition of Dante Bini happened during a period of considerable expansion for the company *Binishell Spa* which, in 1971, was already licensing and constructing pneumatic shells in more than 20 countries. Bini himself describes that moment as a critical one - while he was occupied in promoting his system abroad, a series of changes in the shares of his company's stock turned against him. Michelagnoli, Cappellini and Torniamenti on the board of directors took command and decided to move the *Binishell Spa* head office in Milan [26]. However, the impact of these changes was minimal. . The activity and success of the company continued to increase.Dante Bini's notoriety grew as a consequence and led him, in 1970, to present in person his construction system to the Japanese Taiyo Kogyo. Bini also decided to visit Australia. Since he was no longer part of the company administration, he decided to explore any potential for investment.

In the same period, in 1971, and entirely separately, , the Australian Minister Leon Aston Punch was looking for a rapid system to build multi-purpose centres, schools and libraries in order to fulfil the promises he made during the electoral campaign. For that purpose, he asked Sir Davis Hughes, who was the general agent of New South Wales (NSW) based in London, to investigate the matter. Dante Bini was immediately contacted, as the innovative *Binishell* system was considered by the Government architect of the Department of Public Works of NSW, Mr. Ian Thomson, as the most suitable for their needs.

Before reaching an agreement, Thomson proposed an inspection of the completed domes in Italy, which happened in autumn 1971. Late in 1973, the official offer from the Department arrived. Dante Bini was invited to move to Australia for a year in order to build ten domes with his innovative technique. He would be appointed "as an architectural consultant to design, direct and train staff of the Department in all aspects of constructing the domes" [27]. Bini accepted and decided to move promptly to Sydney with his family.. More details about his relocation can be found in his autobiography [28].

Bini stayed in Australia, for about six years and created more than the initially scheduled ten shells. After completing his work as consultant for the Department of Public Works of NSW in 1976, he continued to work across Australia. Hee built in Victoria and West Australia, and in 1978 he complted the Space City Shopping Centre in Queensland, the biggest complex of intersecting *Binishells* in the world. The structures conceived to serve as public school facilities were in some cases 18-metre diameter domes, a part of which was combined in pairs to increase the internal volume, and 36-metre diameter shells. A first attempt at listing the Australian Binishells is provided in Table 1, in which buildings are classified by state, programme, shell diameter and current state.

The technology that Bini applied for his Australian constructions was published in detail in 1978, in the Italian Journal *L'Industria Italiana del Cemento* [30], but also in the Australian brochure by the Department of Public Works of New South Wales, which was entitled *Construction of Binishell reinforced concrete domes* (Fig. 4) [31]. Recently, in 2013, *The Architectural Review* republished the same details and photographs from the most relevant Australian constructions by Dante Bini [32].

According to the brochure by the Department of Public Works of NSW, erection was completed in fourteen working days, excluding the anchorage beam previously realised [33]. The construction sequence was divided into twelve phases as follows: (0) a reinforced concrete anchorage circular beam was first constructed, and (1) a nylon membrane was securely placed over a floor slab. Then, (2) reinforcing rods were placed and fixed to boundary steel springs and (3) the concrete was poured until they were fully covered. (4) A PVC sheet was used to cover the concrete and was secured to the anchorage. The system was then ready to rise (5) introducing air through ducts in the anchorage beam, and therefore under the membrane. During the process, (6) tension in the springs was used to control the dome shape. The last phases (7-11) generally required one week more and were related to the membrane removal and realisation of shell openings.

In 1979, Dante Bini decided to leave Australia for a new professional challenge in the United States. At present, he is living in the Napa Valley, San Francisco, while his business activity is currently managed by his son Nicolò in Los Angeles [34].

Tab. 1. A first attempt at listing the Australian Binishells.

STATE	NAME	PROGRAM	ADDRESS	DIAM. (m)	CURRENT STATE
NSW	Narrabeen North Public School	Nursery School	6 Namona St, North Narrabeen	18	Existing
NSW	Killarney heights Public School	Library for a school	Tralee Ave, Killarney Heights	18	Existing
NSW	Peakhurst High School	Multi-purpose centre	Rona Street, PEAKHURST	36	Existing
NSW	Randwick Girls High School	Multi-purpose centre	Barker St, Randwick	36	Demolished
NSW	Pitterwater High School	Multi-purpose centre	Mona St Mona Vale, Pittwater	36	Collapsed [29]
NSW	Ingleburn High School	Multi-purpose centre	Oxford Rd Ingleburn	36	Demolished
NSW	Ku Ring Gay High School	Multi-purpose centre	Bobbin Head Rd, Turramurra North	36	Existing
NSW	Fairvale High School	Multi-purpose centre	Thorney Road, Fairfield West	36	Collapsed and rebuilt
Victoria	Monash University	School	Northways Rd, Churchill	?	Collapsed
Queensland	Space City Shopping Centre	Shopping centre	Anzac Avenue, Kallangur, Brisbane	?	Demolished
Queensland	-	Resort	Vaseley Esplanade, Trinity Beach, Cairns	?	Existing
West Australia	-	Recreation centre	Penistone Reserve, Greenwood	36	Demolished
West Australia	-	Recreation centre	Padbury	36	Demolished
West Australia	Phil Renkin Recreation Centre	recreation centre		36	Existing

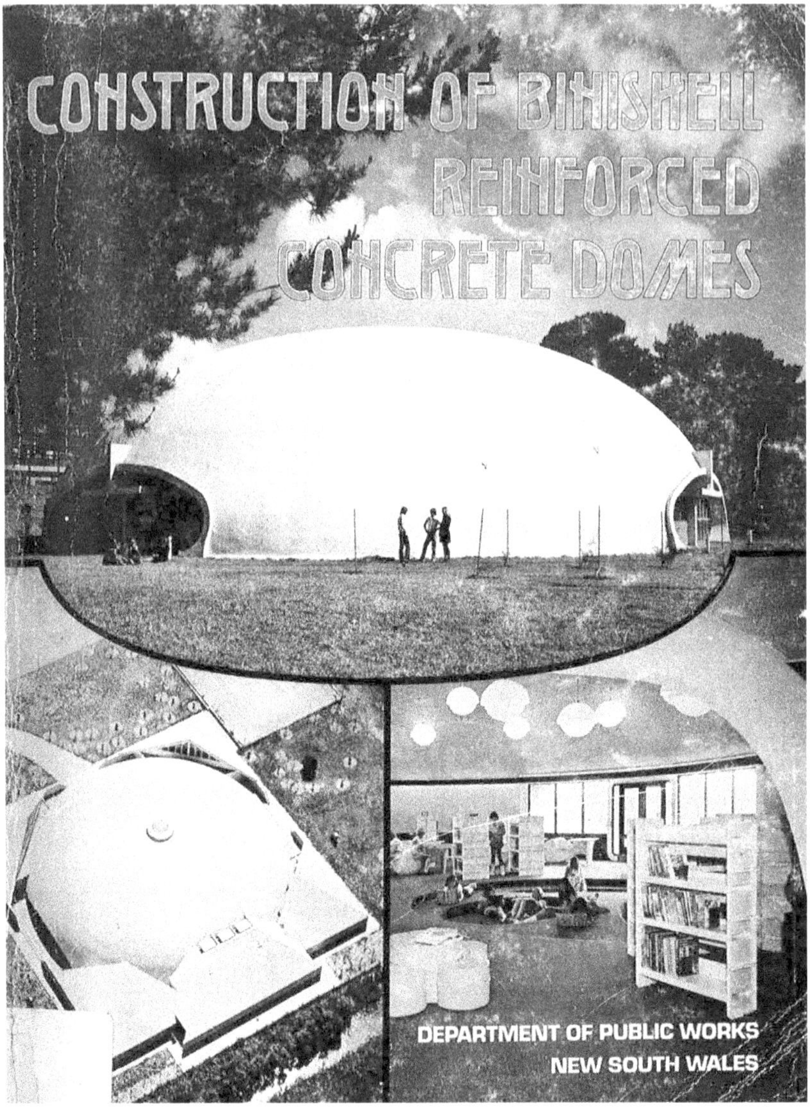

Fig. 4. Cover of the publication by the Department of Public works of NSW on the Binishells built in Australia in 1974 for the Department of Education.

Conclusions

Dante Bini represents one of the last master builders of domes and shells. He has been continuing a purely Italian millenary tradition, an approach to structures which started with the Pantheon and arrives until the realization by Pier Luigi Nervi.

This paper has shown that Bini's invention was the result of on-going research on pneumatic structures that began in the1920s.. Within the field of pneumatic structures, his radical innovation was achieved by combining form-finding and construction phases together. That was achieved by pouring concrete prior to the membrane inflation through the use of a spring system which was able to control the deformation of concrete reinforcing rods during the erection process.

Bini achieved international recognition after realising a prototype at Columbia University in the USA. After that, he moved to Australia in order to build a set of shells with his fast and innovative construction technique. His experience represents, stilltoday, a unique example for Australia, acountry in which form-resistant structures have not yet been widely appliedto consider.

References

[1] This paper is the result of the joint interest of the two authors. Part 1 has been written by Alberto Bologna while part 2 by Alberto Pugnale. Both parts have been revised by the authors and the paper structure has been collectively conceived. This research has been financed by an Early Career Research Project Grant of the Faculty of Architecture, Building and Planning of the University of Melbourne, entitled *Dante Bini's air structures in Australia: construction history and future,* and of which Alberto Pugnale is the chair investigator.

[2] «By the latest system of pneumatic, inflatable and deflatable forms, dock pouring, launching, floating to position and submerging in place, submarine tubes may be built at one-third the cost of suspension bridges», N. W. Mohr, "The city of Saint Francis idealized", *The Architect and Engineer*, February 1927, p. 83.

[3] Patent US 1,064,240, N. W. Mohr, *Combined aeroplane and dirigible balloon,* April 16, 1910.

[4] «Curved in sky line in mass and detail, facilitated by improvement in method of construction (notably the pneumatic form [...]). By this method buildings or circular form will be more economical to construct than our present rectangular system. Also being more closely natural in form, more beautiful», Mohr, *The city*, (note 2) p. 84.

[5] «After plaster shell is set, the rubber form is deflated and is removed with canvas-casing, to be used over and over again. The insulation paper which prevents plaster from adhering to canvas-casing is washed away, leaving a finished blaster interior», W. Mohr, "Thumb Tacks and T-Square", *The Architect and Engineer*, August 1933, p. 4.

[6] W. Neff, Jr. (ed.), *Wallace Neff. Architect of California's Golden Age*. Santa Barbara: Capra Press, 1986, pp. 177-186.

[7] J. Head, *No Nails, No Lumber. The Bubble houses of Wallace Neff.* New York: Princeton Architectural Press, 2011. p. 18.

[8] Patent US 2,270,229, W. Neff, *Building construction,* April 3, 1941.

[9] Patent US 2,335,300, W. Neff, *Building construction,* November 25, 1941; patent US 2,388,701, W. Neff, *Method and apparatus for constructing sheel-form structures,* July 15, 1942; patent US 2,892,239, W. Neff, *Improved method of erecting shell-form concrete structures,* June 19, 1952.

[10] "Balloon Houses Designed for Defense Workers Bloom Under Virginia Trees", *Life*, December, 1 1941, pp. 34-35; "Ballyhooed Balloon", *The Architectural Forum*, December 1941, p. 421.

[11] "Airform House is improved for U.S. market, also going overseas", *The Architectural Forum*, July 1947, p. 15; J. F. Muntz, "Bubbles for defense", in *Wallace Neff 1895-1982. The Romance of Regional Architecture.* San Marino California: The Huntington Library, 1989. pp. 69-89.

[12] "Semitropical Igloo. 'Bubble House' is the new version of a pratical cement dwelling for $6,500", *Life*, February, 22 1954, pp. 75-76; "Air-formed concrete domes. House Hobe Sound, Florida", *Progressive Architecture*, June 1954, pp. 116-119.

[13] L. A. Scipio, *Structural Design Concepts. Some NASA contributions,* Washington DC: Tecnology Utilization Division, National Aeronautics and Space Aministration, 1967, p. 60.

[14] J. Chilton, *Heinz Isler. The Engineer's Contribution to Contemporary Architecture,* London: Thomas Telford Publishing, 2000, pp. 34-37.

[15] D. Bini, "1500 buildings shaped by air", in O. J. Vogl (ed.), *Save Proceedings. International Conference. May 6-9, 1984, Sacramento, California,* Sacramanto: Society of American Value Engineers, 1984, p. 189.

[16] D. Bini, *A cavallo di un soffio d'aria. L'Architettura Autoformante,* Milano: Guerini e Ass., 2009, p. 28.

[17] Bini, *A cavallo*, (note 16) p. 30.

[18] T. Herzog, *Pneumatic Structures. A Handbook for the architect and engineer,* Stuttgart: Verlag Gerd Hatje, 1976, pp. 148-154.

[19] Bini in his writings places his invention in the wake of the domes built by the most renowned designers of the last Century: the use of the term «ferrocemento» to indicate the material of his shells may still underlie his debt to Nervi's work, who patented in 1943 a material called «ferrocemento» with which he realized his well-known prefabricated domes: D. Bini, "Dante Bini, progettista delle stazioni lunari", in F. Kostoris, G. Rossi (ed.), *I leoni della Silicon Valley. Storie geniali di italiani all'estero.* Milano: Guerini e Ass., 2008, p. 124.

[20] Bini, *A cavallo*, (note 16) p. 31.

[21] Bini, *A cavallo*, (note 16) p. 40. Patent IT 18171/65, filed August 10, 1965; DE 1684163, Binishells S.p.A, *Verfahren zum Errichten von Gebäuden,* August 10, 1965; FR 1489329, D. Bini, *Méthode pour l'érection de bâtiments constitués principalement d'une structure portante en une seule pièce, équipement pour réaliser cette méthode, et bâtiment ainsi obtenu,* August 9, 1966.

[22] US 3,462,521, D. Bini, *Method for erecting structures,* December 12, 1966. F. Whitehouse, "Architectural Happening at Columbia University: To Blow a Bubble That Won't Burst, Make It of Concrete", *The New York Times*, May 17, 1967, p. 49; A. Bailey, "The Talk of the Town. Bini Shell", *The New Yorker*, June 3, 1967, p. 28; "Bubble-building", *The Architectural Forum*, June 1967, p. 85.

[23] H. Isler, "Pneumatic shape for concrete shells", in A. M. Haas (ed.), *Proceedings of the 1st International Colloquium on Pneumatic Structures, International association for shell structures,* Stuttgart: University of Stuttgart, 1967, p. 51.

[24] Bini puts the emphasis on the economy of his construction method, explaining that the domes already built are used as barns, offices, one-family houses or multiple dwellings, playground roofs and storage tanks, D. Bini, "A new pneumatic technique for the construction of thin shells", in Haas, *Proceedings* (note 23), pp. 52-55.

[25] See the following patents: DE 1684178, Binishells S.p.A, *Vorrichtung zum Errichten von Bauwerken oder Bauwerksteilen,* April 20, 1967; GR 32683, D. Bini, ΜΕΘΟΔΟΣ ΑΝΕΓΕΡΣΕΩΣ ΚΤΙΡΙΩΝ, April 24, 1967; FR 94811, D. Bini, *Dispositif pour ériger des oeuvres de construction, par gonflage d'un élément expansible,* April 27, 1967; DK 129205, D. Bini, *Fremgangsmade til opførelse af bygningsværker ved hjælp af en oppumpelig ballon,* April 28, 1967; CH 478320, D. Bini, *Structure pour ériger un bâtiment au moyen d'un élément gonflable,* May 18, 1967; IT 853736, D. Bini, *Un metodo per l'erezione di edifici e strutture a membrana perfezionato nel complesso dell'armatura estensibile, nonché nella compattazione del materiale*, November 21, 1968; US 3,686,818, D. Bini, *Expandible reinforcement structure for inflatable domes,* November

20, 1969; CA 875074, D. Bini, *Method for erecting edifices,* July 6, 1971; AT 300284, D. Bini, *Verfahren zum Errichten einer schalenförmigen Baukonstruktion,* November 15, 1971; CY 660 (A), D. Bini, *Method for erecting edifices,* October 31, 1972; AU 7022774, D. Bini, *Expandible reinforcement structure for inflatable domes,* January 8, 1976.

[26] Bini, *A cavallo*, (note 16) p. 56.

[27] New South Wales. Dept. of Public Works, *Construction of Binishell reinforced concrete domes*, Sydney: New South Wales, Australia. Dept. of Public Works, 1978, p. 5.

[28] Bini, *A cavallo*, (note 16) pp. 58-70.

[29] Bini, "The causes of the partial collapse of the Pittwater Binishell", in *Engineers Australia*, 1991.

[30] Bini and Morelli, "Cassaforme Pneumatiche per la Costruzione di Cupole in Cemento Armato", *L'Industria Italiana del Cemento*, June 1978.

[30] New South Wales. Dept. of Public Works, *Construction of Binishell*, (note 27).

[32] McLean, "Domes of Discovery", in *The architectural Review*, February 2013, pp. 86-93.

[33] New South Wales. Dept. of Public Works, *Construction of Binishell*, (note 27). p. 15.

[34] Two websites are currently active and showing the work done by Dante Bini and his son Nicolò: <www.binisystems.com> and <www.binishells.com>.

Lean principles in the design and construction of the Empire State Building

Kenneth F. Robson and Somik Ghosh
College of Architecture, University of Oklahoma, USA

Summary

The Empire State Building was the tallest building in the world when it was officially opened on May 1, 1931. It held that distinction for over 40 years. In 20 months the building was designed, engineered, permitted, demolition of an existing building completed, and the building constructed. At the peak of construction, framework rose 4 ½ floors a week. Starrett Bros. & Eken was the general contractor for the Empire State Building, and the project was completed on time and under budget. In order to complete the Empire State Building under the allotted 18 months schedule, Starrett Bros. & Eken employed innovative construction methods and techniques. Many of these construction methods qualify as tools of lean construction practiced in today's construction industry. Comparing the design and construction processes adopted in the Empire State Building with the underlying concepts of lean construction, it is evident that several lean construction principles were employed in the project, a quarter of a century before lean concepts were formalized by Engineer Ohno at Toyota. Using books, archives and historical records, this paper examines the design and construction processes of the Empire State Building and compares them with the lean construction methodologies that are increasingly employed in the construction industry today.

Introduction

The Empire State Building is arguably one of the most famous buildings in the world. It was the tallest building in the world when it was officially opened on May 1, 1931. It held that distinction for over 40 years. In 20 months the building was designed, engineered, permitted, existing buildings were demolished, and the building constructed. At the peak of construction, framework rose 4 ½ floors a week. The Empire State Building was completed on time and under budget. One year and 45 days from start of construction until completion is extraordinary for a project of this size.

Twenty months from the first signed contracts for design to completion of construction of a building with 1.2 million square feet of rentable space was a remarkable achievement in itself. It required adoption and application of sound planning and control practices by the stakeholders of the project. Interestingly, the Empire State Building was built at a time when many of the management practices that are at our disposal today, were not formalized. The processes adopted for design and construction of the building were models of efficiency and bear striking resemblance with many modern day management practices. One such practice is that of lean construction. This paper analyzes the design and construction processes employed in the Empire State Building and shows how lean principles were being employed on this project much earlier than their formalization by Engineer Ohno in the 1950s.

Key stakeholders in the Empire State building project

The key organizations in the design and construction of the Empire State building were Empire State, Inc., the owners; Shreve, Lamb and Harmon, the architects; and Starrett Bros. and Eken, the contractors. These organizations were comprised of many individuals. There were, however, key individuals within each organization that committed their companies to the team approach that was so influential to the success of the project.

The backgrounds of the key stakeholders were varied and undoubtedly played an important role in the management processes of the Empire State Building's design and construction. John J. Raskob, the driving force of Empire State, Inc., had worked for the du Pont de Nemours Company. He became a close, trusted advisor to Pierre du Pont. In this capacity he convinced du Pont that the profits made from manufacturing gunpowder during World War I could be put to better use by investing in General Motors stock than reinvesting it in the company. This strategy was so financially successful that Pierre du Pont eventually became chairman of GM and Raskob became the vice-president. Pierre du Pont and Raskob made millions in the purchase of GM's stock. Alfred Smith's working career included a variety of jobs until at age 21 he was appointed as a subpoena server for the commissioner of jurors. This was his stepping stone into politics and the eventual governorship of New York. The Empire State Building was Raskob, Pierre du Pont and Al Smith's first entry into the building/real estate market.

Both Shreve and Lamb were college educated, joined the architecture firm of Carrere & Hastings and eventually formed their own firm of Shreve and Lamb. Harmon joined the firm in 1929 and it became Shreve, Lamb and Harmon. "Shreve's proclivities were organizational – his was the genius that solved the operational and administrative problems that had the Empire State Building completed in one year"[1]. Their firm believed that when dealing with the challenges presented by a large office building that "these problems must be dealt with through authority greater than the architect possesses; many of them require ability, experience and organization beyond the scope of a single professional unit, or would, if undertaken by the architect's office, involve a duplication of effort and loss of time too expensive to be tolerated in an operation requiring large capital investment"[2].

Paul Starrett had been employed in the architectural firm of Burnham and Root in Chicago. It was here that he discovered his passion for building while working as a construction superintendent for the firm. Starrett began his construction career with the Fuller Company in New York, starting as a time-keeper. His talents landed him the position of superintendent on the Washington D. C. Union Station project. During World War I, he was responsible for domestic construction for the War Industries Board, becoming well known for his efficiency in building. In 1922, Paul and William joined forces with Andrew Eken, to form Starrett Bros. & Eken. William Starrett quoted Ward P. Christie, a civil engineer, in his book, *Skyscrapers and the Men Who Build Them*, that a contractor's function was 'not to erect steel, brick or concrete, but to provide skillful, centralized management for coordinating the various trades, timing their installations and synchronizing their work according to a predetermined plan, a highly specialized function the success of which depends on the personal skill and direction of capable executives"[3]. The key players' careers and management philosophies were developed during a time of rapid change in America. Continued industrialization, World War I, and the roaring twenties had a great influence on American industry. It was during this time that large corporations, along with management theories to successfully manage them, began to develop.

Evolution of lean construction

Engineer Taiichi Ohno of Toyota developed lean production as a design and production management system in the 1950s. However, the term "lean" was coined by the research team investigating the waste reduction practices of the Toyota production system[4]. The main aim of lean production is to shorten the cycle time to get products to market by minimizing waste. Lean production, advocated by Engineer Ohno focused on optimizing the entire system instead of optimizing a few selected activities, as done in mass and craft production. According to Engineer Ohno, lean production philosophy should provide custom products to the customers, deliver instantly with minimum inventory, and produce with zero waste[5]. Since the 1950s, lean principles have been applied successfully worldwide in the automobile industry. Manufacturers such as Toyota have employed lean practices to achieve production with minimal waste. Lean principles are being increasingly utilized in various industrial sectors with success, especially when used in a repetitive or continuous production environment[6]. Lean philosophies have also been adapted and adopted in the construction industry. Guided by the research efforts to adapt concepts of lean production to the construction industry, a number of companies have embarked on a lean conversion path.

Lean construction accepts Ohno's entire production system design criteria as a standard of perfection. Waste in construction and manufacturing arises from the same activity-centered thinking. In contrast to the traditional form of production management in construction that aims to optimize the project activity by activity, assuming customer value has been identified in design, lean tries to optimize the whole project as a production system. Lean views the entire project in production system terms and brings into practice the concept of "value generation" instead of "activity management"[7]. Unlike traditional production planning, lean construction identifies the concept that planning and control go hand in hand throughout the project. While planning defines criteria for success and produces strategies for achieving objectives, control causes actions to conform to the plan and initiate re-planning when necessary. As defined by Koskela et al. [8] lean construction is a "way to design production systems to minimize waste of materials, time, and effort in order to generate the maximum possible amount of value". Based on the underlying concepts of lean construction, a number of tools have been formally developed to be utilized by practitioners. A study of the management practices and construction means and methods on the Empire State Building clearly shows that lean principles were utilized on the Empire State decades before their formalization by Engineer Ohno at Toyota.

Conceptual phase

From the very outset, Empire State, Inc. under the guiding influence of Raskob, established the vision for the Empire State – it was to be the tallest building in the world, and be built as quickly as possible. Raskob understood the value of assembling a team of experts quickly if they were to meet their May 1, 1931 or October 1, 1931 opening deadline (a final decision had not been made on the opening date). Either deadline would require timely decisions by the owners. On September 9, 1929 the architects were hired. On September 11, 1929 Alfred Smith and Paul Starrett dined together to work out the details of the agreement and on September 13, 1929 the construction company received a letter of agreement confirming their discussions and a signed contract was in place on September 20, 1929[9]. In the July 1931 issue of *The Constructor* John Carmody stated that demolition of the Waldorf Astoria began on September 24, 1929[10].

Empire State Inc. realized the importance of establishing the collaborative relationship with the architects and the contractors, and assembled the team at a pace that is not common in construction (and certainly not common for a project of this magnitude). The owner realized the importance of collaboration in managing the interactions between the various activities of the Empire State Building project in order to deliver it in the shortest time. The fact that Paul Starrett received the signed contract from Alfred Smith within ten days from their initial meeting is evidence to the mutual trust among the key stakeholders. The trust established at the beginning of any project lays the foundation for the exchange of commitments throughout the duration of a project. For such an ambitious project as the Empire State building, it was absolutely critical to establish the trust among the stakeholders so that they remain accountable for their commitments in the later part of the project. The following incident is a testimony to the trust that existed among the stakeholders of the project:

During job negotiations for the project, agreements were reached to change some of the insurance and payment conditions. The contract was ready for signature within a very short time period. At the signing, several of the agreed upon points had not been changed. Paul said "Governor, that isn't what we agreed on" [11][and] Alfred Smith would turn to his lawyer, dictate the changes and the changes were made immediately.

In order to construct the building under the allotted 18 months schedule, Starrett Bros. & Eken employed a method of simultaneous design and construction – a technique now known as fast-tracking[12]. Excavation of the site began before the demolition of the existing structure was complete. The contractor had employed this same technique a year earlier on the Manhattan Building. The concept of fast-tracking was not commonplace in those days. But the fact that Starrett Bros. & Eken had used it in another project shows their inclination of being early adopters of innovations, and gives a hint of 'pursuing continuous improvement'. The innovative construction methods and techniques employed to construct the Empire State Building project are well documented and were important aspects in the speed of construction. These innovative construction methods were part of an overarching management process. Lean philosophy advocates that any new technology has to be introduced only after it is proven through direct experimentation – in this case the contractors experimented with fast track construction on the Manhattan project before the Empire State Building project.

The project delivery method employed in the Empire State Building project bears a close resemblance with the Integrated Project Delivery (IPD) method of today. Having a clear set of objectives for the delivery process is an integral aspect of lean philosophy. Lean construction advocates the importance of collaboration throughout the lifecycle of the project. A delivery method similar to IPD provides a formal platform for collaboration among the stakeholders of any construction project. The owners of the Empire State Building project realized the importance of collaboration and formed the building committee comprised of the architects, engineers and production designers to work collaboratively. Collaboration and teamwork allowed the Empire State Building to be designed and constructed in record setting pace. Friedman's essay "A Story a Day: Engineering the Work," in the book *Building the Empire State*, states that "No one involved in the work was a lone genius; rather the group of architects, engineers, and contractors worked collectively to overcome the problems posed by a building and a building process that in many ways has not been surpassed"[13]. Lean philosophy emphasizes the importance of teamwork and trust building for the success of any construction project. Bringing all the key stakeholders aboard early in the project lead to empowerment of the members and increased sense of belongingness to the team. Koskela [14] indicated that empowerment of participants play a vital role in the rapid response to any problem. The perspective of empowerment and belongingness give rise to a feeling of making valuable contributions as a team. Participant involvement and the resulting increase in communication also establish a base level of trust, which allows participants within a system to shift their attention toward improvement at the system level instead of simply defending their limited interests[15]. The establishment of the Building Committee, including all key stakeholders, immediately after the design and construction contracts had been signed, formed a cohesive team that had the completion of the tallest building in the world in less than two years as its overarching goal.

Design phase

The Building Committee for the Empire State Building project was set up so the key stakeholders could meet on a regular basis. The approach similar to IPD advocated by lean construction facilitated improved communication and the efficient exchange of information without the communication barriers typically experienced in construction projects. According to Tauranac [16], the establishment of the Building Committee made the "cycle of information exchange between disciplines short, enabling rapid decision making and response to mistakes in the design or incompatibility between the design and the reality of the construction site."

The Building Committee met on a continual basis, often several times a week, to make timely decisions on the design and building materials and methods. The program was a simple one. According to Shreve "a fixed budget, no space more than 28 feet from window to corridor, as many stories of such space as possible, an exterior of limestone, and a completion by May 1, 1931, which meant a year and 6 months from the beginning of sketches"[17]. Each new design required input from the Building Committee and revised cost estimates and presentations to the executive committee. After approximately four weeks of 16 design variations, their 17th version, Scheme K was adopted on October 3, 1929 by the executive committee. The builders, engineers and subcontractors had taken just two hours the night before to give their approval to "Scheme K"[18].

The design approach for the Empire State Building project was kept relatively simple and standardized for a project of this magnitude. Standardization, which has been identified by lean construction as one of the methods to increase efficiency was utilized in the design approach. The result was a shortened cycle time for production and less waste. Due to the standardized approach, the cycle times for producing the design alternatives could be kept exceptionally low (the design committee produced four alternatives per week on average). Another aspect of the design process that bears close resemblance with the lean principles is the application of target value design. Target value design evolved from target costing that has been extensively used in the manufacturing industry for management and profit planning. Target value design (cost and quality) requires collaboration among designers and contractors throughout the design phase. In this approach, designers need contractors' assistance to determine viable design alternatives, to estimate costs, and to provide value engineering services as the design evolves. The building committee of the Empire State Building project was instrumental to implement this approach.

John McInerney, the on-site construction manager of the Empire State Building project said "The building presented problems to architects and engineers for which no precedents existed. However, the architects, contractor and engineers united in reducing every question to mathematics, ascertaining to a pound or an inch the complete accuracy of their plans. The building was planned and built, down to the smallest detail, on paper before the first step was taken"[19]. It is evident from the description of the design processes that the resemblances of the management practices of the project with that of lean construction were not sporadic. The planning process followed in the Empire State Building project, as described by John McInerney is very similar to the tenets of production planning of lean construction. Production planning in lean construction involves breaking down each project to the detailed task level – if possible to the level of accounting for every motion (referred to as task explosion). Lean construction acknowledges the interdependence and dynamic nature of construction processes, and thus by breaking down the tasks to a finer detail, it tries to identify the constraints that can impede the successful completion of the tasks. The production planning process in lean construction is carried out in a collaborative fashion by directly involving all the stakeholders. Once the constraints are identified the production planning team tries to find intervention strategies to eliminate, or if not possible, to reduce the constraints.

The importance of collaboration among the stakeholders cannot be emphasized more in lean philosophy. The seed of early collaboration that was engrained in the Empire State Building project through the formation of the building committee was evident in the process of preparing the specifications as well. Shreve said "specifications, if wisely prepared, express the consensus of opinion and experience of the architect, builder, the engineers, and the owner."[20]. Immediately after the preliminary drawings were produced, the architects developed a set of outline specifications. These specifications were sent to the builders, owners, subcontractors and material suppliers for comment and input. Decisions reached from the collaborative input were the basis for the working documents, contracts and the job schedule[21]. By involving the different stakeholders in scrutinizing the specifications, the architects utilized the stakeholders' expertise. In some of the recent literature of lean construction, not utilizing the talents of the team members has been identified as a form of waste. Interestingly, the Empire State Building project team was effective in utilizing the expertise of all the participants.

An essential feature of lean is to pursue continuous improvement. Seeking perfection and continually making modifications to the existing processes to make improvements are emphasized in the lean philosophy. The same attitude of continuous improvement is reflected in Paul Starrett's comment: "the singular unanimity of our thoughts in this instance was one that prevailed throughout the whole job. I doubt if there was ever a more harmonious combination than that which existed on the Empire State, between owners, architects, and builder. We were in constant consultation with both of the others; all details of the building were gone over in advance and decided upon before incorporation in the plans"[22].

Construction phase

Lean construction envisions construction as a "project based production system." Shreve, Lamb and Harmon and Starrett Bros., Inc. viewed the construction of a skyscraper similar to Raskob – as a production process similar to the manufacturing industry. Shreve stated "as far as possible hand work was done away with, for in quantity production with thousands of pieces of each material identical in shape and size, the delay would have been disastrous. Windows, spandrels, steel mullions and stone, all fabricated in various parts of the country, were designed so that they could be duplicated in tremendous quantity with almost perfect accuracy and brought to the building and put together almost like an automobile on the assembly line"[23]. William Starrett stated "given this design, our job was that of repetition – the purchase, preparation, transport to the site, and placing of the same materials in the same relationship, over and over. It was, as Shreve said, like an assembly line – the assembly of standard parts"[24]. In lean philosophy standardized work is thought of as the foundation for continuous improvement and the reduction of errors. On the Empire State Building project, standardization was practiced throughout and was one of the reasons why the process was so efficient.

To effectively manage the construction of the Empire State, Starrett Bros. & Eken had to make an early decision – whether they would approach the management of the job using a centralized or decentralized method. They chose the decentralized method. John Carmody, in charge of production and cost distribution on the project, explains the decentralized method. "The essential nature of the decentralized expediting method places the responsibility of meeting his part of the schedule directly upon each subcontractor and allows him to utilize freely all of this specialized knowledge and skill in coordinating his work with that of other subcontractors and with operation of the general contractor." He goes on to say that "a qualified general contractor who selects only qualified subcontractors and succeeds in establishing mutually intelligent management will find that the system of decentralizing responsibility for expediting will respond splendidly to intelligent leadership"[25]. Additionally, Carmody believed this method encouraged initiative by the subcontractors. This decentralized method of management allowed the individual subcontractors and suppliers to determine the best methods of producing and/or installing their particular materials.

Starrett Bros. & Eken continued the committee approach by establishing the Empire State Builders Club, a club that consisted of all of the subcontractors and suppliers key personnel on the project. They had an open discussion once a month to address coordination issues and foresee any problems on the job [26]. Lean construction advocates a decentralized bottom-up management approach, to utilize the talent and expertise of those doing the work (in this case the subcontractors). The workers are most knowledgeable about their own scope of work with a realistic understanding of their working conditions, and hence their inputs are invaluable. By placing the responsibility of meeting the schedule on each subcontractor, Starrett Bros. & Eken moved away from the traditional approach of directive project control and increased the workers empowerment. On one side this approach made the system more reliable as the commitments of meeting the schedule were actually coming from the subcontractors, and at the same time it increased the accountability of the subcontractors. Starrett Bros. & Eken's management systems would insure that suppliers, subcontractors and tradesmen on the project would also buy-in to the common vision established by Empire State, Inc.

Supporting the production approach to construction, William Starrett believed that construction could be accomplished more rapidly if the materials were delivered in a timely fashion – especially steel. He developed this belief while working as an inspector for Burnham and Root, a prominent Chicago architect [27]. Idle workers (waiting for materials to arrive) or workers having to move their own materials are identified by lean as two of the eight types of wastes (considering underutilization of talent as the eighth waste) prevalent in the construction work sites. On the Empire State Building, Starrett Bros., Inc. made sure that materials were delivered to the craftsmen when and in the quantities needed. A rail system was set up on the first basement and first floor, and upper floors as needed, to deliver materials to the craftsmen in a more efficient manner.

In lean construction, the pull system and just-in-time deliveries go hand in hand – when materials are needed they are ordered and delivered just-in-time. On the Empire State Building project, designers provided designs at the "last responsible moment" and the constructors called for additional materials at the "last responsible moment" as well. Structural steel is a prime example of the pull flow on the project, beginning at the design level. Designers would detail several floors at a time and provide those drawings to the steel fabricators for detailing "just in time." Fabrication of the steel began hours after receipt of drawings and detailing [28]. Steel was reportedly delivered within 80 hours of fabrication in Pennsylvania to the Empire State Building. To avoid bottlenecks, materials were obtained from several different suppliers based on the demand and capacity of the suppliers.

Steel was one of the four pacemakers identified by Starrett Bros., Inc. early in the planning stages of the project. Pacemakers were considered the "four divisions of construction work which had to take the lead and set the pace for other trades that followed" [29]. The pacemakers for the Empire State Building were structural steel, concrete floor arch construction, exterior metal trim and cast aluminium spandrels and exterior limestone (including common brick backup). These four pacemakers were the basis for the entire construction schedule. To reduce variability in workflow, lean construction has adopted the concept of 'takt' time from mass production. The activities were thus grouped into four streams according to the 'takt' time each controlled by the pacemakers. The grouping of the activities were done based on the base 'takt' rate so that the variability of the overall work process could be reduced. During production control, the tasks that took longer than the takt time were identified and multiple crews were assigned for those tasks to be performed simultaneously in multiple floors following a linear schedule [30].

Conclusion

The clear vision established by Empire State, Inc. – build the tallest building in the world and have it completed and ready for tenants by May 1, 1931 – provided the framework for the management systems established on the Empire State Building. To achieve the ambitious goal, the owner and the contractor sought to incorporate innovative techniques and methods from the project's inception through the design and construction phases. As discussed previously, many of these techniques and methods bear close resemblances with the tools of lean construction. Starting with the formation of the building committee comprised of the project's key stakeholders, the design and construction processes employed on the Empire State Building resembled the underlying concepts of lean long before lean concepts were formalized by Toyota. Shreve, Lamb and Harmon made innovative design decisions in consultation with the building committee to speed construction. The design process was kept relatively simple and standardized for a project of this magnitude and target value design was utilized to provide more value to the owner. The key stakeholders, especially the designers, contractor, and the owner viewed the construction of the skyscraper as a production process similar to manufacturing industry. Fast track construction, a relatively new process, was used by Starrett Bros. & Eken on this project to meet the schedule. Starrett Bros. realized the importance of utilizing the knowledge of the subcontractors in order to meet the aggressive schedule of the project and decided on a decentralized method of management thus allowing the individual subcontractors and suppliers to determine the best methods of producing and/or installing their particular materials. Efficient management of delivery of materials that impacted the pace of the construction such as steel was handled by an expeditor. Just in time delivery was adopted in order to increase the efficiency of materials delivery and installation. To reduce workers idle time, a rail system was set up to deliver materials to the craftsmen in a more efficient manner. Other efforts such as utilization of takt time, standardization of installation processes, efficient delivery of materials, and utilization of the expertise of the different stakeholders were utilized in the Empire State Building project in order to achieve the common goal of providing the owner with the tallest building in the world.

As evident in the discussions presented in this paper, many of the processes adopted during the design and construction phases of the Empire State Building project bear close resemblances with the concepts of lean construction. To handle the increasingly complicated projects, the construction industry is developing new delivery methods and adopting novel concepts, lean being one of them. However, it is worth mentioning that the modern day concepts of lean construction were formalized more than 50 years after the completion of the Empire State Building project. While the construction industry is putting more and more emphasis on collaborative teams and speedy construction, studying the management systems and collaborative teamwork environment developed during the design and construction of the Empire State Building provides valuable lessons for today's architecture, engineering and construction market.

References

[1] J. Tauranac, *The Empire State Building: The Making of a Landmark.* New York: Scribner, 1995, p.101.

[2] R. Shreve, The Empire State Building Organization. *The Architectural Forum*. 1930, p. 771.
[3] W. Starrett, *Skycrapers and the Men Who Build Them*. New York: Charles Scribner's Sons, 1928, p.87.
[4] J. P. Womack, D. T. Jones, and D. Roos, *The Machine That Changed the World*. New York, NY: Rawson Associates, 1990.
[5] G. A. Howell and G. Ballard, "Implementing Lean Construction: Understanding and Action", paper presented at the 6th Annual Conference of the International Group for Lean Construction (IGLC-6), Gauruja, Brasil, 1998.
[6] G. A. Howell, "What is lean construction?", paper presented at the 7th Annual Conference of the International Group for Lean Construction (IGLC-7), Berkeley, United States, 1999.
[7] Howell and Ballard, 1998 (Note 5).
[8] L. Koskela, G. Howell, G. Ballard, and I. Tommelein, "The Foundationss of Lean Construction", in R. Best & G. De Valence (eds.), *Design and Construction: Building In Value*. Oxford, UK: Butterworth-Heinemann, Elsevier, 2002.
[9] Anon, "Skyscrapers: The Paper Spires", *Fortune*, 58, 1930, pp.119.
[10] J. Carmody, 1931a. "Expediting The Empire State Building". *The Constructor* 23-27, p.24.
[11] P. Starrett, *Changing the Skyline*. New York: McGraw-Hill Book Company, Inc., 1938, p.292.
[12] D. Friedman, "A Story A Day: Engineering the Work" in C. Willis (ed.), *Building the Empire State*, New York: W.W. Norton & Company, 1998, pp. 43.
[13] D. Friedman 1998, (Note 12) p.46.
[14] L.Koskela, *Application of the New Production Philosophy to Construction*. Standford, CA: Center for Integrated Facility Engineering, Department of Civil Engineering, Stanford University, 1992.
[15] G. A. Howell and G. Ballard, 1998 (Note 5).
[16] J. Tauranac, 1995 (Note 1).
[17] W. Lamb, "The Empire State Building, VII. The General Design", *The Architectural Forum* 54, 1931, p.1.
[18] J. Tauranac, 1995, (Note 1) 157.
[19] J. Tauarnac, 1995, (Note 1) p.181.
[20] Shreve 1930, (Note 2) p.772.
[21] Shreve, 1930 (Note 2).
[22] Starrett, P., 1938, (Note 11) p.293.
[23] W. Lamb, 1931, (Note 17) p.6.
[24] Starret, P., 1938, (Note 11) p.296.
[25] J. Carmody, 1931a. (Note 10), p.24.
[26] J. Tauranac, 1995, (Note 1) p.181.
[27] P. Starrett, 1938, (Note 11) p.72.
[28] R. Sacks and R. Partouche, "Empire State Building project: Archetype of 'mass construction.'", *Journal of Construction and Engineering Management* 136, 2010, pp. 702-710.
[29] J. Carmody, "The Empire State Building. X. Field Organization and Methods", *The Architectural Forum* 54, 1931b, pp.504.
[30] R. Sacks and R. Partouche (Note 28)

Other references

1. Hagley Museum and Library Archives, Catalog Description of du Pont Papers. Empire State Building Papers, 1929-1953. Call Number LMSS: 10.

2. Hagley Museum and Library Archives, Contract between Empire State, Inc. and Starrett Bros. & Eken Exhibit C of Financial Prospectus, Papers of John J. Raskob (Manuscript Accession #473), File 743.
3. Anon., 1950, The Skyline Builders, *Fortune*, 95.
4. Anon., Notes on Construction of Empire State Building in *Building the Empire State*, C. Willis (ed.), pp.1-77.
5. N. Bascomb, *Higher: A Historic Race to the Sky and the Making of a City*. New York: Doubleday, 2003.
6. W. Starrett, Making Buildings From Blueprints. *Engineering News-Record*, 1931, pp. 326-330.
7. D. Wren, *The History of Management Thought.* Hobokan: John Wiley & Sons, 2005.

Acoustic Design in the 1930s: Robinson and Keefe's Church of Corpus Christi, Dublin

Fiona Smyth
School of Architecture, University College Dublin, Ireland

Introduction

On 22nd of December 1915, the *Architects' and Builders' Journal* published an article, written by Hope Bagenal (1888-1979), on the acoustics of three London cathedrals.[1] The article was derived from a thesis which had been submitted to the Royal Institute of British Architects (RIBA) as part of the author's final examinations in architecture a little over a year earlier. Examined under the category of "science as applied to building", that thesis and the published article were supported by a research methodology based on extensive site surveys and analyses. Thesis, article, and their common research methodology constituted the earliest work on architectural acoustics undertaken by Bagenal. Based on the turn of the century advances of physicist Wallace Clement Sabine (1868-1919) in Harvard, that work by Bagenal also represented the first British research on architectural acoustics to have had a quantifiable basis in science.

Bagenal's early work on churches and cathedrals was based on a system of parallel surveys.

These aligned objective findings - based on measured acoustic and architectural data - with the subjective impressions of those who used the spaces for speech and for music. The aim was two-fold: to determine a catalogue of subjective parameters that constituted qualitative acoustics, and to establish objective design criteria for those conditions. The only objective parameter that existed in contemporary architectural acoustics was reverberation time: a measure of the length of time taken for a sound to decay to inaudibility within a given enclosure.[2] Reverberation time had first been mathematically defined by Sabine at Harvard some years previously, and published in *The American Architect*.[3] Intrigued by the scientific elucidation of a hitherto qualitative concept, Bagenal had entered into correspondence with Sabine directly on the topic. From that exchange of letters, dating from 1914, Bagenal had incorporated the physicist's methods of acoustic data collection into his system of parallel surveys.[4]

In Bagenal's work, Sabine's formula was implemented to determine predicted reverberation time values from architectural surveys. A measured reverberation time for different registers of the musical scale was also recorded. Bagenal's comparison of measured versus predicted reverberation time values was to prove invaluable in shaping the course of subsequent research in architectural acoustics. As the surveys progressed, deviations between the measured and predicted values were identified for different vocal registers, prompting further investigative work into the acoustic effects of specific materialities and forms.[5] Much of this later work was undertaken in conjunction with the Building Research Station (BRS).

The acoustic data filed for the various ecclesiastic sites also included a comprehensive survey of all internal materials. These were catalogued in terms of surface area and respective absorption coefficients. The absorption coefficients were derived in some instances from data recorded by Sabine, and in other instances from data recorded by Bagenal and fellow researchers at the BRS. Bagenal was involved in experimental work at the BRS from 1923 onwards when the Station's investigative remit was first extended to include architectural acoustics.[6] Early acoustic research undertaken at the BRS focused on quantifying the absorptive effect of various materials commonly used in the British construction industry. In addition, new materials, particularly new recipes for acoustic plaster were also developed. The data for each material was tabulated in terms of absorption coefficients for different octaves of the musical scale ranging from C1 to C7. This data could then be employed as a reference tool for predicting reverberation time in new-builds. [7]

Bagenal's early survey work was particularly concerned with ecclesiastic buildings. Documenting the acoustic response of each building and its component materials with respect to octaves of the musical scale and to different vocal ranges was a critical aspect of his work. He was adamant that in buildings where music formed an integral function, the practice of averaging acoustic analysis to mid-frequency values was insufficient, arguing that "the whole frequency range must be considered."[8] Thus acoustic behaviour was graphed in a manner which displayed the "reverberation characteristic". This was essentially a time-frequency curve which recorded fluctuations in reverberation time across a frequency range which extended from 60 to 30,000 cycles (Hertz).

The practical implications for acoustic design which emerged from this research path were to find expression in a number of civic and ecclesiastic buildings across Britain and Ireland. One of these buildings is the Church of Corpus Christ on Griffith Avenue in Dublin. Hope Bagenal was contracted as an acoustic consultant for the landmark church by its architect John Robinson in 1938,[9] after the original design for the church had been changed - quite radically - to a form that posed a number of acoustic challenges.

Background to The New Church

The first set of plans for the new Church of Corpus Christi was drawn up in 1937 by the Dublin-based architectural firm of Robinson & Keefe.[10] At the time, suburban development in Dublin, and the creation of a number of new Catholic parishes, had necessitated the construction of new parochial infrastructure including schools, amenity centres, and churches. In north Dublin, the parish of Whitehall had seen a five-fold increase in population in the preceding fifteen years.[11] A temporary church for the area, designed in the 1920s by the same architect, had catered for a congregation of 450.[12] The new church was to accommodate a capacity congregation of 1800.[13]

Since the construction of the earlier church, Dublin Corporation had instigated a development scheme that made provision for the construction of 1,000 new dwellings in the locality. Furthermore, a minimum family-size policy was to be implemented in allocating the new dwellings. Each family to be housed was to have a minimum of four children.[14] Three new schools and a significantly larger church were anticipated requirements of the parish, and the parish priest, Rev. J. Dwyer petitioned the archbishop to authorise construction of these. The pressing need for a new church was outlined in a letter from Dwyer to the archbishop, dated January 17th 1935, "There are 7 Masses on Sundays and each Mass is crowded... there is great difficulty in accommodating the congregations at the Masses. This difficulty is likely to increase in the near future..." [15] Dwyer's requests were duly sanctioned, and permission was obtained in 1937 to begin fundraising over and above weekly Mass collections.[16] Plans for a substantially larger new church were approved by the archbishop in the spring of 1937.[17] Construction was scheduled to begin the following year. Costs were expected to be high, but it was another two months before an estimated sum was disclosed.[18]

Altered design: aesthetics v. economics

In early1938, just before construction had been scheduled to commence, the proposed design for the Church of Corpus Christi was radically altered. Originally intended to be built of granite and brick to a cruciform plan with a tower and connecting arcade (Fig. 1), the new design was for an open-span concrete-shell construction, faced with granite. The new plan was more t-shaped than cruciform, and the church now featured shallow transepts, barrel vaulting, and a central dome clad externally in copper. The freestanding tower was omitted entirely from the design (Fig. 2). Even more curiously, given that the impetus for the new church had stemmed from unprecedented expansion in the population of the area, the new design was expected to accommodate a smaller congregation than the previous proposal: 1,600 as opposed to 1,800.[19] Ultimately, construction of the new design cost almost one and a half times as much as the anticipated expenditure for the original.[20]

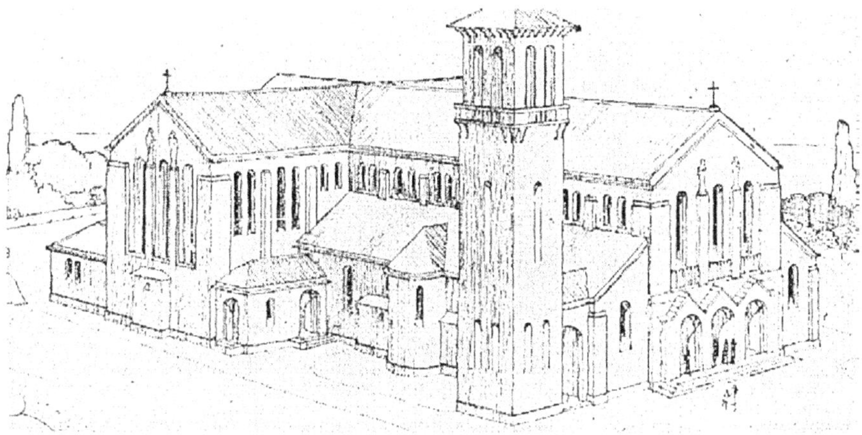

Fig. 1. Architect's sketch of the original scheme published in the Irish Independent in 1937

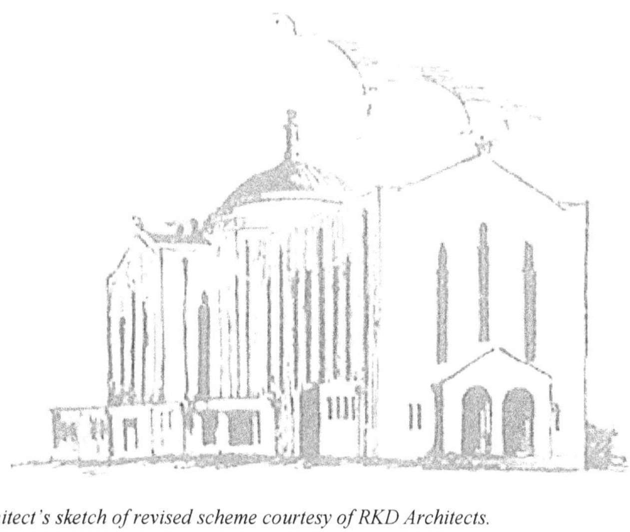

Fig. 2. Architect's sketch of revised scheme courtesy of RKD Architects.

The rationale behind such a sudden, and pronounced, change in design has been the focus of much speculation. However, the reasoning is quite clearly outlined in a series of letters exchanged between the architect, the Parish Priest and the Archbishop. The estimated cost of the original design, when issued to Rev. Dwyer in the summer of 1937 was substantially in excess of that of similar new churches constructed in the locality. The Schedule of Costs showed an anticipated expenditure that ranged from £40,000 to £50,000, depending on the selected construction materials.[21] This was considerably more than was evidenced in a comparative schedule of costs for other newly-constructed churches in the neighbouring parishes of Cabra and Skerries, as well as in the substantially more affluent south-side suburb of Foxrock. These three churches, although smaller in volume, had cost £32,000, £25,000 and £31,000 respectively. The per seat cost of Corpus Christi was anticipated to be £29, in comparison with £19, £20, and £21 respectively for the other three churches. The proposed volume of the original Corpus Christi scheme, at 23,330 cubic metres (820,000 cubic ft)[22] was to be greater than the volume of either of Dublin's two cathedrals.

Correspondence from the architect to the Parish Priest attributed the high cost of the proposal to the depth of the transepts, the consequent additional "façade work" and height, as well as the tower and connecting arcade.[23] A number of modified options, addressing these elements of the design, were put forward by the architect. However, none of them were deemed to possess the monumentality or aesthetic appeal of the original.[24]

While discussions were ongoing, a new chapel in the nearby Mater Misericordiae hospital was officially opened. That chapel had been designed by an architect named Ralph Byrne who was also the architect responsible for the afore-mentioned three new schools which were to be constructed alongside the new church for the parish of Whitehall. Dwyer was so impressed by the new Mater chapel that he suggested to the Archbishop that the new Church of Corpus Christi be *"an imitation…on a larger scale"*.[25] In both form and structure, the new design and its concrete-shell construction were considered to represent a break with tradition in Irish ecclesiastic architecture.[26] Described variously as "Byzantine" and "modernised Romanesque"[27] in style, the church was to be a landmark building, and both Dwyer and Robinson expressed great enthusiasm for the new design.[28] Thus, despite the fact that the proposed new scheme would seat fewer and cost more than the earlier design, aesthetics won over economics, and the new proposal was sanctioned by the Archbishop. With an open internal volume of 15,807 cubic metres (558,223 cubic ft), a large central dome, and smooth unadorned walls, form, scale and materiality all posed significant acoustic challenges. Hope Bagenal, who had recently delivered a lecture on the topic of acoustics to the Architectural Association of Ireland,[29] was contracted as acoustic consultant by John Robinson.[30]

Acoustic challenges: 'designing for reverberation'[31]

The two primary acoustic problems posed by the internal form of the church – specifically the dome and barrel vaults - were eliminating echo and mitigating against unequal distribution of sound. A further issue with respect to the surface finishes was what Bagenal described as "a slightly harsh "ringing" tone imparted to the acoustic by the "hardness" of the material. Over-riding the specific challenges posed by Corpus Christi was the perennial issue associated with all churches: striking a balance between the conditions required for speech and those required for music. Speech requires "clarity" and music requires "warmth". The former tended to be associated with lesser reverberation, the latter with increased reverberation. The balance between the two, in Bagenal's words, was "qualitative reverberation" *viz.* "an intelligent synthesis, a placing of the two side by side"[32]

A set of descriptors for optimum acoustic conditions for music had emerged from Bagenal's earlier survey work. For church acoustics – largely dominated by choral and organ music - these included "fullness of tone", "brightness of tone", and "strength of tone".[33] "Fullness of tone" was considered to be a by-product of sustained reverberation across all registers. "Brightness of tone" was dependent on upper frequency reverberation while "strength of tone" was defined as being independent of reverberation. Translated into the context of Corpus Christi, the acoustic design was to focus on the achievement of clarity for speech combined with these three characteristics specific to music.

Bagenal's earlier surveys had established a bracketed range of 1.8 to 3 seconds as the optimum for mid-frequency reverberation time in churches with a predicted occupancy of two-thirds.[34] This range was deemed to provide the greatest scope (in tandem with other measures) for creating a balance between warmth in music and clarity in speech. Reducing reverberation time, even to the upper level of this range, was a challenging aspiration for an open volume of the scale of Corpus Christi. That challenge was compounded by the impact of the dome, barrel vaulting, and the nature of the internal surface finishes, all of which had the potential to increase reverberation.

Furthermore, achieving the requisite balance for "qualitative reverberation" necessitated the manipulation of reverberation time at different registers. In determining that balance, Bagenal was informed by the findings of the time-frequency data which he had collected for architectural spaces deemed optimum for music. From these findings, he had extrapolated an idealised curve to which the acoustic design would be tailored relative to the optimum mid-frequency reverberation time. This idealised time-frequency curve was a shallow spoon shape (Fig. 3).[35] In effect, achieving a curve of this nature involved reducing lower frequency reverberation to a level consistent with intelligibility while preserving what was termed "brightness" from reinforcement at upper frequencies.

A strategy described as "distributed absorption"[36] was implemented to reduce the anticipated reverberation time in a manner consistent with a flattened overall time-frequency curve. This approach was based on the strategic use of materials with varying capacities to reflect, absorb and diffuse sound at different registers. As Bagenal himself wrote, "...the right frequency curve must depend on providing the right mixture of materials..."[37] In preserving upper frequency brightness while controlling for excessive reverberation at lower frequencies, he wrote that, "The aim is now to distribute a partial absorption over the main bounding surfaces, so as to get controlled inter-reflection from opposite pairs of parallel walls and hence the desirable type of reverberation."[38]

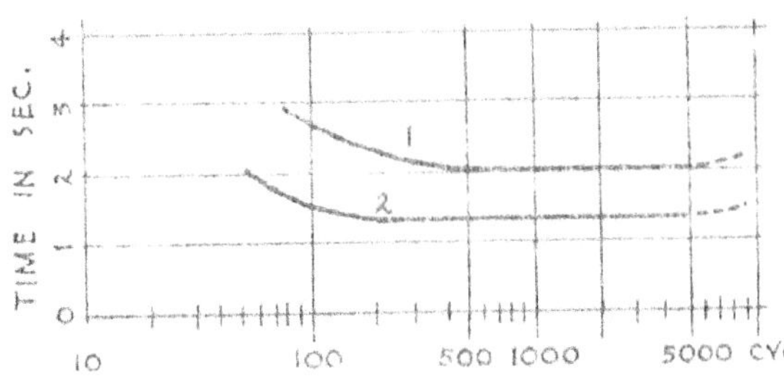

Fig. 3. Hope Bagenal: "Frequency Curves Generalised from Existing Halls"

This is not to imply however, that free rein could be exercised in the choice of materials. A further stricture on the acoustic design, was imposed in the form of finance. In the 1930s Ireland faced a set of challenging economic restraints and in both institutional and personal realms austerity was a necessary measure to maintain solvency. Other public buildings in Ireland in the 1930s which required specialist acoustic input - such as hospitals - had depended on revenue from the controversial "hospital sweepstakes" while landmark ecclesiastic buildings such as Galway Cathedral were, in many instances, funded by overseas philanthropy. Fundraising for Corpus Christi however, was a wholly local effort, led by its Parish Priest and the community it was to serve. [39]

Method: form, material, and economics

Construction work on the new church began in the autumn of 1938.[40] Fundraising continued as the construction work progressed. In the autumn of 1939 anticipated construction costs of £62,500 for the shell were publicly declared.[41] This was already substantially more than the figure, which, two years earlier, had been deemed excessive for the original scheme. By this stage, £30,000 had been raised – through a mixture of charitable donations and a concerted fundraising drive - to underwrite the costs of construction. Dwyer and Robinson were both acutely aware that when finishings were taken into account, costs were going to rise above £62,500. At a meeting of the building committee, just prior to the ceremony to mark the laying of the foundation stone, Rev. Dwyer gave a speech declaring that, "Not a penny will be spent on mere adornment or decoration."[42]

Dwyer remained true to his word: decoration and acoustic measures were integrated in such an exemplary fashion that the acoustic design was the subject of an educational site visit organised for the Architectural Association of Ireland (AAI) on Wednesday the 18th September 1940.[43] Specific reference was made on this occasion to the fact that no ornamentation had been included in the design that did not have a carefully considered acoustic purpose.

Data

The integration of "distributed absorption" into the decoration of the church was very precise. The budget left no room for guesswork, error, or waste. Materials, their locations, and the extent of their respective coverages were carefully calculated from up-to-date absorption coefficient data. Reference data for acoustic design in Britain at the time was a compilation of US and UK work. UK research was undertaken in laboratories at both the BRS and the National Physical Laboratory (NPL). All of the materials data used in Corpus Christi had been derived from experimental work undertaken at the BRS. The fact that all of this data had been derived under the same laboratory conditions added a measure of precision which was endorsed by Bagenal as it eliminated additional variables.[44] The data was used to assess the effect of the materials at different octaves of the musical scale.

Plaster

Two types of plaster were used on the walls of Corpus Christi for two very different purposes. A smooth-finished plaster was applied to the dado at either side of the nave and in the transepts, to act as a reflector for upper frequency sound. The same "glossy surface" was maintained in the plasterwork across the full expanse of wall area behind the altar. This was to compensate for the loss of upper frequency reverberation incurred in meaningful absorption at lower frequencies. It was also designed to act as a reflective sounding board for the spoken word from the altar. Above the dado, in nave and transepts, a proprietary acoustic plaster was applied. This latter plaster had been developed at the BRS from modifications to a recipe initially devised by Sabine. The modifications to the plaster formula - and the thickness of its application - almost doubled its achievable absorbency at the lowest measured frequencies. This type of plaster had also been shown to provide significant absorption at middle-frequencies, but a negligible, or slightly lesser absorbency, than the original at upper measured frequencies. These properties made it a useful tool in achieving the desired frequency curve. The plaster was to be finished with a granular surface and rounded arrises.[45] As the absorptive nature of the material depended on its porosity, a particular, undistempered finish was specified. Furthermore, the porous capacity was to be maximised by allowing each coat to dry out before application of the next coat.[46]

Paint

To an extent, acoustic practicalities even informed the internal painting scheme. In general, paint was known to reduce the efficiency of acoustic plaster as an absorber given that it reduced the porosity. Additionally, the propensity of the very porous plaster to absorb water vapour and condensation caused it to colour with age at a contrasting rate to the surrounding areas of smooth, hard-finished plaster. Thus, contrasting colours in the paint scheme were recommended for all areas of differing acoustic treatment. This was not to highlight the materials used, but to reduce the effect of disparities in colour emerging with age.[47] The selected colour scheme in Corpus Christi, although it has been altered in the intervening years, was originally, "cream and blue-grey".[48] Another aspect of the church which has been altered in recent years is the lighting scheme. In the original acoustic design, loudspeakers were incorporated into the light fixtures.

Transepts and west wall

Whilst the wall behind the altar was to function as a reflector, in as much as was feasible, the opposite wall was to act as an absorber. End walls of transepts were also to act as absorbers, and in Corpus Christi, these three wall surfaces were treated in much the same way.

In general, Bagenal tended to recommend increasing the absorption of an end wall by the use of a wallboard – typically composed of compressed fibre-slabbing with perforations for middle-frequency-absorption. In Corpus Christi, the perforated wallboard featured a profiled surface. This was applied to the apron of the gallery and to the wall behind it. The perforated board had an added decorative value in that – unlike the acoustic plaster - it could be painted without reducing the absorption.[49] The fluting of the profile was decorative in that it mirrored the shape of the organ pipes, and functional in that it acted as a diffuser for sound (Fig. 4).

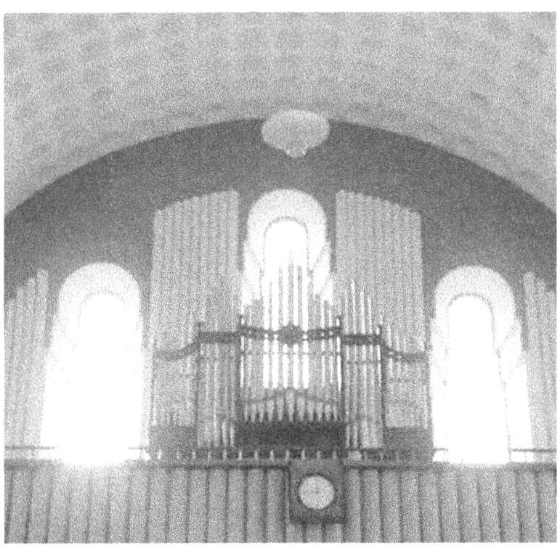

Fig. 4. Perforated Fluting to Gallery Apron and End Wall

The same strategy was applied to the end walls of the transepts where perforated fluting was fixed level with, and in between, the lancet widows.

Flooring

Three contrasting floor surfaces were utilised, to differing acoustic effect. The end wall and floor surrounding the altar were designed to maximise the carrying power of the spoken word. Flooring in this area, by necessity, was to be hard and glossy. Marble was the obvious choice to mark the venerable location and to provide the requisite acoustic reflection. The tendency of marble to increase overall reverberation time was countered with absorbent treatment to the ceiling and baldachin above.

Wooden block flooring was laid beneath the seating areas with rubber flooring reserved for areas of foot traffic. Rubber flooring was utilised for its "noiseless" rather than sound absorbing properties. Bagenal's data charts it as exhibiting 7% absorption at mid frequencies. This is only marginally more than his data for wood block flooring which denotes 5.5% absorption at mid frequencies. The underlying purpose of these floor types was to reduce impact sound, and not the overall reverberation time.

The strategy for advancing clarity in speech was two-fold. The first measure was to capitalise on conditions for early-reflected sound in areas which constituted source points for speech. The second was to reduce later reflection at points of audition by the congregation. This strategy was implemented, not just in the distribution of floor surface types, but also in the use of reflectors above and behind the pulpit.

Pendentives

The concave surfaces of the pendentives were treated with acoustic plaster to absorb sound. Further to this, they each feature plaster relief work to diffuse sound that would otherwise have been concentrated at focal points (Fig. 5). Again, this formed part of the decoration as each pendentive features "representation... of the four Evangelists with attendant angels skillfully carried out in plaster".[50] The fibrous plaster utilised for decorating the dome and pendentives alone was expected to cost £ 1,600.[51]

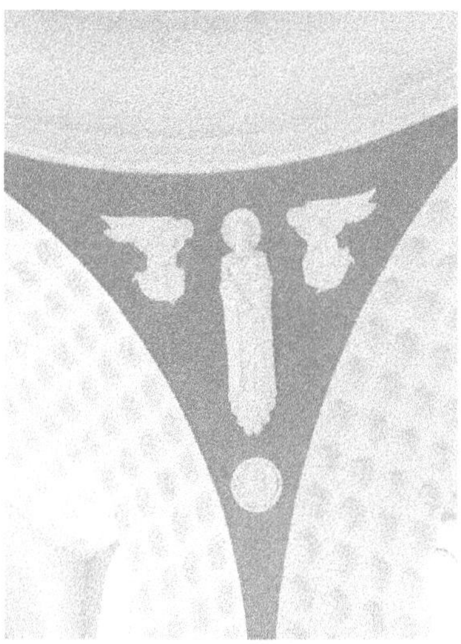

Fig. 5. Pendentive Rendered in Acoustic Plaster with Relief Work

Barrel vaulting

The focusing effect of the barrel vaulting above the nave and transepts was tempered by profiled coffering. The coffers were designated octagonal in shape and a fibrous plaster was utilised. The effect of this was to diffuse sound and reduce focusing. Both depth and profile of the coffering were important to the overall acoustic design. Bagenal's strategy in using relief work and coffering to diffuse sound was to ensure that the depth of coffering equate to at least "one-tenth of the [average] wave length" of the speaking voice." His recommendation was for a minimum depth of 6 ½" in respect of a 5'6" wavelength. [52] The octagonal coffers in Corpus Christi are 8" in depth, measuring 1'10" between extremes. [53]

Dome

The dome (Figs 6a and 6b) presented three primary acoustic challenges. The first two, echo and adding to the overall reverberation time, were tempered by rendering the dome in acoustic plaster. The third, the focusing of sound, was usually addressed by Bagenal subsequent to detailed geometric plots of sound paths and reflected rays. These plots, for Corpus Christi, are no longer in existence. However, it is worth noting that the dome has an angle of curvature which reaches a focal point above the plane of the congregation.

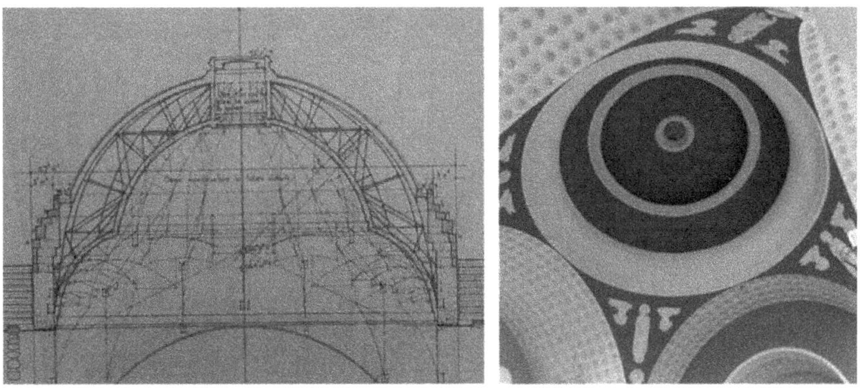

Fig. 6 (a) Section through Dome of Church of Corpus Christi (RKD Architects, held in Irish Architectural Archives); (b) Dome of Church of Corpus Christi

Reception

The Church of Corpus Christi opened its doors to the parishioners of Whitehall in May 1941 with an inaugural mass presided over by the new Archbishop of Dublin, John Charles McQuaid. In subsequent coverage of the event, *The Irish Times* declared the new church to have "the advantage of perfect acoustic properties."[54] Comment on the acoustics was repeated in another article less than a week later, while the *Irish Independent* drew attention to the fact that "the acoustics finely clarified the tones of the chanting and singing"[55] The calibre of acoustics in the new church was also positively viewed by the *Irish Builder and Engineer* with the statement, "That the acoustic properties are particularly good was evident during the opening ceremony, when every word that was uttered could be heard distinctly by every member of the congregation."

Fitness for purpose of the acoustic with respect to music was thoroughly tested with the inclusion in the Solemn Mass of choral works by Mozart, and appropriately, the proper of Sir Richard Terry's Mass of St Dominic. Terry's work in tempering music to acoustics at London's Westminster Cathedral had inspired Bagenal's first acoustic case study some 27 years earlier.[56] It was agreed in media coverage that the strains of Terry's Mass "fill[ed] the edifice nobly."

Conclusion

The Church of Corpus Christi on Griffith Avenue is a landmark building in Irish twentieth-century architecture for a number of reasons. On a hilltop site overlooking the north city, it stands proud as a symbol of urban expansion and planning that followed Irish independence. In a city that is predominantly Catholic, yet possesses no Catholic cathedral, the monumental proportions of Corpus Christi transcend the functional requirements of a new suburban parish church.

Stylistically, it is unlike the other ecclesiastic work of its architect in the post-Independence years. There is no nod to Celtic symbolism in the decoration and no vernacular tradition espoused in the construction methods. Built during a time of economic hardship, and funded to a large extent by local donations and by its parishioners, the church is an expression of community pride and solidarity.

Both design and construction methods are remarkable and at the time were considered to be de rigueur. However, Corpus Christi is also one of the earliest Irish buildings to have incorporated a scientific acoustic consult in its design process. As such, the church is a prime example of the state of play in contemporary acoustic design. When first built, Corpus Christi was cited as an exemplar of modern acoustics - a manifestation of modern building science and integrated design. By the 1960s, the acoustic reasoning behind much of the internal finishing had been forgotten,[57] yet Corpus Christi remains as a landmark not just in the history of architecture and construction, but also in the history of science. The methods, materials, and aspirations which underpin the acoustic design of Corpus Christi are a benchmark of contemporary developments and rationale in that branch of modern building science.

Acknowledgements

The author gratefully acknowledges a Research Support Grant from the Paul Mellon Centre which enabled the archive work undertaken as part of this, and related, research. Special mention must also be made of the families of Hope Bagenal and Wallace Clement Sabine. I am very grateful to RKD Architects for permission to reproduce their original drawings of the church, and especially to Vincent Delany of RKD Architects for the time he took to discuss the work of John Robinson. Thanks are also due to the clergy of Corpus Christi for permitting study and photography of the church, and to the staff of the various archives consulted: The Irish Architectural Archives, Dublin Diocesan Archives, Church of Corpus Christi Archives, the National Library of Ireland, Southampton University Archives, and the RIBA Archives. Finally, this paper is dedicated to the memory of Patience Bagenal.

References

[1] H. Bagenal, "Musical Acoustics in the Three Cathedrals of London", in *The Architects' and Builders' Journal*, 22 December 1915, pp. 272-273.

[2] Reverberation time is still the central objective parameter in architectural acoustics although it has been supplemented by a number of other parameters since then. The standard parameters for assessment currently are listed in ISO 3382.

[3]Sabine's published papers including those on reverberation time are contained in W.C. Sabine, *Collected Papers on Acoustics*. Cambridge Massachusetts & London: Harvard University Press & Oxford University Press.

[4]Private Collection, letter of H. Bagenal to W.C. Sabine, 25 Mar. 1914. Private Collection, letter of W.C. Sabine to H. Bagenal, 13 Apr. 1914.

[5] H. Bagenal, "Reverberation: Some Historical Notes" pp. 34-39, in *Papers and Discussions of the First Summer Symposium of the Acoustics Group, 1947*. London: Royal Institute of British Architects, 1947.

[6] See F.M. Lea, *Science and Building: A History of the Building Research Station*. London: HMSO, 1971.

[7] H. Bagenal, "Planning for Good Acoustics", *RIBA Journal,* vol.32, 22 Nov., pp. 29 - 43. See also H. Bagenal and A. Wood, *Planning for Good Acoustics*. London: Methuen, 1931.

[8]H. Bagenal, *Practical Acoustics and Planning Against Noise.* London: Methuen, 1942, p. 106.

[9] J. Bagenal, *The Best of Bagenal 1930-1939,* Family Papers.

[10] The principals of the practice, John Robinson and Cyril Keefe, were joined by a third partner, Andy Devane in 1946, shortly after the completion of the Church of Corpus Christi. The company, which then became Robinson Keefe and Devane, are now registered as RKD Architects. A number of early drawings have been retained in the offices of RKD. Final and construction drawings were donated by RKD to Irish Architectural Archive and may be consulted there.

[11] "Drumcondra's New Church", *Irish Times,* (9 Oct. 1939) p.2.

[12] Dublin Diocesan Archive, letter of J. Robinson to Rev. J.Dwyer, 20 Oct. 1922. The temporary church, constructed of wood and corrugated iron, was affectionately known by parishioners as "the little tin church".

[13] Dublin Diocesan Archives, letter of Rev. J. Dwyer to Archbishop E. Byrne, undated letter c. 1937.

[14] For discussion on the policy and its envisioned implications for the parish, see DDA, letter of Rev. Dwyer to Archbishop, undated letter c. 1937 (Note 13). For further detail on housing policy under the Cosgrave government, see National Archives of Ireland, Dublin Corporation Housing Programme files. For an evaluation of policy and social context see R. McManus, *Dublin 1910-1940: Shaping the City and Suburbs*, Dublin: Four Courts Press (2002).

[15] Dublin Diocesan Archives. Letter of Rev. J. Dwyer to Archbishop E. Byrne, 17 Jan. 1935.

[16] see DDA, letter of Rev. Dwyer to Archbishop, undated letter c. 1937 (Note 13). See also "Drumcondra's New Church", *Irish Times,* (9 Oct. 1939) p.2.

[17] "New Church for Whitehall", *Irish Times,* (10 May 1937), p.13.

[18] Dublin Diocesan Archives, letter of J. Robinson to Rev. J. Dwyer, 3 Aug. 1937.

[19] Unlisted Author, "Whitehall Church and Schools: An AAI Visit", in *The Irish Builder and Engineer,* vol. 82, 28 Sept. 1940, p. 602. See also Dublin Diocesan Archives, letter of Rev. J. Dwyer to Archbishop E. Byrne, 13 Oct. 1937. Further detail on capacity can be found in the seating schedule contained in Irish Architectural Archive, 86/93, 'RKD Church of Corpus Christi Drawings', 1939.

[20] The Irish press documented the rising costs and fundraising efforts in great detail between 1937 and 1941. The final figure of £70,500 was published by the contractor Messrs Sisk and Son, and also in *The Irish Builder and Engineer.* Unlisted Author, "New Dublin Church: Impressive Building at Glasnevin", in *The Irish Builder and Engineer,* vol. 83, 5 July. 1941, p.325.

[21] Dublin Diocesan Archive, Schedule of Costs for R.C. Churches in Whitehall, Skerries, Foxrock and Cabra, enclosure to letter of Robinson to Dwyer, 3 Aug. 1937 (Note 20).

[22] Volume quoted by John Robinson in comparative schedule of costs issued to Rev Dwyer, 3 Aug. 1937 (Note 18).

[23] DDA, letter of Robinson to Dwyer, 3 Aug. 1937 (Note 18).
[24] See DDA, letter of Robinson to Dwyer, 3 Aug. 1937 (Note 18) and DDA, letter of Dwyer to Archbishop, 13 Oct. 1937 (Note 19).
[25] DDA, letter of Robinson to Dwyer, 3 Aug. 1937 (Note 18).
[26]. *The Irish Builder and Engineer,* (Note 19) p.602 and *The Irish Builder and Engineer,* (Note 24) p.325
[27] DDA, letter of Robinson to Dwyer, 3 Aug. 1937 (Note 18) and *The Irish Builder and Engineer,* (Note 24) p.325.
[28] DDA, letter of Robinson to Dwyer, 3 Aug. 1937 (Note 18).
[29] For transcript of lecture see H. Bagenal, "Practical Acoustics and Modern Planning in *The Irish Builder and Engineer,* vol. 78, March 1936 pp. 290 - 293.
[30] Private Collection, Job Register of Hope Bagenal for years 1933-1939.
[31] Title of Chapter Six in Bagenal, *Practical Acoustics*, (Note 8)
[32] Bagenal, *Practical Acoustics*, (Note 8) pp.135.
[33] For definitions of these see Bagenal, *Practical Acoustics*, (Note 8) pp.118 - 119
[34] Bagenal, *Practical Acoustics*, (Note 8) p. 90
[35] Image from Bagenal, *Practical Acoustics*, (Note 8) p. 111, fig 23.
[36] Technique discussed in Bagenal, *Practical Acoustics*, (Note 8) pp. 111-116
[37] Bagenal, *Practical Acoustics*, (Note 8) p. 109.
[38] Bagenal, *Practical Acoustics*, (Note 8) p. 113.
[39] Funds for the church were raised through a mixture of charitable donations, largely received from the local community and the catholic clergy, and a concerted fundraising drive undertaken by a committee of parishioners and led by Rev. Dwyer. The fundraising drive was a very high profile one which received much contemporary media coverage. An annual event known as the Whitehall Carnival was instigated which was very successful. "Whitehall Carnival", *Irish Times,* (24 Oct. 1936), p.4. "Carnival in Aid", *Irish Independent*, (29 Apr. 1937), p.12. As a measure of support for the new construction invested by the citizens of Dublin as a whole, the fundraising drive was publicised and supported by both the GAA and the Dublin Tram Company. Substantial sums of money were also raised through weekly mass collections. "New Dublin Church", *Irish Independent,* (19 Oct. 1940), p.5. A list of benevolent donations received up until 1939 was published in "New Church of Corpus Christi", *Irish Times,* (28 Oct. 1939), p.3.
[40] "New Church of Corpus Christi, Parish of Glasnevin", *Irish Times,* (7 Oct. 1939), p.4
[41] *Irish Times* p.2 (Note 16), "New Dublin Church", Irish Times, (14 Oct. 1939), p.17. See also, "News in Brief", in *The Irish Builder and Engineer,* vol. 81, 14 Oct. 1939, p.870
[42] Dublin Diocesan Archives, transcript of speech delivered by Rev. J. Dwyer at meeting of Building Committee for Church of Corpus Christi, undated, c. Oct. 1939.
[43] *The Irish Builder and Engineer,* (Note 19) p.602. See also "Notes: Notification of AAI Site Visit", in *The Irish Builder and Engineer,* vol. 82, 14 Sept. 1940, p.576
[44] For a contemporary discussion on differential data resulting from altered lab conditions see Bagenal and Wood *Acoustics* (Note 7) pp. 239-40.
[45] Irish Architectural Archive, Bill of Quantities for the Church of Corpus Christi, p. 129.
[46] This is discussed in Bagenal, *Practical Acoustics*, (Note 8) p. 96.
[47] This is discussed in Bagenal, *Practical Acoustics*, (Note 8) p. 96.
[48] "New Glasnevin Church Opened", *Irish Times,* (26 May. 1941), p.2
[49] Bagenal and Wood *Acoustics* (Note 7) p. 248.
[50] *The Irish Builder and Engineer,* (Note 19) p.602.
[51] IAA, Bill of Quantities (Note 45) p. 131.

[52] Bagenal and Wood *Acoustics* (Note 7) pp. 71-2.

[53] IAA, Bill of Quantities (Note 45) p. 130.

[54] 'New Glasnevin Church Opened' in *The Irish Times,* May 26th 1941, p. 2. See also 'New Dublin Church' in *the Irish Times,* May 31st 1941, p. 12.

[55] ''New Church Opened at Place where St Mobhi Preached' in *Irish Independent,* May 26th 1941, p. 5

[56] F. Smyth, *"Scales of Interaction"* in G. Adler et al, (eds.) *Scale: Imagination, Perception and Practice in Architecture*. London: Routledge pp. 88-98.

[57] This is clear from the maintenance reports which are held in the Dublin Diocesan Archives.

New technologies for a new society: German museum architecture after 1945

Benedetta Stoppioni
Independent scholar, Italy

Summary

WWII reduced Germany to a pile of rubble. The situation got worse with the painful division of the country into two sister, but antagonistic republics. In both states a spiritual search into a new identity was propelled, an identity that would be able to erase the shameful Nazi past and the misery of the present.

In the German Federal Republic (GFR) these issues became reflected within the design of the

museums that were being built since the fifties, during the so-called "economic miracle". The new requirements of exposition and conservation of the artworks induced a virtuous alliance between social-anthropological needs and technical innovations, with which architects and engineers were experimenting in the numerous construction sites of this period.

The aim of this paper is to inquire how the museum architecture in the GFR – spanning between 1945-1968 – could be considered a mirror reflecting the rebirth of this country, both in the exterior shape of the museums and in the technological solutions adopted.

The museum Wilhelm Lehmbruck in Duisburg, the subject of my doctoral thesis, represents a useful instrument to explore this field. Its rich and unpublished archival documentation can explain the motivations behind the choice and adjustment of some construction systems, that were typical of German architecture after 1945, such as raw concrete and large glass walls.

This page of the history of architecture has been mainly investigated by German scholars since the eighties. My intention is to look at these studies with a new perspective, that could highlight the role of the new technologies in this context.

New technologies for a new society: German museum architecture after 1945

WWII reduced Germany to a pile of material and spiritual rubble. The two young German republics - born in 1949 from the painful division of the country imposed by the Allies - both found themselves faced with the necessity of rebuilding their territories and cities devastated by the bombings, and at the same time with the questions of guilt and responsibility for the German people for the war and the crimes perpetrated by the Nazis. These newly-born countries had to strive to rise from the ashes of the Reich with new identities, capable of ferrying them forward into to the future.

In the German Federal Republic (GFR) the new democratic identity was translated by joining the free market and the prosperity assured by the new happy course of the economy on one hand; and on the other, in the search for a "founding legend" that could stand as a guarantor of this democracy. [1] Such a legend was to be found in the glorious Weimar Republic.

With the so-called *Wirtschaftswunder* [2], society in the GFR intended to move on to the next chapter quickly and leave its past behind. Condemned by the world for their blind obedience to Hitler's immoral orders, West Germans were able to transform their subdued industriousness into national virtue. During the 1950s, they earned international respect thanks to an equally genuine and absolute loyalty to efficiency, industriousness and the quality of their industrial productions. Factory smokestacks resumed smoking, industries in the *Ruhrgebiet* [3] severely affected by the Royal Air Force air raids reopened their doors to meet the pressing demands of raw materials by the many building sites working on the reconstruction.

German architects working in this period trained during the 1920s, called this the fervent Golden Age of Weimar democracy. Those who set on the road to exile during the Nazi dictatorship mostly remained in their country of choice, and only few got involved in the reconstruction of their homeland. For the most part the reconstruction architects had not gone into exile: some of them opted for the so-called internal emigration, substantially giving up the perspective of moving on in their career; some others, without much resistance, chose employment in public offices. In the regime years, in the building sites of the *Reichsautobahn* [4] and in those of the military industry, a certain culture of the Modern was enforced, and this led to an evolution from the radical precepts of the Weimar years. After 1945, these were the architects who guided the reconstruction, rather than the "great emigrants" (e.g. Mies van der Rohe and Walter Gropius). This is the reason why the *Nachkriegsmoderne* [5] did not take inspiration from the *Moderne* [6] but, instead, from the art of those professionals who remained in Germany: the continuity of their activity, from the regime to the post-war years, represented the broadest contradiction to the legend of the direct derivation of post-war German architecture from the classic Modern. [7]

But the Federal Republic found this reference to Weimar convenient when it attempted to reaffirm its legitimacy as a democratic nation. Declaring itself heir of the Weimar Republic meant it could suddenly delete and forget the twelve and gloomy years of National Socialism and that it could show itself to the world in a new light: transparent and cleaned from inconvenient references. It was in this disguise that the newly-born country made an appearance at the first post-war world exhibition, held in Brussels in 1958. The German pavilion - a transparent steel and glass structure - was clearly inspired by Mies van der Rohe's work and therefore recalled the tradition of the Modernism of the 1920s. [8]

The issues of identity and collective memory found a peculiar application in the development of post-war museum architecture. In its territory bent by WWII, patching the artistic and cultural fabric back together was one of the primary objectives pursued by the Federal Republic. The end of the war marked the beginning of the rebuilding of theaters and concert halls; during the 1950s, artwork collections were also rebuilt, together with new suitable spaces to accommodate them, since the oldest museums had largely been destroyed or damaged by the war.

The museum architecture in West Germany during the post-war period on the one hand became an expression of the need for collective oblivion and therefore chose the Weimar language of the Modern; on the other hand, however, it did not reject the need for individuals to accept their own personal history and painful memories. Post-war museums became the place for intimate and introverted reflection with their humble and collected character. They assumed a Janus-faced appearance: a face dedicated to the collective oblivion and the other to individual memories.

The refurbished museums and those built from scratch in the post-war period had a number of elements in common: grand windows, inner courtyards often used as ground for sculpture installations, elements of small cube-shaped pavilions that made up the entire structure and plain doorways either at street-level or raised by just a couple of steps. Any reference to monumentalism was avoided: a museum was not to convey a sense of prestige and magnificence and had to give up the character of *Musentempel* [9] but was rather placed in a relationship with and at the service of people, of visitors.

The new democratic society, driven by the *Wirtschaftswunder* and eager to forget past miseries, was craving life and abundance. Consumerism not only pervaded economy and trade, but also dictated the everyday connection that people had with the city, its services and its institutions. So museums could no longer afford to appear like treasure chests for valuable and unapproachable works of art. They were now becoming the heritage of the community: from sacred object to product of mass consumption. And for this reason they had to be easily accessible to the people, i.e. the "consumers".

The new museums opened to the city with large glass walls, with windows that allowed passers-by to savour the treasures held within, even when just walking on the city streets. The entrance was more appealing through doors that were all but grand: the museum was no longer a destination where visitors could "elevate" themselves with reverential fear, but rather one of the possible and natural locations of the new *flâneurs.*

The large stained glass boxes enclosed open and flexible inner spaces that would accommodate the various possible interior settings. This model was imported to Germany mainly from the USA. Moreover, the newly-born Federal Republic saw the United States as a real reservoir from which it could draw an entire symbolic universe where liberalism, democracy and consumerism intermingled until they became, perhaps ironically, synonyms. When the GFR joined the West Front, not limitedly to the diplomatic and political sphere, it led to the complete adoption of the so-called "American way of life".

The renewed demands that museums had to fulfil triggered a series of in-depth field studies, most of which concerned the field of building technology. [10]

German architects were all set for the market demands as they had mostly trained, as previously mentioned, in areas where technical and construction research played a major role. Furthermore, immediately after the war the Allies supported the democratic re-education of the German population by sponsoring trips abroad for young architects, where they could certainly gather fresh ideas and designs that could inspire them. [11]

They first had to identify structural devices that were suitable for the installation of large glass surfaces, which, used in the context of museum architecture, marked a strong German emancipation from traditional models. [12] The building site for the Folkwang Museum in Essen was one of the first that tested innovative solutions in this respect, so much so that it became a valuable and familiar reference. [13]

Getting rid of the divide between art and life and of the idea that museums were a "device" functional to enjoying the works also entailed rethinking of lighting and conservation techniques. The American model was popular once again when artificial lighting was introduced; this however was not promoted enough in Germany to replace natural light.

Designers at the Focke Museum in Bremen used a combination of artificial and natural light: the first was on for a few minutes over the permanent collection of pieces dating back to the Middle Ages, while the rest of the exhibits were lit with natural light flowing from the large glass façades that looked out onto the nearby park. [14]

The Federsee Museum in Bad Buchau on the other hand, is a good example of solutions chosen to bring together a more unrestricted enjoyment of art and at the same time the suitable preservation of the objects displayed. The museum displays ancient prehistoric artefacts, whose easy perishability required particular care in preservation techniques. Manfred Lehmbruck [15], the architect that oversaw its design, chose to store the objects in large glass displays, inside which they were kept at constant temperature and humidity and lit by skylight windows and spotlights. Their peculiarity was that they extended from floor to ceiling and blended in with the entire architectural structure - again featuring large windows - until they disappeared altogether.

For the increasingly growing number of diversified temporary exhibitions offered to the eager German "consumer", special pavilions were often created, designed almost exclusively on the basis of the open museum model thus far described. However, the permanent collections also required suitable spaces. They collected testimonials from the past that were properly catalogued and organised, and by their very nature existed to tell a story, to which the building had to adapt, for example through the arrangement of successive spaces that flowed into one another or through a marked separation from the outside, i.e. from reality. These parts of the museums were coloured by an intimate atmosphere and didn't quite open out on the road but rather on the inner courtyards or on the adjacent parks, if available. [16] Here the other face of post-war Germany was unveiled, the face that did not look to the future and the new-found wealth, but rather to the past and gave space, voice and breath to distant pictures and memories.

From the outside, these pavilions sometimes showed more traditional walls, other times actually displayed more innovative solutions such as the *béton brut* or stone cladding details, designed in a more modern approach in their stereotomy and often built on metal structures. [17]

The case of the Wilhelm Lehmbruck Museum in Duisburg is actually paradigmatic as it includes all the above features of post World War II museum architecture in the GFR and is useful here to gain an insight into the specificity of the technological innovations adopted. [18] The building in question was one of the first museums that had been built *ex novo* in the Federal Republic. It was designed by Manfred Lehmbruck – the son of the sculptor to whom it is dedicated – and built between 1957 and 1964. [19]

The Lehmbruck Museum is a relatively small construction within Kant Park, a small green oasis at the heart of the industrial area of Duisburg. The park was a treasure that needed to be safeguarded and the museum would have to be a quiet addition, so as not to create an awkward, unfitting object. Its designer planned the building as divided into three rather low sections, which expanded considerably in the basement and ran inwardly to make up a sort of central courtyard (Fig. 1). [20]

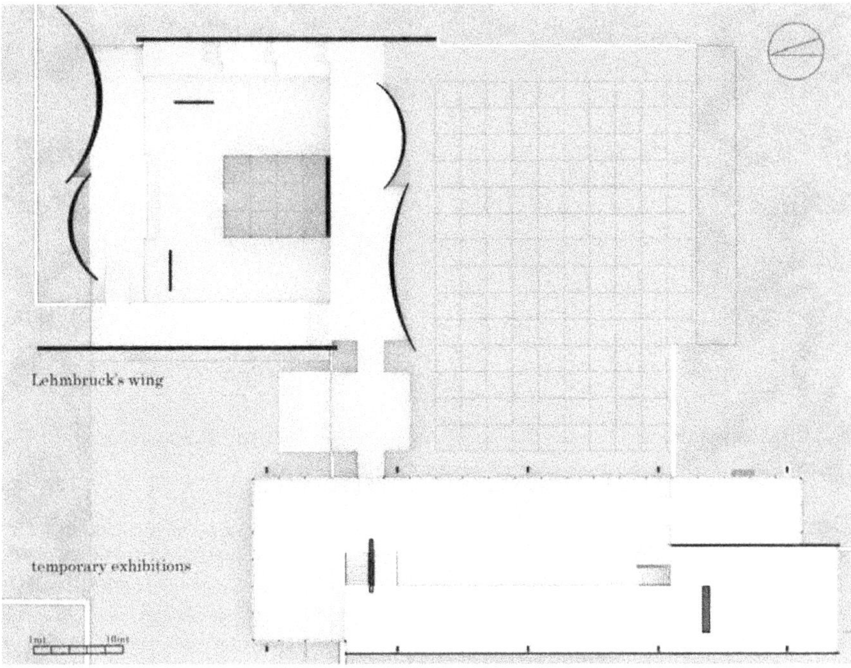

Fig.1. Lehmbruck Museum: the pavilions designed and built between 1957 and 1964. (First floor. Graphic elaboration by Andrea Sbarbati)

The wing that hosts the works by Wilhelm Lehmbruck is a low cubical building topped by a thick flat canopy. It is empty in the middle leaving a sort of small courtyard, also cubic. The outside features low walls of exposed concrete, sometimes bowed and interrupted by openings sealed by glass windows. The small central courtyard radiates natural light into the inner space, which is arranged on two levels and continues partially below ground. A small gallery connects the wing of the museum where temporary exhibitions and paintings are on display. This building is a glass parallelepiped without an inner bearing structure. In its place there are five large external steel frames, which support the flat cover (Fig. 2). On the ground floor there is a mezzanine that runs along the length of the room. The paintings are displayed in the basement, which also has no pillars, with totally artificial lighting to protect the works from the damaging effect of sunlight.

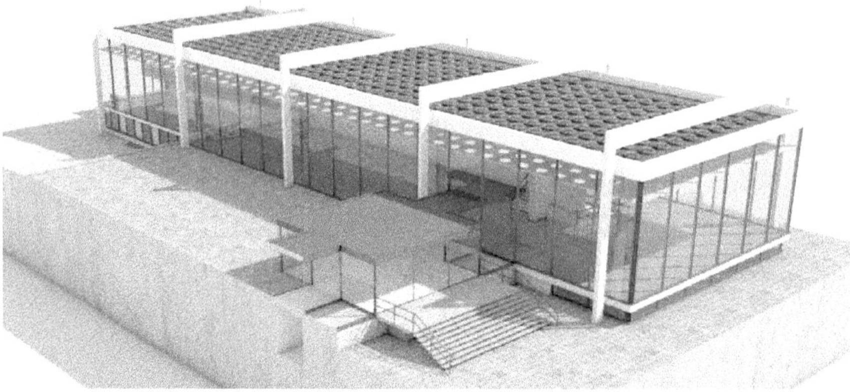

Fig. 2. Lehmbruck Museum: the wing for the temporary exhibitions. (Virtual model elaborated by Andrea Sbarbati)

In this wing of the museum the reference to Mies is obvious. [21] We can mention the Museum for a small town (1942), the Crown Hall (IIT) in Chicago (1950-1956), the Cullinan Hall (1958, part of the Museum of Fine Arts in Houston), which have similar structures to that chosen by Manfred Lehmbruck for the Duisburg Museum with mighty frames outside the building to hold up the glass box. [22] These materials were so innovative for Germany in those years that Lehmbruck met many difficulties in finding the firms that would produce them; in addition, the Duisburg town hall planning application office expressed their concern with respect to such choices as they certainly would have caused installation difficulties and also affected the cost of construction.

Lehmbruck opted to hang the glass walls from the roof, a decision that was almost forced due to the possible movements of the steel structure, but also functional to the achievement of large whole glass surfaces, chosen to avoid unsightly subdivisions. The installation was entrusted to the company Hahn from Frankfurt, who had previously worked at the construction site of the Focke Museum. Despite their expertise in the field, the company needed the advice of the American firm Boston Blacking Company on some construction details.

In addition to the glass panes, mobile *brises-soleil* had to be anchored to the roof (and therefore to the outer steel frames that supported it) arranged in such way to yield abundant natural light, so important to Manfred as it was seen as one of the vital factors of the works of art on display. The *brises-soleil* had to be placed inside the building so that air pollution would not compromise its working. The company Fries & Sohn of Duisburg was in charge of the construction of the steel support structure; they immediately expressed their concern for the peculiarity of the chosen solutions.

Lehmbruck was also adamant to have lighting from above to light up the works, so he also designed a see-through flat roof made up of about 400 small acrylic glass domes. Again, it was difficult for the architect to reach an agreement with the construction companies. Eventually, the chosen company was Düsseldorf-based company Esser that, however, immediately stated they could not supply any guarantee on the chosen material, in particular with regard to its resistance to weather conditions, fire and wear and tear. [23]

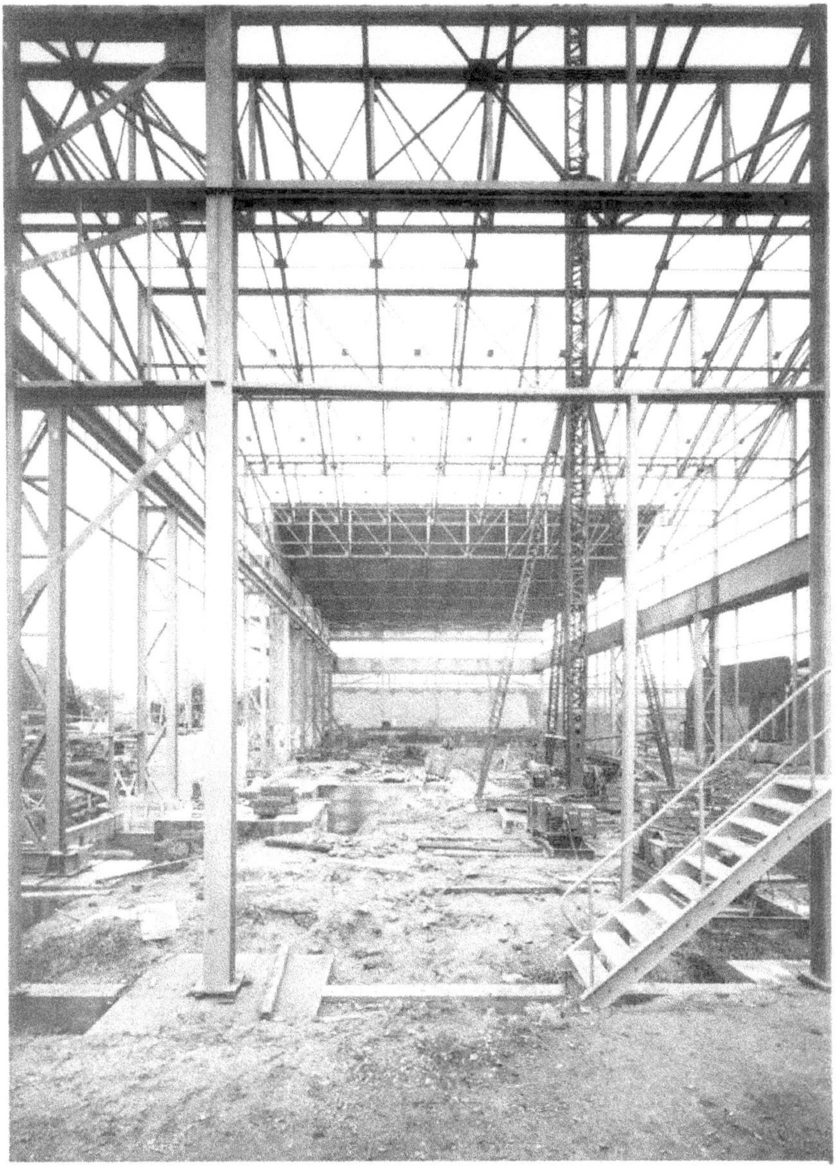

Fig. 3. Lehmbruck Museum: the wing for the temporary exhibitions during its construction, 1962. (Photo by Bernd Kirtz)

The structure that supports the roof at its intrados is made with a light reticular mesh that barely affects the rate of light rays going through (Fig. 3). Naturally, there are hanging horizontal clear acrylic panes that can be adjusted in height and replaced with opaque sheets to use according to the amount of light required for the internal displays. In addition to natural light, we find ceiling spotlights, as well as a system of diffused lighting, mainly turned on at sunset.

This wing of the museum was designed with maximum flexibility in mind, for it was conceived as a large open space left neutral and that could be divided as required depending on the different exhibition settings. For this purpose, moveable partitions were provided, mounted on aluminium frames that sit on the floor and are inserted in special slots placed in line with floor tile joints. These same slots - when not used to accommodate the partitions - are used for the technical systems necessary for the preparations (e.g. electricity cables), carefully concealed by removable steel bars.

In contrast with the buoyant and wide space of this building, the other area, dedicated to the work of Wilhelm Lehmbruck is an "amniotic" kind of architecture, surrounded by a partial shade that promotes a meditative and collected state (Fig. 4). The building materials are the same ones used by Wilhelm Lehmbruck in his work: concrete, stone, sand and steel or wood sections.

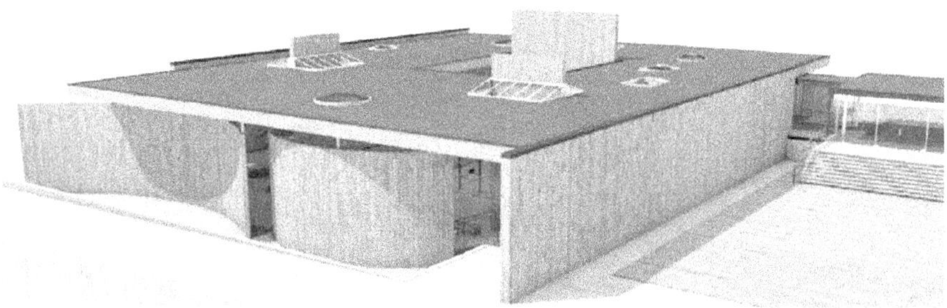

Fig. 4. Lehmbruck Museum: Lehmbruck's wing. (Virtual model elaborated by Andrea Sbarbati)

The load bearing structure is made up of *béton brut* side walls and interconnected steel beams for the roof. (Fig.5) These lean on the dividing walls only in a few points: the impression is such that the ceiling is suspended like a piece of fabric over the works scattered unsystematically in the space below. The large windows that surround the inner courtyard are also suspended from the roof and are fastened to the floor via special slots. In addition to the courtyard, the main source of natural light is provided through a few skylights that pierce the dark false ceiling made with a thick metal grille installed to conceal various technical facilities placed in its hollow. Here we also have some spotlights directed downward to light up the individual sculptures. The two larger skylights are bisected by slabs of reinforced concrete that continue beyond the ceiling, and help to convey the light within and serve as a backdrop for the works of art. A similar structure is also located on the west side of the small courtyard.

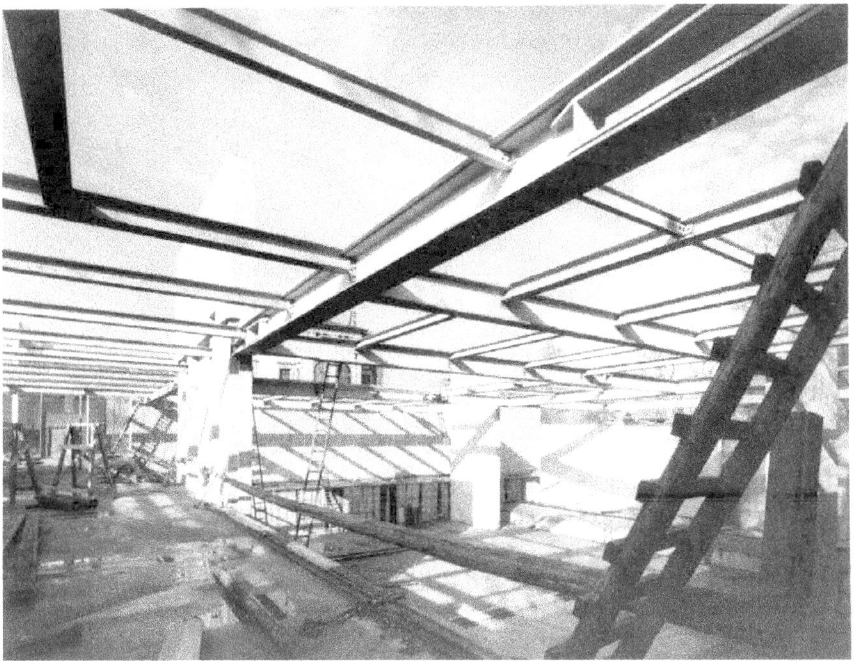

Fig. 5. Wilhelm Lehmbruck Museum: the Lehmbruck's wing during its construction, 1963. (Photo by Bernd Kirtz)

The use of *béton brut* was not an obvious choice for the architectural landscape of Germany in those years. One of the first German building sites where it was used was the *Hochschule für Gestaltung* in Ulm (1955), which was designed - not by chance - by the Swiss designer Max Bill. Swiss architecture indeed boasted a long tradition in the use of reinforced and bare concrete, and just after World War II pioneered interesting experiments in this field. [24]

With regard to the *béton brut*, Manfred Lehmbruck had to argue with the construction bureau of Duisburg, which again disagreed on this choice because of technical implementation difficulties and also a significant expense aspect. [25]

The construction site of the Lehmbruck Museum stayed open for about six years. Work proceeded slowly because most of the construction details were innovative and decidedly uncommon and had to be studied and redesigned during the process, in a constant - and often difficult - dialogue between the architect and the construction companies. Just as drawn out was the relationship between Manfred and the municipality, mainly due to the continuous increase in the budget required. Throughout the entire construction phase, Lehmbruck continued to stand by his decisions, saying that a museum was not an ordinary building, but architecture with special requirements that therefore involved extraordinary technological and design solutions.

The museum was inaugurated with grand celebrations and accolades from the critics and press in June 1964. Soon the *scenario* of German museum architecture would undergo a sharp turn. When at the end of 1960, a new generation pushed its way on the stage of history claiming its own identity, museums also changed their face and their architecture spoke a new language, finally freed from the reference to the *Moderne*. [26]

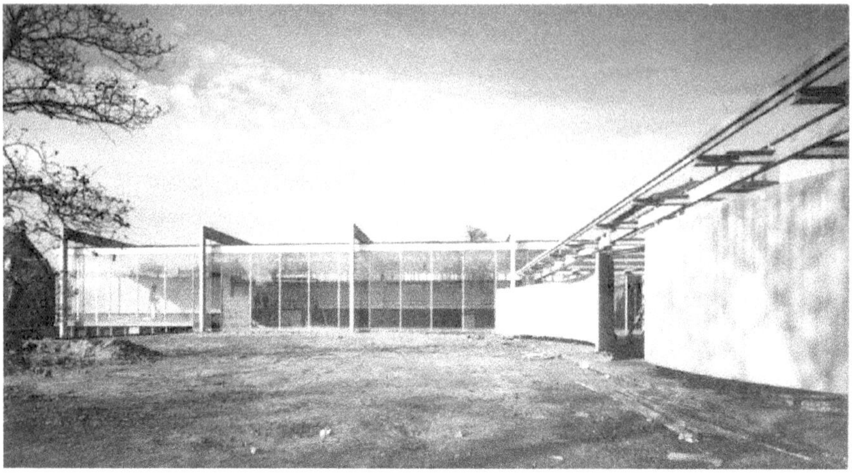

Fig. 6. The construction site of the Lehmbruck Museum, 1964.(Photo by Bernd Kirtz)

References

[1] My paper focuses on the German Federal Republic. In the German Democratic Republic - that joined the Soviet bloc after the end of WWII - the new democratic identity resulted essentially in anti-fascist ideology.

[2] The "economic miracle", i.e. the recovery of the economy in the German Federal Republic after WWII.

[3] The *Ruhrgebiet* (Ruhr region) is an urban area located in the German Nordrhein-Westfalen federal state. This large urban agglomeration comprises some important industrial cities (Düsseldorf, Duisburg, Essen, Dortmund, and Bochum among others) that developed near the banks of the river Ruhr. Since the XIX century the region become a stronghold of the coal mining and of the heavy industry. During both World Wars the cities of this area were the driving force of German war industry and as such they were subjected, more than others, to Allies bombings.

[4] The Reich's motorway.

[5] German modern architecture of the second post-war period.

[6] German modern architecture of the 1920s.

[7] See: H. Frank, "La tarda vittoria del Neues Bauen. L'architettura tedesca dopo la seconda guerra mondiale", *Rassegna: problemi di architettura dell'ambiente*, no.54, 1993, pp. 58-67.

[8] The architects of the German Pavilion at the Brussels World's Fair were Egon Eiermann and Sep Ruf.

[9] Temple of the Muses.

[10] This kind of research was promoted, among other institutions, by the *Deutscher Museumsbund* (league of German museums), founded in 1917 and closed during the Nazi dictatorship. The league opened again in 1960 and two years afterwards it also started the review "Museumskunde" that became almost like its press service. The objectives of this association were the reorganisation of

German museums from a scientific and administrative perspective as well as opening up to the community.

[11] During the Nazi dictatorship the architects could not have a great deal of contact with foreign countries. After WWII the Allies - especially the Americans - sponsored young German architects to travel and study abroad, with the aim to divulge a new "democratic" architectural education. Among the favorite destinations were the United States and Scandinavia.

[12] The large glass walls and the copious light they let into the buildings were a *Leitmotiv* of the German architecture after WWII also in other areas, like residential housing. Light played the same role as nature in the architecture of this period: these should be like an instrument of "purification" from the shameful German past. Transparency became the metaphor of the new and democratic GFR.

[13] The "first" Folkwang Museum in Essen was built in 1929, designed by architect Edmund Körner. The museum was damaged during WWII and rebuilt at the end of the 1950s (1956-1960). The new building was designed by the architects Werner Kreutzberger, Erich Hösterey and Horst Loy.

[14] The Focke Museum of Bremen was designed and built between 1959 and 1964. Its architects were Heinrich Bartmann and Reinhold Kalger.

[15] Manfred Lehmbruck (1913-1992) - son of the renowned sculptor Wilhelm Lehmbruck (1881-1919) - was a German architect particularly well-known for his projects of a few museums in Germany after WWII. The most famous include the Reuchlinhaus in Pforzheim (1954-1961), the Wilhelm Lehmbruck Museum in Duisburg (1957-1964) and the Federsee Museum in Bad Buchau (1954-1968). Since 1962, Manfred Lehmbruck was a member of the ICOM (International Council of Museums) and as such he traveled around the world to take part in many important conferences on museum architecture. From 1967 to 1978 he taught at the TH Braunschweig.

[16] This kind of ideas about museum architecture was initially divulged by the German art historian and museum director Alexander Dorner (1893-1957) since the 1920s.

[17] Some examples are the Folkwang Museum, the Focke Museum, the Theodor Heuss Bau (Germanisches National Museum, Nürnberg, 1958. arch. S. Ruf and H. Roth).

[18] The Lehmbruck Museum is the subject of my doctoral thesis discussed in June 2013 at the Alma Mater Studiorum-Università di Bologna (Italy), in co-direction with the Karlsruher Institut für Technologie (Germany).

[19] The Lehmbruck Museum is located in Duisburg, a German city of the Nordrhein-Westfalen federal state. Its main purpose is to host the permanent collection of the works of Wilhelm Lehmbruck, but it also includes an important selection of European sculptures of the early twentieth century and the post-war period and a number of paintings, mainly by German Expressionists.

[20] The project conceived by Manfred Lehmbruck at the end of the 1950s included three pavilions: the wing that hosts the works of Lehmbruck, the wing for temporary exhibitions and an auditorium that was imagined as a half dome partially underground. This last pavilion was not built because of a shortfall in funding. Currently, the third pavilion incorporates an extension built in the 1980s also designed by Manfred Lehmbruck. My paper only focuses on the project of the 1950s.

[21] Mies van der Rohe had been one of Manfred's mentors, as well as a dear friend of his father's. In 1932, Manfred got in touch with Mies, when he chose to become an architect. He worked as an apprentice for the construction of Haus Lemcke and he studied at the Bauhaus. In 1933, advised by Mies (the Bauhaus closed when the Nazis were coming into power), he enrolled at the TH Berlin-Charlottenburg and then at the TH Stuttgart, where he graduated with Professor Paul Bonatz in 1936.

[22] Similar structures were employed by Mies also in other projects: the Farnsworth Haus (Plano, Chicago, 1945-1951) and the 50/50 Haus (1950/1951).

[23] In the correspondence between Manfred Lehmbruck and the company Klaus Esser (kept in the SAAI archive, Karlsruhe) this company pointed out to the architect that only some factories

producing this kind of products - like the aircraft factories - could provide all the warranties for this material. In August 1959 the company produced a prototype of the acrylic glass dome for Manfred.

[24] After his doctoral dissertation (TH Hannover, 1942), Manfred was called to fight on the frontline and then he was taken prisoner in Russia. In 1945, he took refuge in Switzerland, where he remained until 1951 and worked in the studios of Roth, Brechbühler and Salvisberg. The Swiss experience enriched the architectural panorama of Lehmbruck and to him, this also led to a deeper understanding of the architecture by Le Corbusier, the Master of the expressive use of *béton brut.*

[25] The elements in *béton brut* were made by the firm Kiefer of Duisburg.

[26] See the Neue Staatsgalerie of Stuttgart (arch. James Stirling and Michael Wilford; 1977-1984) and the Städtisches Museum Abteilberg of Mönchengladbach (arch. Hans Hollein; 1972/1981), that could be defined as "postmodern" architecture.

Stone barrel vaulting in late medieval churches in Scotland

Dimitris Theodossopoulos
ESALA, University of Edinburgh, UK

Summary

This paper explores the construction characteristics and structural design of plain stone barrel vaults built to roof a new type of single-nave churches in Scotland in the 15th century, such as Dunglass Collegiate Church in 1423, Seton Collegiate Church in 1492, St. Salvator's Chapel in St. Andrews in 1450, Corstorphine in 1429 and Ladykirk in 1500. Four case studies are surveyed (Dunglass, Seton, Corstorphine, Ladykirk) and their geometry is analysed to assess their design, construction quality and structural efficiency. The analysis demonstrates a very good quality control and a consistent approach to the simplification of the pointed profile developed in cross vaults.

Introduction

Fire-proof stone vaulting in Scottish churches developed a simpler configuration of barrel vaults over single naves in the 15th century, in parallel with cross-vaulting used in the new contemporary burgh churches. These vaults are found in royal or aristocratic foundations such as Dunglass Collegiate Church in 1423, Seton Collegiate Church in 1492, St. Salvator's Chapel in St. Andrews in 1450, Corstorphine in 1429, and Ladykirk in 1500, away from the earlier large scale monasteries. Similar vaults also characterise votive or burial aisles in existing churches (Cockburnspath parish church C15, Wardlaw Vault in Dunfermline in 1617, Dirleton Aisle in 1664, Abercorn in 1727).

Cross-vaulting is usually defined by a system of ribs, which guarantees the quality of construction and creates a strongly linear architecture as they have to fully integrate with the decorative programme (concordance) and make larger openings in the envelope possible. Cross-vaults require however complex construction, especially for exposed stonework, and are not widely encountered in Scotland away from large churches. When compared, barrel vaults are simpler in design and execution, and relate to the lateral walls in a very different manner because of the single curvature and continuous support along the wall-head.

A study of the development of the technical culture of this type is presented here. The vaults have a straightforward structural scheme which will be analysed to interpret the efficiency of their design and containment of thrusts, but also to explore the masons' understanding of load paths. Their stonework is usually dressed internally but overall the structures are robust as they are roofed in flagstones and the space of the pitch is filled with rubble. The quality of the masonry bond will also be discussed in relation to the structural design to reveal the extent to which the simplicity of the structural scheme permitted a better tectonic control.

Four case studies from the 15th century were chosen (Dunglass, Seton, Corstorphine and Ladykirk), which demonstrate a certain chronological development in bolder forms. They also represent a re-elaboration of pointed profiles and ribs, key Gothic characteristics, in the barrel vault context, and the study explores whether these were constructional devices or stylistic affiliations.

Scottish late medieval vaults

The C15 vaults studied here are elaborate and consistent designs, comparable to earlier examples such as St. Margaret's chapel in Edinburgh Castle, the C12 Dalmeny church or Sutra aisle. This new design over single-nave churches is in line with the emergence of different types of smaller and more compact, still significant, churches that deviate from previous large scale constructions such as cathedrals and abbey churches. These were founded by aristocrats as colleges of priests (Dunglass, Seton, Bothwell, Corstorphine, Crichton, Roslin, Carmelite Priory in S. Queensferry) or prestigious chapels such as St. Salvator's college chapel in St. Andrews, while royal foundations later followed the pattern too (Ladykirk). The origin of these forms seems to be more civic than religious, following techniques developed and tested in basements or banqueting halls in castles [1]. The barrel vaults built in the cloister and additions in the Augustinian Inchcolm Abbey during the 14th C are an example of robust, relatively inexpensive and durable roofing in a compound that has to provide or at least show off its ability to defend itself.

From the beginning of the 15th C, following the assertion of James I's rule, Scottish castles developed a stronger residential character combined with the need to resist the effects of artillery assault. Prominent examples that may have an influence are Bothwell, Strathclyde (imposing form with architectural qualities, refurbished in C15); Dirleton (in East Lothian, where many collegiate churches were founded), characterised by several vaulted ceilings serving various purposes (storage, kitchen, private apartments) along the face of defence and refined ashlar dressing of the elevation (Fig. 1).

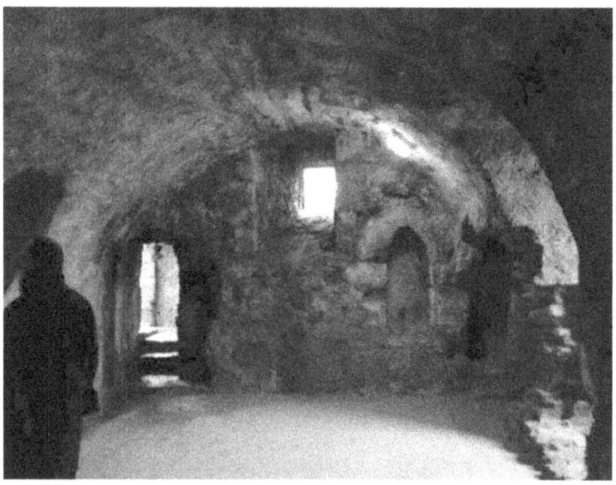

Fig. 1. Vaults in Dirleton Castle.

If these vaults are viewed as an evolution or diversification of earlier Gothic vaults, it is worthwhile to examine how their characteristics were adapted. The cloister vaults at Inchcolm already present a smooth pointed profile, which becomes sharper (Corstorphine 1429) and much more clearly defined later (Carmelite Priory in S. Queensferry 1441,Crichton 1449), eventually presenting an ashlar finish (Ladykirk 1500).

Ribs take a different character in this context. In the case of Bothwell (1398) or Ladykirk (1500) they divide the barrel vault into bays and are pronounced (Fig. 2A), while they spring immediately above the windows. In some cases, ribs were planned to be attached, possibly to confer some rhythm or even prestige to the design, but they would spring from corbels as surviving post holes show (Dunglass – Fig. 2B, Seton) [1].

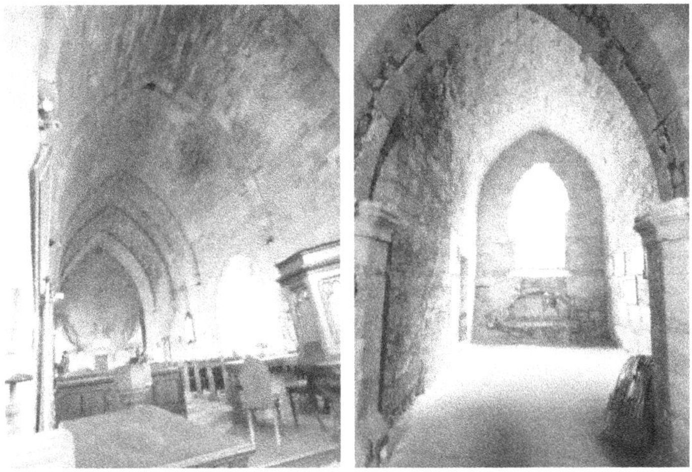

Fig. 2. A) Ladykirk; B) Dunglass.

The only case of concordance is Roslin (1450) where ribs are heavy and integrated in a tripartite scheme along the shafts of the piers of the aisles, but usually the smooth barrel is continuous with the walls. They are often used however to guide difficult junctions like apses, especially in the case of Seton (Fig. 3A). Closely spaced ribs are found in the apse of Holy Rude, Stirling (the only stone-roofed part of the choir) and they are more like a flat reinforced mesh, dating before mid C16 (Fig. 3B). There is also a unique external system of ribs above the cross-vaults in the single nave of the earlier but very elaborate Lincluden Collegiate Church (1389).

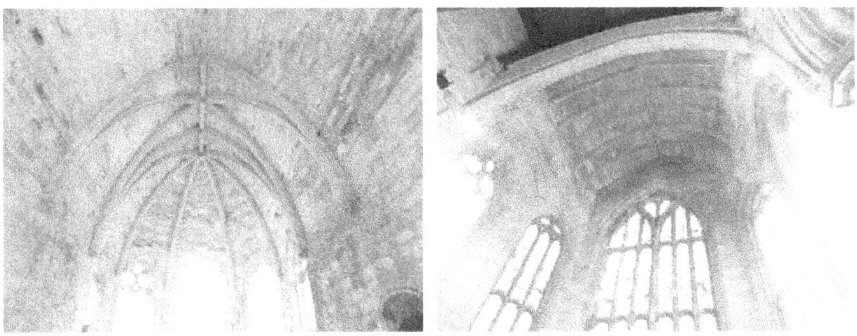

Fig. 3. A) Radial ribs framing the apse of Seton; B) Orthogonal mesh roofing the apse of Holy Rude, Stirling.

What characterises these vaults is that they all have exposed stonework now, although it is possible that Corstorphine or Dunglass were planned to be rendered, and in most cases the finish is ashlar: Ladykirk and Bothwell (red sandstone) or Crichton (buff sandstone).

Regarding spans and proportions, these will be examined in more detail in the four case studies. It is interesting that most span 6m to 7m while St. Salvator in St. Andrews (1450) is the biggest (ca. 9m), which had substantial amount of windows and consequently daylight, and was possibly the weakest as it was demolished in 1773. The survey of the shell itself in the case studies shows a span to rise ratio L/F between 1.4 and 2.3, but uncertainties in the springing level often indicate that the design was generated by an equilateral triangle (L/F=1.15).

The case studies will also evaluate the structural efficiency of the vaults' design. Pier buttresses are employed to resist thrust and their application varies: Dunglass has few and irregularly spaced; Corstorphine has a sufficient number; Ladykirk has regularly spaced, deep piers that are integral to the overall quality of the design. Empirical rules that relate span and buttress size will be tested in the discussion of the case studies [2, 3].

Focusing on the buttresses at the apses, some are perpendicular (like Dunglass, Bothwell, Crichton, Pencaitland) and others are diagonal, as in Corstorphine. Seton and Ladykirk however have buttresses normal to their polygonal apse bays, while Roslin applies flying buttresses over the aisles. The diagonal ones may indicate some stronger experience with the behaviour of the vaults and an attempt to provide a more efficient bracing. Polygonal apses also improve the stiffening of the barrel vault at its ends but the majority show straight apse ends. The polygonal ones in Seton (1492), Ladykirk (1500) or Holy Rude (Fig. 3) coincide with the latest period of this typology but are not fully integrated with the barrel vault.

Walls in medieval churches are characterised by the quest to balance daylight with structural support. The majority have a main tracery E window at the apse and very few lateral openings. Many windows appear only along the walls of St. Salvator or Bothwell but the fabric in-between does not receive the treatment of skeletal piers as traditionally occurs in the Gothic style.

Finally, an interesting aspect to be considered is the propagation of the system beyond C15, the long term safety and durability, and also its legacy. Few barrel vaults were built after 1500. The plain, ashlar vault of Dirleton Aisle is the closest direct connection but it is smaller, less deep, with a semi-circular profile spanning 4.65m and belongs to a more neoclassical context (1664) that may even have rejected pointed profiles. St Salvator's vault (the largest span) was deemed unsafe in 1772 by the neoclassical architect James Craig and, possibly because of lack of knowledge of its behaviour and construction (there was no debate or ideas recorded about a repair), was subsequently demolished [4, 5]. This may have also been precipitated by the earlier collapse of Fearn Abbey church in Ross (1742). The complete detachment from the technology of these plain structures can even characterise the reconstruction of the vaults in the nave of Corstorphine in 1903-5, where solid prefabricated concrete panels were used, strengthened with ribs.

Four case studies

The design aspects and technical culture of this typology will be explored in four representative case studies. The main tool is a measured survey using a Trimble S6 total station to perform a medium density laser-scan. The point-cloud allowed the geometry of a variety of areas within the church and each vault to be analysed and eventually key cross sections across the barrel were taken and compared (Fig. 4) - the ordinate axis starts at what is assumed to be the level of springing.

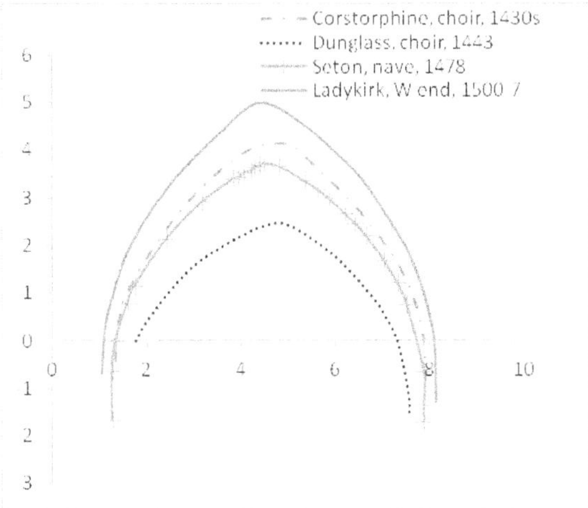

Fig. 4. Comparison of representative cross sections between the four case studies.

It is not always straightforward to establish the true springing of the vault. The discussion of the proportions and angles that define the vaults of each case study focuses on the clearly observed curved parts of the shell. Very often however the crown angle φ_1 is very close to 60° or the centre of the circular segment of the pointed vault lies at a distance from the springing that is equal to the radius R. This gives hints that the proportions of the vault were defined by an equilateral triangle which was then adjusted during construction.

Evaluation of thrust and min. thickness are based on shell solutions for masonry domes calculated by J. Heyman based on thrust line and limit-state principles [6], as in the graphs in Fig. 5 for example that relate thrust and min. thickness to the crown angle φ_1. Thrusts Nf are normalised over wR, where w is self weight and R the radius of the pointed arch. To define w a uniform, reasonable thickness t=0.4m is assumed in the absence of survey data - probably the real thickness varies from a max at the base (wall head) to a min. at the crown. The unit weight of stonework γ is taken as 2 t/m^3 =20 kN/m^3, therefore w $=\gamma$ t =20·0.4 =8 kN/m^2.

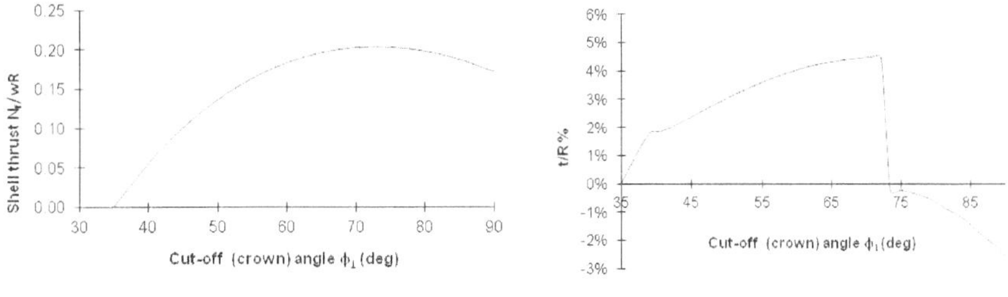

Fig. 5. A) Shell thrusts Nf/wR for a pointed shell with a springing angle φ_0= 35; B) Required thickness t for a pointed shell with φ_0 =35.

A) St. John's Collegiate Church, Corstorphine, Edinburgh

The church built before 1158 became a collegiate one possibly in 1429 by Sir John Forrester [7]. The building was transformed to the new form of worship, allowing space for the parish, and this included the absorption of earlier Gothic features and the erection of the barrel vaults, which may have been completed by 1436. The E choir (Fig. 6) and the adjacent sacristy are the only original roofs that survived, and the vaults of the nave and aisle were re-made in precast concrete ribbed panels during the restoration in 1903-5 by George Henderson [8].

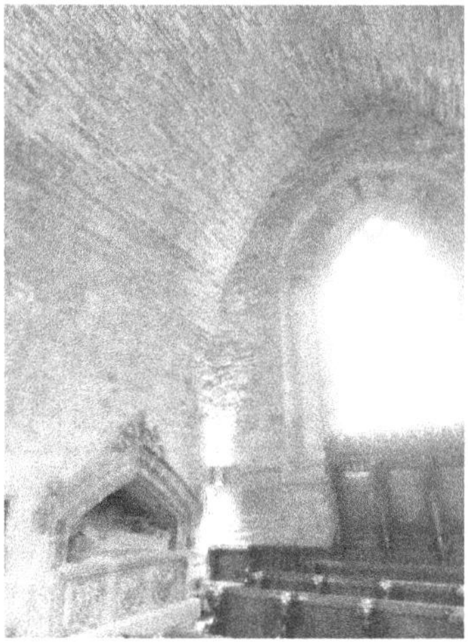

Fig. 6. Corstorphine Collegiate Church, vault at E choir.

The survey of the E choir demonstrates the very good quality of the construction of the vault as there are almost no differences in various sections drawn along the bay. The span from the springing is measured as L=6.55m (21.5ft) and the rise F=3.97m (13ft), resulting in L/F =1.65. The corresponding circle has a radius R =5.3m (ca. 17ft) and there are minor deviations around the spandrel, supporting the observation of good quality control of construction. The springing angle $\varphi_0 = 9^\circ$ and at the crown $\varphi_1 = 63^\circ$, therefore the normalised thrust of a representative tranche [6] is Nf =0.6471 wR or 0.6471·8·5.3 =27.4kN/m. The minimum thickness seems to be very small and hinges may form close to the crown, so the form appears to be almost a catenary and develops mainly axial compressive forces. These large thrusts are contained by widely projecting buttresses and rather thick walls.

The proportions of this vault give a hint that an equilateral triangle of sides equal to R defines the profile, which then was adjusted to the thickness of the walls or extended with stilting.

The walls of the choir have a rubble finish currently and at the springing of the vault the masonry is made of oblong thin blocks laid regularly, which makes the construction of the vault feasible without specialist skills or experience. There was no apparent intention of an architectural finish to the stonework and the surfaces were possibly unified by lime-wash, to match the quality of the tombs and tracery in the window.

It is very interesting how a more sophisticated, pre-existing vault like that of the S transept was treated in this remodelling. This vault was absorbed into the remodelled church in 1429 and at the N end the ribs were truncated without apparent concern for integration into the new design. Their design does not appear as a dominant structural scheme to work around as it shows a structurally slightly arbitrary pattern of a semi-quadripartite arrangement of ribs rising from a springing point at the centre of the bays [1].

B) St. Mary's Collegiate Church, Dunglass, E. Lothian

The church may date from after a college of priests was founded in 1443 [9]. The plan is quite articulated and all the spaces are vaulted, creating a continuous stone surface between the walls and the vaults, which are not at present divided by any permanent decoration or cornices (cf. Bothwell). There are two main barrel vaults at the E choir and W nave (taller, but narrower), and three minor ones over the two transepts and a burial vault at the S side. Stylistic analysis of the arches where these vaults meet the nave and traces of foundations of walls show these vaults were added later [9].

The E window is smoothly pointed (Fig. 2A) and the profile of the barrel vault is similarly gentle, especially when compared to the other cases (Fig. 4). The vault of the nave was surveyed and spans L = 5.5m at the springing (or 18 ft), while the walls at footprint divert slightly to a span of 6.1m (20ft). The rise of the vault (from the springing) F= 2.43m (8ft), producing L/F =2.25. The radius R that generates the pointed profile was evaluated as 6m (20ft) and the centre lies about 10ft below springing.

The springing angle φ_0 was measured as 37° and crown φ_1 is 69°. The normalised thrust is evaluated from Fig. 5A as Nf=0.202 wR (for φ_0=35°, φ1 =70°), or 0.202·8·6 = 9.7kN/m. Regarding the min. thickness to withstand self-weight, Fig. 5A shows it 0.045R or 0.27m and if below, a hinge can form at a 46° from springing. These calculations show the design of the choir has a comfortable safety margin. Flagstone roofing covers all bays which indicates each vault successfully carries a substantial load.

If the circle segment that corresponds to these proportions is superimposed, there is a vertical deviation at the crown but it is small, almost 8cm, i.e. the quality of construction is of efficient level to maintain the original form as planned.

The vault of the nave has a shorter span and its overall height is taller than the choir. The proportions therefore are not too different, but the pier buttresses are only normal to the walls of the choir and are shallow, so either the masons became confident of the stability of the design of such a vault after the choir was built or they relied on the later addition of the S burial vault to abut the nave. It would be interesting to survey the nave vault as well to verify any differences in the design or deformation.

The quality of the stonework bond is variable and improves to a firmer ashlar in the later transepts and nave. In the choir, the blocks of the wall are rectangular up to a height of 4.5m where they become thin and oblong for a further meter until the springing and continue as such along the barrel vault. A visual inspection shows different proportions and springing points at each bay. There are no ribs or intentions for them, and the sockets on the barrel were made when the church became a stable and decks were added.

C) St. Mary and Holy Cross Collegiate Church, Seton

The development of the church is well documented in the history of the founders, the Seton family, written by G. Maitland around 1560 [9]. Today the (original) W nave is missing but there remain the very simply and uniformly designed vaults over the choir and the two transepts. This documentation is summarised in [9]: the choir (apparently the ribbed apse) and the sacristy were built by the third Lord

Seton (before his death in 1478) but the plain W part of the choir was added by the fourth Lord, who also founded a college in 1492, while the choir was roofed later with a flagstone roof by the fifth Lord (+1513). The transepts were added later, first the N one (1541) and then the S (1545).

Sections along the W choir taken from the survey are very similar, showing high construction quality control. Minor sinking or deviations from a circle were measured. A similar profile however was more difficult to obtain for the apse, probably due to the presence of the ribs. The irregular profile (Fig. 7A) and the layout of the blocks (Fig. 7B) show the ribs were built first as a geometric guide and the oblong blocks of the shell were then attached.

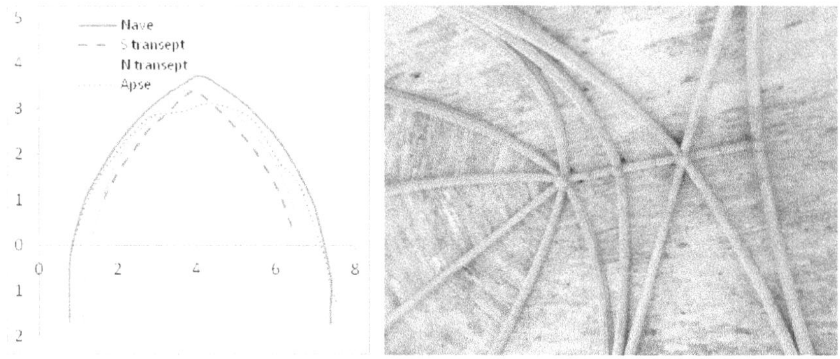

Fig. 7. A) Profiles of the barrel vaults in Seton Collegiate Church; B) The apse.

Focusing on the choir, the generating radius R=4.85m (ca. 16ft), the span L = 6.46 m (ca 21 ft) and rise F=3.75m, so L/F=1.7. The springing angle φ_0 was measured as 9.5° and cut-off (crown) φ_1 is 68°. The normalised thrust is calculated as Nf=0.647wR (for φ_0=10°, φ1 = 65°) =0.647·8·4.85 =25.1kN/m. The calculation of the thickness [6] shows a similar phenomenon as in the right-hand section of Fig. 4, i.e. there is no min. thickness as the profile of the vault may correspond to a catenary, hence no bending develops.

All divisions are braced with substantial pier buttresses, built contemporarily with the fabric and vault, and at some point would have been highlighted with a pinnacle. Several geometric rules existed around the time to calculate the depth of the buttress, as the later manuscript of Rodrigo Gil de Hontañon indicates (1544-54) but there is no evidence recorded of such rules in Scotland. Since they are all empirical in nature, one that is quite intuitive is considered here, the rule of Martinez de Aranda [2], which evaluates the buttress depth as ca. 1.5m. This matches well the real depth of the buttress 5ft or 1.52m.

The bays of the choir vary 16ft -18 ft in spacing and apart from a regular division of the area they do not seem to be in any proportion with the profile of the vault. There are several large openings, each with a tracery, invariably on each division and the survey shows some minor deformations in their vicinity. The (later) S transept shows a minor distortion of the vault towards its S end and the N transept shows a small irregular deformation at the N window.

Study of the stonework shows a visual and tectonic continuity as the rectangular ashlar of the wall changes quite smoothly around the springing to the exposed oblong blocks of the vault. The plain and almost continuous aspect of the stonework (affected slightly today by humidity and salts around the spandrels of the vault) provides a uniform, subtle and peaceful treatment of the interior, which may not have been altered significantly by the furnishings of the period.

When compared with the choir, the transept's vaults are smaller (Fig. 7A), which shows the masons were not capable of (or interested in) copying the vaults of the choir. Even the construction quality between them varies, with the N transept being slightly more irregularly built, although they followed the same materiality principles.

D) Ladykirk Parish Church (or Kirk of Steill), Scottish Borders

This is a brand new design that had no constraints from pre-existing structures. It is considered as a votive church founded by James IV and was built between 1500 and 1507, the year when the master-mason Nicholas Jackson (who had worked in Linlithgow Palace) was at work on the roof [10]. The church has a unique pattern combining a three-sided E end apse (instituted earlier in St. Salvator's from 1450) with two transverse ones, each roofed with a domical ribbed vault which is a variation of the strongly pointed vault of the nave.

The nave vault is the highest and most imposing of the four case studies and is prominently pointed (Fig. 4). The section is constant throughout the nave, indicating high control during construction, which is also manifested in the ashlar finish of the shell. The vault profile matches fully the corresponding circle of a radius R =7.3m (24 ft). L/F = 7.2/4.99 = 1.44, the springing angle φ_0 was measured as 18° and cut-off (crown) φ_1 is 62°. The normalised thrust is evaluated as Nf=0.374wR (for φ_0=20°, φ_1 =60°) =0.374·8·7.3 =21.8 kN/m. Thickness is evaluated t= 0.04R = 0.29m, and there is an apparently large safety margin.

To test the efficiency of the profile, it was compared with a catenary that corresponds to the proportions of the vault, calculated using a readily available spreadsheet [11]. There is a visible deviation from this profile so some bending develops, which is absorbed by the thickness of the shell.

The vault carries a substantial load from the flagstone roof and the infill or diaphragm walls that form the pitch. The thrusts are contained by deep buttresses, topped with a pinnacle, which shows some awareness of their function. These piers and the corresponding bays are widely and regularly spaced every S=13ft. The divisions are marked with ribs, springing from corbels at a height of ca. 22 ft = 6.6m, which is not far off the radius.

To complete the structural scheme, the transeptal apses abut the barrel nave at mid-height and their edge is finished with a solid arch. There is no full assimilation of the visual aspects of this function [1], which is in contrast with earlier achievements of Gothic as have been manifested in the cross-vaulted churches of the period (St. Giles, Trinity College).

Discussion

Gothic masons' structural design (*scientia*) was firmly based on geometry as the form guarantees a safe state of equilibrium [2]. This initial study from the survey and the structural analysis of the forms reveals some things about the designers' choices and their relation to the construction process. The profiles in the four case studies are consistently pointed and there appears to be an underlying scheme of equilateral triangle that defines the overall dimensions of the vault itself. The comparison of the profiles show a slow increase of the sizes (Fig. 4) and Ladykirk shows a confident and safe design of the pointed shell

(6.5m span), with a low normalised thrust 0.374wR, something that may have failed earlier in the larger span of St. Salvator (9 m).

The construction is solid and of good quality, and not many defects or deviations from a circular profile are observed (Fig. 7). This shows the structural form is not too sensitive to the construction process (whether bottom to top or end to end).

The impression of the structures, even within the context of large stone roofing of the time, is one of uniformity, maturity and simplicity, as can be compared with Trinity College, Haddington or St. Giles. In a way is a return to more archaic, Romanesque forms (see for example Tranni Cathedral, Puglia), but with a tectonic impression of a continuous surface of stone adjusted here for the absence of aisles and lateral walls. In conclusion, the simple form of single curvature vaults shows Scottish masons were capable of achieving high quality of construction and control of the forces (both the load path and the containment of the thrusts). Among them only St Salvator (and possibly Fearn) failed but the effectiveness of the scheme was undermined by the rigidity of the space it created, which curiously enough was more suitable for the different spaces created after the Reformation later in C16.

Acknowledgments

AOC Archaeology (especially Graeme Cavers and Gemma Hudson) gave fundamental support by providing the total station and technical guidance; Gergana Staneva, MA Architecture student in ESALA provided essential help during the surveys; Peter Ranson and Nira Ponniah are thanked for allowing access to properties in Historic Scotland's care; Fr. Alan Cartwright, priest of Ladykirk, and Christine Mill and the community at Corstorphine are warmly thanked for valuable access and insight to the history of their churches.

References

[1] R. Fawcett, *The architecture of the Scottish medieval church 1100-1560.* London: Yale University Press, 2011.

[2] S. Huerta, "Geometry and equilibrium: The gothic theory of structural design". *The Structural Engineer,* vol 84(2), pp. 23-28.

[3] S. L. Sanabria, "The mechanisation of design in the 16th century: The structural formulae of Rodrigo Gil de Hontañón", *J. Soc. Architectural Historians*, 41, 1982, pp 281-293.

[4] R. G. Cant, *The college of St. Salvator*. Edinburgh: Oliver and Boyd, 1950.

[5] D. Theodossopoulos, "Limitations in the understanding of the performance of complex structures and vaults as manifested in the fatal repairs of the Holyrood Abbey church roof in 1760". Forthcoming publication, 2014.

[6] J. Heyman, *The Stone skeleton. Structural engineering of masonry architecture.* Cambridge: Cambridge University Press, 1995.

[7] C. McWilliam, D. Walker and J. Gifford. *The Buildings of Scotland: Edinburgh*. London: Penguin, 1984.

[8] Fergusson, Dickson, James and Traquair, "The Church of Corstorphine and its recent restoration" *The Scottish Ecclesiological Society*, vol.1 (3), 1905-1906, pp. 286-298.

[9] C. McWilliam *The Buildings of Scotland: Lothian.* London: Penguin, 1978.

[10] K. Cruft, J. Dunbar and R. Fawcett *The Buildings of Scotland: Borders* London: Penguin, 2005.

[11] Catenary function (2009): <http://newtonexcelbach.wordpress.com/2009/06/25/a-catenary-function/> [posted 25 June 2009]

An alpine hydroelectric construction site

Works for the utilisation of hydraulic resources in Alta Valtellina 1907-1930

Francesco Carlo Toso
Department of Architecture and Urban Studies, Politecnico di Milano, Italy

Summary

In 1906, in the wake of the municipalisation of the electric supply for urban lighting and tramway networks in the city of Milan, the municipality opted to expand the city's productive capabilities by buying the rights for the utilisation of the hydraulic resources of the river Adda, in the Alta Valtellina valley. For the aptly founded company, Azienda Elettrica Municipale (AEM), it was the start of a century-long process of infrastructure building, which shows the evolution of technical and construction systems in hydraulic engineering and has had a defining impact on the landscape of the valley. The research has involved examining the surviving correspondence, technical and photographic documentation, offering an overview of the construction techniques and of the organisation of the construction site. It shows the rapid introduction of new technology in relation to the hydraulic works, especially under the supervision of the first national committee on large dams, while highlighting a reluctance to the introduction of new languages for the architecture of the power station.

Introduction

As with most valleys in the Alps, the beginnings of electrification in the mountains of Lombardy between the 1880s and 1890s was linked to the supply of lighting to the small industries and workshops and of public lighting, by means of small low head power plants built by private companies on the tributaries of the Adda river. However, by the beginning of the twentieth century, the adoption of three-phase current for distance transportation allowed the bigger industrial enterprises in the city of Milan to look at the hydraulic resources in the Alps as their new source of motive power. In the first three decades of the century, the bigger electric companies, such as Imprese Elettriche Conti, Edison and Adamello acquired the rights for the utilisation of the main alpine drainage basins, setting up the schemes for the development of true "hydroelectric districts". [1] The liberal-led administration of the municipality of Milano was an important factor in this early phase of development since 1906, securing the rights for the utilisation of the northernmost section of the Adda river and setting up a municipal company (AEM) in 1910, for the purpose of building and managing the municipal power plants and networks. [2]

These early developments clearly show the relation between the articulation of the facilities, the level of technological advancement and the morphological specificities of the different alpine valleys. Even technical handbooks did not offer fixed criteria on how to devise a utilisation scheme, preferring to illustrate specific technical solutions with a choice of examples. [3] The correct interpretation of the hydraulic problem, which involved a careful recognition of the environmental features of the area, was entirely up to the ability of the designer, due to the great variability of factors that came into play. The process of accumulating knowledge and expertise becomes evident with a closer examination of the technical office historical documentation and correspondence.

Early AEM hydroelectric facilities

The earliest facility, including the Grosotto power plant, was designed by three engineers from the Royal Technical School of Engineering in Milan, Gaudenzio Fantoli, the later AEM director Tito Gonzales and Carlo Mina. The latter was a central figure in the development of a number of projects in Valtellina, including the power plants built by the major steelworks company in Milan, Società Acciaierie e Ferriere Lombarde.

The Grosotto facility derived the needed waters from the river Adda, by means of a 32 m long concrete and granite weir, equipped with a stilling basin and a surveillance house. A 12 km long tunnel connected the intake works with the forebay tank, adjusting to each different slope condition and including sections drilled through solid rock, partly with the use of a supporting concrete vault, and sections with masonry walls and a concrete roof, with the eventual consolidation works needed when crossing alluvial cones. The works were reportedly carried out with minimum use of explosives. [4] Finally, an open air segment of the channel led to a sedimentation forebay and to the loading tanks hidden under the personnels' housing. The anchorage of the steel penstocks, aligned on a concrete bed, required further masonry consolidation works as the slope revealed a tendency to slide. The penstocks themselves, as with the majority of the alpine facilities of the time, were produced and assembled by the Brescia and Genova-based company Tubi Togni. [5] They meet the power house orthogonally, an arrangement that was to be dropped in later years due to the risks connected with an explosion of water under pressure against the building.

The power plant included the turbine hall and a three floor building housing the transformers at the ground floor, the switchboard and buses at the first floor, and the high tension switches and power line at the now demolished second floor. Concrete foundations supported the artificial stone walls, cladded in coloured cement brickwork, which were prepared on site with pedal operated machines. [6] Often attributed to Gaetano Moretti, the eclecticist architect of the celebrated Taccani power plant in Trezzo d'Adda, the simple architectural features may actually have been designed by Mina himself. Mina's works in the preceding and following years showed a similar use of a simple, symmetrical façade and a preference for regular, round-arched openings. The facility was completed and running in 1910 (Fig. 1).

In 1913, the director of the company, Tito Gonzales, devised a project for the construction of a 10,000,000 m^3 reservoir situated in Val Fraele, near the river springs at an altitude of approximately 1800 m above sea level. Despite the project being put on hold during World War I, the increased demand for power due to the shortage of imported coal made the construction of a second facility a necessity. The Roasco facility was built between 1918 and 1921 under the direction of engineer Silvio Apostoli, then at the head of the Valtellina department of the AEM technical office. It was the first facility to include an arch-gravity dam.

Built in a narrow rocky gorge, the 59 m high dam of Fusino is a masonry dam with a stone and concrete core. The cross section is triangular, with the profile of the downstream wall joining the top in the shape of an arch, while the upstream wall is slightly sloping. Both walls are cladded in regular granite blocks, while the core filling is made of stone blocks and concrete. Next to the dam, a 38 m long spillway lets the exceeding water fall directly into the gorge, while four discharge steel pipes pass through the foundations. The top of the dam is carefully crafted with a granite decoration and the surrounding buildings, including the housing and the sluice gate operation room, are designed with similar care.

The construction was carried out by raising the outer granite masonry walls by 1 m increments, then transporting the rock filling by means of light railway cars (decauville). Two Blondin aerial ropeways were used to move the buckets over the dam and a Ransome-Canniff pneumatic machinery equipped with movable pipes was used for pouring the concrete. [7] As the construction was rising, the whole equipment had to be dismantled and reinstalled at a higher level. On completion of a new layer the reservoir level was raised; as soon as the intake works could feed the channel the power plant was put into operation (Fig.2). The channel and penstocks were realised in a similar fashion to the previous facility, with the addition of a cement-gun sprayed internal cladding.

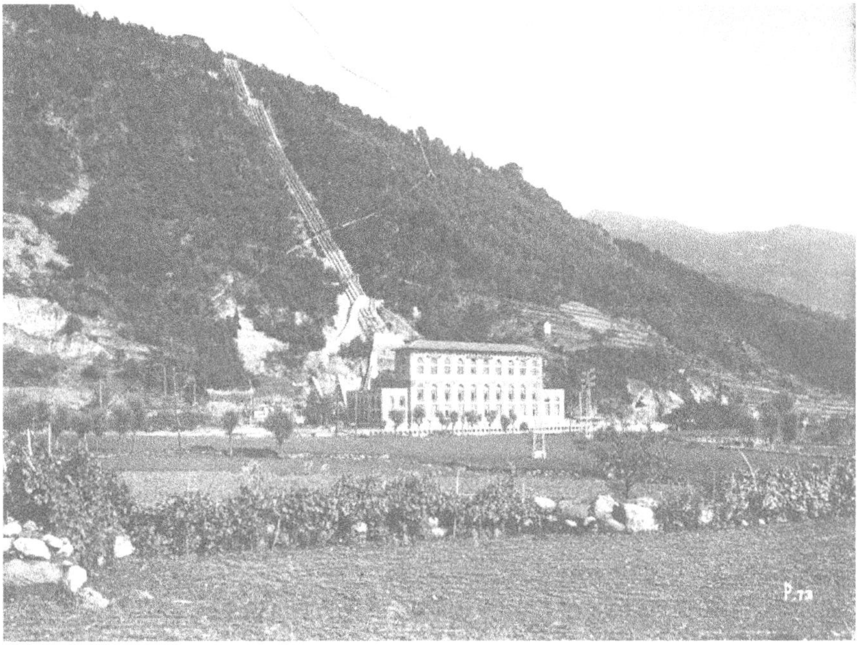

Fig.1. Grosotto power plant. Photo by Antonio Paoletti, early 1920s (Photographic Archive Fondazione AEM, Milano)

Notes on Italian dam building in the 1920s.

The Roasco dam was built in a period of major changes in Italian dam practice. The first comprehensive Italian regulation on dams was issued in 1921 [8] and it was the first to include a classification by type, including gravity dams, multiple arch and buttresses, earth and dry masonry dams. It also provided proper dimensional and calculation prescriptions for each dam type, as well as guidelines for the quality control of materials. Until then, the structural design of a gravity dam relied entirely on weight and adequate anchorage to solid rock, while its impermeability relied solely on the accurate execution of the stone and cement mortar cladding. The greatest novelty in the new regulations was the inclusion, for stress calculation purposes, of uplift, seepage pressure and ice pressure on the crest, along with specific prescriptions for assessing the conditions of the foundation soil, arranging drainage in the dam core and providing documentation of the materials employed.

The further development of regulations followed the infamous Gleno dam disaster, which took place in Val di Scalve (Bergamo) in 1923. [9] The multiple arch dam failed few months after its completion, causing more than 500 victims. Numerous reports were issued in major international journals to assess whether the accident was caused by inadequacy of calculations. [10] However it was later ascertained that the major flaws could be attributed to the inadequate monitoring of the quality of materials and supervision of the construction. The client and construction firm had not followed all design specifications, the concrete was poorly mixed and the core included highly irregular stones. In the aftermath of the accident, a scientific committee, Commissione Grandi Dighe (Large Dams Committee), was formed to assess the state of all built barrages in Italy according to the new criteria and act as a consultant for the major active construction sites, often introducing considerable changes in the design, as was the case with the AEM Cancano dam.

The Gleno accident also led to arch dams having limited diffusion across Italy. In his speech at the World Power Conference in London in 1924 [11], the electrotechnical engineer Luigi Mangiagalli gave an overview of the major dams built since 1910. Earlier, the majority of hydroelectric developments were based on run-of-the-mill facilities and no comprehensive plan for water reservoirs was made until the exploitation of high altitude alpine hydraulic resources gave the major impulse in dam building. There were few examples of multiple arch dams, the most important one being the Tirso dam in Sardinia. Arch gravity dams, both non-overflow or with a spillway at the crest, were the most used typology in the alpine developments. [12] Mangiagalli remarked the prevailing single arch planimetric arrangement and the little variability in the cross sections adopted nationwide, acknowledging that Italian practice was mainly indebted to American dam building technology. [13] In an important article published in the journal *L'Energia Elettrica* in 1924, Gaudenzio Fantoli stated the arch-gravity typology to be a safer, more appropriate solution for the alpine developments than the multiple arch or the thin section dams. He cited numerous examples from the neighbouring countries to support his observations, including the Swiss Wäggital, Barberine and Grimsel dams, the French Eguzon and Chavanon barrages, the Schwarzenbach and the Waldecker dams in Germany and the Spullersee dam in the Austrian Voralberg. His technical opinion was supported by the words of his correspondent, the Engineer-in-Chief in the municipality of San Francisco, O'Shaugnessy: "I am throughly convinced of the sound logic of your position in favor of the modern type of arch gravity dam. Where the foundation and other characteristics are favorable I believe this is the only type to employ where human life and property lower down are involved". [14]

The Fraele-Viola facility and the Cancano Dam.

In its earliest design, the Cancano dam in Val Fraele was conceived as a double arch-gravity dam, joining on a single central buttress. After an accurate inspection by geologist Dal Piaz, professor at the Univeristy of Padua and a regular consultant for AEM, various other solutions were taken into consideration, including a hollow gravity dam and a single arch-gravity dam. [15] The main auxiliary buildings, workers' accommodation and warehouses were built at the valley bottom, about 500 m below the dam construction site and in proximity to the power station site. A system of cable cars connecting the two sites was installed, as well the extension of the power lines coming from the Grosotto power station.

The works started in 1924, with the preliminary excavations and consolidation of the foundation rock and the drilling of the pressure conduit that should connect the reservoir, through the mountain, to the nearby valley where the power plant was to be built. The overall scheme of the facility also included a diversion channel ("Canale Viola") intercepting all the minor rivers on the valley slope, joining the main conduit at the surge tank. The main pressure conduit was drilled in solid limestone and cladded with cement and smooth mortar. During the construction unexpected fractures in the limestone and incoherent sections were found, the former secured with reinforced concrete rings, the latter requiring the construction of a whole tunnel. A well for the operation of the sluice gate at the intake works was dug nearby the dam, where the surveillance house was to be built. The 2 km long pressure conduit joined the surge shaft on the other side of the mountain, where a second sluice gate was installed to shut the penstock. The main steel penstock itself, 1.8 m in diameter, was installed in an inspectable tunnel. This first section, almost 800 m in length, ended with a junction subdividing the main pipe in three penstocks anchored on an open air concrete trench (Fig. 3). The trench was equipped with an inclined plane for the installation and maintenance of the piping.

In 1926, well into the construction works, it was clear that the storage of concrete in the valley bottom was insufficient. The high altitude site was equipped with cement storage, crushers, gravel washers and two Ransome towers, installed at the opposite ends of the valley, which allowed the buckets to be positioned at any place along the dam, considerably accelerating the pouring process and eliminating the need for most of the scaffolding, a major technical improvement on the former Roasco dam (Fig. 4). In the winter of 1926-27, the Large Dams Committee further reviewed the design, specifying measures against water seepage. Gaudenzio Fantoli, member of the commission and rector at Politecnico di Milano, advised the AEM technical office to adopt a similar system to the one applied to the Spullersee dam for the construction of the core, including an internal drainage network and the adoption of expansion joints. [16] The dam was subdivided into 5 sections, which were raised alternately. As the sections were ready, cement was injected along the joints. The building sites became true open air workshops for the Large Dams Committee, through which observations and further knowledge on dam execution was gathered, allowing for the development of future regulations.

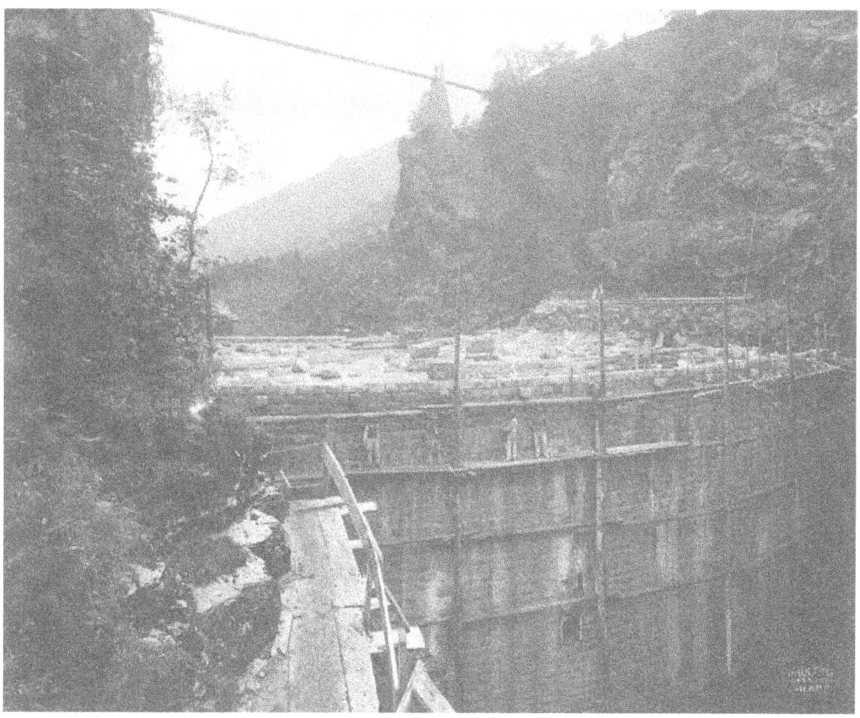

Fig. 2. The Fusino dam under construction. Photo by Antonio Paoletti circa 1922 (Photographic Archive Fondazione AEM, Milano)

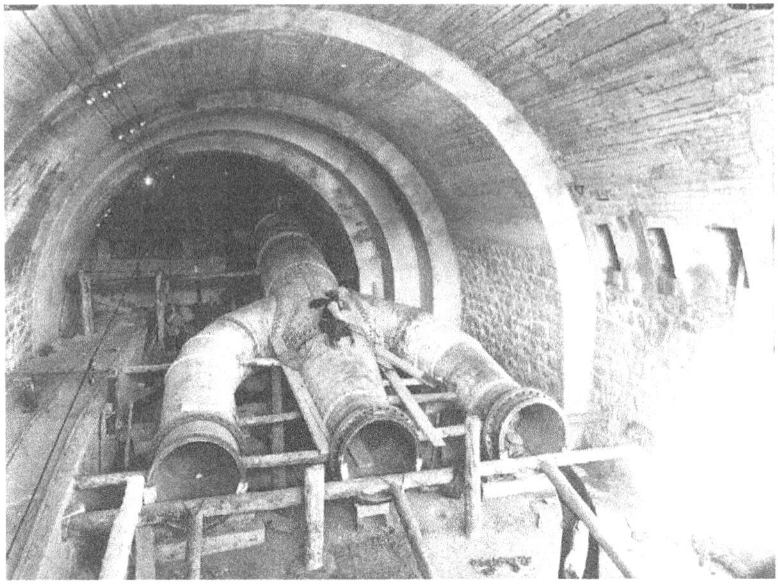

Fig. 3. Penstocks installation. Photo by Antonio Paoletti, circa 1927 (Photographic Archive Fondazione AEM, Milano)

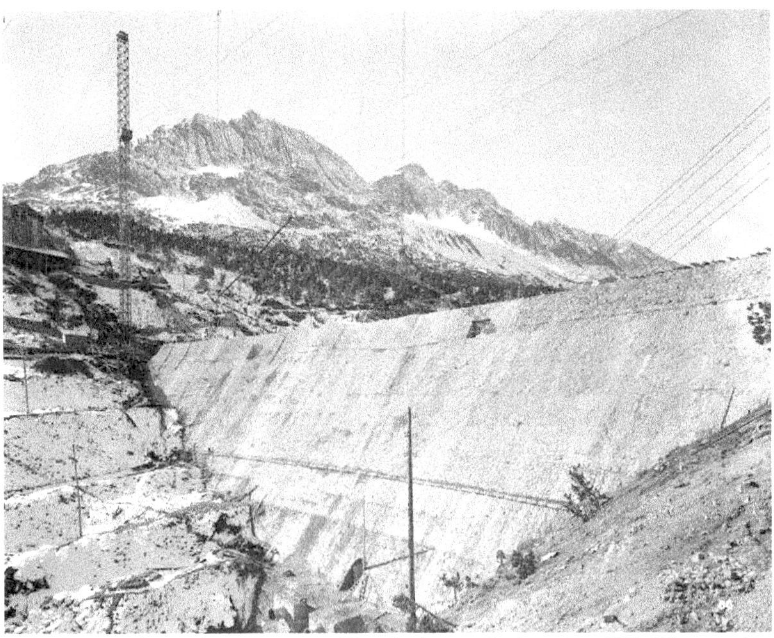

Fig. 4. The Cancano dam under construction. Photo by Antonio Paoletti circa 1928 (Photographic Archive Fondazione AEM, Milano)

Power station design and construction

In both the Roasco and the Fraele facilities, the power plant typology and technical characteristics were defined well before an external architect was invited to provide a suitable, representative shell for the building. For Roasco, AEM employed Piero Portaluppi, the later head of the architecture faculty in Milan, whose continuing collaboration with the Conti electrical company made him the most famous "hydroelectric architect" of the time [17]. He was presented with a fixed layout and cross sections [18], which he adapted to the shape of an inventive art-deco castle, in a dialogue with the nearby mediaeval remains of the Visconti castle (Fig. 5).

The more complete documentation of the later Fraele power plant (Fig. 6), built from 1924 to 1928, offers a more precise insight of the overall design process. The number of generating units was fixed from the beginning according to the available stream flow and head – thus being mostly fixed in number and power from the preliminary utilisation schemes, before the building typology was discussed. In high head developments the main constraints to the location of the power plant were the availability of access routes and a suitable foundation ground, therefore the building was conveniently situated in the valley bottom near the river Viola, which would collect the water from the tailrace. Its single row layout was decided without the involvement of an architect, and as soon as all the technical details on the turbine and generator bays were provided by the Riva and Tecnomasio Brown Boveri companies in Milan, the foundation works could start. The foundations of the building are laid on solid rock, with walls partly built in reinforced concrete, partly in dry stone (Fig. 7). Where the foundation walls couldn't directly reach the solid rock, they were supported by reinforced concrete arches, transferring the load to the rock through concrete piles made using removable steel tube jackets (Wolfsholz system).

Unlike the earlier Grosotto power plant, the superstructure of the new building is a hybrid construction, with outer walls making use of a more common masonry technique, in limestone blocks and cement mortar. The overhead crane was supported by a continuous reinforced concrete shelf. The switchgear and controls room was supported by concrete slab floors over reinforced concrete beams, themselves supported entirely by the peripheral masonry walls. The roofing system was supported by iron Polonceau trusses, used in the machine hall to leave more space for operating the crane, while "English trusses" [19] were used in the transformers and switches hall, where they served also as a support for the electrical equipment. The whole roof was covered with Perret hollow tiles, bituminous paper waterproofing and local slate tiles.

Between the building of the foundations and the start of the superstructure works, architect Giulio Paleni from Bergamo was commissioned to design the façades and interiors. Paleni received the plans and cross sections of the building and had to provide it with a suitable appearance - the eclectic design, where round arch windows and pilasters reflect the internal arrangement of engines and transformers, included the use of noble materials such as cut granites and marble, the majority of which were dropped by the company in favour of less expensive artificial stone and stucco decorations. In fact, while the cement was bought from the Bergamo Italcementi facility, most of the remaining materials came from nearby quarries, since the company preferred to cut on transportation costs, and the only stone imported from outside the valley was the green marble of the interior staircase. [20]

As soon as the definitive dimensions and layout were fixed, the superstructure of the turbine hall was built in its entirety, including the roof trusses temporarily covered with wooden benches to allow for the machinery and generators to be safely installed, while the remaining part of the building was still to be built. The same pattern was followed for the second hall, where a reinforced concrete frame structure, independent from the outer masonry shell, was built to support the electrical equipment and divide the room into separate cells for the transformers.

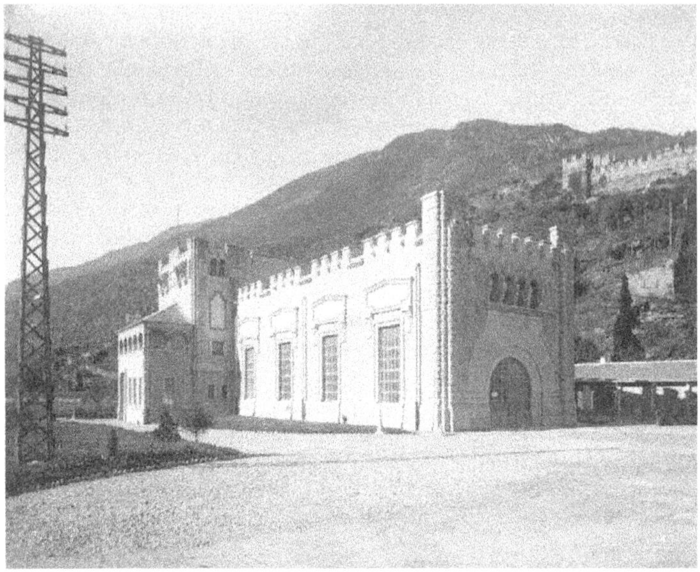

Fig. 5. Roasco power plant. Photo by Antonio Paoletti circa 1930 (Photographic Archive Fondazione AEM, Milano)

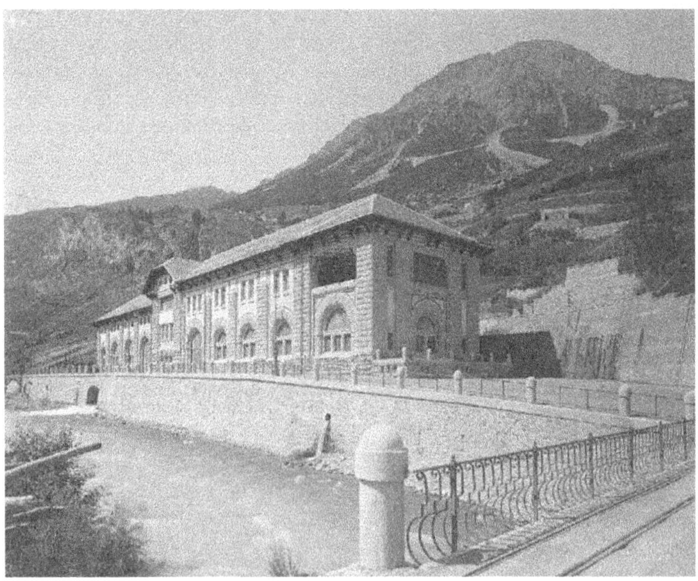

Fig. 6. The completed Fraele power plant. Photo by Antonio Paoletti circa 1930 (Photographic Archive Fondazione AEM, Milano)

Conclusions: hydroelectric heritage between construction and landscape

The documentation of the central AEM technical office in Valtellina shows an overall utilisation scheme, which was strategically devised at the beginning of the twentieth century and refined through thirty years of development. The construction progressed through subsequent adjustments, and a single technical office was managing resources on multiple sites. The sources offer insight on the construction process of the single power plant, showing how the increasing standardisation of the mechanical and electrical industry met the construction industry in a process of rapid development.

The technological contents of the buildings, which have in part survived retooling to this day, provide evidence of the relations with the steadily growing Milano industries. The architectural works themselves, often praised for their aesthetic qualities, are late examples of an eclectic and ornamental design that was to be abandoned in facilities built later in the 1930s, both due to the development of new languages and to the introduction of tunnel power plants, whose prominent outer architectural features are limited to the access gate. From the construction point of view, they present elements of industrial innovation within common building practice, making them site-specific examples which only partly match the construction typologies presented in handbooks. [21] While it is difficult to retrace the diffusion of construction procedures through archival documentation alone or assess how much the building techniques owe to local practice, further studies on both company archives and the built heritage could offer a wider picture. It is however clear that this phase of alpine history not only greatly contributed to the industrial development of Milan, but also had a strong impact on the landscape of whole valleys and on local construction practice, enhancing the infrastructure networks which aided the later tourism-oriented development of the alpine region.

References

[1] Examples of these early configurations can be found in *Impianti idroelettrici della Società imprese elettriche Conti nelle valli ossolane. Note della Società Costruzioni Meccaniche Riva*, Milano: Associazione Elettrotecnica Italiana, 1922; C. Bonomi, *Gli impianti idroelettrici dell'Alta Valle Camonica (iniziative della Societa generale elettrica dell'Adamello)*. Roma: Tipografia del Senato, 1923.

[2] C. Pavese, *Un fiume di luce. Cento anni di storia della AEM*. Milano: Rizzoli, 2011.

[3] Most handbooks of the time hardly try to classify the development schemes beyond the head range criterion. See e.g. F. Marzolo, *Utilizzazioni di forze idrauliche. Impianti idroelettrici*. Padova: CEDAM, 1926.

[4] A.Manfredini, "Gli impianti idroelettrici di Valtellina e le installazioni elettriche del Comune di Milano", *Il Monitore Tecnico*, no.8, 1910, p.144.

[5] *Tubi-Togni Brescia. Condotte forzate 1903-1923*. Milano: Arti Grafiche Alfieri e Lacroix, 1926.

[6] Historical Archives Fondazione AEM, folder A101, "Società Idroelettrica Italiana: Valutazione dei mezzi d'opera ceduti dall'Ing Carlo Mina fino a tutto Maggio 1908 all'Ente costruttore dei Lavori Idraulici per l'impianto idro-elettrico di Grossotto".

[7] Historical Archives Fondazione AEM, folder A271, file "Direzione Lavori Impianti Idroelettrici 1920", 20 Dec. 1920.

[8] Ministry decree 2 Apr. 1921 "Norme generali per i progetti e per la costruzione di dighe di sbarramento per serbatoi e laghi artificiali".

[9] See U. Barbisan, *Il crollo della diga di Pian del Gleno: errore tecnico?* Mantova: Tecnologos, 2007.

[10] See, among others, "La rupture du barrage à voltes multiples sur le Gleno (Lombardie), Rapport Officiel des experts" in *Le Génie Civil*, no.16, 1924, pp. 346-348; J. Boudet, "La rupture du barrage du Gleno, étude descriptive, critique et analytique", in *La Houille Blanche*, no. 2 ,1924, pp.34-47;

A. Ludin, "Der Bruch der Gleno-Talsperre nach dem amtlichen und anderen Gutachten", in *Zeitschrift für Bauwesen*, no. 7, 1925, p. 65.

[11] L. Mangiagalli, "Sviluppo recente nella costruzione delle dighe italiane (comunicazione alla World Power Conference, Londra, luglio 1924)", in *L' Elettrotecnica*, no. 36, 1924, pp. 1006-1013.

[12] M. Dolcetta Capuzzo, R. Pavia, "Le grandi dighe e i serbatoi artificiali", in R. Pavia (ed.), *Paesaggi elettrici. Territori, architetture, culture*. Venezia: Marsilio, 1998, pp.221-224.

[13] While German handbooks such as Ziegler's and Sympher's works (P. Ziegler, *Der Talsperrenbau nebst Berechreibung ausgeführter Talsperren*. Berlin: W. Ernst, 1911; L. Sympher, *Der Talsperrenbau in Deutschland*, Berlin: W. Ernst, 1907) are to be found in the historic library collections of both Padua University and Politecnico di Milano, the cross section typologies used in gravity dams are mostly similar to those found in the available American handbooks (E. Wegmann, *The design and construction of dams: including masonry, earth, rock-fill, timber, and steel structures*. New York: John Wiley & sons, 1911; C. W. Smith, *Construction of masonry dams*. New York: McGraw-Hill, 1915; W. P. Creager, *Engineering for masonry dams*. New York: Wiley & Sons, 1917).

[14] G. Fantoli, "Intorno ai problemi delle Dighe per Serbatoi e del loro tipo nelle applicazioni italiane", in *L'Energia Elettrica*, no. 1, 1924, p. 5

[15] Historical Archives Fondazione AEM, folder A302, file "Direzione Lavori Impianti Idroelettrici 1923. 2° trimestre", 17 apr. 1923.

[16] Historical Archives Fondazione AEM, folder A332, file "Ufficio Lavori Valtellina: Roasco e Fraele: Gennaio 1927", 22 Jan. 1927.

[17] Selvafolta, "Figura Multiplicavit: le centrali idrolettriche di Ettore Conti, Umberto Girola e Piero Portaluppi", in Bigazzi, R. Poletti, *Accoppiamenti giudiziosi. Storie di progettisti e costruttori*. Milano: Skira, 1995.

[18] Historical Archives Fondazione AEM, folder A252, file "Portaluppi Arch. Ing. Piero", 21 Sept. 1918.

[19] "Capriata inglese", a term found both in contract specifications and in Italian construction handbooks, is the iron equivalent of a Queen-rod truss.

[20] Technical Archives A2A Group Grosio, folder "Fraele Muraria 1", file "Imprese costruzione fabbricato Centrale" 16 Jul. 1927

[21] See e.g. G. Revere, *Elementi delle costruzioni. Costruzioni industriali. 1909-1910*. Milano: R. Istituto Tecnico Superiore di Milano, 1910, a course book on industrial building technology written after the course held by the author at Politecnico di Milano.

The emergence of a new generation of building products in post-war Belgium

The case of lightweight concrete

Stephanie Van de Voorde, Inge Bertels and Ine Wouters
Department of Architectural Engineering, Vrije Universiteit Brussel, Belgium

Introduction

According to a survey in 1969, only two per cent of the houses in Belgium were constructed using industrialized building techniques. This percentage is significantly lower than in other countries such as the UK, France, Germany and Eastern Europe, where 10 to 30 per cent of the houses were constructed with industrialized building methods.[1] The reason for this low percentage in Belgium is often sought in 'the habits, the everyday routine and the scepticism of the people towards everything that is new and hasn't proven its worth yet.'[2] If we take a closer look at the Belgian building industry, in particular the house building sector, we can find underlying and related causes to account for this. As we will explain in the first part of this paper, one of the main reasons can be found in the general structure and organization of the Belgian building industry; in addition, also the government policy, the traditional Belgian housing typology and the interaction between public bodies, owners, architects and builders lead to the use of certain types of building products and methods. The second part of this paper will elaborate on how knowledge of the new or improved building materials and techniques was disseminated through architectural journals, being one of the most appropriate mediums to reach a wide audience. Finally, we will focus on one specific category of this new generation of building products, namely lightweight concrete. Starting from articles in the periodical press, we will elaborate on how, why, when and where lightweight concrete was developed, marketed and used in the Belgian house building sector.

A. Post-War building in Belgium: actors and organization

1) Organization of the building industry

In the post-War period, the Belgian building industry was characterized by a very large proportion of small and medium-sized companies: 85 per cent of building companies employed only one to four workers.[3] Additionally, the sector was not very well provided with capital, hence the frequent reproaches of technical inertia, lack of specialized labor and low productivity. Indeed disorganized, yet not solely responsible for this, the sector had no perspective or incentive to invest in large-scale building techniques, equipment and training, unless the public authorities set up large-scale building programmes. The vicious circle created by the industry's focus on small, traditional and 'safe' contracts on the one hand and the lack of large-scale projects supported by the government on the other hand, left little room for alternative building methods to fully develop. It is not until the 1960s that we see a first sign of professionalization, scientific application and technical upgrading of the sector. In 1960 the Belgian Building Research Institute was established by the Federation of Belgian building contractors. Ten years later a technical quality label for building products was introduced. .[4] Yet during the 1970s, the first oil crisis weakened this positive trend.

2) Public commissioners and private owners

Immediately after WWII, a limited number of innovative initiatives and large-scale projects was set up by public bodies, yet for several reasons (technical, financial, aesthetic, etc.) none of them was continued beyond the experimental stage.[5] The authorities generally adopted an ad hoc policy, operating independently from each other, with little attention to the transfer of knowledge. Seeking to solve the housing problem in a swift and easy way (limiting the need for technical innovation and supervision by the government to the minimum), Alfred De Taeye, representative for the Catholic People's Party, submitted a law in 1948 stimulating private initiative.[6] The famous/notorious 'Law De Taeye' offered grants to private individuals to build a modest family house. It was an immediate success: in only six years', 100,000 De Taeyegrants were awarded. However, the grant was indirectly detrimental to collective housing projects. The outcome of the law is also very apparent in the street scene, at least in the height and typology of buildings. If we look at the number of houses built between 1946 and 1981, only five per cent are apartment buildings, 23 per cent are terraced houses, 29 per cent are semi-detached houses and 42 per cent are detached houses. If we look at the number of floors, all categories are best represented in the two- or three-storey buildings (even the apartment buildings), except for the detached house (mainly bungalows with only one floor).[7]

3) Improved traditional building methods

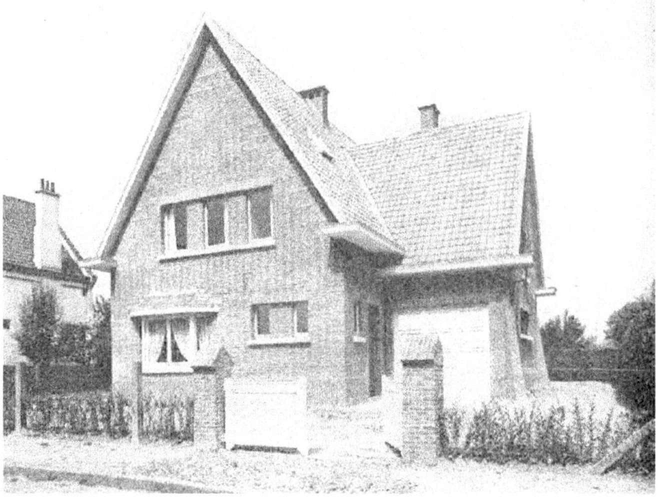

Fig. 1. A typical De Taeye-house with red bricks and ceramic roof tiles (P.-L. Flouquet, La petite maison familiale, Brussels: Art et Technique, 1950, p.60).

Although the De TaeyeLaw did not specify construction techniques or architectural styles, it (unintentionally) gave rise to conservatism. As it did not provide stimuli for a rational, efficient and economical building practice, owners and builders rarely abandoned the traditional idiom of the small house with red bricks, wooden window frames and a saddleback roof with ceramic roofing tiles (Fig. 1).[8] Nevertheless, the building industry did not come to a standstill: even within the traditional idiom, there was room for improvement, with materials and products that could be integrated without complicated adjustments or details nor heavy equipment. Steering a middle course between innovation and tradition, 'le traditionnel évolué' or 'improved traditional building methods' made use of often common materials but with enhanced technical, economical or architectural characteristics. Examples of these improved traditional building materials are lightweight building blocks; hollow core slabs; new (finishing) materials such as rubber, vinyl or linoleum; prefabricated panels in different (composite) materials like cement, asbestos, wooden fibres and plastics; new types of glass and various materials for acoustic or thermal insulation. These materials and products did not imply high-tech developments or fundamental changes in the building process, yet it was necessary to develop the (practical and theoretical) scientific equipment: manufacturers needed to deliver full data on the characteristics of the materials, which in many cases were already developed during the first half of the twentieth century, but were only used on a large scale during the 1950s. In addition, commercial strategies were needed to sell these materials and products to architects, commissioners and builders, for example through architectural magazines and journals.

B. New construction techniques and materials in the periodical press

Architectural and technical periodicals are an interesting and rich source of information for research in construction history.[9] As a direct, heterogeneous and topical means of communication, the periodical press was a perfect medium to promote new building materials and techniques, either by means of articles on specific applications or by means of advertisements and supplements. For this study, we have selected three Belgian post-War journals with a different profile and discourse, namely *La Maison*, *Architecture* and *Bouwen&Wonen*. A close-reading of these journals corroborates the typological importance of low-rise housing in Belgium, in which improved traditional building methods are used increasingly. We will focus on how these three journals report on the new building materials and techniques, to elaborate on specific characteristics and applications further on.

La Maison was the first and one of the longest running post-War architectural periodicals in Belgium. Published every month between March 1945 and February 1970, it focused on the design, decoration and equipment of the (private) house, not only in Belgium but with a broad international perspective. Putting both modernism and traditionalism or regionalism on the agenda, it aimed to reach a wide audience. With a discourse that is rather informative and educational than polemic or progressive, *La Maison* is considered as one of the conservative architectural journals in Belgium, at least in comparison with for instance *Bouwen&Wonen* or *Architecture*. Maybe surprisingly, because not consistent with its (supposed) traditional image, *La Maison* gave relatively substantial information on new building materials and construction techniques. Even in the first volumes, reflecting the reconstruction and housing needs of that time, *La Maison* wrote about experiments in house building such as the international exposition on experimental temporary houses and the national building yards for alternative building methods. In general, the editorial voice took on a critical-positive tone towards these reconstruction experiments. From the 1950s onwards, we see a diversification of the articles on construction and techniques. Next to more general or reflective articles on broader themes such as industrialization, normalization and rationalization of the building industry, we can distinguish three main categories. A first series of articles deals with building services, mainly heating and ventilation, but also electricity, lighting, building acoustics, heat recovery and kitchen and bathroom equipment. Published with roughly the same frequency as the first topic, were articles on specific building systems for houses in different materials (e.g. houses in monolithic or lightweight concrete, a bungalow in aluminum, wooden houses or houses in plastics). Perhaps more representative of/relevant for the daily building practice are the articles reporting on building materials and products, with new types of glass and thermal or acoustic insulation as very popular topics, next to wood, cement, plastics, wooden frames, gas concrete (Fig. 2), fire-resistant boards, and asbestos.

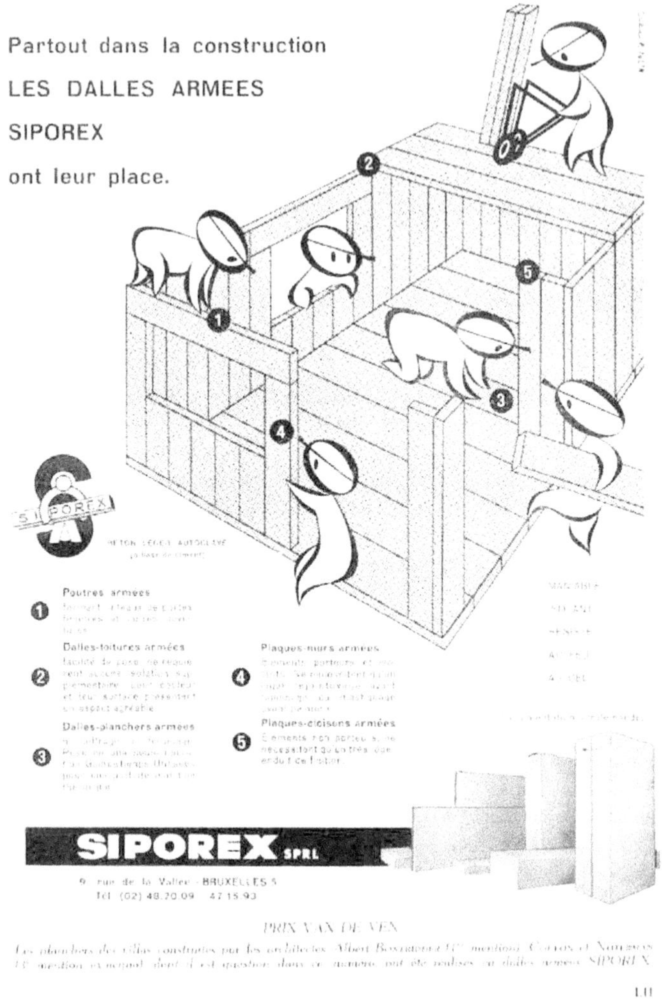

Fig. 2. A typical advertisement for gas concrete: lightweight, easy to handle and low thermal conductivity (La Maison, vol. 16, no.3, 1960).

The monthly periodical *Architecture* roughly covers the same period (January 1952 until April 1970), but holds a very different stance: founded by a very large group of progressive architects, the so-called new generation of modernists, *Architecture* reports almost exclusively on modernist architecture. Next to news items on current affairs, a few monographic studies or editorial pieces on architectural trends, almost all articles in *Architecture* dealt with recent completions, yet within this one type of article almost every architectural typology was covered, from small family houses, offices, churches and schools, to high-rise apartment buildings and industrial complexes, often brought together in thematic issues.

Information on building materials and construction techniques was mainly confined to the technical information given in the description of completed buildings, supplemented with an occasional article on technical issues (e.g. heating systems, acoustics, industrialization of house building, prefabricated schools, vacuum concrete or the Preflex-beam) and a few thematic issues on technical topics and materials (prefabrication, prestressed concrete, wood, glass and heavy prefabrication). In all, technical information was mainly inserted in descriptions of practical applications, as an inherent part of the architecture, making *Architecture* (at least its table of contents) less technical than one would expect from the editorial board which included a few architects with a distinct technical interest such as Willy Van Der Meeren and Renaat Braem.

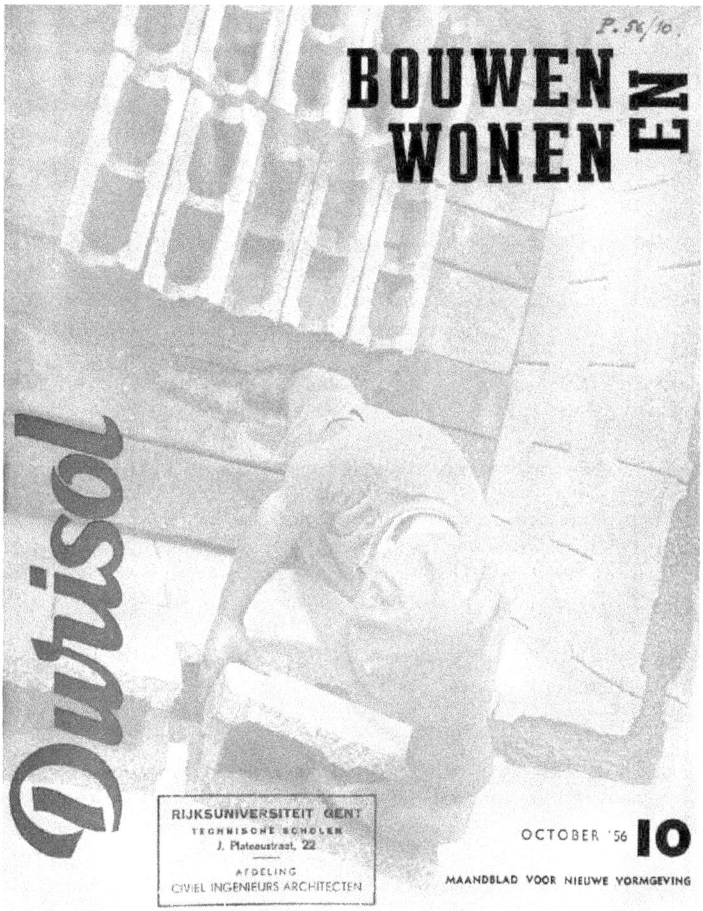

Fig. 3. Cover of the October issue of Bouwen&Wonen in 1956: an advertisement for Durisol-blocks.

The latter, Renaat Braem, one of the key figures of post-War modern architecture in Belgium, was also founder and editor-in-chief of *Bouwen&Wonen*, a monthly periodical that ran from October 1953 until March 1962, with a distinct technical character. Not that it reported exclusively on construction techniques and materials: *Bouwen&Wonen* covered a broad range of topics, with regular features on, for example, interior architecture and urban planning as well as art, but these topics and articles were strongly related to the field of interest of the members of the editorial board. The technical thematic issues of *Bouwen&Wonen* dealt with wood (several times), concrete, bricks, new finishing materials and boards, insulation and the National Building Centre in Antwerp (several times). As in *La Maison*, among the technical articles we find more general pieces (building costs, industrialization, cheap housing) and articles on building services (heating, ventilation, lighting, electricity). Far more important in *Bouwen&Wonen* though are pieces on specific materials: many articles on wood (including complete building systems, new types of beams, calculation methods,...), steel and aluminum, cement and concrete, insulation materials, plastics and synthetic materials (e.g. polystyrene, paint and resins). Prominently present are new types of prefabricated boards and panels (e.g. Novopan, Ultrawood, Eternit, Formica, Triplex, Soundex, Mureclair,...) and products in lightweight concrete (gas concrete and wood fibre concrete such as Durisol, Ytong, Siporex and Durox) (Fig. 3). Often these articles on products and materials were related to an event in the National Building Centre, e.g. the erection of the Durox- and Eternit-pavilion or an exhibition on thermal and acoustic insulation. The centre was created in 1958 to provide information and advice on building and to lift the contemporary building practice to a higher level.[10] Although these objectives sound very estimable and altruistic, some doubts about the objectivity and independence exist. What applies to the articles and the advertisements in *Bouwen&Wonen*, also applies to the information and the pavilions at the National Building Center: all four often carry the same signatures, both from architects and building companies as well as manufacturers and material suppliers. Not only did the two main figures running the journal and the National Building Centre, namely architect Braem and contractor Victor Van Coillie, take on different roles simultaneously, the centre could also rely on the financial or material support of many companies who appeared in the advertisements and articles in *Bouwen&Wonen* quite regularly. This rather small but dense network of actors calls for prudence when interpreting the data.

C. The case of prefabricated elements in lightweight concrete

If we want to look at the facts and figures, commercial strategies and practical applications of one type of material, lightweight concrete is a very interesting case in point. *La Maison, Architecture* and *Bouwen&Wonen* each report on lightweight concrete in a very different way, expressing different degrees of approval and advocating different applications for the material.

Within the class of lightweight concrete, we can distinguish lightweight concrete made with wood fibres (e.g. Durisol) and lightweight concrete with a foaming agent, also called gas concrete (e.g. Ytong, Siporex and Durox).

Ytong is made of lime and silica, mixed with aluminum powder, provoking a chemical, foaming reaction. After the mixture has hardened and the concrete is sawn into pieces, it is put in autoclaves under steam pressure to finalize the chemical reaction and give the material its full strength. With only approx. one third of the specific gravity of concrete and very low heat conduction, the material was very popular in Sweden (where it was developed in 1929 and had become one of the most important post-War building materials) and Germany. In Belgium, the first Ytong-factory was erected in 1954-55. Even by 1955, in its thematic issue on concrete, *Bouwen&Wonen* predicted a swift development for Ytong in Belgium, as the material was well tested elsewhere and the benefits were numerous.[11]

In the same category of gas concrete, Siporex was developed in Sweden in 1935 and was produced in Belgium from January 1956. In October that year, in a thematic issue on Ytong, Siporex and Durisol, *Bouwen&Wonen* listed the advantages of Siporex: solid, resistant to fire and moist, isothermal, easy and quick to handle with less equipment, modulated and enabling architectural diversity (Fig. 2).[12] These advantages are not exclusive to Siporex and are valid for other lightweight concretes as well. In fact, Siporex is manufactured in the same way as Ytong, except for using cement instead of lime as its base material. One of the differences, albeit superficial, was the shape and size. The range of Ytong was mainly limited to blocks, while Siporex was delivered in different forms: elongated blocks (50cm long, 25cm high and 15 or 30cm thick), entire wall panels for insulation (from 7 to 15cm thick), reinforced floor or roof plates (up to 6m long, between 12.5 and 25cm thick to carry a maximal load of 400kg/m^2), reinforced lintels, etc.

For typical applications of steam-cured lightweight concrete, Durox provides an interesting comparison. Durox was given a lot of attention in *Bouwen&Wonen*, especially with respect to the experimental Durox-house by Braem at the site of the National Building Centre (Fig. 4).[13]

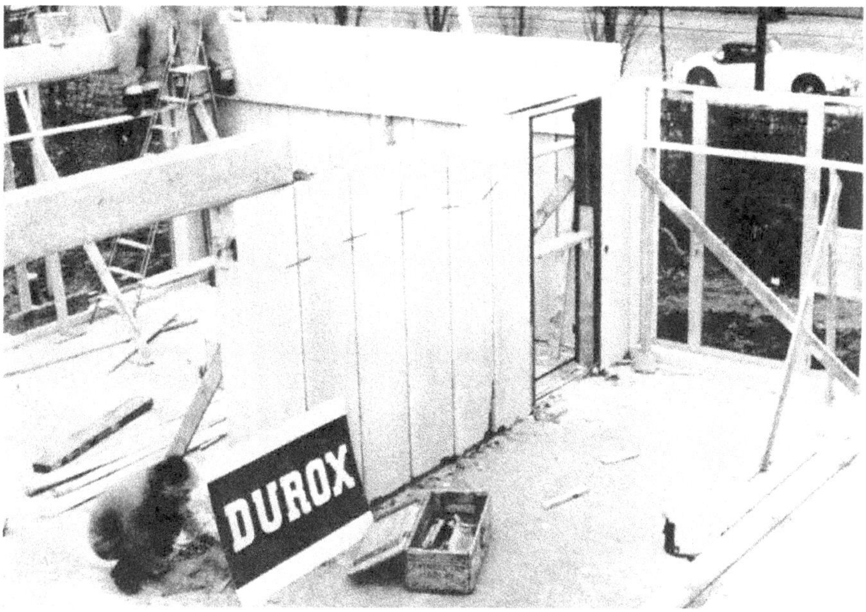

Fig. 4. The erection of the Durox-house at the National Building Centre took only 5 days (R. Braem, "Durox-huis type Braem", Bouwen&Wonen, vol.7, no.5, 1960, p.188).

The goal of the experiment was to show that it was possible to construct a house quickly and cheaply, with the use of prefabricated elements. Why Braem choose Durox for this experiment (and not another brand of gas concrete), is not clear: the literature rather lets us believe that the different brands were interchangeable. The Durox-pavilion was constructed in March 1960, in a record time of only five days (not counting the two days extra for electricity and finishing). It was a single-storey house, based on a module of 50cm, being the standard width of Durox-panels. The external walls were 21cm thick, and 11 or 15cm for the inner walls and 15cm for the roof plates. The joints between the Durox panels were flushed, after which the external walls were painted. While Braem did not succeed in executing this prototype on a larger scale, architects Constant and Godart did manage to construct a social housing project in Seraing with 60 houses, with inner and outer walls made with standard Durox-panels of 2.50m by 0.50m (Fig. 5).[14]

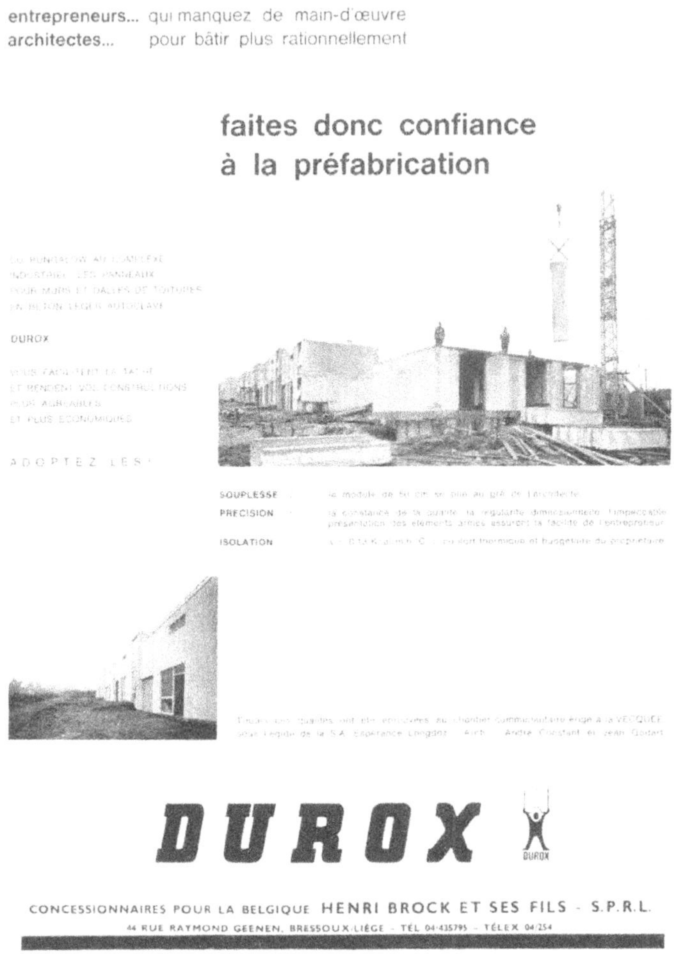

Fig. 5. An advertisement of Durox, depicting the housing project in Seraing (Architecture, no.52, 1963).

Next to these examples, in which the manufacturer was closely involved from the outset, there is also another type of application. In a country with an individualistic architectural tradition, the external appearance was no less important than the technical advantages. Accused of being dull and too prone to damage because of its softness, different options for its external treatment were elaborated, from simple painting over plaster to a complete wall in bricks.[15] Although it was perhaps not completely in accordance with the philosophy of the material's inventors, the option with an outer brick wall seemed to be the most popular. For example this was used for the 289 single family houses of the housing project

Ban Eik in Wezembeek-Oppem by the architects of Groupe Structures: the inner walls in Ytong are covered with external walls in traditional bricks (Fig. 6).[16]

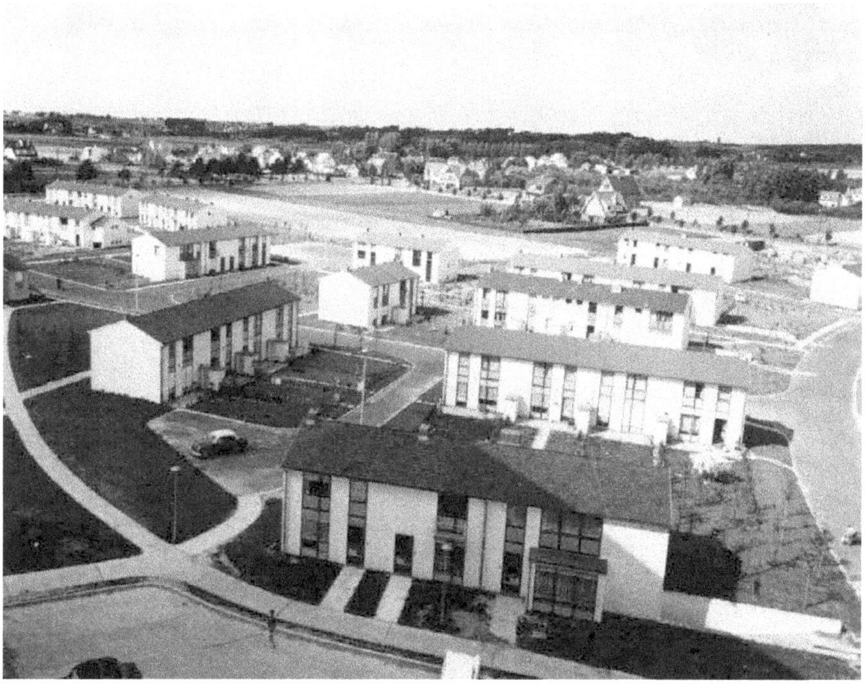

Fig. 6. Aerial view of the Ytong- and brick-houses in Wezembeek-Oppem (Private collection, Van Coillie).

In the second category of lightweight concrete, namely concrete with wood fibres, there is only one main popular brand, namely Durisol. Developed in Switzerland during the 1930s, Durisol was manufactured by mixing cement and water with mineralized scraps of wood which were resistant to moisture and chemicals.[17] Durisol is not just a different material from gas concrete (yet with similar characteristics such as a low specific gravity, easy to handle and easily adjustable at the construction site by sawing), but is also used in different ways: either as panels, planks and slabs (similar to gas concrete panels) or, the more popular way, as hollow blocks and slabs to become a permanent mould for loadbearing walls and floors in reinforced concrete (for buildings up to 15 floors) (Fig. 3). The Durisol-blocks were used in numerous projects (e.g. the large-scale housing projects 'Kiel' in Antwerp by Braem, or 'Plaine de Droixhe' in Liège by EGAU, but also in a large number of small-scale constructions), yet as they are plastered or clad afterwards, they are hardly ever visible.

The way the architectural journals report on the various kinds of lightweight concrete is very different, but this is more related to the journal than to the manufacturer. The manufacturers came up with many initiatives and events to convince potential clients, directly or indirectly through the journals as a communication medium, for example by inviting the editorial board to the factory, organizing technical meetings, cooperation on practical projects, organizing an abstract design competition amongst architects to get them acquainted with the material,...[18] The way journals reacted to these initiatives is of course very different: *La Maison* and *Architecture* for example were rather 'economical' with technical information, while *Bouwen&Wonen* was not afraid to give specific data on material characteristics and physical properties. Incidentally, in many cases the information is incomplete and not always sufficient to set up an appropriate and respectful renovation strategy for those post-War houses in urgent need of renovation.[19]

Summing up how these new or 'improved traditional' building materials have influenced post-War design and building practice in Belgium, one of the findings is that the use of these materials has hardly ever had heavy implications for the architectural design. As for lightweight concrete products, they were mainly used because of economical and pragmatic advantages (lighter, quicker, cheaper), yet still within a relatively traditional concept: the properties which are attributed to the materials (prefabricated, rationalized, innovative) are not, or not necessarily, transmitted to the buildings in which they are applied. Although a few architects made an honourable effort, the use of lightweight concrete did not (successfully) lead to an inherent architectural aesthetic or idiom: the building practice might have changed, yet the architectural design practice was relatively unaltered.

Acknowledgements

This paper fits within the framework of a research project (2013-2015) on post-War housing in Brussels, funded by the Brussels Capital Region through the Innoviris Strategic Platform Environment 2012 'Brussels Retrofit XL'. See also www.brusselsretrofitxl.be.

References

[1] N. Schmitz, M. Ducuroir, "Vers l'industrialisation du bâtiment en Belgique", *Cahiers économiques de Bruxelles*, no.49, 1971, pp.5-50.

[2] C. Crappe, "Kijkdorp Limal en prefabricatie", *Landeigendom*, no.297, 1972, p. 37.

[3] E. Buyst, *An economic history of residential building in Belgium between 1890 and 1961*. Leuven: Leuven University Press, 1992, p.134.

[4] S. Van de Voorde, *Bouwen in beton in België (1892-1975)*. Ghent University (PhD), 2011, pp.469-471.

[5] For example the International Exposition on Temporary Houses in Brussels in 1945-46, the so-called 'chantiers nationaux' (1946-1949) and the large-scale housing projects for miners (1946-48).

[6] K. Van Herck, T. Avermaete (eds.), *Wonen in welvaart. Woningbouw en wooncultuur in Vlaanderen, 1948-1973*. Antwerp: VAi, 2006, pp.67-77.

[7] <http://statbel.fgov.be/nl/statistieken/cijfers/economie/bouw_industrie/gebouwenpark/>[accessed on 31 December 2013]

[8] P.-L. Flouquet, *La petite maison familiale*. Brussels: Art&Technique, 1950; R. Braem, "Alle De Taeye-woningen zijn niet lelijk", *Bouwen&Wonen*, vol.1, no.3, 1954, pp.110-113.

[9] A. Sornin et al., Revues d'architecture dans les années 1960 et 1970. Montréal: Institut de recherche en histoire de l'architecture, 2008; J.-M. Leniaud, B. Bouvier, *Les périodiques d'architecture, XVIIIe-XXe siècle: recherche d'une méthode critique d'analyse*. Paris: Ecole des chartes, 2001.

[10] "Nationaal Bouwcentrum", *Bouwen&Wonen*, vol.5, no.9, 1958, pp.248-260.

[11] P. Chapeaux, "Ytong", *Bouwen&Wonen*, vol.2, no.10, 1955, pp.383-386.

[12] "Lichte betonelementen Siporex", *Bouwen&Wonen*, vol.3, no.10, 1956, pp.463-466.

[13] R. Braem, "Durox-huis type Braem", *Bouwen&Wonen*, vol.7, no.5, 1960, pp.183-186.

[14] "Maisons en éléments préfabriquées à Seraing", *Architecture*, no.52, 1963, pp.169-171.
[15] "Technische bijeenkomst 'Gasbeton'", *Bouwen&Wonen*, vol.7, no.8, 1960, pp.367-368.
[16] Cité 'Ban Eik' à Wezembeek-Oppem", *Architecture*, no.33, 1960, pp.442-447; "Le quartier 'Ban Eik', à Wezembeek-Oppem", *La Maison*, vol.16, no.8, 1960, pp.261-266; B. De Grunne, R. Stenier, "L'expérience du quartier 'Ban Eik'", *Habiter*, no.26-27, 1964, pp.2-41.
[17] "Durisol", *Bouwen&Wonen*, vol.3, no.10, 1956, pp.450-450/12.
[18] "De fabricatie van Ytong. Een bezoek aan de fabriek te Burcht", *Bouwen&Wonen*, vol.3, no.10, 1956, pp.459-462; "Concours Ytong", *La Maison*, vol.25, no.6, 1969, p.233.
[19] Additional information can be found in publications by the manufacturers or in secondary literature, e.g. R.M.E. Diamant, *Industrialised building*. Londen: Iliffe books, vol. 1, 2 & 3, 1964/1965/1968.

Bibliography

1. "Cité 'Ban Eik' à Wezembeek-Oppem", *Architecture*, no.33, 1960, pp.442-447.
2. "Concours Ytong", *La Maison*, vol.25, no.6, 1969, p.233.
3. "De fabricatie van Ytong. Een bezoek aan de fabriek te Burcht", *Bouwen&Wonen*, vol.3, no.10, 1956, pp.459-462.
4. "Durisol", *Bouwen&Wonen*, vol.3, no.10, 1956, pp.450-450/12.
5. "La petite maison familiale dans le cadre de la loi De Taeye", *La Maison*, vol.5, no.4, 1949, pp.95-96.
6. "Le quartier 'Ban Eik', à Wezembeek-Oppem", *La Maison*, vol.16, no.8, 1960, pp.261-266.
7. "Le rôle des produits DURISOL dans l'isolation acoustique et thermique des bâtiments", *De Ingenieur-Architect*, no.3, 1937.
8. "Lichte betonelementen Siporex", *Bouwen&Wonen*, vol.3, no.10, 1956, pp.463-466.
9. "Maison expérimentale, à La Hulpe", *La Maison*, vol.13, no.4, 1957, pp.118-119.
10. "Maisons en éléments préfabriquées à Seraing", *Architecture*, no.52, 1963, pp.169-171.
11. "Maisons en éléments préfabriqués de béton cellulaire", *La Maison*, vol.19, no.5, 1963, pp.157-158.
12. "Nationaal Bouwcentrum", *Bouwen&Wonen*, vol.5, no.9, 1958, pp.248-260.
13. *Nouvelles techniques et matériaux nouveaux. Actes du 6e Congrès de l'UIA à Londres*. s.l.: UIA, 1961.
14. "Technische bijeenkomst 'Gasbeton'", *Bouwen&Wonen*, vol.7, no.8, 1960, pp.367-368.
15. "Wandpanelen van gasbeton", *Cement*, vol.20, no.10, 1968, p.385.
16. J. Braeken et al., *Renaat Braem 1910-2001 architect*. Brussels: ASA, 2010.
17. R. Braem, "Alle De Taeye-woningen zijn niet lelijk", *Bouwen&Wonen*, vol.1, no.3, 1954, pp.110-113.
18. R. Braem, "Durox-huis type Braem", *Bouwen&Wonen*, vol.7, no.5, 1960, pp.183-186.
19. N. Bullock, *Building the post-war world. Modern architecture and reconstruction in Britain*. London: Routledge, 2002.
20. E. Buyst, *An economic history of residential building in Belgium between 1890 and 1961*. Leuven: Leuven University Press, 1992.
21. B. Campbell (ed.), *Housing from the factory*. London: Cement and Concrete Association, 1962.
22. P. Chapeaux, "Ytong", *Bouwen&Wonen*, vol.2, no.10, 1955, pp.383-386.
23. C. Crappe, "Kijkdorp Limal en prefabricatie", *Landeigendom*, no.297, 1972, pp.36-39.
24. B. De Grunne, R. Stenier, "L'expérience du quartier 'Ban Eik'", *Habiter*, no.26-27, 1964, pp.2-41.
25. R.M.E. Diamant, *Industrialised building*. Londen: Iliffe books, vol. 1, 2 & 3, 1964/1965/1968.
26. M. Eleb, L. Engrand, *Préfabriquer la tradition: la troisième voie de la maison individuelle 1920-1960*. Paris: ACS, 2002.

Recording the 'building world': how oral history transforms construction history

Christine Wall
Department of Architecture, ProBE (Centre for the Study of the Production of the Built Environment), University of Westminster, UK

I mean, ...the inspection system on this job! ...we used to do the setting-out, we used to put gridlines down, and offset lines down, and then the chippie foreman used to take over, and then he would set-out the actual wall. And if it was an awkward bit, you'd actually do it so that you might set the actual thing out, you'd scabble the floor for the kicker...

The engineer came along – not the clerk of works, but his engineer equivalent – he used to check the scabbling. And making sure that the cover on the starter bars was correct. He then allowed you to go ahead and concrete the kicker. When the kicker was concreted, you then had to scabble the concrete on the kicker, and then the back shutter got put up. The reinforcement was then fitted in position. Once all the reinforcement was in place, the engineer then came round and checked all the reinforcement – and I don't mean just looking at it to see if it looked pretty. He actually came out with the reinforcement drawing and checked every reinforcement bar and position, and he would even check the dimensions between the bars. He would then check to make sure the cover was right. ... You'd put all the concrete spacers, where we used to make concrete spacers in trays, and then put wires in so you could wire them onto the vertical work. Once everything was in, he would then check it off and he would agree, and then the front shutters could go up. Once the front shutters was up, the following inspections had to take place before concrete could commence: the clerk of works came out, and the chippies had a plumb-bob down one end, a plumb-bob lay on the bottom end, and why in heaven's name they had to do this, I do not know, but they put a string line along the bottom, hard up against the kicker, and a strong line across the top, offset, so that the clerk and works could go along and check the dimensions and make sure everything was perfect. He would then say, "Yes, it's okay to concrete," provided the engineer signed it off. *Peter Day, site engineer QEH*

Everything was levelled and done within 5mm, which is a quarter of an inch. And that was done on a certain step on Waterloo Bridge, we made a datum, so that it would never be disturbed, and everything was levelled – the tops of a shutter being poured was levelled to within 5 millimetres. And that is exceptional, believe me! ... If you had something ready to pour in concrete, you got the clerk of the work's permission actually to concrete it. He would then check it, from this datum on Waterloo Bridge, to within five mill! And then you'd concrete, and say a wee prayer when you struck the shutters to make sure it was okay! *Ted Newbury, site foreman Hayward Gallery*

In order that we didn't get any grout loss at all and spoil the finish, all the [shuttering] boards were tongued and grooved. They were butted up tight, they were six inch wide boards, sawn-finish, and then the next one would be an eighth projection, like that, and that would setback like that, and they were tongued and grooved to avoid any grout loss. And all the bolts, bolting, to bolt on the back shutters, all had to be done in a symmetrical pattern. So each wall elevation, the company had to produce the drawing. We ended up producing over 1,200 drawings for this contract, and employed up to 12 draughtsmen. *Rod Bond, assistant project manager QEH*

But being that the CHAG is unique in its finish, that it is all reinforced concrete, which had a timber effect on the concrete, we had to decide which way to handle it. It wasn't like a normal foreman doing the shuttering. We employed draughtsmen because the architect wanted the pattern of the concrete in panels, and the only way we could do this was by drawing out all the plans of the outline of what they wanted each wall or each section of the building to look like.

And this was done by using rip-sawn timber that had been cut, not by a circular saw, because a circular saw marks like curves from the size of the blade, and these had to be cut square off the cuts. And this was done with inch and a half diameter thick timber, six inches wide, and alternate board was one-eighth thicker than the others. So, every six inches, there was an eighth of an inch step-in and out again. These were joined together with a tongue and groove system to make sure that they were properly sealed because, if there was a grout leak the effect of the sawn board wouldn't come out.

We had to find a way of doing these panels so that the concrete was lowered by a crane on a chute into the panels, on a very slow basis, so that we vibrated the concrete slowly to make sure that all the air-pockets were completely out of them, and finished on a perfectly straight line, with a timber screwed onto the shutters, so that we had a perfect line to work off when we started off with the next layer.

... But we did it slowly, we didn't rush it, and we carefully monitored, with torches, looking down inside the shutter as the concrete was placed in and vibrated. What you had to be very careful is, with these heavy vibrators that you use internally in concrete, if it got on the face of the shutter, it would take the markings of the sawn board off, and that is what they didn't want. On an ordinary wall, on ordinary shuttering, that didn't matter so much, but we had to be very, very careful on that. And it was a slow operation.

Now, the Queen Elizabeth Hall was a vast concrete structure, and the whole of the underside of the Queen Elizabeth Hall was built on a cantilever basis, and it was laid down and supported on some temporary steel structures, some tubular, some steel-framed, and, first of all, we built the columns up to it, and then we built the auditorium floor, which, structurally, could not support itself, and then we built the walls up from there, and eventually, we put the roof on, and it's only when the roof tied the auditorium floor, the walls and the roof together, structurally complete, with cube results for the strength of the concrete, that we could then dismantle the supports underneath the auditorium floor. *Glan Davies, general foreman, QEH*

Working life on the Queen Elizabeth Hall and the Hayward Gallery site 1961-7

The recollections above are extracts from over fourteen hours of recordings made with former Higgs and Hill site staff, and building workers, engaged on building the South Bank arts complex between 1961 and 1967.[1] They give immediacy to the building process only guessed at from examining archival photographs and drawings (Fig. 1), and the last paragraph sums up the entire process with succinctness only possible from someone who has seen the entire job through from start to finish. All the men named above were direct employees of Higgs and Hill, three of them having come up through the tools to white collar site manager positions, and all considered that the slow and painstaking process of creating a perfect sawn-board finish to the concrete specification was possible only because Higgs and Hill, founded in 1874, was a traditional firm. By the end of the contract firms like this were beginning to disappear from the British construction industry, but in 1961, when Higgs and Hill won the tender for the South Bank Concert Hall and Art Gallery (CHAG) scheme, they still employed an extensive labour force including bricklayers, carpenters and joiners, steel-fixers, plant operatives, scaffolders, painters etc. For the duration of the CHAG contract, the firm was based in Vauxhall at the Crown Works where it ran an extensive timber store, joinery workshop, metalworking shop and stonemason's yard. Higgs and Hill ran exemplary apprenticeships and career progression up the company hierarchy was encouraged with suitable candidates moving from skilled occupations into management positions. The firm held annual apprentice dinners, gave awards for excellence and long-service and ran an in-house magazine, *The Crown Journal*. The South Bank site was typical in accommodating drawing offices, canteens, offices for bonus clerks and wages clerks as well as on-site joinery workshops for the manufacture of the shutters, all of which were operated by the main contractor.

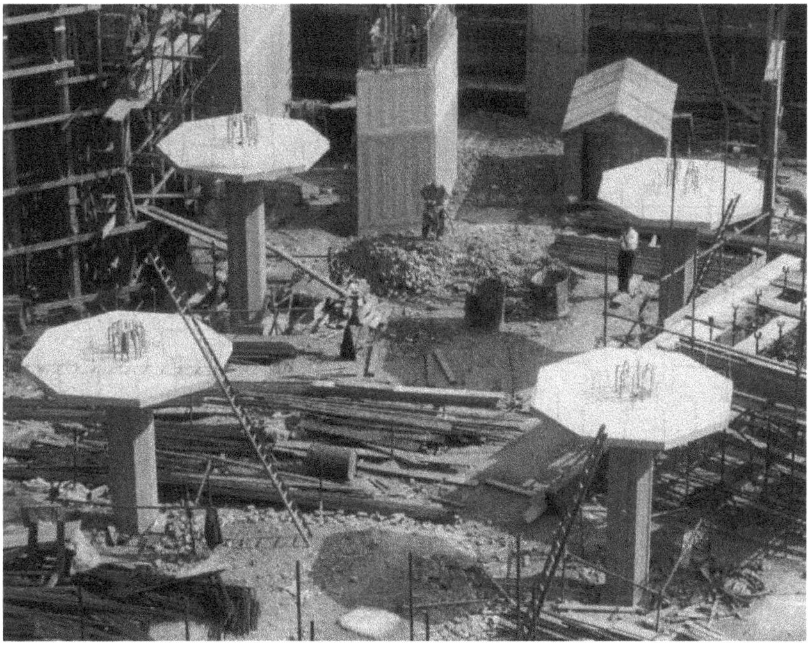

Fig. 1. Mushroom columns under construction on the Concert Hall and Art Gallery scheme, South Bank, London. Photograph reproduced courtesy of Rod Bond.

What emerges with great clarity from the oral histories is the high esteem given to site managers who were able to turn their hands to solving practical problems on site. All the interviewees remembered, and praised, J.P. Morgan, the project manager, as someone who spent most of his time out on site with the men and who frequently grabbed tools from people if he thought he could make a better job of it. He was also renowned for his skill in 'faking' the grain on concrete in order to cover up the very few areas which were spoilt. By using the life-history approach to interviewing it was possible for each respondent to recount the way in which they entered the building industry, again revealing the importance of early years spent as manual workers in a trade and education received from technical college. It was this specific skill and knowledge that enabled them not just to become effective site managers but also to innovate at site level, for as Glan Davis put it, the architects and engineers were responsible for, 'where the walls were, they designed the cavity, they designed the pattern, but not *how* to do it. We had to find the way of doing it' (Fig. 2).

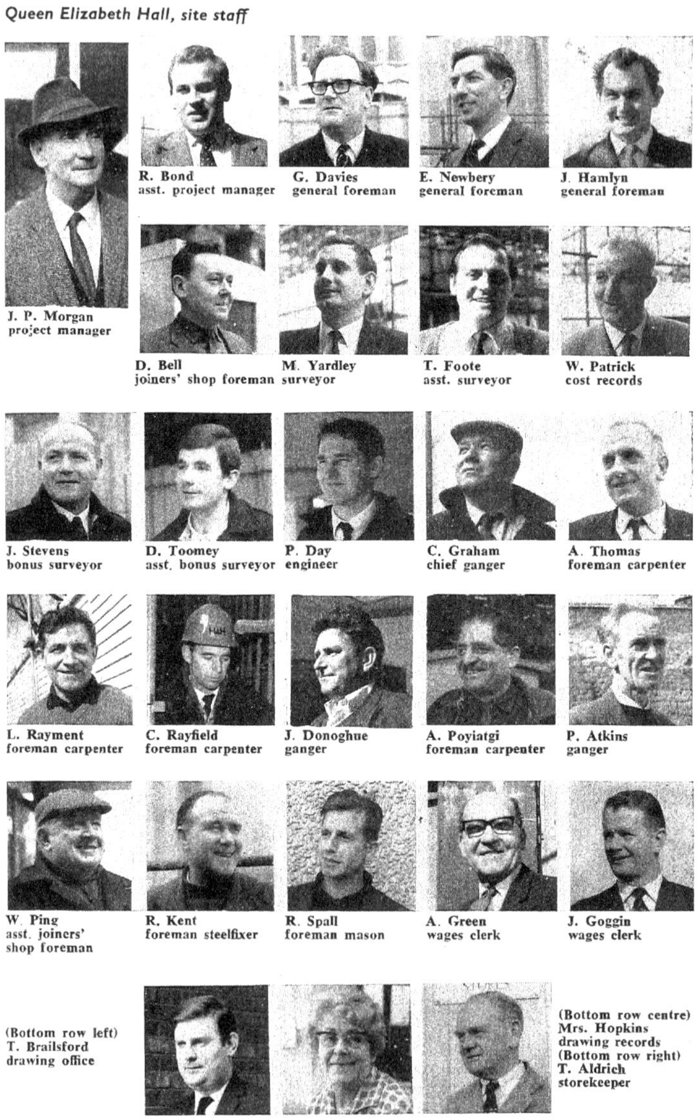

Fig. 2. Staff employed by Higgs and Hill on the Concert Hall and Art Gallery scheme. Source: The Crown Journal 142, March-April, 1966.

Among the solutions to the everyday practical problems of building to the exacting standards of the GHAG specification was a system, much praised by Construction News, consisting of mild steel 'U' channels that were bolted to the ply shutters.[2] The channels 'were held away from the concreted walls with greased mild steel rods that could be blocked to the required gap width, and easily withdrawn to allow striking'. This particular system of 'U' channels was devised by Glan Davis during the construction of the Purcell Room and he described it in precise detail many years later when interviewed in 2012.

Taken together these testimonies provide insight into a particular form of social organization of the building process found in mid-twentieth century Britain at the very point at which it was poised to disappear with the imminent change from direct to self-employment, and increased use of labour-only subcontracting.

> The CHAG site developed during the time when most contractors employed their own labour, and it is some years afterwards that subcontractors developed into the industry in the way that is done today. In that respect, I think that's the worst thing that's ever happened to the industry because they don't have apprentices, where [before] the main contractor employed his own labour and had his own apprentices. *Glan Davis*

The social 'building world'

Antoine Picon described construction history as existing in a 'strategic position between the history of technology and cultural history' with the potential to inform both disciplines.[3] While the history of technology was transformed in the 1980s by the work of Bijker, Hughes and Pinch (1987), much construction history, with some notable exceptions, e.g. Clarke (2011), Bullock (2002), and Swenarton (2008), has clung to a technological imperative which refers to political and socio-economic forces merely as 'context' and peripheral to the core narrative.[4] This is despite John Summerson's early (1973) evocation of the 'building world' to describe the complex social organisation required for the massive building projects taking place in London of the 1860s.[5] Summerson (1985) suggested that the social world of building practice, and the ramifications of this 'world' within wider society, rather than the study of innovations in materials and structural design, should be the direction for the new discipline of construction history.[6]

This paper follows on from this position in arguing for examination of the centrality of social relationships to the study of historic building practice, that is, in the words of Summerson:

> the total process of getting a building up on the site, including everything from the recruitment of labour, selection of materials, transport of materials and equipment on the site, down to the supply of drawing materials for the office, the method of payment to builder and architect and so on and so on. It is the history of a complicated process always in a state of change, which it is the historian's business to investigate and expound.[7]

While many construction historians now examine the building process, and to a greater or lesser extent the social relationships it encompasses, it is noticeable that very few include the contribution made by workers at the point of production, on the building site. This has the effect of making invisible a large part of the social 'building world' and their contribution. As the accounts above reveal, it is the collective activities on site that enable the success, or failure, of any structure imagined by architects or devised by engineers.

Even before the phrase 'building world' was coined William Lethaby's account of the thirteenth century craftsmen workers of Westminster Abbey set out to rectify the omission of any master painters, sculptors, goldsmiths or masons from the newly published National Dictionary of Biography.[8] Lethaby considered that architects had much to learn from building workers and considered the optimum approach to building to be one of a 'community of interests' between all concerned with the building process. But both Summerson and Lethaby were constrained by their sources: none of which originate from the building workers themselves.

One of the strengths of construction history is that it ignores the arbitrary distinctions between architecture, building and civil engineering and approaches all the outputs, and participants in the building process as worthy of investigation. But this does not overcome the central problem of lack of sources emanating from the working men and women who make up the largest part of the industry. In the case of civil engineering there are very few autobiographical accounts by workers dating from the nineteenth century, mainly because many men were illiterate, but Patrick MacGill's *Children of the Dead-End* (1914) is notable. Coleman's well-known book, the *Railway Navvies* is an excellent, if somewhat sensationalist, introduction to the social worlds of civil engineering but more detail is found in David Brookes' account.[9] Photographic archives have added to our knowledge of the social and working lives of these nomadic workers in for example S.W.A. Newton's photographs of the men who worked on the construction of the Great Central Railway.[10] But few of these accounts give insights into the precise nature of the work done by the manual workers or indicate the extent of knowledge and skill needed: this information is acquired obliquely from other sources than the men themselves and is not infrequently that authors mention that 'good men' were invaluable assets to contractors.[11] Much the same is true for the nineteenth century building industry, with the exception of Tressell, but for both civil engineering and building this begins to change with the repeal of the Combination Acts in 1824 and the gradual emergence of trades unions as social and political forces with the funds to publish their own newspapers and journals.

By the early twentieth century there were a large number of publications reflecting the numerous trade unions, which, after a series of amalgamations during the 1880s, became federated in 1918 into the National Federation of Building Trade Operatives (NFBTO).[12] These publications continue throughout the twentieth century and are a valuable historical source for understanding industrial change in terms of wages, working conditions, training, occupational change and industrial disputes. They rarely, however, provide detailed accounts of the building process. This is occasionally referred to in the Annual Reports of the NFBTO conference in contributions by union representatives to key debates on working with new materials and techniques but these are carefully prepared arguments in relation to submissions to conference.

The twentieth century, in a similar way to the nineteenth century, produced very few autobiographical accounts of work on building sites and apart from Ultan Cowley's books on the Irish in Britain, no comprehensive portrayal of building workers during the twentieth century.[13] There are then, few detailed descriptions of the historic building process as experienced by working men and women.

To compound these difficulties it is usually the case that contractors' records are dumped not archived and this is particularly true when a firm has lost its identity through a takeover. Most of the traditional firms who were active in the 1960s no longer exist, Peter Lind and Higgs and Hill, to name but two of many, have not left behind any accessible archives. The current National Archives mapping project seeking to identify archive sources for architecture, construction and building has resulted in the usual bias towards architecture and surveying with contractors trailing behind.[14] There are however former employees of these and similar firms who are prepared to give accounts of working on many prestigious post-war architectural and engineering schemes.

Oral history and construction history

Although it emerged as a distinct discipline in the 1970s oral history as a method is an offshoot of the earlier *'History from below'* movement, initiated in the early twentieth century by Marc Bloch and the Annales School, which sought to incorporate social phenomena, including ordinary people as subjects, into historical narratives. In Britain George Ewart Evans began recording the memories of elderly agricultural workers in the 1940s and 50s, just as post-war mechanisation made many of the old practices of working on the land redundant. Taken up in the 1960s by History Workshop, with Ralph Samuel as an advocate, it was closely aligned with labour historians who undertook studies of mining, fishing and other industries.

Oral history's original aims sought to give voice to groups and individuals who had been neglected by mainstream historians. The rich accounts provided by the different perspectives of ordinary people were deemed to enlarge and widen the scope of historical writing so that history itself became 'more democratic'.[15] Oral history is commonly described as emancipatory, allowing less privileged people to tell their own histories, and also claims to diversify history through recovering the voices of groups whose experiences would otherwise have been lost.[16] The original concern of many oral historians, therefore, has been to give voice to people enabling them to tell their story in the way they want to tell it, and according to their own 'agenda'.[17]

However, historians' interpretation of oral history testimony as a positivist record of 'lived experience' has been critiqued by Louisa Passerini who stressed the roles of subjectivity and memory.[18] For Passerini oral history testimonies are not just simple repositories of facts but are constructed stories that give meaning to lives and contribute to the formation of identity. The role of the researcher and interviewer is also acknowledged as a participant in the shaping of any narrative, thus, while the interview is a joint activity, the researcher has a particular responsibility to acknowledge their own position, and the ways in which this might influence and shape the interview. While the oral history approach goes some way towards addressing the power relationship between researcher and narrator, it is still the former who initiates, frames and shapes the research process and outcome through their interpretation.[19]

Oral history no longer needs defending as a discipline within history but has become an expanding territory within academia, as well as a popular method for local and community studies. Over the last twenty years it has also become a means for recording the histories of elite groups and the archive of the British Sound Library has become the repository of recordings of a large number of eminent twentieth century architects, artists and a smaller number of engineers. As Robert Proctor has pointed out oral history interviews have frequently been used uncritically by architectural historians even though architects are known to be unreliable narrators of their design intentions.[20] Proctor is instructive on the architect interview as a source for writing architectural history, seeing it as a form, which requires careful interpretation, together with a critical awareness of the autobiographical retrospective of an architect reflecting on his work coupled with his desire for esteem. To some extent these concerns do not apply to the testimonies of building workers who are usually perceived as being at the bottom of the hierarchy of relationships found on a building site and without the social esteem and power of professionals. However this is not a qualification to exclude critical analysis of the oral testimonies we recorded: the boasting of laying extraordinary numbers of bricks a day, and the disappearance of team members to enable a narrator to become the singular hero of his anecdote were noted in two different testimonies.

On the recent project described here we approached the oral testimonies with the same care as we would any other historical source and with the awareness that how people create meaning out of life experiences and how they historicize themselves is a complex process relying on memory and reflection. However, by foregrounding the subjects' accounts, and using a life-history approach, we hoped to achieve what Paul Thompson noted: [that]… it is a primary merit of oral history that, to a much greater extent than most sources, it allows the original multiplicity of standpoints to be recreated'.[21] For construction history the accounts of knowledgeable voices from building sites supply a valuable counterpoint to accounts of the 'evolution' of structural forms, and foregrounding these voices also avoids the use of oral testimony to add 'colour' to largely technological narratives of the building process.

The oral histories we recorded for *Constructing Post-War Britain* are an important tool, not just for answering the question, how did this structure get built but also through revealing the centrality of the social process in production. They provided accounts of working conditions, technical details, pay systems, forms of employment, training and lack of training, camaraderie, mutual support, industrial relations, career progression, and social mobility. Frequently they also, through direct contact and visits to participants' homes, provided documentary evidence in the form of photographs, certificates, drawings, plans, tools and materials dating from the building process.

Conclusion

Historians of building practice prior to the twentieth century must of necessity use documentary sources in their explorations but oral history has not been developed, as a method or a source, to any great extent in construction history. This paper has argued, using evidence from a recent research project, that oral history is one of the key means to understanding the nexus of social relationships, which underpin the construction process. It provides information on the social structures and informal connections between participants in the building process both within and outside contractual relationships. Most importantly it provides information on knowledge and skill evident at site level enabling an analysis, which in this case, has transformed our understanding of the mid-twentieth century construction process and its outputs. It remains an under-utilised resource for historians of the late twentieth century built environment.

References

[1] The three year Leverhulme Trust funded study, '*Constructing Post-War Britain: building workers' stories 1950-1970'* was carried out by Christine Wall and Linda Clarke, Centre for Research into the Production of the Built Environment, (ProBE) at the University of Westminster. The findings can be viewed at the project website: <http://www.westminster.ac.uk/probe>

[2] Complex structure for South Bank Art Centre', *Construction News*, 18 March, 1965.

[3] Antoine Picon, "Construction History: between technological and cultural history", *Construction History Journal*, vol. 21, 2005-6, pp.5-21.

[4]W. Bijker, T. Hughes and T. Pinch (eds.), *The social construction of technological systems*, London: MIT Press, 1987; Linda Clarke, *Building Capitalism*. London: Routledge, 2011; Nick Bullock, *Building the post-war world: modern architecture and reconstruction in Britain*. London: Routledge, 2002; Mark Swenarton, *Building the New Jerusalem: Architecture, Politics and Housing 1900-1930*. Garston: IHS-BRE, 2008.

[5]J. Summerson, *The London building world of the eighteen-sixties*. London: Thames and Hudson, 1973.

[6] John Summerson, "What is the history of construction?", *Construction History Journal*, vol 1, 1985, p 1.

[7] Ibid.

[8] William Lethaby, "Education in Building", *RIBA Journal*, vol. VIII 3rd series, 1901, pp. 385-395.

[9]P. MacGill, *Children of Dead-End Street, The Rat-Pit*; T. Coleman, *Navvies*; D. Brooke, 1983, *The Railway Navvy.*

[10] These can be viewed at: <http://www.railwayarchive.org.uk/stories/pages.php?enum=le123&pnum=0&maxp=8>[accessed on October 20 2012]

[11] Keith Middlemas, *The Master Builders: Thomas Brassey, Sir John Aird, Lord Cowdray [and] Sir John Norton-Griffiths*. London: Hutchinson,1963, p.40.

[12] W.S. Hilton, *Foes to Tyranny: A History of the AUBTW*. Amalgamated Union of Building Trade Workers, 1963.

[13] See Behan, Brian, *With Breast Expanded*. London: MacGibbon and Kee, 1964; D. MacAmhlaigh, *An Irish navvy. The diary of an exile*. London: Routledge and Kegan Paul, 1964. The few sociological investigations include: J.M. Sykes, Work attitudes of Navvies. *Sociology*, Vol. 3, 1969, pp. 21-34.

[14] See the interim results, revealing that out of twenty-two participating firms only four are building contractors, at: <http://www.nationalarchives.gov.uk/archives-sector/architecture-building-construction-records.htm>

[15] Paul Thompson, *The Voice of the Past*. Oxford, 1988, p.9.

[16] Joanna Bornat, "Reminiscence and oral history: parallel universe or shared endeavour", *Ageing and Society*, 2001, 21(2): pp.219-241; Robert Gildea, James Mark, Anette Warring (eds.), *Europe's 1968 : Voices of revolt*. Oxford, 2013.

[17] Bornat, 2001, p.226.

[18]Luisa Passerini, *International yearbook of oral history and life stories. Vol.1, Memory and totalitarianism*. Oxford, Oxford University Press, 1992.

[19] A. Portelli, *The Death of Luigi Trastulli and other Stories*. New York: State University of New York Press, 1991,

[20] Robert Proctor, "The Architect's Intention: Interpreting Post-War Modernism Through the Architect Interview", in *Journal of Design History*, vol. 19, no. 4, 2006.

[21] Thompson, 1988, p. 6.

»Ich traue der reinen Rechnung nicht, wenn diese nicht am Modell überprüft werden kann (I cannot trust the pure calculation, if this cannot be verified with a model)«

Frei Otto and model testing

Christiane Weber
Arbeitsbereich Baugeschichte und Denkmalpflege, Leopold-Franzens-Universität Innsbruck, Austria

Introduction

The Südwestdeutsche Archiv für Architektur und Ingenieurbau at the Karlsruhe Institute of Technology (KIT) took over the collection of architect, designer, researcher, and visionary Frei Otto in June 2011. The inventory comprises more than 400 models, and is thus among the most comprehensive twentieth-century model collections in Germany.[1] Since 1980, approximately twenty of Otto's models have been the property of the Deutsches Architekturmuseum. This abundance of models is evidence of Otto's passion for model making, which was so greatly admired by his colleagues.[2] Since it was nearly impossible to draw the forms of the structures that he developed with precision, he needed the models as a medium "to materialize the idea".[3] Models were thus employed at all phases of the design process: as presentation models, as form-finding models, and finally as measurement models on which the forces within the structure could be measured. Yet in what way did models define Frei Otto's work? His model-based designs meant that he was just as much at home in the art of architecture as he was in the science of civil engineering. The following will examine the extent to which he adopted and applied methods developed in the field of civil engineering.

Frei Otto's work is usually viewed from the perspective of "natural design".[4] His structures are defined by the transfer to architecture of the forms and design principles that prevail in the world of nature. His approach entails a considerable amount of self-discovery, meaning that as long as the general conditions are conducive, the form of his objects is quasi-self-optimizing.[5] Otto's engagement with this theme began in the 1950s, specifically in 1957, when he opened his development lab for lightweight design in Berlin following his appointment to the Institute for Light-weight Structures (IL) that had been set up specifically for him. From 1964, he continued his work at the Technische Hochschule in Stuttgart and at his studio in Warmbronn.[6] The aim of all of his experiments was to develop structures capable of bearing loads with minimal material input.

Soap-film Experiments

One area in which Otto worked extensively from 1958 onwards was soap-film models.[7] His starting point here was his realization that soap solutions were invariableand of their own accord form the smallest possible surface within a given frame[8], taking on a form with equal surface tension in all directions. This discovery enabled him to generate minimal surfaces whose forms served as models for the numerous tent structures realized in the course of the 1950s and 1960s. It was thanks to his soap-film experiments that he was able to develop the cable loops that starting in 1965 were used for the IL pavilion, for the German pavilion at Expo 1967 in Montreal, and for the roofs of the Olympic stadium in

Munich.[9] In 1963, Otto built a *Minimalwegegerät* (minimum path machine) in Berlin as part of his series of experiments on minimal networks.[10] The machine underwent further development at the IL in Stuttgart, where it became the "minimum surface machine" or "soap-film machine" used for photographing the transient forms of soap-films and for obtaining exact photo grammetric images in parallel projection (Fig. 1) The team of engineers at the Institute for Geodesic Applications in Civil Engineering (IAGB) headed by Klaus Linkwitz worked out a method for the photogrammetric measurement of soap-films.[11] Added to the experimental set up was an optical bench which generated the parallel light needed to project the contours of the minimum surface so that they could be traced or recorded by cameras.[12] The discovery that the shape of the soap-film reacted quite strongly to the climate of the surrounding space meant that in 1973, the soap-film machine had to be adapted so that the soap-film could be developed in a climate chamber with the greatest possible atmospheric humidity.[13] An add-on wind tunnel allowed the testing of soap-film forms exposed to wind.

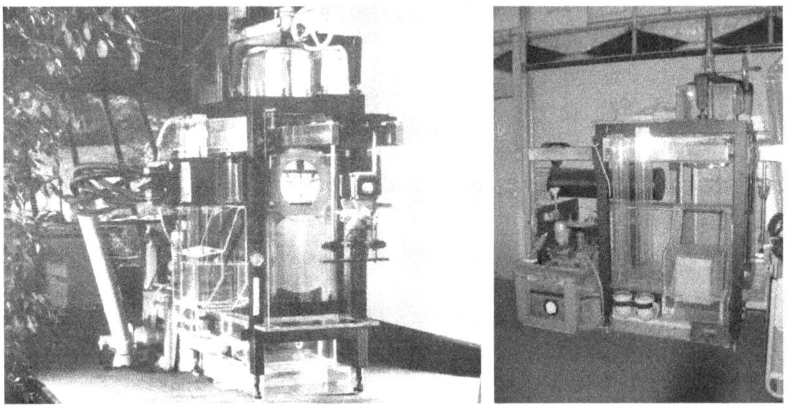

Fig. 1. Institute for Lightweight Structures (today ILEK, Stuttgart University), soap-film machine – historic photo and today.[14]

The researchers saw another potential use for the minimal surfaces being developed experimentally with soap-film in pneumatic structures.[15] These air-dome structures brought Otto very close to his goal of roofing-over large areas with minimum effort and expense and of creating climatic domes. In contrast to the tent structures, however, very few of these ideas were ever actually implemented.

Hanging Models

Otto and his team took a different approach to optimizing supporting structures with the most diverse forms of hanging models. In this case, researchers were able to consult previous studies, such as those by Heinrich Hübsch or Antoni Gaudí.[16] These rest on the physical phenomenon that a hanging mesh will always take on the form corresponding to the line of thrust of the inverse structure. In other words, if the hanging mesh were to be "frozen" and inverted, that is, mirrored on the horizontal plane, it would produce an arch with ideal statics, whose form was perfectly aligned with the line of thrust.[17] In 1946, applying this law of physics, Otto began experimenting with hanging cloths and later with a wide range of chains and meshes.[18]

In 1968, for example, a hanging mesh made of machine-made hooks and eyes was used for optimizing the grid shell of a congress center that was to be built in Saudi Arabia, although it was never actually realized in this form.[19] For the preliminary design of the Mannheimer Multihalle for the Bundesgartenschau (Federal horticultural show) in 1975, the participants — the architectural office Carlfried Mutschler and the Atelier Frei Otto in Warmbronn — modelled wire mesh on the freely formed landscape of the horticultural show grounds.[20] The amorphous shape canopied a multifunctional events space and restaurant. For the structure, Otto suggested a grid shell of wooden slats, thus, a light construction using minimal materials.

In order to define the form of the shell, after measuring the pre-design model, a large hanging model was built in Otto's studio in Warmbronn based on the inversion principle.[21] The grid of the hanging model (scale of 1:98.5) (Fig. 2) was created by hand from 15-millimeter-long chain links and rings with a diameter of 2.5 millimeters (Fig. 3). In order to reduce the complexity, only every third mesh of what would be the grid shell was depicted. The knotting occurred in separate parts, which were hung on a temporary frame made of balsa wood.[22] The form of the grid shell was examined based on this model. For example, the half-dome widening of the entrances arose from the model's process of finding its own form.[23] In a further work step, this model was photometrically measured by the IAGB under Klaus Linkwitz,[24] which means that the model was photographed from above from two different angles, thus generating a digital, spatial model. Based on this, the form of the slat grid was then calculated. Especially in the border area, the kinks, in part, strong ones, had to be balanced out and the intersections, missing in the model, had to be added in. The data, created with a CDC 6600 mainframe computer, could be distributed with a so-called "Zeichenautomat" (drawing machine) and served as base for the planning of the work, among other things, for the determination of the pattern (pattern: system of length of all rods of the grid as well as of all edges).[25]

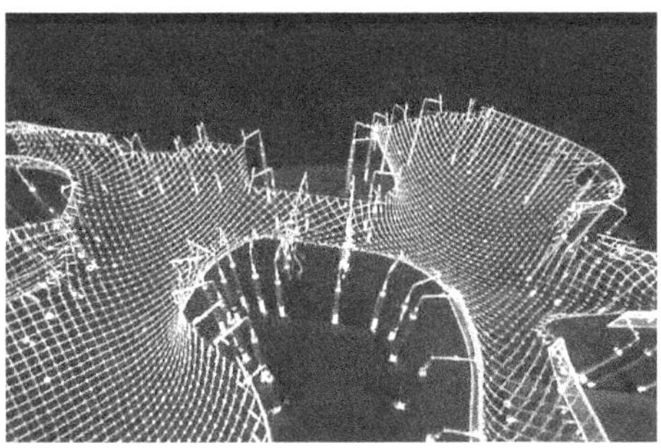

Fig. 2. Studio Warmbronn, Hanging model at a scale of 1:98.5

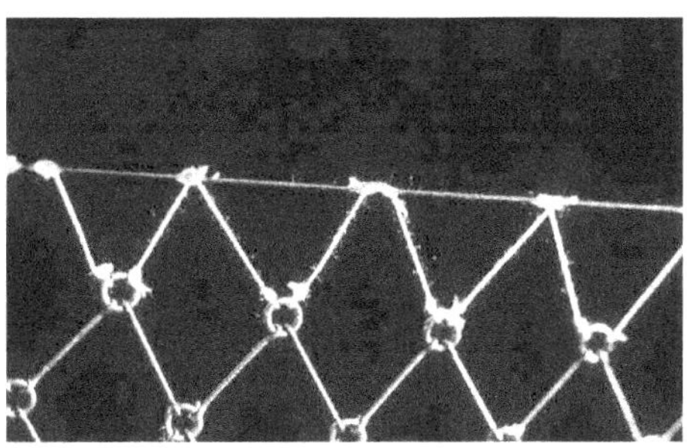

Fig. 3. Hanging model, detail

»Ich traue der reinen Rechnung nicht, wenn diese nicht am Modell überprüft werden kann« Frei Otto and model testing

Until this time, only two other wood grid shells had been built worldwide. A trial shell at the *Deutsche Bauausstellung*, 1962 in Essen, and the roof of the lecture hall in the German pavilion at the Expo 1967 in Montreal. In order to gather an idea of the load bearing behavior of the construction, the engineers from the office of Ove Arup & Partners built a first static measurement model based on the Essen dome.[26] The model, at a scale of 1:16 was made of narrow perspex strips, which were tied to the intersections with small pins. Stress tests were then carried out with suspended nails. A model of the Multihalle was subsequently built at a scale of 1:60, to test the significantly more difficult geometry of the Mannheim design (Fig. 4).[27] Strains could be examined on it with the help of dial gauges. The trials showed that the assemby of tension members (steel cables) and the doubling of the grid shell were necessary. Since the amorphous form could not be calculated with the usual specificatons for wind loading, the British Hydrodynamics Research Association carried out wind tunnel experiments.[28] For this, several geometrically similar measuring models (scale of 1:200) were built, which were equipped with 150 measuring pressure taps, the total pressure at each tap location was measured by manometer (Fig. 5).[29] The wind pressure and suction could thus be investigated at a wide range of wind directions and strengths. In order to develop details of the grid construction, several models were built at a scale of 1:5, which showed the planning state of the single-layer grid. Parts of the construction were even built at a scale of 1:1.[30] The largest model had a surface of 80 square meters and was used by the company Koitwerk H. Koch in Rimsting, to test the displacement of the roof skin and instruct the carpenters.[31]

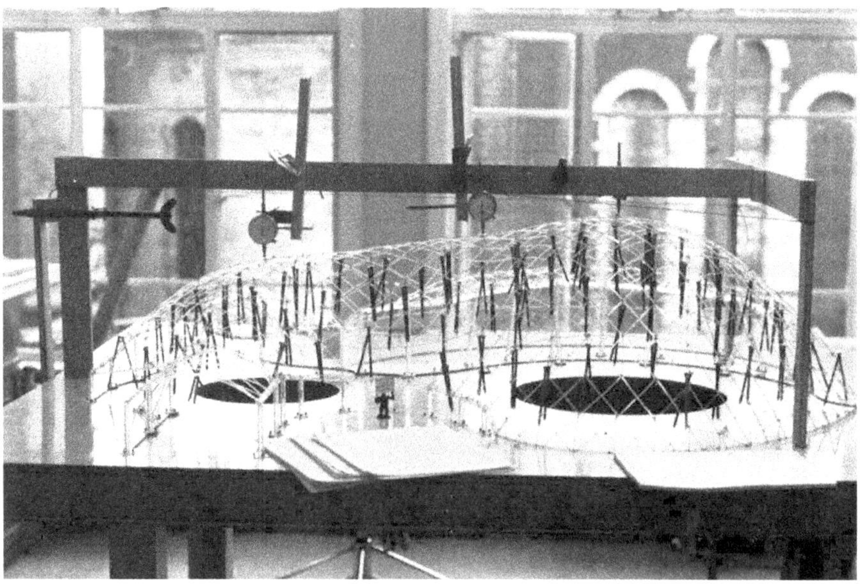

Fig. 4. Ove Arup & Partners, measuring model at a scale of 1:60

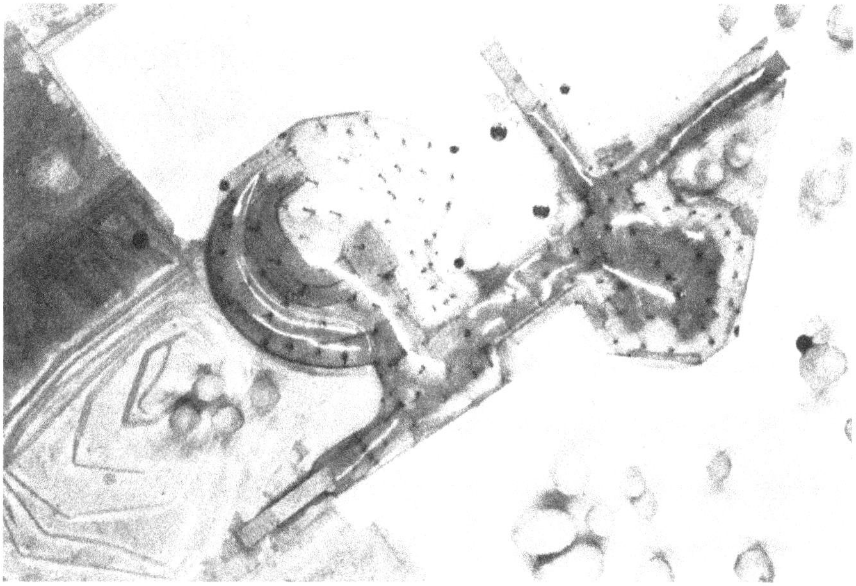

Fig. 5. British Hydrodynamics Research Association, Model for wind tunnel experiments, 1:200

In addition, the experimental set up for fire proofing consisted of a part of the roof structure, just about 5 square meters.[32] The grid shell could finally be set up over the foundation in winter 1974/75. The engineers discussed the various possibilities for setting up the shell beforehand, based on a model at a scale of 1:60.[33] The decision was made to successively lift the flat, laid out grid with the help of forklifts and scaffolding elements into the final position. The test engineer Fritz Wenzel from the Technische Hochschule Karlsruhe demanded a stress test after completion of the construction. On 30 January 1975, the roof of the Multihalle became its own 1:1 measuring model: 205 garbage cans, hung on the grid shell, were filled with water to simulate snow loads.[34]

Capturing Form

One fundamental difficulty presented by Otto's lightweight load-bearing membrane structures was how to capture them graphically. The spatially complex geometries are difficult to represent using the standard representational techniques of floor plan and section. Since the forms were based on models, the challenge was to translate these into two-dimensional drawings (to produce cutting plans for the membranes, for example).

A measuring table designed specifically for this stage in the design process was built at the Institute for Lightweight Structures. Using this apparatus, the xyz coordinates of the models could be taken by means of a plummet (Fig. 6). The measurements could be displayed directly and exported onto punched tape.[35] Photogrammetry, introduced for the German pavilion at Expo 1967 in Montreal, presented an alternative to this tracing method. The large measurement model on a scale of 1:75 was measured stereophotogrammetrically by taking photos from above at specific distances with the help of a splint.[36]

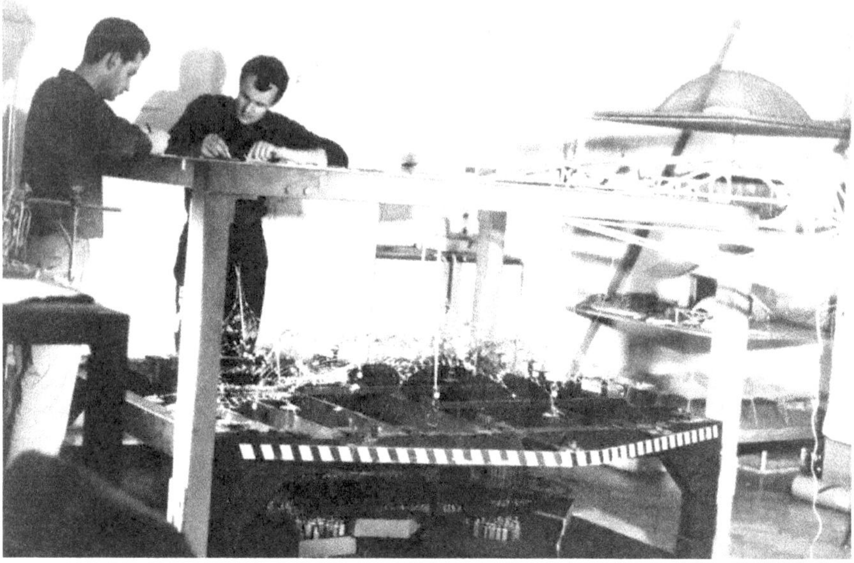

Fig. 6. Institute for Lightweight Structures, measuring table

A computer-assisted process for the evaluation of the measurements was introduced in 1966. Under the leadership of Linkwitz, the IAGB at the Technische Hochschule Stuttgart created a computer program that could translate the data recorded on punched tape into drawings. This made it possible to draw random cuts in the free-tent form. The "Zuse-Graphomat" output device — one of the first computer-based drawing systems ever used in the building industry — was displayed as an exhibit in the German pavilion in Montreal. This non-contact photogrammetric method was also used for the soap-film experiments, as the delicate forms were liable to burst on the slightest contact. Another process used was measurement with parallel light, which the optical bench could generate as large as 120 mm & 120 mm and project as a vertical line grid onto the model.[37]

Measuring Forces

The form-finding models described were used for developing and optimizing load-bearing structures. Measurement models in the strict sense of the term, on the other hand, were used when the load-bearing structures were too complex to be analysed with arithmetical methods. How forces and distortions are measured on the models will be explained in what follows.[38] In structural engineering, the term "model analysis" refers to a method in which the load-bearing structure is reconstructed in models that are similar in scale, geometry, and elasticity.

The testing of load-bearing structures with the help of models has long been standard practice among engineers,[39] "but in the twentieth century, model analysis was applied systematically only after similitude fluid mechanics, metrology, and material research created the necessary preconditions for it".[40] In the early twentieth century in Europe and the USA several research institutes and research teams were working on model testing, i.e. Georg E. Beggs, who published his results since the 1920s.[41] After this, model analysis as a method spread rapidly, and was soon being used in Stuttgart, where the engineer Karl Schaechterle applied it in bridge construction.[42] Thus, the Material Testing Institute (MPA) at the Technische Hochschule Stuttgart conducted model analyses of suspension bridge models for important *Autobahn* projects, such as the highway bridge over the Rhine in Cologne-Rodenkirchen or the bridge over the Elbe in Hamburg.[43] Hermann Maier-Leibnitz compared the mathematical results with those obtained from model analyses in an article serialized in the magazine *Die Bautechnik* of 1941 and 1942.[44] Due to extensive financial support provided for these model experiments in National Socialist building industry, measuring statics became an established method. After World War II, the MPA Stuttgart conducted model measurements to test the new, cable-stayed Nordbrücke built in Düsseldorf in the years 1952 to 1957 according to plans by Leonhardt + Andrä. The engineers also conducted a series of model analyses for the single-cable suspension bridge which Fritz Leonhardt proposed in 1961 for a bridge across the Rhine near Emmerich.[45]

However, owing to its efficiency and economy, model analysis was deployed in more than just bridge building. Another important area of application was shell structures. The architects Niessen & Störmer began planning an expressive, cantilevered pre-stressed concrete shell for the indoor swimming pool on Sechslingspforte in Hamburg in 1956. Jörg Schlaich of Leonhardt + Andrä was respon- sible for designing the load-bearing structure[46] and it was for this shell roof that a Plexiglas measurement model was delivered to the Faculty of Civil and Environmental Engineering for the performance of stress tests using strain gauges and the automatic measuring apparatus available at the Institute of Model Statics[47] at the Technische Hochschule Stuttgart (Fig. 7).[48] Internationally renowned architects, such as Heinz Isler in Switzerland or Pier Luigi Nervi in Italy worked with professional engineering laboratories — in Nervi's case the *Laboratorio Prove modelli e costruzioni dell'Istituto di Scienza delle Costruzioni* in Milan — to measure and test such shell structures.[49]

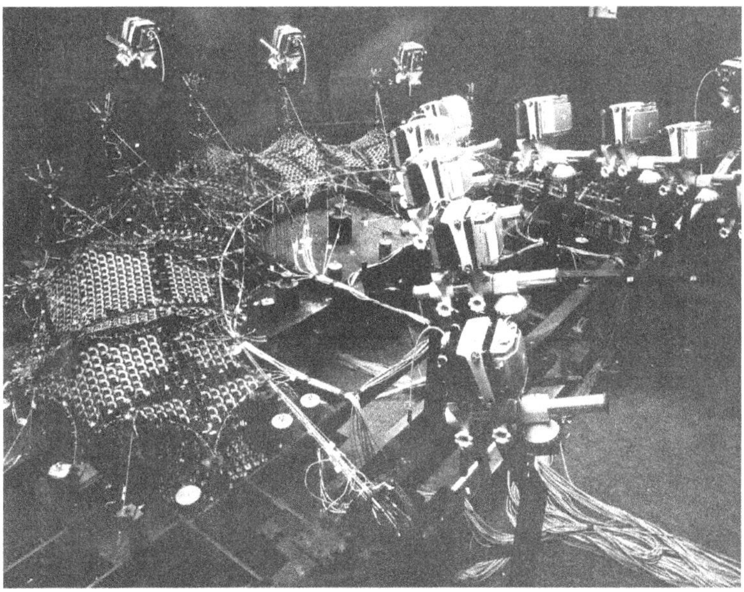

Fig. 7. Institute for Model Statics at the TH Stuttgart, model measurements for the hypershell roof of the indoor pool on the Sechslingspforte in Hamburg, 1:26.67

Measurement Models for Determining Forces in the Works of Frei Otto

As Otto planned his tent and tensile structures with increasingly large spans, he had to rely more and more on model analysis. The first extensive use of measurement models was during the planning phase of the German pavilion at Expo '67 in Montreal. For this project, Frei Otto's Institute for Lightweight Structures (IL) built a measurement model on a scale of 1:75 based on the principle of geometric and static similarity (Fig. 6). Thin, rust-free steel cables that were seventy-five times thinner than the steel cables built in Montreal were used.[50] Tensions and changes in tension under stress were measured by hanging small weights from the nodes.[51] The "Montreal micrometer," a small gauge mounted on three points developed specifically for this purpose, was used to determine the elongation that became evident in the cables.[52] In 1968, by which time Otto's involvement in the planning of the Olympic stadium in Munich based on the design by the architects Behnisch & Partner was definite, the task assigned to IL comprised "producing measurement models for all building parts, including for the measurement of forces and distortion".[53] The team of engineers at Leonhardt + Andrä headed by Jörg Schlaich did a lot of work on the models and "exerted a certain control function during the phase of model construction."[54] The measurement models, whose scale of 1:125 resulted from the size of the measuring table at IL,[55] allowed for form to be optimised and the measurement of the tension in the cables and in the load-bearing properties of the structure when under pressure.[56] The "multimedia experiments" in which several measuring techniques were applied simultaneously also involved the Institute for Model Statics, which used strain gauges to determine the changes in forces in the border cables (Fig. 8).[57] As in Montreal, the cutting pattern was determined in close collaboration with the IL and the IAGB.

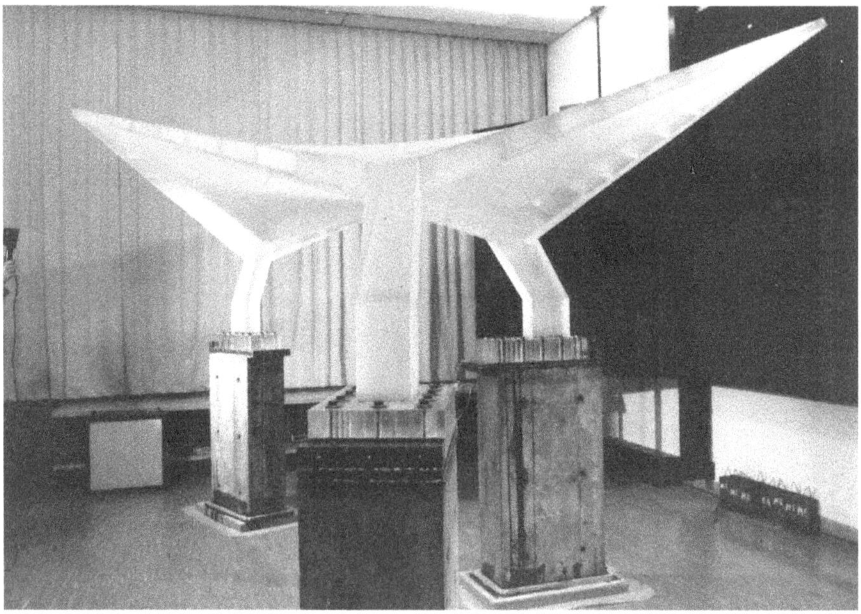

Fig. 8. Institute for Lightweight Structures with IAGB, roof stress tests on the model of the stadium for the 1972 Olympics in Munich, 1:125

Since the measurement model for the cut was not sufficiently precise for all models, Leonhardt + Andrä involved the Technische Hochschule's Institute for the Statics and Dynamics of Aerospace Engineering in Stuttgart headed by John Argyris. Argyris is considered a cofounder of the Finite Elements Method.[58] This method made it possible to calculate the pattern for the sports hall mathematically using the IAGB's data.[59] Fritz Leonhardt was among the first to recognize the potential of computer-aided measuring processes and as early as 1961 launched the company for *Rechneranwendungen im Bauwesen* (RIB) together with Volker Hahn and Friedrich Wilhelm Bornscheuer. It was this company that brought the first Finite Elements Program (FEM) onto the market in 1972. In the following years, the development of ever faster and ever cheaper computers made them a more economical measuring method than model analysis. Robert Müller, head of the Institute for Model Statics at the Technische Hochschule Stuttgart, saw this coming in his *Handbuch zur Modellstatik*[60] of 1971. Economic pressure ultimately rendered model analysis redundant. When Müller retired in 1993, the institute he had headed was dissolved and the professorship renamed Information Processing in Engineering Design, marking the end of the era of model analysis in structural engineering.[61] Otto, however, followed his own path. Although he worked quite successfully with engineers on the realization of his projects — after the Multihalle in Mannheim primarily with Ted Happold of the Structures 3 depart- ment at Ove Arup & Partners[62] — Otto himself remained true to model-based design. In the following decades he stuck doggedly to the practice of testing forms and the flow of forces on physical models, producing more than 120 models alone for the Stuttgart 21 project. As Frei Otto himself said recently, "I do not trust pure calculations if they cannot be tested on models."[63]

References

[1] KIT press release no. 109 from June 29, 2011, p. 1.

[2] Walker, Addis: Happold. The Confidence to Build, 1997, p. 57.

[3] Lennart Grut in: "Lennart Grut, Ted Happold, and Peter Rice discuss Frei Otto and his work", in: *Architectural Design*, no. 3, 1971, p. 150 (quoted from Christian Brensing: "Frei Otto und Ove Arup. Szenen einer gegenseitigen Inspiration", p. 106).

[4] See the exemplary book and exhibition: Nerdinger Winfried (Hrsg.): Leicht Bauen – Natürlich Gestalten. Frei Otto – Das Gesamtwerk, Basel, 2005.

[5] Gaß: Experimente, IL 25 (Form – Kraft – Masse 5), 1990, p. 0.15.

[6] Ibid.

[7] Bach, Burkhardt: Seifenblasen, IL 18, Stuttgart, 1987, p. 15.

[8] Ibid., p. 6.

[9] Ibid., p. 18.

[10] Burkhardt: Experimentelle Ermittlung von Minimalnetzen, IL. 1, 1969, pp. 12–14. Bach, Burkhardt: Seifenblasen, IL 18, Stuttgart 1987, p. 320.

[11] Ibid., p. 327.

[12] Ibid.

[13] The collaboration with an engineering !rm in the building of the soap-skin machine had to be terminated as their solutions proved too elaborate and expensive. Ibid., p. 328.

[14] Ibid., pp. 322–331.

[15] Otto, Trostel: Zugbeanspruchte Konstruktionen, 1962, pp. 10–168.

[16] Otto: Natürliche Konstruktionen. Formen und Konstruktionen in Natur und Technik und Prozesse ihrer Entstehung, 1982, p. 48. Galileo Galilei already made the first experiments with chain lines in ca. 1600. Karl-Eugen Kurrer: Geschichte der Baustatik, Berlin 2002, p. 126; Tomlow: Das Modell: Antoni Gaudis Hängemodell und seine Rekonstruktion. Neue Erkenntnisse zum Entwurf für die

Kirche der Colonia Güell, IL 34, Stuttgart 1989 (Mit einem Vorw. von Rainer Graefe und Frei Otto und einem Beitr. von Harald Szeemann).

[17] Graefe: "Gewölbe und Kuppeln, Schalen und Stabtragwerke, Gitterschalen. Beispiele aus der Baugeschichte", 1974, pp. 20–25.

[18] Ibid, pp. 22, 32–44.

[19] Otto: Gitterschalen, IL 10, Stuttgart 1974, p. 294f.

[20] Otto: Multihalle Mannheim, IL 13, Stuttgart 1978, p. 33.

[21] The model is part of the collection of the Deutsches Architekturmuseum (DAM Frankfurt), who thereby has available the second largest museum collection of the work of Frei Otto. The founding director of the Deutsches Architekturmuseum, Heinrich Klotz, first presented Frei Otto's designs and models at the exhibition *Vision der Moderne* (1986).

[22] Otto: Multihalle Mannheim, IL 13, Stuttgart 1978, pp. 34–39.

[23] Ibid, p. 39.

[24] Gründig, Hangleiter, Preuss: „Berechnung des Hängenetzes und Zuschnittsermittlung". Ibid, pp. 41–50.

[25] Otto: Multihalle Mannheim, IL 13, Stuttgart 1978, pp. 41.

[26] Ibid, p. 76f.

[27] Ibid, p. 80f.

[28] Ibid, p. 66f.

[29] Ibid, p. 68.

[30] Ibid, p. 98.

[31] Ibid, p. 150f.

[32] Ibid, p. 131f.

[33] Ibid, p. 142f.

[34] Ibid, p. 156f.

[35] Otto: Gitterschalen, IL 10, Stuttgart 1974, p. 132.

[36] Bundesbaudirektion (ed.): Expo '67 Montréal. Deutscher Pavillon. Dokumentation über das Bauwerk, Düsseldorf [1967], p. 18.

[37] Otto: Gitterschalen, IL 10, Stuttgart 1974, p. 134.

[38] Otto: Netze in Natur und Technik, IL 8 Stuttgart 1975, p. 276.

[39] E.g., Leonardo da Vinci's vault models. Cf. Geschichte der Baustatik, 2002, p. 133. Addis, „'Toys that save millions' – a history of using phisical models in stuctural design", in: The Strucutral Engineer, 9 (2013), 4, p. 12.

[40] Müller: Handbuch der Modellstatik, 1971, pp. 1–2.

[41]Addis, „'Toys that save millions' – a history of using phisical models in stuctural design", in: The Strucutral Engineer, 9 (2013), 4, p. 17.

[42] Leonhardt: Baumeister in einer umwälzenden Zeit. Erinnerungen, 1984, p. 78.

[43] Materialprüfungsanstalt (Material testing facility) at the Technischen Hochschule Stuttgart. Institut für die Materialprüfungen des Bauwesens. "Prüfbericht. Antragsteller Direktion der Reichsautobahnen, Berlin. Modelluntersuchungen zur Rheinbrücke Rodenkirchen. 3. Bericht mit den Ergebnissen der Untersuchungen an der Modellbrücke III. Fortsetzung unserer Berichte vom 22. April und 30. Juni 1938," January 15, 1939. (Universitätsarchiv Stuttgart/Bestand 33/1/KFM 8).

[44] Maier-Leibnitz: "Grundsätzliches über Modellmessungen zur Klarlegung der Formänderungen und Spannungen von verankerten Hängebrücken", in: *Die Bautechnik*, 19, 1941, no. 46/47 pp. 508–515, no. 48, pp. 522–526, no. 53/54, pp. 573–583, and 20, 1942, no. 53/54, pp. 457–463 and no. 54/56, pp. 486–490.

[45] Kleinmanns, Weber: Fritz Leonhardt 1909–1999, 2009, p. 210.

[46] Ibid.

[47] Fritz Leonhardt, who was appointed to the Lehrstuhl für Massivbau in 1957, personally supported the furnishing of the Institut für Modellstatik. The trained physicist and mathematician Robert Müller was appointed as director in 1960. Later, endowed with a C3-professorship, he led its establishment as a separate institute, "Institut für Modellstatik." Cf. Leonhardt: Baumeister in einer umwälzenden Zeit. Erinnerungen, 1984, p. 78; Müller: "Modellstatik an der Universität Stuttgart", in: *Südwestdeutsche Monatsschrift für Kultur, Wirtschaft und Reisen*, 19, 1969, p. 102.

[48] Kayser: "Das Hyperschalendach des Hallenbades Hamburg Sechslingspforte", in: *Die Bautechnik*, 65, 1970, pp. 245–247.

[49] Chiorino: "L'expérimentation dans l'œuvre de Pier Luigi Nervi", 2010, pp. 61–83; Addis, „'Toys that save millions' – a history of using phisical models in stuctural design", in: The Strucutral Engineer, 9 (2013), 4, pp. 18–19.

[50] Frei Otto: Protokoll über die Arbeiten des Instituts für Leichte Flächentragwerke 1967 [3. Messmodell Montreal] (Privatarchiv Bertold Burkhardt, Braunschweig); Leonhardt: "Der deutsche Pavillon auf der Expo '67 in Montréal. Eine vorgespannte Seilkonstruktion", in: *Der Stahlbau*, 37, 1968, no. 4, p. 103.

[51] Otto: Protokoll über die Arbeiten des Instituts für Leichte Flächentragwerke 1967 [2.9 Belastungsversuch] (Privatarchiv Bertold Burkhardt, Braunschweig); Müller: "Modellstatik an der Universität Stuttgart", in: *Südwestdeutsche Monatsschrift für Kultur, Wirtschaft und Reisen*, 19, 1969, p. 103.

[52] Otto: Protokoll über die Arbeiten des Instituts für Leichte Flächentragwerke 1967 [3. Messmodell Montreal] (Privatarchiv Bertold Burkhardt, Braunschweig).

[53] Otto: IL 8 Stuttgart 1975, p. 275.

[54] Ibid, p. 276.

[55] Ibid, p. 285.

[56] Ibid, p. 276.

[57] Ibid, p. 291.

[58] Spieker: "Die Planung des Olympiadachs in München", 2009, p. 122.

[59] Otto: IL 8 Stuttgart 1975, p. 287

[60] Müller: Handbuch der Modellstatik, 1971, p. 8

[61] Senatsprotokoll der Universität Stutt- gart, November 3, 1993 (Universitätsarchiv Stuttgart), p. 18.

[62] Dickson: "Frei Otto und Ted Happold. 1967–1996 und darüber hinaus", 2005, pp. 111–123.

[63] Frei Otto in conversation with the author, December 12, 2011.

Bibliography

1. Addis, Bill: „'Toys that save millions' – a history of using phisical models in stuctural design", in: The Strucutral Engineer, 9 (2013), 4, pp. 12–27.
2. Bach, Klaus; Burkhardt, Berthold: Frei Otto, Seifenblasen, IL 18, Stuttgart 1987.
3. Brensing, Christian, „Frei Otto und Ove Arup. Szenen einer gegenseitigen Inspiration", in: Winfried Nerdinger (Hrsg.), Leicht Bauen – Natürlich Gestalten. Frei Otto – Das Gesamtwerk, Basel 2005, S. 102–109.
4. Burkhardt, Berthold (Hrsg.): Experimentelle Ermittlung von Minimalnetzen, IL 1, Stuttgart 1969.
5. Bundesbaudirektion (Hrsg.), Expo '67 Montréal – Deutscher Pavillon. Dokumentation über das Bauwerk, Düsseldorf [1967].
6. Cachola Schmal, Peter; Elser, Oliver (ed.), Das Architekturmodell. Werkzeug, Fetisch, Kleine Utopie, Zürich 2012.

7. Chiorino, Mario Alberto, „L'expérimentation dans l'œvre de Pier Luigi Nervi", in: Carlo Olmo, Cristiana Chiorino (Hrsg.), Pier Luigi Nervi. L'architecture comme défi, Mailand/Brüssel 2010, S. 61–83.
8. Dickson, Michael, „Frei Otto und Ted Happold. 1967–1996 und darüber hinaus", in: Winfried Nerdinger (Hrsg.), Leicht Bauen – Natürlich Gestalten. Frei Otto – Das Gesamtwerk, Basel 2005, S. 111–123.
9. Gaß, Siegfried: Experimente, IL 25 (Form – Kraft – Masse 5), Stuttgart 1990.
10. Graefe, Rainer; Hennicke, Jürgen; Otto, Frei: Aufgaben, Probleme und Fragen in Forschung, Entwicklung und Anwendung des Leichtbaus, IL 20, Stuttgart 1979.
11. Graefe, Rainer: „Gewölbe und Kuppeln, Schalen und Stabtragwerke, Gitterschalen. Beispiele aus der Baugeschichte", in: Frei Otto (Hrsg.), Gitterschalen, IL 10, Stuttgart 1974.
12. Graefe, Rainer; Otto, Frei u.a.: Natürliche Konstruktionen. Formen und Konstruktionen in Natur und Technik und Prozesse ihrer Entstehung, Stuttgart 1982.
13. Janke, Rolf: Architekturmodelle. Beispielsammlung moderner Architektur 1, Stuttgart 1962.
14. Kayser, Rolf; Lehmitz, Karl; Leonhardt, Fritz; Müller, Robert K.; Schlaich, Jörg; Voßbein, Hartmut, „Das Hyperschalendach des Hallenbades Hamburg Sechslingspforte", in: Die Bautechnik, 65, 1970, Teil I (Leonhardt/Schlaich): „Entwurf und Tragverhalten", Nr. 9, S. 207–215, Teil II (Müller/Kayser): „Modelluntersuchung", Nr.10, S. 245–249, Teil III (Voßbein/Lehmitz): „Bauausführung", Nr.11, S. 261–264.
15. Kleinmanns, Joachim; Weber, Christiane (Hrsg.), Fritz Leonhardt 1909–1999. Die Kunst des Konstruierens/The Art of Engineering. Stuttgart/London 2009.
16. Kurrer, Karl-Eugen, Geschichte der Baustatik, Berlin 2002.
17. Kurrer, Karl-Eugen, The History of the Theory of Structures. From Arch Analysis to Computational Mechanics, Berlin 2008.
18. Leonhardt, Fritz; Egger, Harald; Haug, E., „Der deutsche Pavillon auf der Expo `67 in Montréal – eine vorgespannte Seilkonstruktion", in: Der Stahlbau, 37, 1968, Nr. 4, S. 97–106, Nr. 5, S. 138–145.
19. Leonhardt, Fritz, Baumeister in einer umwälzenden Zeit. Erinnerungen, Stuttgart 1984.
20. Maier-Leibnitz, Hermann: "Grundsätzliches über Modellmessungen zur Klarlegung der Formänderungen und Spannungen von verankerten Hängebrücken", in: Die Bautechnik, 19, 1941, no. 46/47 p. 508–515, no. 48, p. 522–526, no. 53/54, p. 573–583, and 20, 1942, no. 53/54, p. 457–463 and no. 54/56, p. 486–490.
21. Müller, Robert K.: „Modellstatik an der Universität Stuttgart", in: Südwestdeutsche Monatsschrift für Kultur, Wirtschaft und Reisen, 19 (1969), Sonderausgabe Dez. 1969, S. 101–102.
22. Müller, Robert K.: Handbuch der Modellstatik, Berlin/Heidelberg/New York 1971.
23. Nerdinger Winfried (Hrsg.): Leicht Bauen – Natürlich Gestalten. Frei Otto – Das Gesamtwerk, Basel 2005.
24. Otto, Frei; Trostel, Rudolf, Zugbeanspruchte Konstruktionen, Band 1, Frankfurt/Berlin 1962.
25. Otto, Frei (Hrsg.): Netze in Natur und Technik, IL 8 Stuttgart 1975.
26. Otto, Frei (Hrsg.): Gitterschalen, IL 10, Stuttgart 1974.
27. Otto, Frei (Hrsg.): Multihalle Mannheim, IL 13, Stuttgart 1978.
28. Otto Frei u. a., Natürliche Konstruktionen. Formen und Konstruktionen in Natur und Technik und Prozesse ihrer Entstehung, Stuttgart 1982.
29. Spieker, Elisabeth: "Die Planung des Olympiadachs in München. Fritz Leonhardts Mitwirkung und Impulse", in: Joachim Kleinmanns, Christiane Weber (Hrsg.), Fritz Leonhardt 1909–1999. Die Kunst des Konstruierens/The Art of Engineering. Stuttgart/London 2009, S. 118–125.

30. Tomlow, Jos: Das Modell: Antoni Gaudis Hängemodell und seine Rekonstruktion. Neue Erkenntnisse zum Entwurf für die Kirche der Colonia Güell, IL 34, Stuttgart 1989 (Mit einem Vorw. von Rainer Graefe und Frei Otto und einem Beitr. von Harald Szeemann).
31. Tomlow, Jos: „The hanging model - Some historical applications and its use in experimental historical survey“, in: Proceedings of the International Conference on Studies in Ancient Structures 1997, Istanbul 1997.
32. Derek Walker, Bill Addis (ed.), Happold. The Confidence to Build, London 1997.
33. Weber, Christiane: »Frei Otto – Experimentelle Modelle«, in: Peter Cachola Schmal, Oliver Elser (Hrsg.), Das Architekturmodell. Werkzeug, Fetisch, Kleine Utopie, Zürich 2012, S. 45–50.
34. Weber, Christiane: »Frei Otto – Formfindungsmodelle«, in: Peter Cachola Schmal, Oliver Elser (Hrsg.), Das Architekturmodell. Werkzeug, Fetisch, Kleine Utopie, Zürich 2012, S. 174–195.

Archive Sources

- Bestand Frei Otto, Südwestdeutsches Archiv für Architektur und Ingenieurbau, Karlsruher Institut für Technologie.
- Modellsammlung des Deutschen Architekturmuseums Frankfurt.
- Sammlung des ehemaligen Instituts für Leichte Flächentragwerke (heute ILEK), Universitätsarchiv Stuttgart.
- Privatsammlung, Bertold Burkhardt, TU Braunschweig.

Figures

- Ill. 1.1: Bach, Klaus; Burkhardt, Berthold: Frei Otto, Seifenblasen, IL 18, Stuttgart 1987, p. 331.
- Ill. 1.2: photo Christiane Weber 2011.
- ill. 2, 3: collection Berthold Burkhardt, Braunschweig.
- ill. 4, 5: Otto, Frei (Hrsg.): Multihalle Mannheim, IL 13, Stuttgart 1978, pp. 66, 81.
- ill. 6: Frei Otto, Protokoll über die Arbeiten des Instituts für Leichte Flächentragwerke an den Modellversuchen und Auswertungen am Projekt „Deutscher Pavillon, 1967, Weltausstellung Montreal“ (Stuttgart 1967), collection Berthold Burkhardt, Braunschweig.
- ill. 7: Frei Otto (Hrsg.), Netze in Natur und Technik, IL 8 Stuttgart 1975, p. 292.
- ill. 8: Abb. 9 Rolf Kayser, Robert K. Müller, „Das Hyperschalendach des Hallenbades Hamburg Sechslingspforte. Modelluntersuchung“, in: Die Bautechnik, 65, 1970, Nr.10, p. 245.

The vault with curvilinear ribs in the "Hall of Arms" in the Albrechtsburg Meissen: studies on the concept, design and construction of a complex late Gothic rib vault

David Wendland, María Aranda Alonso and Alexander Kobe
Technische Universität Dresden, Germany

During the 14th and 15th centuries, the challenge of creating vaulted ceilings lead to ever more complex solutions in late Gothic architecture. These ambitious, astonishing and sometimes daring constructions rank amongst the finest masterpieces of architecture – extremely demanding from the structural point of view and particularly challenging in their geometric design. Their builders managed to overcome the difficulties of planning the complicated meshes of ribs soaring along spatial curves, providing instructions for the production of their single components and their assembly on the building site, and achieving a curved vault surface which corresponds to the equilibrium condition of shell structures.

The discussion on how the design and planning of these structures was performed and how their construction process was organized, has been so far largely based on sources (some of them dating from long after late Gothic architecture was practised), and in particular on their interpretation as established in the 19th century. However, this current state of research can be shown to be in contrast with the evidence of many of the actual built objects.

At this point, it appears necessary to formulate hypotheses on the design directly from the built artefact, and on this basis attempt a re-interpretation of the known drawings and treatises [1]. For this approach it is also necessary to deal with the methodological challenges of using the building as source. One of the objects analyzed in this on-going research is the vault with curvilinear ribs (Fig. 1) in the Albrechtsburg at Meissen (Germany), which is presented here as work in progress. [2]

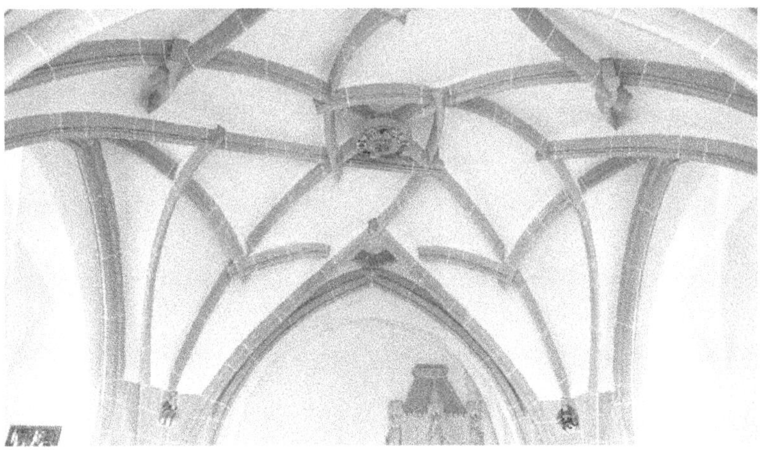

Fig. 1. The vault in the "Hall of Arms" in the Albrechtsburg at Meissen (Ventas / Wendland)

The vault with curvilinear ribs in the "Hall of Arms" in the Albrechtsburg Meissen: studies on the concept, design and construction of a complex late Gothic rib vault

The vault is located in the second floor of the north wing of the great palace of the Electors of Saxony, which had been under construction since 1471. While the walls and roofs were completed in the 1480's, part of the internal structure including some of the vaulted ceilings at first remained unfinished, until only decades later the construction works resumed and the vaulted ceiling in the room which is now called "Hall of Arms" was built. It can be dated to shortly after 1521 both by documentary sources [3] and archaeological evidence [4]. The room has a square plan with four annexes, which are carved in the volume of the perimeter walls and separated by profiled stone arches. These pre-existing arches were integrated as *formerets* into the design of the new vault, as well as the diagonal walls that crop the corners of the square plan and contain the springers of the main vault.

Most of the ribs in the vault, in addition to their curvature in elevation, are curved also in plan, thus describing three-dimensional curves. The design is attributed to Jakob Heilmann, a master mason working in Saxony and Bohemia whose major works were built in the first decades of the 16th century and who in that time was working on St-Anne's Church at Annaberg where he built the remarkable vaults with curvilinear ribs [5]. It is supposed that Heilmann was pupil of Benedikt Ried at Prague, who built the first vault with curvilinear ribs on monumental scale in Vladislav's great hall in Prague Castle, completed in 1502. This vault, together with the adjacent "Knights' Stairway", led to a whole series of vaults with curvilinear ribs. Initially, the invention of the double-curved rib is associated with the master masons Madern Gerthener in Frankfurt and Hans von Burghausen in Landshut in the beginning of the 15th century [6]. These early examples incorporate only short segments of double-curved ribs within a grid of the usual single curved ribs, while double–curved ribs that dominate the entire figurations became more frequent only after the middle of the century [7].

The design and geometric concept of the rib system

Fig. 2. Diagonal cross section of the vault in the "Hall of Arms": over the main shell of the vault four additional thin vaults support the floor, leaving void spaces in the spandrels (CAD M. Aranda Alonso).

The vault spans c. 9.40 m in transversal and 11.30 m in diagonal, the height at the summit is 6.72 m and the fleche of the diagonal arch 3.12 m. The ribs, keystones, members of the *tas-de-charge* and the boss in the summit are made of sandstone from the Elbe valley. While the rib voussoirs generally remain underneath the vault surface, as far as it could be observed, the keystones penetrate through the whole thickness of the shell – their plane horizontal upper surfaces in some cases show engraved marks [8]. The main structural feature of the vault is the shell [9] built in brick masonry, which is 30 cm thick and has reinforcing ribs on the extrados. Four additional barrel vaults, only 15 cm thick, support the upper floor in the area of the spandrels, providing empty spaces instead of the usual filling (Fig. 2). This is surely motivated by the aim of reducing the weight of the structure, considering the enormous vertical height of c. 30 m between the foundation level and the springing of the vault.

The spatial system of ribs that intersect on different levels consists of four plane diagonal arches meeting at the central keystone and several groups of curvilinear ribs (Fig. 3). Two sets of curvilinear ribs start from each of the four springers located in the corners of the room: one of them loops around the summit, running through two of the four triple intersections in the central part of the vault and then ending with a clean cut, while the course of the other is interrupted much earlier, again with a clean cut, just above the intersection with its long counterpart. Further, from the vertex of each of the four *formeret* arches a pair of double-curved ribs descends towards the corners of the hall, crossing the course of the long double-curved ribs and ending just after the intersection with the diagonal arches. Apart from the pre-existing *formerets*, all ribs have a uniform profile.

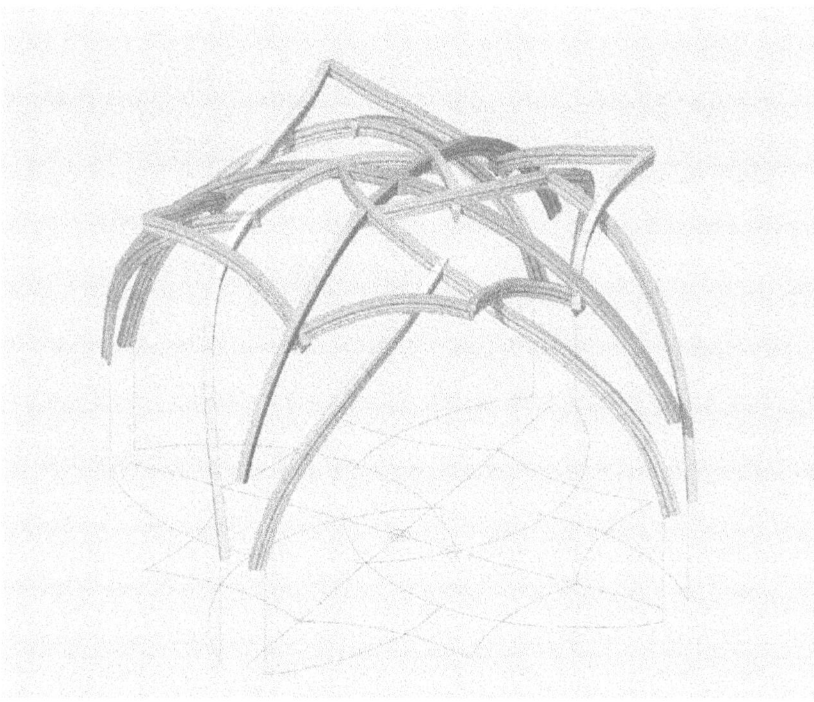

Fig. 3. CAD model of the spatial system of the ribs, showing only the profiled parts (CAD M. Aranda Alonso).

Especially eye-catching are the cut-off ribs that are suddenly truncated and virtually resting in the void. In particular, the way the lower ends of the arches descending from the *formerets* are attached underneath the profile of the continuing diagonal rib frankly contradicts the idea of tectonics. Likewise, a portion of rib is literally cut out between each long curvilinear rib that rises from the springer and the shorter one that comes from the next corner of the hall. In consequence, the figure of the ribs becomes rotation-symmetric.

As we know, generally in late Gothic vault design the curves of the ribs are not derived from the vault surface, but on the contrary, pre-determined: they are the primary feature of the vault, forming a spatial system of curves to which the surface is adapted [10]. Therefore, the phenomenon of ribs intersecting on different levels and springing on different heights is usually explained by the geometric definition of the ribs as autonomous curves. Often it is considered to be a consequence of a supposed design routine that is based on the use of one singular radius for all ribs, determined according to certain rules, called "principal arch" (*Prinzipalbogen*) [11].

While there is no doubt on the characterization of the ribs as autonomous curves and that they are a primary feature of the vault design in late Gothic, the validity of the "principal arch" concept turns out to be questionable. The term is definitely not associated with the late Gothic building practice [12]; the characterization of the design process as "algorithm" is doubtful [13]; moreover, the evidence on the built vaults (as far as geometric information is available) does not confirm the use of uniform arch radii to have been general practice [14].

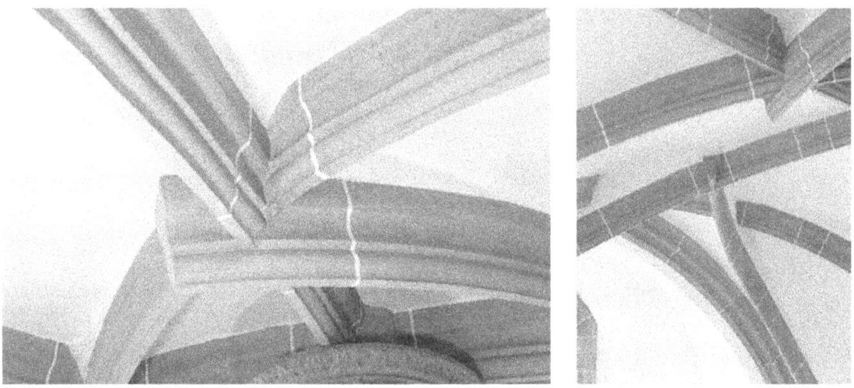

Fig. 4. Details of the rib system. Left, keystones with triple intersection near the summit. Right, the gap between the curvilinear ribs, showing that their course is discontinuous in the third dimension (Ventas / Wendland).

In the particular case discussed here, we find that the superposition of the ribs in the upper part of the vault is not caused by the autonomous development of the curves described by the ribs, but that the intersection on three different levels in the four keystones around the summit is given *a priori* (Fig. 4, left). Each of the four long curvilinear ribs passes through two of these triple crossings: the first time above the diagonal rib, and the second time below, while the diagonal ribs remain in the intermediate level. At least the two curvilinear ribs could intersect straightforwardly in one point, because both their springing and their first intersection occur at the same level. But the arrangement of the intersection in three different levels determines the situation at the gap between the long and the short curvilinear ribs: while in the plan the course of one long and one short curvilinear rib in principle could be inscribed in one single arch (cf. Fig. 3), this is not the case in the spatial development, where the two ends of the ribs clearly miss each other (Fig. 4, right).

Hence, the asymmetry of the figure in plan is related to the vertical development of the curves – it depends on the choices made in the layout of the four triple intersections where the multi-layer development in the spatial system of curves is deliberately imposed. Another interesting aspect in the spatial arrangement of the ribs can be observed at the intersections of the diagonal rib with two curvilinear ribs descending from the *formerets*, which have two levels: the two curvilinear ribs intersect in one point underneath the diagonal rib. The vertical distance corresponds to the height of the rib profile, suggesting their level at the intersection point to be given *a priori* and not resulting from the development of the curve. Consequently, these ribs are determined by their starting and ending point and by the horizontal tangent at the upper end – in plain contradiction to the "principal arch" concept. This is an interesting parallel to the geometric concepts exposed in the schemes of Gothic vaults in Philibert de L'Orme's Treatise (a source usually not considered relevant for the German speaking area [15]), where the positions of intersection points are determined prior to the curves themselves, and the radii of the rib curves constructed subsequently. In the "Hall of Arms", in principle such type of procedure seems to correlate well with the actual configuration of the vault.

The analysis of the geometric definition of the curves described by the looping ribs clearly confirms the idea of autonomous curves. We exposed elsewhere how the curve geometries can be characterized on the basis of the survey of the built object [16]. In Gothic vaults with plane arches, the curves are generally defined with elementary geometric terms, as circle segments (or composed of more than one circle segment with tangential continuity) inscribed in a vertical plane, the curves of the looping ribs are generated as circle segments "enrolled" according to the curvature in the plan, which is also circular (Fig. 5). In geometric terms, this can be referred to as a spiral, with the vertical development defined by a circle segment [17]. This geometric definition is in complete accordance with what we know about the procedures of constructing curves used by the late Gothic architects, who never constructed curves by projection or intersection of geometrically defined surfaces, but always operated with real lengths.

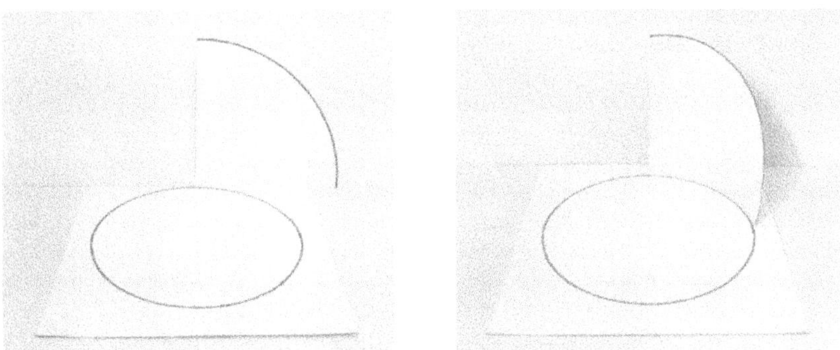

Fig. 5. Visualization of the curve generation for the looping ribs: their development in the third dimension is determined bending a circle segment in the vertical plane according to the circle in the plan (Ventas / Wendland).

As a consequence of this curve definition, by principle the single members are not interchangeable – in contrast to ribs in plane arches, in curvilinear ribs every member will fit only exactly in the position for which it was made (Fig. 6). Moreover, the profile of the rib is maintained vertical throughout the course of the spiral curve. This implicates additional difficulties for the geometric design of the single members and for the organization of their production and assembly.

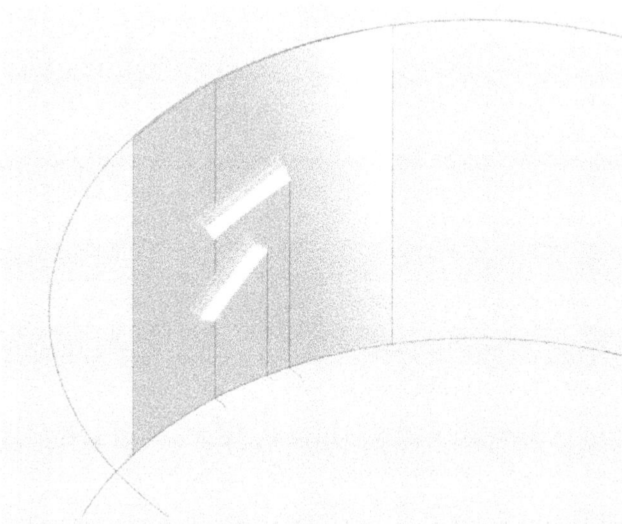

Fig. 6. Two segments of the same rib: The different voussoirs cannot be interchanged, every member fits only in the exact position along the curve for which it was produced (CAD M. Aranda Alonso).

Because the curves of the ribs are not derived from a defined surface, but developed as autonomous curves, by principle they are not inscribed in a continuous surface. Moreover, the multilayer development of the rib system with the intersections in different levels, as shown above, leads to a more complex relation between the curve system of the ribs and the surface of the shell. In consequence, at many positions we can observe considerable vertical distances between the extrados of the rib profile and the shell, which are bridged by vertical diaphragms.

Although the phenomenon of vertical portions above the rib profile has its origins in early Gothic architecture in the Île-de-France, the crucial steps towards the characteristic solutions of the later 15th century were made in the Middle-Rhine region. The master mason Madern Gerthener played a particularly import role in the development in looping ribs, and for the development of these so-called ploughshare webbings (Fig. 7). Until the end of the century, both motifs became dominating elements of German late Gothic vaults, producing dynamic rib figurations and an increasing degree of spatiality of the rib and web structure that required a high degree of rationalization in planning and communication processes to guarantee a perfect realization of the design on the building site.

Investigations on the stone cutting of the rib elements

Since the 1980s, scholars have emphasized the rationalization of building procedures as an important precondition and key feature for the development of Gothic architecture [19]. In relation to stonecutting, research has focused especially on the problem of standardization and pre-fabrication of piers, attached shafts and walls in 13th century churches [20]. Surprisingly enough, its importance in relation to the most discussed element of gothic architecture – the rib vault – has not yet been studied in detail. Moreover, captivated by the apparent analogy between aesthetical and technological logic in French *Rayonnant*, for a long time researchers were not interested at all in the role of rational design strategies in its supposedly ´bizarre´, ´decorative´ and ´anti-rational´ late Gothic counterpart [21]. Until now, the strategies and range of possibilities of late medieval masons to organize the stonecutting and joint system in such a multilayered rib vault with ploughshares have not yet been studied and constitute one of the central issues of the project.

In the "Hall of Arms", the survey of the positions of the joints in the ribs (detecting their actual position under the modern paint on a scaffolding and measuring with a full station) reveals that the members of all double-curved ribs have a regular sequence of long and short pieces, with only slight differences between the lengths of the corresponding members of the different ribs. This suggests fixed lengths for the members in every rib type of the vault (except for the plane diagonal ribs). Given that the position of the single parts of the double curved ribs is unchangeable, as pointed out, such an establishment of consistent joint systems for every type of rib must not only have facilitated the prefabrication of the stone members, but also helped to provide structured design information, assuring the correct transfer of information during the process of planning, carving and assembly of the single elements of the vault.

In relation to the stonecutting, the thin membrane-like elements or ploughshares pose similar problems that extend to the question to which degree the shape of the shell in such late-medieval vaults has been planned in advance. With regard to the design of the joint system, in the "Hall of Arms" the approach is essentially different from that in the earlier examples, like for instance in the western wing of the big cloister in Basel Minster (circa 1465). At the springers of the vault in Basel (Fig. 7), these portions are treated as thin walls made of dressed stones with horizontal beds, sitting on the rib voussoir that comprises only the rib profile and is separated by a continuous joint along its extrados.

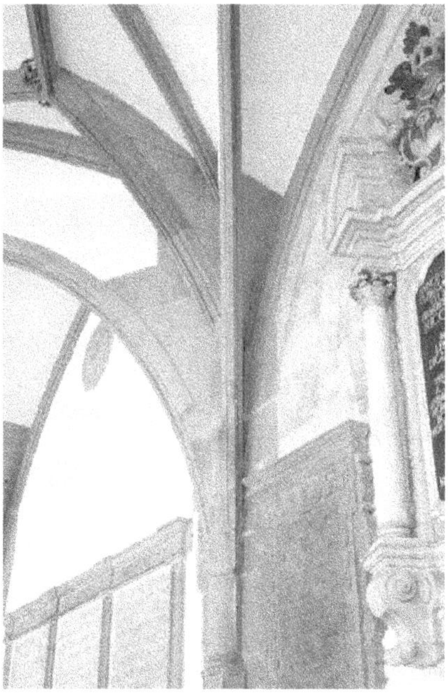

Fig. 7. Ploughshare webs in the vault of the cloister in Basel Minster, c. 1465 (A. Kobe).

In Meissen, on the contrary, at least in the lower parts the stone members of the ribs clearly comprise both the rib profile and the diaphragm reaching to the vault surface: they are worked in single pieces with radial joints that extend all the way up to the shell (Fig. 8, far left, shows the upper end of such an element). As the intrados and the extrados of these voussoirs follow different curves (that of the rib and that of the shell intrados), which are not parallel but have variable distances, the shape and size of these stone members are subject to constant change all along the course of the rib. Again, in respect to the design for the joint system we underline the challenge of creating voussoirs which are not interchangeable due to the double curvature of the ribs and the unity of rib and ploughshares.

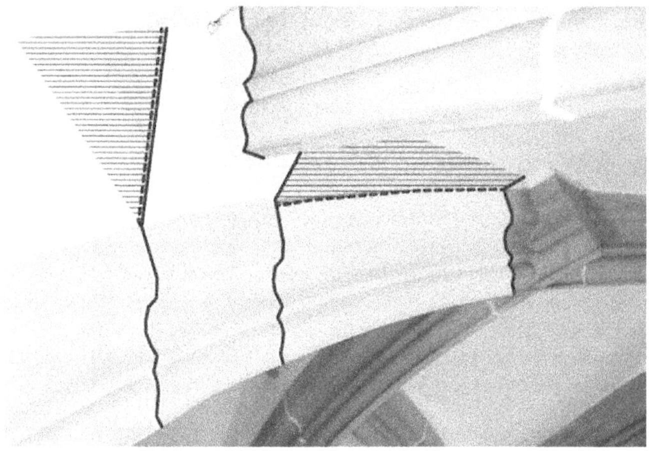

Fig. 8. Detail of the stone members of the vault in the "Hall of Arms". For clarity, the stone surfaces are highlighted, plastered portions marked as hatched areas, and the joints retraced (A. Kobe).

Finally, this design of the rib members also implicates the shape of the shell of the vault to be determined in the design process, i.e. together with the tracing of the stone elements. We can show that at least under some aspects and in some portions the shape of the vault surface was anticipated in the design. This contrasts to the current belief that the masonry of late Gothic vaults was built ad hoc. However, also the principle requirements of shell structures, stable by their form, suggest that there must have been at least some sort of a strategy for the development of the surface. Therefore we assume that the brick masonry structure of the shell is not only well executed, but also carefully designed [22].

Acknowledgements

This research is carried out within the ERC Starting Grant "Design Principles in Late-Gothic Vault Construction – A New Approach Based on Surveys, Reverse Geometric Engineering and Reinterpretation of the Sources", which is receiving funding from the European Research Council under the European Union's Seventh Framework Programme, ERC grant agreement n° 284373. Taking part in the research are D. Wendland (P.I.), M. Aranda Alonso, A. Kobe, K. Küster, C. Mai, K. Schröck and M.J. Ventas Sierra. For information on the earlier investigations and restoration of the vault we thank E. Grotegut and B. Eckoldt; for their friendly permission and support for the work on site special thank is due to the Saxon Castles Administration and in particular U. Michel and the staff of the Albrechtsburg museum.

References

[1] D. Wendland, "Design Principles in Late-Gothic Vault Construction – A New Approach Based on Surveys, Reverse Geometric Engineering and Reinterpretation of the Sources", 2011, research project ERC Starting Grant n° 284373.

[2] A first campaign of geometric survey and analysis was performed by D. Wendland in 2009/2010. The current surveys and analyses are carried out by M. Aranda Alonso, A. Kobe, C. Mai, K. Schröck and D. Wendland.

[3] Sächs. Hauptstaatsarchiv Dresden, Kopial 137, Schellenbergisch Copial, 1521-1527: fol. 4 (alt) / 12 (neu) and fol. 48 (alt) / 57 (neu), containing instructions from Count George. Analysis and transcription by C. Mai.

[4] E. Grotegut, Report on the archaeological investigations in the Albrechtsburg, 3rd floor, during May and September-December 2007, Archive of the Landesamt für Denkmalpflege Sachsen, Dresden.

[5] In 1515 Heilmann took over the direction of the lodge at St. Annen in Annaberg and was appointed in 1517 to build the parish church in Most. Before 1515 he apparently worked under the direction of Benedikt Ried at St. Barbara in Kutná Hora. Despite the lack of written sources in relation to his previous domains and apprenticeship, the close formal relationship between his works and the architecture of the Hradchin palace in Prague is generally recognized. Cf. G. Fehr, *Benedikt Ried. Ein deutscher Baumeister zwischen Gotik und Renaissance*. München: Callwey, 1961; E. M. Kavaler, *Renaissance Gothic. Architecture and the arts in northern Europe, 1470 - 1540*, New Haven, Conn. etc.: Yale Univ. Press 2012, p. 141.

[6] Cf. N. Nußbaum, S. Lepsky, *Das gotische Gewölbe. Eine Geschichte seiner Form und Konstruktion*, Darmstadt: WB 1999, pp. 259-268. Kavaler (Note [5]), pp. 140-146.

[7] Kavaler (Note [5]), p. 140. Some examples are the main nave at St. Stephan Cathedral in Vienna (1446-1459), the choir at St. Jakob in Wasserburg (1445-1448) by Stephan Krumenauer, and the looping vault in the cloister of Basel Minster supposedly designed by Jodok Dotzinger around 1465.

[8] S. Bürger and G. Donath, "Zeugnisse werkmeisterlicher Betätigung – Die Werksteine des Jakob Heilmann im Wappensaal der Albrechtsburg", in S. Bürger, B. Klein and K. Schröck, (eds.), *Werkmeister der Spätgotik: Personen, Amt und Image*, Darmstadt: WB, 2010, pp. 232-243. Similar evidence is also found on other late Gothic vaults and was first described by R. Willis, "On the construction of the vaults in the middle-ages", *Transactions of the Royal Institute of British Architects*, Vol.1, 1842, pp. 1-69.

[9] D. Wendland and M.J. Ventas Sierra, "Designing a Masonry Shell in the Reconstructed Vault of the Palace Chapel at Dresden – an Attempt to Recover the Forgotten Art of Late Gothic Vault Construction", *Informes de la Construcción*, vol.65, N° extra-2, 2013, pp. 49-63.

[10] This has been largely discussed since the early 19th century, in particular Willis (Note [8]). Cf. W. Müller, N. Quien, *Virtuelle Steinmetzkunst der österreichischen und böhmisch-sächsischen Spätgotik: Die Gewölbeentwürfe des Codex Miniatus 3 der Österreichischen Nationalbibliothek in Wien*. Petersberg: Imhof, 2005, and elsewhere. D. Wendland, *Lassaulx und der Gewölbebau mit selbsttragenden Mauerschichten: Neumittelalterliche Architektur um 1825 – 1848*. Petersberg: Imhof, 2008 and elsewhere.

[11] Cf. Müller and Quien, *Virtuelle Steinmetzkunst* (Note [10]). Also W. Müller elsewhere.

[12] On the sources and the epistemology of the *Prinzipalbogen*: D. Wendland, "Johann Claudius von Lassaulx' Gewölbe 'aus freier Hand' – Die Wiedererfindung der gotischen Architektur und die Entwicklung der technischen Literatur", in U. Hassler, C. Rauhut and S. Huerta (eds.), *Construction techniques in the age of historicism*, München: Hirmer, 2012, pp. 92-117. D. Wendland, *Lassaulx* (Note [10]).

[13] This view was suggested by Werner Müller. Cf. Müller and Quien, *Virtuelle Steinmetzkunst* (Note [10]).

[14] This doubt was put forward by Nussbaum and Lepsky, *Das gotische Gewölbe* (Note [6]), p. 180. Our surveys of late Gothic vaults in Saxony in most cases reveal multiple radii. Cf. D. Wendland, "Cell Vaults – Research on Construction and Design Principles of a Unique Late-Mediaeval Vault Typology", vol.3, in *Proceedings of the Third International Congress on Construction History*, Cottbus, 2009, pp. 1501-1508.

[15] Philibert de L'Orme / J.-M. Pérouse de Montclos (ed.), *Traités d'architecture* ('Novelles Invention pour bien bastir et à petits fraiz', 1561, 'Premier Tome de l'Architecture', 1567 / 1648). Paris: Léonce Laget, 1988; W. Müller, "Zum Problem des technologischen Stilvergleichs im deutschen Gewölbebau der Spätgotik", *Architectura*, 1973, pp. 1-12.

[16] D. Wendland, "Arches and Spirals: The Geometrical Concept of the Curvilinear Rib Vault in the Albrechtsburg at Meissen and Some Considerations on the Construction of Late-Gothic Vaults with Double-Curved Ribs", vol.1, in R. Carvais et al. (eds.), *Nuts and Bolts of Construction History: Culture, Technology and Society*, Paris: Picard, 2012, pp. 351-357. The geometric concept of the curves described by late Gothic looping ribs has been first described for the case of the Vladislaw Hall in Prague by J. Muk, "Die Gewölbe des Benedikt Ried", vol.1, in J. Tomlow and R. Graefe (eds.), *Geschichte des Konstruierens IV, Wölbkonstruktionen der Gotik*, Stuttgart/Tübingen: SFB 230, 1990, pp. 193-205.

[17] Muk, *Gewölbe* (Note [16]).

[18] The first example introducing a discontinuity between vertical and curved portions of the webs which would finally lead to thin vertical membranes, appears in the choir of St. Leonhard in Frankfurt, designed by Gerthener and dedicated 1434. The later development was strongly influenced by the work of Nikolaus Eseler the Elder and his son Nikolaus Eseler the Jounger in Nördlingen, Dinkelsbühl and Rothenburg o.d. Tauber. In Saxony, first examples occur around 1460. On the Eseler: A. Pelizaeus, *Die Eseler von Alzey: Werktätigkeit und Wirkungsstätten einer spätmittelalterlichen Bau- und Werkmeisterfamilie*, Alzey 2010.

[19] See especially the studies of Dieter Kimpel. For a summary: D. Kimpel and R. Suckale, *Die gotische Architektur in Frankreich 1130-1270*. München: Hirmer, 1985.

[20] E.g. A. Hartmann-Virnich, "Préfabrication, module et 'standardisation' dans l'architecture de pierre de taille médiévale. Quelques exemples du Sud-Est de la France (XIIe – XIVe siècles)", in *Terres et hommes du Sud, 126e congrès national des sociétés historiques et scientifiques, Toulouse, 9-14 Avril 2001*, Paris 2004, pp. 187-204.

[21] However see recently: M. Hurx, *Architect en aannemer. De opkomst van de bouwmarkt in de Nederlanden (1350 - 1530)*, Diss. Phil., Nijmegen, 2012.

[22] Wendland, *Arches and Spirals* (Note [16]); Wendland and Ventas, *Designing a Masonry Shell* (Note [9]).

Production and assembly marks in wrought iron structures

Ine Wouters and Inge Bertels
Department of Architectural Engineering, Vrije Universiteit Brussel, Belgium

Introduction

Literature covers carpentry marks in timber structures from medieval times to the twentieth century. In timber structures one can distinguish marks related to the shippers and importers who would add marks referring to the quality or the cubic content of the balk. The master adds assembly marks and direction tags to ease erection. Although information on marks in iron is scarce, we can find a similar pattern: marks related to the place of fabrication indicating the quality, marks related to the ironworks and to the assembly of the structure.

Iron brands

In the nineteenth century, iron ores were smelted in the blast furnace to produce pig-iron. Depending on the chemical composition of the ores and the temperature in the furnace, different qualities of pig-iron were obtained. Swedish pig-iron, produced from pure magnetic ore, was known for its excellent quality. Figure 1 illustrates engravings of Swedish marks produced in the 1860's from magnetic ores in Upland, Dalecarlia and Wermland. [1]

The different varieties of pig-iron were classed under different heads as they served different purposes. Henry Fidler mentions in his book, 'Notes on building construction' in 1879, that in the UK, three general heads were applied: Bessemer iron, foundry iron and forge iron. [2] Forge iron consisted of pigs, which were almost free of uncombined graphitic carbon and were suited for conversion into wrought iron. The pig-iron of commerce was further divided into varieties. The number of categories changed over time and was different from country to country. At the end of the nineteenth century, the pig-iron suitable to produce wrought iron was indicated as N°4 (white pig-iron) in English and French literature. [3]

The pig-iron was then sold to iron manufactures. To produce wrought iron, several bars of pig-iron were combined and reheated in a smaller furnace, the puddling furnace, to get rid of the carbon and other impurities. The quality of the resulting puddled bar depended on the quality of the pig-iron selected. The iron manufactures would therefore carefully select and compose the mixtures. [4] The hot bloom that came out the puddling furnace was then rolled into bars. These puddled bars consisted of wrought iron of the lowest quality. To further increase the mechanical properties of the puddled bar they were piled, reheated and rolled several times. In the UK, the following three categories were used to describe the quality of wrought iron applied in structures: best (B), best best (BB) and treble best (BBB).

In the French nomenclature, best iron was indicated as *très bonne* or *n°3* and applied for ordinary structures, roofs and bridges in iron. The best best iron was called *supérieur* or *n*°4 and suitable for bolts, rivets and swing bridges. The treble best iron was referred to as *supérieure extra* or *n°5* and applied for cables. [3]

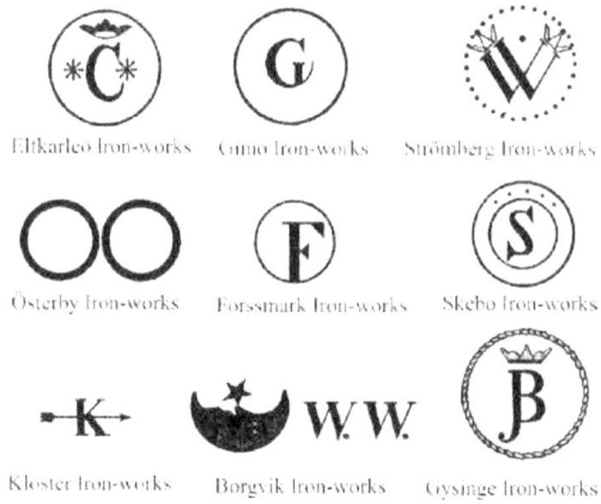

Fig. 1. Engravings on pig-iron produced by Swedish iron-works, known for their purity and superior quality. (Adapted by the author from Percy, 1864, p.737)

The final wrought iron elements were stamped near their end with a mark indicating the ironworks. According to Fidler, the mark generally consisted of the makers' initials or name and a device indicating the quality (Fig. 2). In the UK, the characters B, BB or BBB could be added to refer to the three qualities described above. The crown, which was introduced in many brands, had no special signification as it was also found on bad irons. Some brands were known for their superior quality, like Lowmore in the UK, so they added no mark relating to quality. [2]

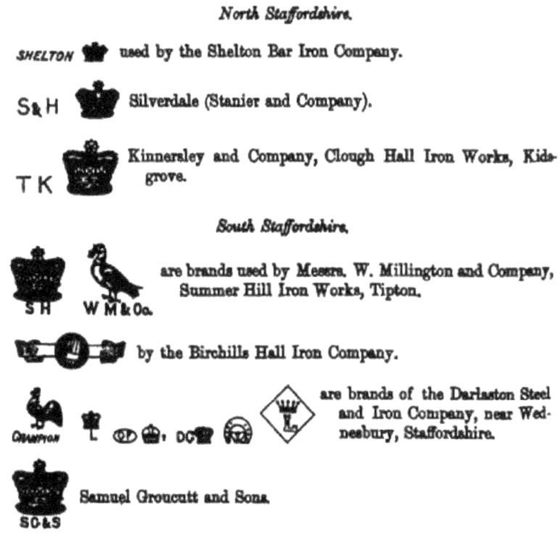

Fig. 2. Brands on wrought iron used by ironworks in the Staffordshire area in the United Kingdom (Fidler, 1879, p. 284)

When discovering stamps of brands on built historic structures, these stamps help to indicate the iron manufacturer and provide indirect information about the quality. However, today it is very difficult to trace back the iron manufactures on the basis of the brand marks as a repertory is missing. For example, the stamp mentioning the characters 'ST' (Fig. 5) was found on an 1824 wrought iron suspension bridge in Kruibeke (Belgium). [5] Sources mention the bridge elements were manufactured in the UK. Relating the stamp to a manufacturer in the UK could confirm this idea. Some stamps contain even more information. The iron roof construction of the Hermitage in St. Petersburg contained stamps on iron, not only indicating the name of the Russian Demidov ironworks but also mentioning the date of production. [6] Documenting marks of iron brands and developing an international database is essential to trace the iron manufactures and to get information about the place and time of fabrication. In 2009, an online database with a repertory of forge marks in Belgium was composed by the Museum for Old Techniques (MOT), but up until now the authors are not aware of a repertory of iron-works brands. [7]

Assembly marks and direction tags

A second group of marks refers to signs, which make the assembly process easier. As the wrought iron elements of an iron structure were forged one by one, they are not identical. The iron master used to assemble the structure in his iron-works to check if all elements fit and adapt them, if necessary. Before dismantling the structure, elements were marked to ease re-assembly. The elements could be marked by drawing, painting, carving or hammering signs onto the elements. As most of the iron structures were painted to protect them against corrosion, mainly carved and hammered marks are visible when assessing a nineteenth century historic structure. The assembly marks of three early nineteenth century structures are discussed.

Iron roof truss in Walhalla, Germany (1838)

In the iron roof structure of the Walhalla, built in 1838 near Regensburg, Germany, the assembly marks were recently documented. [8] Hammered marks were discovered on the trusses. As the roof consists of six iron roof trusses, all elements within one truss were stamped with the same Arabic number. Then, at the connections, a second sign was added (a dot or line) to indicate which elements fit together. Figure 3 illustrates how the individual wrought iron elements of the connection are marked with the signs '5 IIII' and '5 ···'. The Arabic numeral 5 refers to the number of the roof truss. The signs 'IIII' and '···' refer to a specific joint within truss '5'. The documentation of the iron marks revealed that truss number 5 and 6 were switched in the built structure. [8]

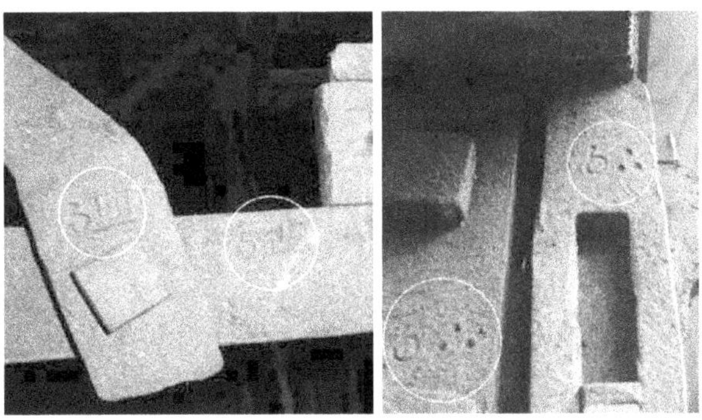

Fig. 3. The assembly mark '5' is hammered onto the main elements of roof truss number 5 of the iron roof of Walhalla. The hammered symbol next to '5' indicates the connection: '5 IIII' and '5 ···'. (Photos by Karen Veihelmann, 2010)

Suspension footbridge, Belgium, 1824

When the structure consists of a sequence of elements, the structure can be numbered consecutively. This was the case for the suspension bridge erected in Kruibeke near Antwerp, Belgium in 1824. During renovation works, which took place in 2012, assembly marks became visible. However, not all marks could be traced back due to deterioration of the surface. The elements of the suspension chain were marked with Arabic numerals, hammered in the iron, going from 1 to 32. Some bars and chains were numbered with Roman numerals (from I to XII), by carving lines into the iron. It is unclear why both systems are combined. Figure 4 (left) illustrates how chain '14' is connected to its corresponding eye-bar '14'. At several connections, however, the marks do not match. Chain '7' is wrongly connected to eye-bar 'XI' (figure 4, right). These confusions hinder the understanding of the applied numbering system.

On the south-west railing, a 'triangle' is hammered next to each Arabic numeral to be able to distinguish this railing form the north-east railing. This triangle therefore serves as a direction tag. Figure 5 (middle) illustrates the combination of the Arabic numeral '3' and the triangular direction tag. A second circular tag is added to the segments 8 to 22 of the handrail. If the circular tag started from segment 11, it would begin in the centre and help to distinguish the southern from the eastern elements. For now it is unclear why this circle was added. (Fig. 4, left)

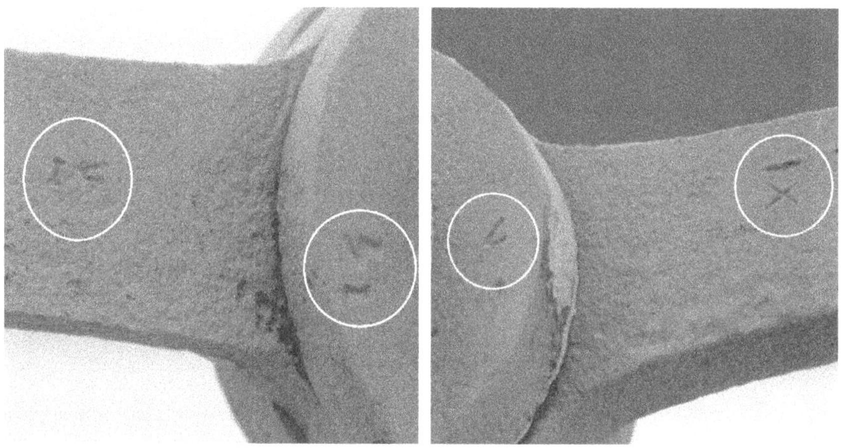

Fig. 4. During renovation of the 1824 wrought iron suspension bridge in Kruibeke (Belgium) several marks became visible on the eye-bars and chains. Assembly marks with Arabic numerals were combined with Roman numerals. (Photos by Ine Wouters, 2012)

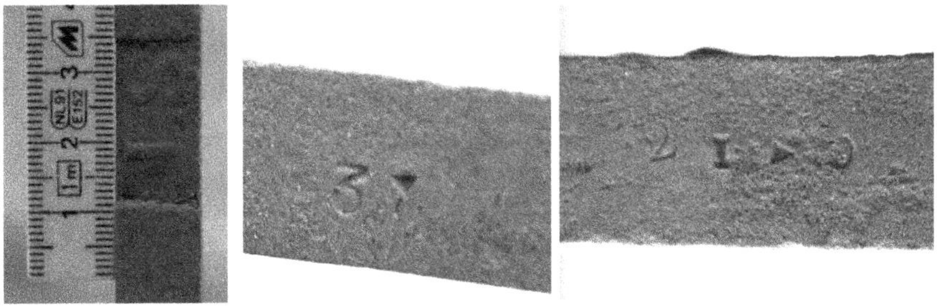

Fig. 5. Marks and stamps on the railing of the 1824 wrought iron suspension bridge in Kruibeke (Belgium): stamp of iron-works mentioning 'ST' (left), triangular direction tag hammered next to Arabic numeral '3' (middle), direction tags hammered next to Arabic numeral '21' (left). (Photos by Ine Wouters, 2012)

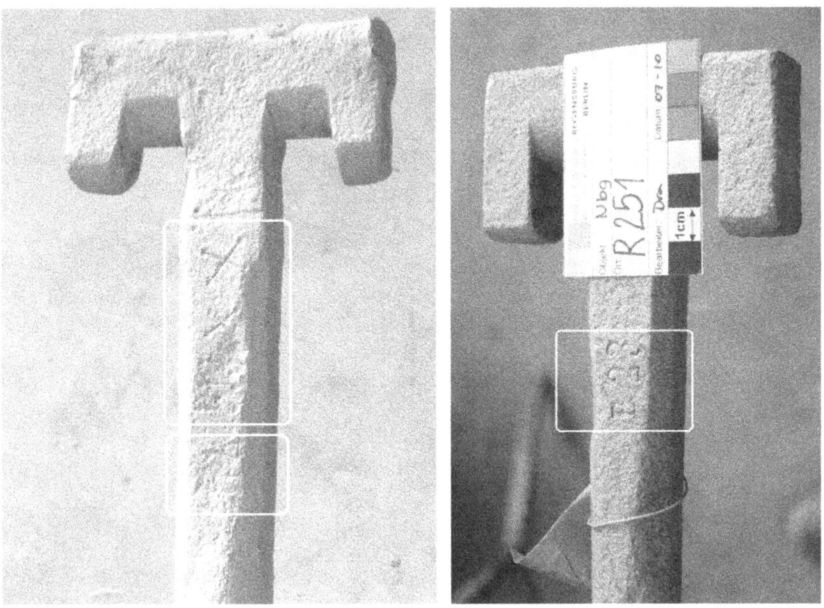

Fig. 6. The assembly marks 'XXIII' and '7B' (left) and 'B23' (right) are visible on the wrought iron bars of the suspension bridge 'Der Kettensteg' in Nuremberg (1824). (Photos by Haber & Brandner, 2010)

Suspension footbridge 'Der Kettensteg', Germany, 1824

The wrought iron suspension bridge 'Der Kettensteg' in Nuremberg, Germany, built in 1824, was renovated in 2010. [9] [10] After sandblasting the structural elements, assembly marks became visible on bars, hangers and connection rings. Roman numerals were found on the original bars. Figure 6 shows the Roman number 'XXIII' carved in the original bar. More recent iron elements, which date back from a restoration campaign in 1909, could be distinguished by their geometry. These 1909 bars carry different marks: a letter is combined with an Arabic numeral. Figure 6 shows the mark 'B23' hammered onto the 1909 bar. This numbering system is also present on the original bar, which contains the mark '7B'.

Conclusion

Looking for marks of iron brands on wrought iron structures can deliver information to identify the iron master and interpret the iron quality. An international repertory on iron brands, which is essential for the interpretation, is however still lacking today.

The study of the assembly marks shows that both carved Roman numerals and hammered Arabic numerals were applied in the early nineteenth century in Germany and Belgium. In all three cases the built structures were not erected as marked in the workshop: elements were switched, the order of trusses inversed. Caution is thus called for when interpreting the assembly marks. At the same time marks are meaningful during the assessment of existing structures and can add arguments to questions about the construction, the place and date of production and the authenticity of parts.

References

[1] J. Percy, *The art of extracting metals from their ores. Volume 2*, London: J. Murray, 1864. *<https://archive.org/details/americana>*

[2] H. Fidler, *Notes on building construction: Part III. Materials.* London: Rivingston, 1879. *<https://archive.org/details/americana>*

[3] M. de Bouw, I. Wouters, J. Vereecken, L. Lauriks, "Iron and steel varieties in building industry between 1860 and 1914 – A complex and confusing situation resolved", *Construction and building materials*, 8, Vol 23, 2009, pp. 2775-2787.

[4] P. Belford, "Extra special best best: Black Country iron puddling and wrought iron manufacture", *Historical Metallurgy*, 38(1), 2004, pp. 47-59.

[5] I. Wouters, M. de Bouw, S. Adriaenssens, A. Verdonck. "Upgrading mainland Europe's oldest iron suspension footbridge", *Steel construction. Design and research.* 1(2), 2009, pp. 36-41.

[6] B. Heres, "The iron roof trusses of the New Hermitage in St. Petersburg", in J. Campbell (eds.), *Proceedings of the Second International Congress on Construction History, Cambridge 2006*, Cambridge: University Press, 2006, pp. 1555-1569.

[7] Museum voor Oudere Technieken-MOT, Forge marks in Belgium <http://www.mot.be>

[8] K. Veilhelmann, "Der eiserne Dachstuhl der Walhalla – ein herausragendes Zeugnis der frühen Stahlbaus", *Stahlbau*, 82, 2013, pp. 494-501.

[9] *Der Kettensteg schwingt! Denkmalgerechte Sanierung und Wiederherstellung des Kettenstegs in Nürnberg als Hängebrücke.* Nürnberg: Baulust, 2011.

[10] Haber & Brandner, *Nürnberg Kettensteg. Dokumentation der Restaurierungs-und Konservierungsmassnahmen.* Unpublished technical report, 2011.

The construction and structure of medieval gates

David Yeomans[1] and Hugh Harrison[2]
1. Vice-Chair Construction History Society, UK; 2. Hugh Harrison Conservation, UK

Summary

The construction of doors and large solid gates from the medieval to the Tudor periods has received little attention except for a brief survey by Cecil Hewett. These are sometimes very heavy timber structures and the structural problem is to prevent their racking. Early doors were of plank construction and the change to framed structures seems to be associated with an understanding of the performance of frames and how they might be strengthened by the addition of bracing members. Three types of door can be distinguished: early planked doors, framed doors and triple skin doors. This paper provides an outline of their development and discusses their structural behaviour.

Introduction

This paper deals with both doors and gates with solid faces, which serve the same function and so require similar structures. A door separates external from internal spaces while the gates dealt with here separate two external spaces but are housed within the archway of a gatehouse, which provides them with some shelter. As the function, structure and to some extent the decorative treatment of both doors and gates was the same throughout the period considered, we may treat them the same, the term used simply distinguishing the function of a particular example, or in many cases pair of examples. The carpentry of a door or gate must be strong and stiff enough in plane to transmit its weight back to the hinges without racking, i.e. to prevent its toe from dropping, and it must be stout enough out-of-plane to resist wind loads and attempts at undesired entry.

There has been no typological study of doors in England similar to that carried out by Rachel Touzé in France [1], and the best that we have are collections of drawings by Cecil Hewett, one study largely of cathedral doors [2] and one of church doors in Essex. [3] A substantial study by Geddes, *Medieval Decorative Ironwork in England*, is largely about doors but, as the title indicates, is concerned more with their ironwork than their carpentry construction. [4] A large number of unpublished studies have been carried out by one of the present authors in the course of repairing and restoring doors and these have been used as a prelude to developing a typological study of their carpentry. A preliminary hypothesis on the development of door types is that in England prior to about 1300 their carpentry was based on the use of planks reinforced with ledges or grids and it was only at this date that frames began to appear. [5] While framed doors might seem to be the most advanced construction, triple skin doors (see below) are another development, possibly in response to the structural problems associated with frames.

A door is the sum of a number of elements: the boarding that forms all or part of the outer surface, the structural system on the back (where this is separate from the boarding), the pattern of nailing and the hinge straps. There may also be purely decorative features, either front or back. Note that structural elements may have also been treated decoratively and that what was, at the time, simply considered to be decorative, might in some instances perform structurally. The survival of the door depends on the success in combining of all these elements such that they contribute to its structural integrity. Naturally all doors include boarding but in the earliest doors the boards formed the structure whereas later doors have frames that form the principal structural elements and the boarding is entirely secondary. However, throughout the period under consideration there is a third category which we may call double or triple skinned doors in which the boarding was in more than one layer.

This last category of doors has the greatest in-plane stiffness and as they present no structural problems may be simply dealt with. A double skinned door has only two layers in different directions, vertical on the outside and horizontal or inclined within. An example is Queen Camel Church, Somerset. As shrinkage in unrebated boards would result in gaps where the joints overlap a third layer may be added to cover the joints in the outer boards usually consisting of decorative ribs. The fifteenth-century church at Tattershall, Lincs is a triple skin door with a central wicket. In the church at Chugford, Devon, either late fifteenth or early sixteenth century, triple skin construction is used for a folding door, which is clearly a testament to its in-plane stiffness. These doors clearly provide the greatest resistance to racking, although at the expense of some weight and they avoid the need for the complex carpentry used in other types of doors.

We need to be clear that these terms 'board or plank door' and 'framed door' are being used in relation to their structure, rather than simply to describe them visually. That is because all doors will have boarding and there are also doors in which a frame has been applied for visual effect to what is essentially a plank door. Although it may look as if framed, such a door would not be regarded as a framed door because the frame serves no structural function. Possibly such applied work might in some cases have added some out of plane stiffness but it is unlikely that it was seen in that way.

The weight of the door has to be transmitted to at least one of the hinges and, as the door is essentially a cantilever, these hinges must also provide horizontal forces. The door itself also has to be constructed in order to prevent its racking, i.e. its toe must not drop. These may seem like obvious points but the forces can be considerable as a large gate or door may weigh as much as three quarters of a ton, but the same considerations apply to the smallest doors. Concern for appearance was generally concentrated on the outside face, so that the internal face shows the carpentry more clearly. However, in doors where framing was added to a single layer plank door it was added to the inside face.

Hinges and pivots

A door may be har-hung. In this arrangement the hanging stile is extended above and below the body of the door to form pivots that engage with sockets in the cill and lintol. The weight is then carried by the cill. The gates of the cathedral close at Peterborough (discussed in more detail later - Fig. 6) are har-hung and although the door has been shortened the original top pivot remains in the masonry.

In hinged doors, either two or three hinges were used. Then the weight may be carried by any of the hinges although wear of that hinge might eventually lead to a more even distribution of weight. Note also that three hinges require accurate setting of the middle hinge so that it is directly between the other two. If it is not additional horizontal forces will be produced in the hinges as the door swings.

Some doors were constructed with a bottom pivot in the cill, which would carry the weight, but also with a hinge or hinges above. The so-called Queen's Gate at Trinity College, Cambridge has this arrangement although there the bottom plates of the frames are extended beyond the hanging stile and the pivots set into those.

Nails

The weight of the door and concomitant horizontal forces must be transmitted to the hinges by their fixing nails and in many doors the hinge straps are an important component of the door's structure as they serve to hold boards together. In the large decorative hinges described by Geddes there was the opportunity to distribute the forces over a large number of nails each of which might carry a comparatively small force. However, the decoration was generally rather delicate extensions of a heavy hinge strap where the forces would have been concentrated. When simple hinge straps came into use with framed doors it was essential that the nails be driven into the muntins or ledges of the internal framing members.

In some cases the nails were driven through the door and then either clenched on the back or riveted over with a small plate, usually a diamond shaped plate, against which the rivet head was formed. Both clenching and riveting seem to have been used throughout the period, but the longer process of riveting combined with the extra cost of the plates is more usually found on earlier doors or later more elaborate doors. In other cases the nails might not be driven through the door, relying simply upon friction to hold them in place. At Holy Trinity, Tattershall, Lincs, the doors were nailed from both sides, and not through nailed. In all four pairs of doors oak pins can be seen which appear to been used to hold the doors together temporarily while nailing.

Wicket gates

In large double gates or doors it is convenient to incorporate a smaller door, a wicket, for daily use. These need to be considered because they interrupt the structure of the gate. They were built into both plank and framed doors and the most common arrangement was to have a small door in one leaf of a pair of doors, often next to the hanging stile. However there are instances when the decorative design of the front of the doors requires that the wicket is also formed as a pair of doors at the centre. The triple skin doors at Tideswell, Derbyshire, have a central wicket gate like that as do the south porch doors of St Nicholas, King's Lynn. Arrangements for framing wickets are too varied to trace any pattern of development but some observations on structure can be made.

The opening for the wicket introduces a weakness into the structure of the door and in a framed door the simplest means of providing a wicket is for it to be framed above the bottom rail, which therefore has to be stepped over. This arrangement can be seen in the door at Peterborough (Fig 6) and the Queen's Gate at Trinity College, Cambridge. However there are also examples where the wicket extends to the bottom of the main door so that in framed doors the cill is shortened and where it is at the centre so are the meeting stiles. If hinges are used in the main door then the wicket may also limit the length of the bottom hinge strap.

Board or plank doors

The simplest way of making a door is to assemble a number of boards side by side and that was the preferred method for the earliest examples. The boards were held together by ledges, which might be fixed to the boards by simple nailing. Two nails per board would be sufficient to prevent relative rotation of ledge and board and so ensure rigidity of the whole door. Thus even in this simple arrangement the nailing pattern was critical to structural integrity. In more skilful carpentry rigidity was achieved by having a pair of comparatively thin horizontal ledges let into the broad thick vertical boards. According to Hewett this was the technique used for surviving early doors at Westminster Abbey and Durham Cathedral but the illustration here (Fig. 1) is of a door of the church at Little Hormead, Herts, dendrochronologically dated to c.1150. This shows clearly the dovetail slot cut into the boards to take the ledge and also shows that skilful carpentry like this was not confined to high status buildings.

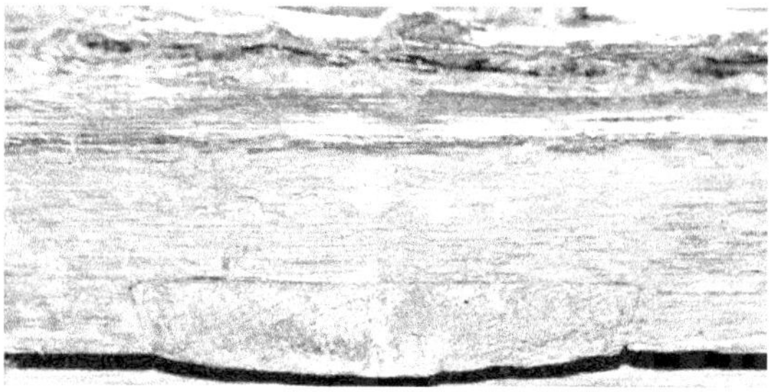

Fig. 1. Edge of door at St Mary's, Little Hormead, Herts, showing a strap let into the board.

The boards were nearly always rebated along their edges so that shrinkage should not result in open joints. By using counter rebates along the length of the boards vertical forces could be transmitted from one board to another (Fig. 2) and that arrangement enabled diaphragm action to be developed in the door. Figure 2 shows this technique used at Bristol Cathedral, but this illustration also shows the rather casual spacing of the nails in the ledge. The carpenters seem to have been aware that the ledge now only served to hold the boards together rather than providing any resistance to racking of the door. In contrast Little Hormead with its let-in ledge also has a detail that achieves the same effect as counter rebating but using V-shaped grooves in the edges of the boards with corresponding projections on the adjoining board (Fig. 3).

The door at York Minster chapter house and the west doors at Peterborough Cathedral, are both plank doors, combining carpentry along the edge of the boards to transmit shear forces, one with the V joints and the other rebated, with a pattern of diagonal timbers on the back. At Peterborough this pattern is combined with three ledges and as the hinge straps do not extend across all the boards this grillage serves to hold the door together. One can see from the details at Peterborough that the diagonal members were seen as decorative but the illustration also shows that they are fixed with two nails per board and so could provide some structural rigidity had the boards not been counter rebated. It follows that any door with a decorative diagonal grillage on one face, and Hewett shows several such doors, may well derive some stiffness from that grillage. In contrast, doors at Salisbury, and in particular the vestry doors, are crudely constructed with diagonal grids that appear to have been conceived as purely structural.

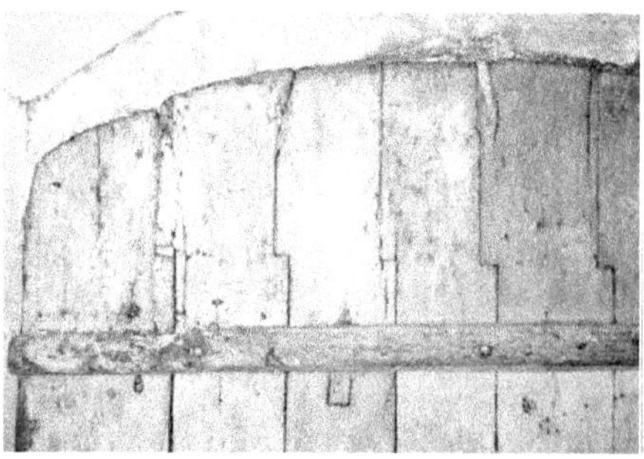

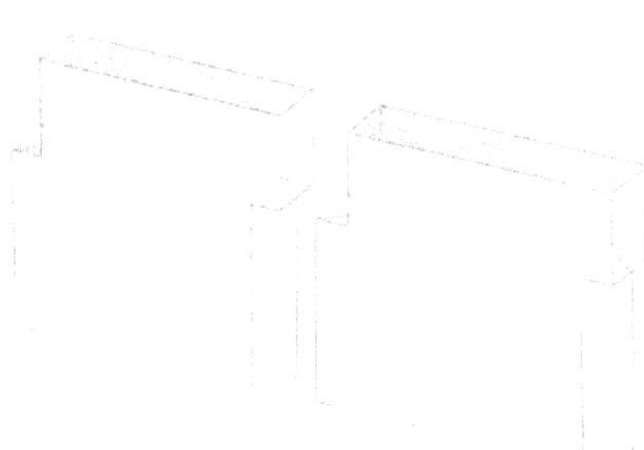

Fig. 2. (left) Night stair door, Bristol Cathedral c. 1140, showing the joggles at the top of the boards.
Fig. 3. (right) Detail of the jointing at St Mary's, Little Hormead.

The examples at York and Peterborough suggest that the grid pattern and/or the outer frame was wanted for visual as much as structural reasons and there are other examples of board or plank doors to which frame elements have been added. At Bradworthy, Devon, for example, a simple outer frame had been planted onto the back of a plank door and used to carry carved decoration.

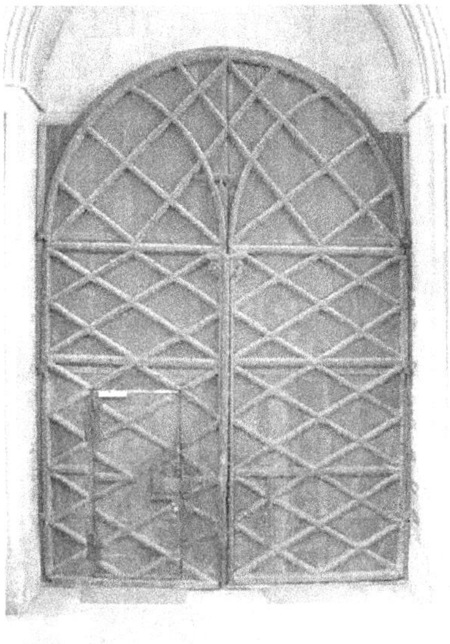

Fig. 4. West doors at Peterborough Cathedral.

Framed doors

The frame is ideal for a rectangular structure, but in a door where either one or both long side stiles have curved tops, the weakness of the joint(s) needed to form the curve reduces the inherent strength of the frame. This was avoided in small single-leaf doors by using grown durns, components taken from timber with naturally grown curves at one end so that the only joint required was at the apex. For the single leaf door of Westminster Abbey Treasury, Hewett shows the door with separate frames front and back with grown durns on one face but separate curved top members jointed to the stiles on the other. The massive construction of this door was for security, but it might have been simpler to reinforce the door with an iron grillage. Such an iron grillage survives in the thirteenth-century door of Dunster Castle, Somerset.

If grown durns were not used there were a number of ways of joining the hanging stile to the curved top rail: either they could be seen as a continuous member with the top of the hanging stile tenoned into the bottom end of the curved top rail, or the curved top rail could be tenoned into the inside edge of the hanging stile. The former was used for the recently reconstructed gates at Hampton Court, the latter for the Great Gate at Trinity College, Cambridge (Fig. 5).

Within the frame of stiles, curved top rail and bottom rail the body of a framed door was then divided into a grid of muntins and ledges (Fig. 5), or vertical muntins and inclined ledges (Fig. 6) or by both. It might also be divided by a mid-rail and while the Trinity College gate is so divided, this seems to have been unusual for gates with a single grid but was, of course, necessary where there was a square grid at the bottom and a grid with inclined ledges at the top. The simple way of joining the members of the grid into the frame was with notched lap joints because the frame could then be assembled first and the internal members let into it. These internal members would be halved over each other with muntins normally interrupted on the back face. That would provide continuous muntins on the outside face to which the boarding might be fastened.

With a heavy gate weighing something like ¾ ton it is difficult to imagine that the external frame would be rigid enough to prevent racking. Besides which, the bottom rail would in many cases be subject to wetting and eventual decay. Thus we must look to the internal structure for stiffness. In a gate with only a square grid of muntins and ledges any tendency of the toe of the gate to drop might be resisted either by nailing of the external boards to the muntins or by resistance to rotation of the halving joints between muntins and ledges. Examination of the gates at Trinity College, Cambridge, shows that it was the latter that provided their stiffness. There was noticeable rotation of the timbers at these joints. In these gates the boards were set into rebates in the muntins and held in place only by the nailing through the outer decorative ribs and could have provided no contribution to the gates' stiffness. Calculations suggest that in spite of the crushing observed stresses at the joints would not today be considered excessive.

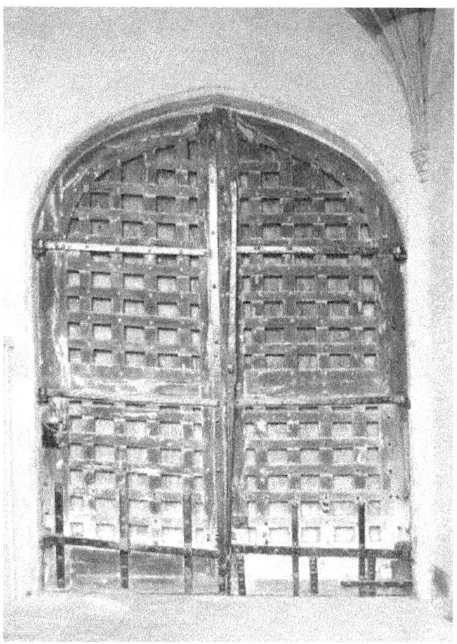

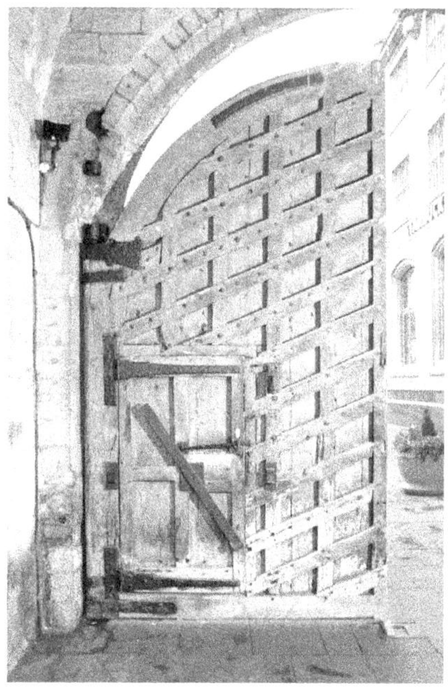

Fig. 5. (left) Great Gate at Trinity College, Cambridge. Fig. 6. (right) Gate at Peterborough Cathedral.

The same resistance to rotation might be provided by inclined members and muntins halved over each other. However, this kind of framing results in some triangulations of the door panel, especially if a mid rail is incorporated into the frame. This is seen in gates at Winchester Cathedral with a square grid below the mid rail. This triangulation would be sufficient to give the gate its necessary rigidity. In contrast a door constructed as a single panel even if with inclined members, as in the Peterborough close gates, can hardly rely upon any such triangulation because it would depend upon the bottom rail and be compromised when that member decayed. Once that happened, the gate could only rely upon the resistance to rotation of the grid of members and/or the pattern of nailing between boards and grid of timbers.

The impression is that the structural properties of the grid of diagonal members and muntins was recognised while the square grid was wanted for its visual effect. It was the visual effect that was provided by the example found at St Nicholas, King's Lynn (Fig. 7). There the construction of this door suggests some confusion about the nature of the structure. Only the ledges are continuous and have bird's beak ends let into recesses cut into the styles. The muntins on the back of the door are complete imitations, in fact made of simple blocks nailed in between the ledges. It is possible that some stiffness is provided by the nailing of the boards, which are laid over this apparent grillage, and the decorative muntins on the outer face of the door. Note that at King's Lynn the boards were applied over the muntins and nailed in place whereas at Cambridge (Fig. 8) they were set into rebates in the muntins and held in place by the decorative scheme applied to the front.

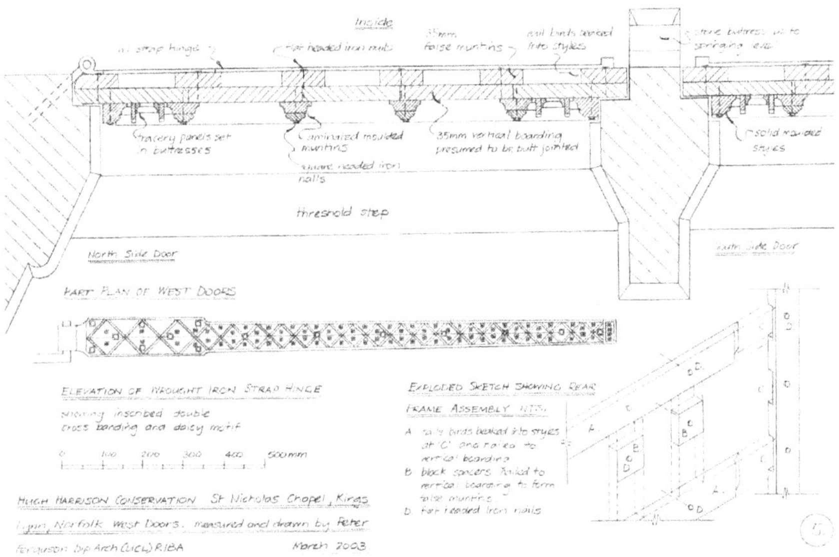

Fig. 7. Drawing of the west doors at St Nicholas, King's Lynn.

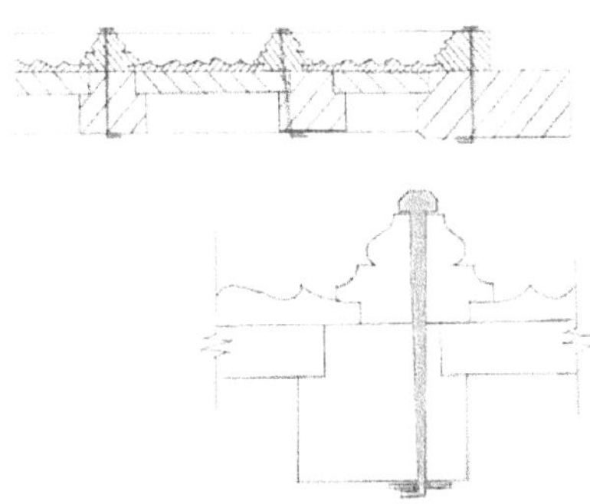

Fig. 8. Construction of the Great Gate at Trinity College, Cambridge.

Conclusions

Double skin doors provided the simplicity of plank construction without the need to incorporate stiffening ledges. Triple skin doors added to that decoration on the front and covered the gaps created by the double skin. However, where elaborate framing was used it seems that the square grillage was wanted for its visual effect. However, the late gates at Trinity College, Cambridge, and also the doors of the chapel at King's College have rectangular grillages that did not provide sufficient stiffness to prevent them from racking. At King's College Chapel, Cambridge, the rectangular grid of the doors to the chapel has been used as a feature on their front face. It seems that by the time these structures were built the visual effect had come to dominate the design and carpenters had ceased to understand the principles essential to their successful framing. Of course, it was many generations since doors or gates of this scale had been wanted for castles or the great cathedrals and it may well be that the understanding of their structural needs had been lost in the meantime.

This analysis, based as it is on a small sample of doors and gates, can only be tentative. It ignores the variations that exist in each of the types described and a fuller understanding of the development of their carpentry must await a much more extensive survey of surviving examples together with accurate dating.

References

[1] Rachel Touzé, *Vantaux de Porte à Planches Jointives du XIe au XIXe Siècle*. Paris: Éditions du Patrimoine, 2009.
[2] Cecil Hewett, *English Cathedral and Monastic Carpentry*, Chichester: Philimore, 1985, pp. 155-187.
[3] Cecil Hewett, *Church Carpentry: a study based on Essex examples*. Chichester: Philimore, 1982.
[4] Jane Geddes, *op. cit*, London: Society of Antiquaries, 1999.
[5] Hugh Harrison, "Church woodwork in England", *Regional Furniture*, vol. 21 (2007), pp. 53-66.

www.ingramcontent.com/pod-product-compliance
Ingram Content Group UK Ltd.
Pitfield, Milton Keynes, MK11 3LW, UK
UKHW051450290926
15471UKWH00008B/1124

9 780992 875107